Bacteria in Agrobiology: Plant Nutrient Management

# Already published volumes:

Bacteria in Agrobiology: Crop Ecosystems
Dinesh K. Maheshwari (Ed.)

Bacteria in Agrobiology: Plant Growth Responses
Dinesh K. Maheshwari (Ed.)

Dinesh K. Maheshwari
Editor

# Bacteria in Agrobiology: Plant Nutrient Management

*Editor*
Prof.(Dr.) Dinesh K. Maheshwari
Gurukul Kangri University
Department of Botany and Microbiology
Faculty of Life Sciences
249404 Haridwar (Uttarakhand)
India
maheshwaridk@gmail.com

ISBN 978-3-642-21060-0 e-ISBN 978-3-642-21061-7
DOI 10.1007/978-3-642-21061-7
Springer Heidelberg Dordrecht London New York

Library of Congress Control Number: 2011935049



*Cover illustration*: Optical micrograph showing cross sections of intercellular colonization rice calli and regenerated plantlets by *A. caulinodans*: CS view of root uninoculated control; magnified cross section view of leaf colonized by *A. caulinodans* in regenerated rice plant; possible sites of infection and colonization of rice root (from left to right); see also Fig. 3.1 in "Endophytic Bacteria – Perspectives and Applications in Agricultural Crop Production", Senthilkumar M, R. Anandham, M. Madhaiyan, V. Venkateswaran, T.M. Sa, in "Bacteria in Agrobiology: Crop Ecosystems, Dinesh K. Maheshwari (Ed.)"

*Background*: Positive immunofluorescence micrograph showing reaction between cells of the rhizobial biofertilizer strain E11 and specific anti-E11 antiserum prepared for autecological biogeography studies; see also Fig. 10.6 in "Beneficial Endophytic Rhizobia as Biofertilizer Inoculants for Rice and the Spatial Ecology of this Bacteria-Plant Association", Youssef G. Yanni, Frank B. Dazzo, Mohamed I. Zidan. in "Bacteria in Agrobiology: Crop Ecosystems, Dinesh K. Maheshwari (Ed.)"

*Cover design:* deblik, Berlin

Printed on acid-free paper

Springer is part of Springer Science+Business Media (www.springer.com)

# Preface

Sustainable crop production should switch from growing plants to the cultivation of plant–microbial communities, which can reach high productivity under minimal energy and chemical investments along with minimal pressure on the environment. Much effort and cooperation among experts in different fields of science is, therefore, needed to be successful in attaining microbial-based technologies. Many of the today's environmental problems for crop production are due to applications of chemical fertilizers that provide imbalance of mineral nutrients.

The organization of the book (14 chapters) is from basic to applied aspects of nutrient management via microorganisms in general and bacteria in particular in support of plant growth and development. To begin with, various strategies and different genera involved in amelioration of nutrients for growth promotion of major oil seed crops have been suitably described. The mineral–microbial interactions lead to their solubilization or rhizoremediation of nutrients such as phosphorus, sulfur, zinc, and iron that are optimally required for different metabolic pathways. For example, low soil phosphate availability is a major constraint for soil fertility. To overcome this problem, use of phosphate-solubilizing microorganisms provides the phosphorus in available form more efficiently. Similarly, sulfur, zinc, and iron nutrition is mediated through specialized plant growth-promoting bacteria (PGPB), making these minerals available in adequate amount to crop plants. The book gives in-depth insight into agronomical and physiological impacts of phytohormones secreted by Azospirilla and other PGPB. A due account is given on the application of ACC deaminase-containing bacteria, which act as a sink for ACC and thus protect the developing seedlings from deleterious effects under various adverse conditions.

Some of the chapters highlight themes regarding the nature and diversity of bacterial biofilms and elucidate their potential as a rich source of novel biologically active compounds for our agro-ecosystems. But their effectiveness involved quorum sensing (QS) systems in PGPB that offer important networks in enhancing their beneficial performance. For such a phenomenon, microbial metabolites intend to play a role in QS besides combating plant diseases. Further, there is mounting evidence of the involvement of *Pseudomonas* spp. that produce 2–4 diacetylphloroglucinol and other phenazine derivatives in signaling function, induction of systemic resistance, and reduction of minerals in soil in relation to plant health. The contents lay stress on Microbial world that synthesizes and emits many volatiles besides depicting that denitrification in soils leads to sustainable agriculture ecosystems.

I wish to acknowledge all the subject experts who are instrumental by providing the valuable scientific piece of work for making this book a success. I am indebted to the many individuals who generously assisted me in reviewing process. Thanks are also due to my research team members namely, Dr. Sandeep Kumar, Abhinav, Rajat, and Pankaj.

The credit also goes to my wife Dr. (Mrs.) Sadhana Maheshwari and son Ashish (B.E.) especially for language corrections.

The production of the book is greatly supported by Dr. Jutta Lindenborn, Springer, Heidelberg, Germany. I owe my thanks.

Haridwar, Uttarakhand, India Dinesh K. Maheshwari

# Contents

1 **Role of PGPR in Integrated Nutrient Management of Oil Seed Crops** ........ 1
R.C. Dubey and D.K. Maheshwari

2 **Mechanisms Used by Plant Growth-Promoting Bacteria** ........ 17
Elisa Gamalero and Bernard R. Glick

3 **Microbial Zinc Solubilization and Their Role on Plants** ........ 47
V.S. Saravanan, M. Rohini Kumar, and T.M. Sa

4 **Effectiveness of Phosphate Solubilizing Microorganism in Increasing Plant Phosphate Uptake and Growth in Tropical Soils** ........ 65
Nelson Walter Osorio

5 **Sulfur-oxidizing Bacteria: A Novel Bioinoculant for Sulfur Nutrition and Crop Production** ........ 81
R. Anandham, P. Indira Gandhi, M. SenthilKumar, R. Sridar, P. Nalayini, and Tong-Min Sa

6 **Role of Siderophores in Crop Improvement** ........ 109
Anjana Desai and G. Archana

7 **Basic and Technological Aspects of Phytohormone Production by Microorganisms: *Azospirillum* sp. as a Model of Plant Growth Promoting Rhizobacteria** ........ 141
Fabricio Cassán, Diego Perrig, Verónica Sgroy, and Virginia Luna

**8 ACC Deaminase Containing PGPR for Potential Exploitation in Agriculture** ........ 183
Venkadasamy Govindasamy, Murugesan Senthilkumar, Pranita Bose, Lakkineni Vithal Kumar, D. Ramadoss, and Kannepalli Annapurna

**9 Quorum-Sensing Signals as Mediators of PGPRs' Beneficial Traits** ........ 209
Leonid S. Chernin

**10 Management of Plant Diseases by Microbial Metabolites** ........ 237
M. Jayaprakashvel and N. Mathivanan

**11 The Role of 2,4-Diacetylphloroglucinol- and Phenazine-1-Carboxylic Acid-Producing *Pseudomonas* spp. in Natural Protection of Wheat from Soilborne Pathogens** ........ 267
Dmitri V. Mavrodi, Olga V. Mavrodi, James A. Parejko, David M. Weller, and Linda S. Thomashow

**12 Plant Root Associated Biofilms: Perspectives for Natural Product Mining** ........ 285
Salme Timmusk and Eviatar Nevo

**13 Bacterial, Fungal, and Plant Volatile Compounds in Reducing Plant Pathogen Inoculums** ........ 301
W.G. Dilantha Fernando and Vidarshani Nawalage

**14 Denitrification Activity in Soils for Sustainable Agriculture** ........ 321
Leticia A. Fernández, Eulogio J. Bedmar, Marcelo A. Sagardoy, María J. Delgado, and Marisa A. Gómez

**Erratum to Chapter 1: Role of PGPR in Integrated Nutrient Management of Oil Seed Crops** ........ E1

**Index** ........ 339

# Contributors

**R. Anandham** Department of Agricultural Microbiology, Agricultural College and Research Institute, Tamil Nadu Agricultural University, Madurai, Tamil Nadu 625 104, India; Department of Agricultural Chemistry, Chungbuk National University, Cheongju, Chungbuk 361-763, Republic of Korea

**Kannepalli Annapurna** Division of Microbiology, Indian Agricultural Research Institute, New Delhi 110012, India, annapurna93@yahoo.co.in

**G. Archana** Department of Microbiology, M.S. University of Baroda, Vadodara 390002, India

**Eulogio J. Bedmar** Departamento de Microbiología del Suelo y Sistemas Simbióticos, Estación Experimental del Zaidin, CSIC, Apartado Postal 419, 18080 Granada, España

**Pranita Bose** Division of Microbiology, Indian Agricultural Research Institute, New Delhi 110012, India

**Fabricio Cassán** Laboratorio de Fisiología Vegetal y de la Interacción planta-microorganismo, Universidad Nacional de Río Cuarto, Campus Universitario, CP 5800, Río Cuarto, Córdoba, República Argentina, fcassan@exa.unrc.edu.ar

**Leonid S. Chernin** Department of Plant Pathology and Microbiology and The Otto Warburg Center for Biotechnology in Agriculture, The Robert H. Smith Faculty of Agriculture, Food and Environment, The Hebrew University of Jerusalem, Rehovot 76100, Israel, chernin@agri.huji.ac.il

**María J. Delgado** Departamento de Microbiología del Suelo y Sistemas Simbióticos, Estación Experimental del Zaidin, CSIC, Apartado Postal 419, 18080 Granada, España

**Anjana Desai** Department of Microbiology, M.S. University of Baroda, Vadodara 390002, India, desai_aj@yahoo.com

**R.C. Dubey** Department of Botany and Microbiology, Faculty of Life Sciences, Gurukul Kangri University, Haridwar 249404, Uttarakhand, India

**Leticia A. Fernández** Departamento de Agronomía, Universidad Nacional del Sur, San Andrés s/n (8000), Bahía Blanca, Provincia de Buenos Aires, República Argentina, fernandezletic@yahoo.com.ar

**W.G. Dilantha Fernando** Department of Plant Science, University of Manitoba, Winnipeg, MB, Canada R3T 2N2, d_fernando@umanitoba.ca

**Elisa Gamalero** Dipartimento di Scienze dell'Ambiente e della Vita, Università del Piemonte Orientale, Viale Teresa Michel 11, Alessandria 15121, Italy, elisa. gamalero@unipmn.it

**P. Indira Gandhi** Krishi Vigyan Kendra, Tamil Nadu Agricultural University, Vriddhachalam, Tamil Nadu 606 001, India

**Bernard R. Glick** Department of Biology, University of Waterloo, Waterloo, Ontario, Canada N2L 3G1

**Marisa A. Gómez** Departamento de Agronomía Universidad Nacional del Sur, San Andrés s/n (8000), Bahía Blanca, Provincia de Buenos Aires, República Argentina

**Venkadasamy Govindasamy** Division of Microbiology, Indian Agricultural Research Institute, New Delhi, 110012, India

**M. Jayaprakashvel and** Biocontrol and Microbial Metabolites Lab, Centre for Advanced Studies in Botany, University of Madras, Guindy Campus, Chennai 600025, India

Department of Biotechnology, AMET University, Kanathur, Chennai 603112, India

**Lakkineni Vithal Kumar** Division of Microbiology, Indian Agricultural Research Institute, New Delhi 110012, India

**M. Rohini Kumar** Division of Industrial Biotechnology, School of Bio Sciences and Technology, Vellore Institute of Technology, Vellore 632014, India

**Sandeep Kumar** Department of Botany and Microbiology, Faculty of Life Sciences, Gurukul Kangri University, Haridwar 249404, Uttarakhand, India, sandeepchokar@gmail.com

**Virginia Luna** Laboratorio de Fisiología Vegetal y de la Interacción planta-microorganismo, Universidad Nacional de Río Cuarto, Campus Universitario, CP 5800, Río Cuarto, Córdoba, República Argentina

**D.K. Maheshwari** Department of Botany and Microbiology, Faculty of Life Sciences, Gurukul Kangri University, Haridwar 249404, Uttarakhand, India, maheshwaridk@gmail.com

**N. Mathivanan** Biocontrol and Microbial Metabolites Lab, Centre for Advanced Studies in Botany, University of Madras, Guindy Campus, Chennai 600025, India, prabhamathi@yahoo.com

**Dmitri V. Mavrodi** Department of Plant Pathology, Washington State University, Pullman, WA 99164-6430, USA, mavrodi@mail.wsu.edu

**Olga V. Mavrodi** Department of Plant Pathology, Washington State University, Pullman, WA 99164-6430, USA

**P. Nalayini** Division of Crop Production, Central Institute for Cotton Research, Regional Station, Indian Council of Agricultural Research, Coimbatore, Tamil Nadu 641 003, India

**Vidarshani Nawalage** Department of Plant Science, University of Manitoba, Winnipeg, MB, Canada R3T 2N2

**Eviatar Nevo** Institute of Evolution, University of Haifa, Mt. Carmel, Haifa, Israel

**Nelson Walter Osorio** Facultad de Ciencias, Universidad Nacional de Colombia, Calle 59 A No. 63-20, Medellin, Colombia, nwosorio@unal.edu.co

**James A. Parejko** School of Molecular Biosciences, Washington State University, Pullman, WA 99164-4234, USA

**Diego Perrig** Laboratorio de Fisiología Vegetal y de la Interacción planta-microorganismo, Universidad Nacional de Río Cuarto, Campus Universitario, CP 5800, Río Cuarto, Córdoba, República Argentina

**D. Ramadoss** Division of Microbiology, Indian Agricultural Research Institute, New Delhi 110012, India

**Tong-Min Sa** Department of Agricultural Chemistry, Chungbuk National University, Cheongju, Chungbuk 361-763, Republic of Korea, tomsa@chungbuk.ac.kr

**Marcelo A. Sagardoy** Departamento de Agronomía Universidad Nacional del Sur, San Andrés s/n (8000), Bahía Blanca, Provincia de Buenos Aires, República Argentina

**V.S. Saravanan** Department of Microbiology, Indira Gandhi College of Arts and Science, Pondicherry 605007, India

**M. SenthilKumar** Department of Agricultural Microbiology, Centre for Plant and Molecular Biology, Tamil Nadu Agricultural University, Coimbatore, Tamil Nadu 641 003, India

**Murugesan Senthilkumar** Division of Microbiology, Indian Agricultural Research Institute, New Delhi 110012, India

**Verónica Sgroy** Laboratorio de Fisiología Vegetal y de la Interacción planta-microorganismo, Universidad Nacional de Río Cuarto, Campus Universitario, CP 5800, Río Cuarto, Córdoba, República Argentina

**R. Sridar** Department of Agricultural Microbiology, Centre for Plant and Molecular Biology, Tamil Nadu Agricultural University, Coimbatore, Tamil Nadu 641 003, India

**Linda S. Thomashow** USDA-ARS, Root Disease and Biological Control Research Unit, Washington State University, Pullman, WA 99164-6430, USA

**Salme Timmusk** Department of Forest Mycology and Pathology, Uppsala BioCenter, SLU, Uppsala, Box 7026, SE-75007, Sweden, salme.timmusk@mykopat.slu.se

**David M. Weller** USDA-ARS, Root Disease and Biological Control Research Unit, Washington State University, Pullman, WA 99164-6430, USA

# Chapter 1
# Role of PGPR in Integrated Nutrient Management of Oil Seed Crops

**R.C. Dubey and D.K. Maheshwari**

Please note the Erratum to this chapter at the end of the book

## 1.1 Introduction

High oil prices contributed to price rise of most agricultural crops by raising input costs on the one hand, and by boosting demand for agricultural crops used as feedstock in the production of alternative energy sources on the other. In the past few decades, it has been increasingly recognized that plant nutrient needs in many countries can best be provided through an integrated use of diverse plant nutrient resources. An integrated plant nutrition system (IPNS) or integrated nutrient management (INM) enables the adaptation of the plant nutrition and soil fertility management from farming systems to site characteristics, taking advantage of the combined and harmonious use of organic, mineral and biofertilizers, nutrient resources to serve the concurrent needs of food production and economic environmental and social viability (Roy et al. 2006). An increase in corn and wheat prices leads to shifting of land under edible oil seeds to grain, and this trend would intensify in 2010–2011 due to heavy speculation in grain (Anonymous 2010a). The total demand for edible oils is expected to increase from the current level of 156 lakh ton (2010) to 208 lakh ton by 2015 at global level. This assumes a modest per capita consumption increase of 4% a year and population growth of 1.8% a year, which translates to an overall growth in demand at the rate of 6% per annum. However, if the per capita consumption growth turns out to be higher at 5% or 6%, then demand will be much higher, approximately touching the figures of 226 lakh ton or 246 lakh ton by 2015 (Anonymous 2010b).

However, the incidence of plant diseases is a major reason for decline in cultivation of oilseeds. The prominent ones among them are charcoal rot of root

R.C. Dubey • D.K. Maheshwari (✉)
Department of Botany and Microbiology, Faculty of Life Sciences, Gurukul Kangri University, Haridwar 249404, Uttarakhand, India
e-mail: maheshwaridk@gmail.com

D.K. Maheshwari (ed.), *Bacteria in Agrobiology: Plant Nutrient Management*,
DOI 10.1007/978-3-642-21061-7_1, © Springer-Verlag Berlin Heidelberg 2011

and stem, fusarial wilt, alternaria, necrosis, rust and leaf spot. Union Government is investing much money for addressing this problem, but still the problem persists due to poor technology transfer as one of the main reasons. Although quality technologies are available, they do not reach to the farmers readily. The proof of this is the huge gap existing between the demonstrated plot yield and the farmers' yield. In addition, edible oil availability in the international market was declining because 20% of the world vegetable oil is being diverted for production of bio-fuels.

Three oilseed crops, i.e., groundnut, soybean and rapeseed/mustard, together account for over 80% aggregate of cultivated oilseed outputs. World's largest edible oil consuming countries are the USA, China, Brazil and India. India contributes about 8% of the world oilseeds production and about 6% of the global production of oils and fats, and currently is the fifth largest vegetable oil economy in the world, after the USA, China, Brazil and Argentina (Ramesh and Hegde 2010). A growing population, increasing rate of consumption and increasing per capita income are accelerating the demand for edible oil globally. The demand is increasing for newer oilseed crops such as sunflower, soybean, other vegetable oil and rice bran oil which are used along with traditionally used sesame, mustard, groundnut and coconut oil, because of increased awareness of health and dietary benefits.

In intensive cropping system, supplementing soil with all nutrients using chemical fertilizer is considered inevitable for obtaining optimum yield of crops, but their utilization efficiency remains low, due to loss by volatization, denitrification, leaching and conversion into unavailable forms (Ayala and Rao 2002). On the other hand, Adesemoye et al. (2009) reported that fertilizers are essential components of modern agriculture because they provide essential elements for growth and development. However, overuse of chemical fertilizers may cause unanticipated environmental impacts. Chemical fertilizers are becoming too expensive day by day and their excessive uses proved hazardous to both human and soil health. Alternative to chemical fertilizers are biofertilizers which help in enhancing the crop yield quality and play a role in imparting resistance to environment stress such as drought, extreme heat, early frost, pests and pathogen problems (Kumar et al. 2006). Plant growth-promoting rhizobacteria (PGPR) in the form of biofertilizer (Cakmakci et al. 2006) and biopesticides (Arora et al. 2008) have been proved to be a boon to agricultural crops. Integration of PGPR with traditional inorganic fertilizers in the field proved to be effective means to increase the availability of nutrients to plants with simultaneous reduction in diseases incidence of oil seed crop has been reported (Kumar et al. 2009).

During last three decades or so, plant growth promotion and biological control of pest and pathogen have been widely applied and now well established and intensively investigated and used commercially in a number of different countries worldwide in sustainable agriculture, silviculture, horticulture and environmental remediation (Kloepper et al. 1989; Jeffries et al. 2003; Reed and Glick 2004; Fravel 2005). Organic substances that are capable of regulating plant growth included oil seed crops produced either endogenously or applied exogenously which regulate growth by affecting physiological and morphological processes at very low concentrations (Arshad and Frankenberger 1991). Involvement of plant growth regulators produced by bacteria such as indole acetic acid (IAA) (Park et al. 2005; Mordukhova et al. 1991; Gupta et al. 1999; Kumar et al. 2005a), gibberrelic acid (Mahmoud et al. 1984), cytokinins (Tein et al. 1979; Garcia de Salamone et al. 2001) secreted from microorganisms has significant effects. Growth regulators such as IAA and cytokinin-producing PGPR were observed in growth promotion of non-leguminous plants (Noel et al. 1996; Hirsch et al. 1997; Patten and Glick 2002; Kumar et al. 2005b).

## 1.2 PGPR vs. Oilseed Crops

Biofertilizers, microbial inoculants that can promote plant growth and productivity, are internationally accepted as an alternative source of fertilizer (Mia et al. 2010). In the past few decades, field and greenhouse inoculation studies with PGPR have been shown that these rhizobacteria are able to promote yield of agriculturally important crops grown under climatic conditions and different soil (Okon and Labandera-Gonzales 1994). However, information gathered on the inoculation of PGPR on oilseed crops as biofertilizers, biocontrol agents and bioenhancer activities has been summarized here and the possibilities for commercial utilization have been explored. Inoculation of oilseed with PGPR has been reported to increase plant biomass, palmitic acid, nitrogen and protein content. The PGPR inoculation increased oil content along with biological control of fungal pathogens (Chandra et al. 2007; Kumar et al. 2009).

Agrochemicals including the chemical fertilizers reduce the population of beneficial microorganisms, due to inhibitory effect on bacterial growth (Smiley 1981). The abnormal morphological changes in *Rhizobium meliloti* cells have been observed by Strzeleowa (1970), by growing *R. meliloti* on throton's agar medium containing 2.5 mg/ml urea. Excessive nitrogenous fertilization increases the generation time and disrupts protein synthesis in *Acetobacter diazotrophicus* (renamed as *Gluconoacetobacter diazotrophicus*) as visualized by Becking (1995). Further, Muthukumarasamy et al. (2002) examined higher population of *G. diazotrophicus* from low fertilized soil in rhizosphere as compared to excessive N-fertilized soil. Such effects of different chemicals on microbes have been documented by a number of researchers (Kantachote et al. 2001; Saraf and Sood 2002; Joshi et al. 2006). Similarly, the application of pesticides on oil seed crop resulted in the accumulation of their residues in seeds, oil and oil cake (Bhatnagar and Gupta 1998). On the other hand, Gricher et al. (2001a) reported a reduction in the height of sesame up to 66% and plant stand (biomass) 8–98% as compared to control, after the application of different herbicides. The seedlings damage and reduction in the crop stand due to application of various chemicals including fungicides, weedicides and pesticides has been reported by several researchers (Shukla 1984; Bansode and Shelke 1991; Gricher et al. 2001b).

Although the exact biocontrol mechanisms are still unknown in many microorganisms (Vassilev et al. 2006), biological methods demonstrate a number of environment-friendly and economically advantageous results over chemical-based control application. A few researchers reported that many plant growth promoters used for inoculation in cropping systems might serve as to enhance nutrient uptake and biocontrol agents for dreaded pathogens. Recently, Arora et al. (2009) reported that biological formulations containing fast-growing *Sinorhizobium meliloti* RMP5 and slow-growing *Bradyrhizobium* BMP1 enhanced the process of symbiotic nitrogen fixation in the soil supplemented with Mo and Fe.

Fluorescent pseudomonads, rhizobia and a few others showed phosphate solubilizing activity (Gupta et al. 1999, 2001, 2002; Arora et al. 2001; Deshwal et al. 2003; Negi et al. 2005; Kumar et al. 2005a, b) that has substantially affected the growth and productivity of oil seed crops, such as sunflower (Reddy et al. 2005), peanut (Parasuraman and Mani 2003; Dey et al. 2004) and mustard crops (Penrose and Glick 2003; Joshi et al. 2006). Arora et al. (2001) isolated siderophore-producing strain of *S. meliloti* from the rhizosphere of *Mucuna pruriens*. Deshwal et al. (2003) reported plant growth-promoting and biocontrol potential of *Bradyrhizobium* sp. on peanuts. Attachment of rhizobia with non-leguminous crops such as maize, wheat, rice, oat, sunflower, mustard and asparagus has been observed (Planziski et al. 1985; Triouchi and Syono 1990; Biswas et al. 2000; Peng et al. 2002; Chandra 2004) that led to plant growth promotion and enhancement of yield (Shanmugam et al. 2002; Yadav et al. 2002). Interactive effect of biofertilizers and INM has been studied on oil seed crops such as mustard (Chauhan et al. 1995; Glick et al. 1997; Vyas 2003; Gudadhe et al. 2005; Singh and Sinsinwar 2006; Joshi et al. 2006), soybean (Dubey 1998; Singh and Abraham 2001), groundnut (Meena and Gautam 2005), sunflower (Rao and Soren 1991) and sesame (Kumar et al. 2009; Aeron et al. 2010).

### *1.2.1 Sesame*

Sesame (*Sesamum indicum* L) is one of the oldest oilseed crops in the world. Archeological records indicate that it has been used in India for more than 5,000 years (Bedigian 2004). Adverse effects of different groups of chemicals have been observed on sesame by Gricher et al. (2001a, b). Accumulation of pesticides in sesame seeds, oil and oil cake is a serious concern as reported by Bhatnagar and Gupta (1998). Sesame oil is good for health due to the presence of antioxidants such as sesamolin, sesamin and sesamol (Suja et al. 2004) and the presence of low level of saturated fatty acids and antioxidants. In fact, sesame oil reduces the incidence of cancer (Hibasami et al. 2000; Miyahara et al. 2001), hypertension and the cholesterol level in human beings (Lemcke-Norojarvi et al. 2001; Sankar et al. 2004). International demand for sesame seeds and oil is continuously increasing with the growing health consciousness and increasing knowledge on the dietary and health benefits.

Wherever sesame is grown, it is liable to attack by various pathogenic fungi (Abd-El-Ghany et al. 1974). Among the fungal diseases, charcoal rot of sesame caused by *Macrophomina phaseolina* is the most devastating disease (Dinakaran and Manoharan 2001) of the crop in India with reports of about 50% disease incidence resulting in heavy yield losses (Chattopadhyay and Kalpna 2002) and quarantine processing in more than 2,000 germplasm samples of sesame observed the interception of 12 pathogenic fungi of high economic significance (Agarwal et al. 2006). Integrated use of organic, inorganic and biofertilizers may sustain productivity of sesame by improving soil physical conditions and may also reduce the costly inorganic fertilizer needs (Deshmukh et al. 2002; Joshi et al. 2006; Kumar et al. 2009). Kumar et al. (2005b) reported an inoculation of *P. fluorescens* along with chemical fertilizers is an effective way to reduce the *Meloidogyne* spp. of sesame. Recently, Aeron et al. (2010) reported differential response of sesame under the influence of indigenous and non-indigenous rhizosphere-competent fluorescent pseudomonads.

### *1.2.2 Canola*

The *Brassica* oil seed crops including *Brassica campestris*, *B. napus* L., *B. rapa* L. and *B. juncea* are the world's third most important source of oil seed and edible oil (Raj et al. 1997). Largely in response to the continuing increase in world edible oil demand, unfortunately the crops suffer with about 26 diseases which caused due to fungi, bacteria, viruses, insects etc. at different stages of development and during storage (Mehta et al. 1996). Out of which, fungal diseases stem rot (*Xanthomonas compestris*), fusarium wilt (*Fusarium oxysporium*), charcoal rot (*Macrophomina phaseolina*) and damping off (*Pythium debaryanum*) are important and impart significant change in the field of international crop protection research. Chandra et al. (2007) reported that *Mesorhizobium loti* MP6 is an efficient root colonizer of *B. campestris* and not only increases growth and yield but also controls white rot disease significantly.

Earlier, Penrose and Glick (2003) reported bacterial strains that could utilize ACC as sole source of nitrogen from the soil. All of these PGPR strains induced root elongation in canola seedling under the gnotobiotic conditions. On the other hand, Lifshitz et al. (1987) reported that *P. putida* stimulated the growth of root and shoot and increased $^{32}$P-labeled phosphate uptake in canola. Noel et al. (1996) reported direct involvement of IAA and cytokinin production by PGPR in the growth of canola. Asghar et al. (2002) showed significant correlation between auxin production by PGPR in vitro and vegetative and yield attributes i.e., the number of pods per plant and grain yield of *B. juncea*. Recently, Maheshwari et al. (2011) reported co-inoculation of *S. meliloti* RMP1 and *P. aeruginosa* GRC2 with urea and diammonium phosphate (DAP) enhanced biomass and yield of *B. juncea* as compared to control. The yield was better with application of half and full doses of recommended fertilizers (Table 1.1).

**Table 1.1** Effect of integrated use of chemical fertilizers and co-inoculants (RMP1 + GRC2) on growth and yield of *Brassica juncea* after 120 DAS

| | | Root | | | Shoot | | | Yield | |
|---|---|---|---|---|---|---|---|---|---|
| Treatments | Seed germination (%) | Length (cm) | Fresh wt. (g) | Dry wt. (g) | Length (cm) | Fresh wt. (g) | Dry wt. (g) | No. of siliquae per plant | Seed yield per hectare (kg) |
| RMP1 | 85 | 15.2 | 13.7 | 8.4 | 180 | 86 | 41 | 246 | 631 |
| $GRC_2$ | 87 | 15.9 | 14.1 | 8.9 | 184 | 89 | 43 | 249 | 639 |
| $N_{25+25}$ $P_{20}$ | 89 | 16.8 | 15.3 | 9.2 | 190 | 96 | 52 | 291 | 781 |
| $N_{50+50}$ $P_{20+20}$ | 88 | 19.8 | 20.1 | 11.3 | 201 | 106 | 62 | 337 | 981 |
| RMP1 + $GRC_2$ | 90 | 19.5 | 20.0 | 10.1 | 196 | 101 | 60 | 315 | 978 |
| RMP1 + $GRC_2$ + $N_{25+25}$ $P_{20}$ | 95 | 20.2 | 21.8 | 12.4 | 204 | 108 | 69 | 341 | 989 |
| Control | 79 | 11.2 | 10.7 | 5.9 | 146 | 47 | 28 | 146 | 467 |
| SEM | 1.46 | 0.86 | 0.75 | 0.37 | 0.51 | 0.29 | 0.43 | 0.79 | 0.83 |
| CD @ 1% | 6.53 | 3.85 | 3.36 | 1.65 | 2.31 | 1.33 | 1.94 | 3.59 | 3.71 |
| CD @ 5% | 4.6 | 2.71 | 2.36 | 1.16 | 1.62 | 0.93 | 1.36 | 2.50 | 2.61 |

Values are mean of ten randomly selected plants from each set, $N_{50+50}$ $P_{20+20}$ full doses of chemical fertilizers, $N_{25+25}$ $P_{20}$ half dose of chemical fertilizers (*Source* Maheshwari et al. 2011)

### *1.2.3 Ground Nut*

Groundnut (*Arachis hypogaea* L.) is a major oilseed and food crop of the semi-arid tropics. The late leaf spot disease of groundnut, caused by the fungus, *Cercosporidium personatum*, almost co-exists with the crop and contributes to significant loss in yield throughout the world. Leaf spots can cause up to 53% loss in pod yield and 27% loss in seed yield (Patel and Vaishnava 1987). Smith (1992) reported pod loss of 10–50% by late leaf spot disease. Control of this disease mainly depends on fungicides, although considerable effort has been invested in developing biocontrol methods (Meena et al. 2002). On the other hand, Jadhav and Desai (1996) reported a *Rhizobium* isolate, increased plant growth and chlorophyll content in groundnut. This observation might explain in part the rhizobia-enhanced mineral uptake in groundnut tissues (Howell 1987). Gupta et al. (2002) found reduced disease incident, better vegetative growth parameters and ultimately enhanced grain yield in peanut by the addition of *P. aeruginosa* $GRC_2$ in *M. phaseolina*-infested field soil. Dey et al. (2004) reported fluorescent *Pseudomonads* isolates increased the growth and yield of peanut. Recently, Bhatia et al. (2008) reported increased seed germination, growth promotion and suppression of charcoal rot due to *M. phaseolina* with fluorescent pseudomonads. Earlier, Arora et al. (2001) observed enhanced seed germination, seedling biomass and nodule weight with reduced disease incidence in groundnut. Similarly, Meena et al. (2006) applied *P. fluorescens* for plant growth and in biocontrol of late leaf spot caused by *C. personatum* in groundnut. Seed treated with *P. fluorescens* strain Pf1 recorded the highest seed germination percentage and the maximum plant height with significantly controlled late leaf spot disease of groundnut resulting in an increased pod yield. Effect of INM with PGPR studied in peanut and reported significant positive effect over the control (Parasuraman and Mani 2003).

HCN-producing fluorescent *P. aeruginosa* $GRC_1$ and *P. aeruginosa* $GRC_2$ (Gupta et al. 1999, 2001; Bhatia et al. 2003, 2005) inhibited charcoal rot of peanut and sclerotial rot of *Brassica* caused by *M. phaseolina* and *Sclerotiorum sclerotonium*, respectively. Plant growth-promoting fluorescent pseudomonads applied through seed bacterization checked charcoal rot due to *M. phaseolina* in peanut. About 45–68% reduction in disease symptoms has been recorded in comparison with peanut crop raised in *M. phaseolina*-infected soil (Bhatia et al. 2003). This clearly indicated the role of fluorescent *pseudomonads* in the charcoal rot disease suppression caused by *M. phaseolina*. Similarly, Gupta et al. (2002) observed a reduction in disease incident, better vegetative plant growth and ultimately enhanced grain yield in peanut by application of *P. aeruginosa* $GRC_2$ in *M. phaseolina*-infested field soil. Shanmugam et al. (2002) co-inoculated peanut seeds with *Pseudomonas* and *Rhizobium* and reduction in root rot incidence and significant improvement have been observed on plant growth and health in comparison with control.

### *1.2.4 Sunflower*

Shehata and El-Khawas (2003) examined the effect of two biofertilizers on sunflower (*Helianthus annus* L. cv. Vedock) and showed an increase in seed yield, nutrient content of seeds, nitrogen and all nitrogenous compound, mineral and seeds' oil contents and also showed a decrease in the saturated fatty acid as palmitic and stearic, while an increase in the main unsaturated fatty acid (oleic, linoleic and linolinic) and they induced the synthesis of two new protein of low molecular weight (2.1 kDa for biogien and 14.9 kDa for microbien). Application of phosphorous and phosphate solubilizing bacteria (PSB) has been reported promising in seed yield, oil content and oil yield in sunflower as compared with uninoculated control (Patel and Thakur 2003; Jones and Sreenivasa 1993). Integrated use of N, P and K together gave better seed yield in sunflower as compared with their individual application (Rao and Soren 1991). Fertilization pattern of different fertilizers (N, P, K and S) has been developed for the sunflower and observed that application of $N_{80}$, $P_{17.5}$ and $K_{33.5}$ resulted in a higher yield (Mandal et al. 2003).

Application of biofertilizers, *Azospirillum*, *Phosphobacterium* along with inorganic fertilizers and nutrient sources (NPK) resulted in higher growth parameters, stalk and seed yield in sunflower than sole application of nutrient sources (Reddy et al. 2005). Application of phosphorous and PSB has been reported promising in seed yield, oil content and oil yield in sunflower as compared with uninoculated control (Patel and Thakur 2003; Jones and Sreenivasa 1993). Some pseudomonads namely, *P. fluorescens* and *P. putida* produced HCN, inhibited the wilt of sunflower due to *Sclerotinia* spp. (Epert and Digat 1995). Forchetti et al. (2007) isolated endophytic bacteria *B. pumilus* and *Azospirillum xiloxosidans* form sunflower which produced jasmonates and abscisic acid in culture medium.

### *1.2.5 PGPR vs. Chemical Fertilizers for Oil Crops*

Amir et al. (2003) reported that inoculation of the rhizobacteria in combination with inorganic N enhanced oil-palm seedlings' stimulation. Similarly, integrated use of *Rhizobium*, phosphorous-solubilizing microorganism (PSM) along with recommended dose of inorganic fertilizer showed maximum dry matter, seed oil content and seed yield of soybean (Lanje et al. 2005). Nowadays, it is an endeavor to blend ecology and economy in a cost–benefit frame work. Hence considerable attention has been paid to PGPR as the best combination with chemical fertilizer (Mohiuddin et al. 2000).

The integrated approaches achieve the health of soil and plants' environment. N, P, K and S are the basic nutrients for the plant. As nitrogen is required from tillering to flowering, two splits of fertilizers, N is more responsive to production than a single dose prior to planting. Phosphate must be applied prior to planting in one dose in the nutrient and required at the beginning, and it has the ability to spread at the start and spread over the needed areas. Potash may be applied at any stage prior to flowering. Throughout the world a very few agriculturists have the luxury of production of crop (yield) organically using microbial fertilizer and or pesticides. Although the increase in food production in the past 50 years has been associated with the use of synthetic chemicals, especially inorganic fertilizers and pesticides, yet it has disturbed soil environment, contaminated underground water, resulted in the development of resistant races of pathogens, and caused health risks to humans. In addition, chemicalized farming is costly and causes the loss of inherent soil fertility and disturbance in microbial ecology. The ecological and economic constraints require this biological alternative approach. Development of crop and eco-specific bioinoculants (biological control agents, bio-fertilizers and bio-degraders) is perceived as one of the most exciting areas owing to its potential of solving multifarious agricultural and environmental problems concurrently.

## 1.3 PGPR vs. Integrated Nutrient Management

Fertilizers are the carrier of poisonous heavy metals such as lead, cadmium, etc. Some of the synthetic fertilizers may contribute a significant quantity of heavy metals to the soil due to their use for a long time. Consequently, the amounts of the heavy metals in soil ecosystem may buildup up to toxic levels for plants, animals and human health (Ram 1994). For maintenance and enhancement of soil fertility, appropriate sustainable technology is required in order to replenish the nutrients so as to maintain and build up the soil nutrient status (Hera 1996). Long-term studies have been carried out at several locations on different cropping systems (Belay et al. 2002; Katayal et al. 2003) indicated that the application of all the required nutrients through chemical fertilizers has deleterious effect on soil health, leading to unsustainable yield. On the other hand, biofertilizers in combination with chemical fertilizers (NPK) or alone have been reported to improve growth and yield of *Brassica* crop (Vyas 2003). Large-scale application of PGPR to crops as inoculants seems to be attractive as it would substantially reduce the use of chemical fertilizers and pesticides (Singhal et al. 2003).

Nutrient management is the key issue in sustainable soil fertility. Integrated use of N, P and K together gave better seed yield in sunflower as compared with their individual application (Rao and Soren 1991). Fertilization pattern of different fertilizers (N, P, K and S) has been developed for the sunflower and observed that application resulted higher yield (Mandal et al. 2003). Alagawadi and Gaur (1992) reported inoculation of rhizobia with reduced doses of nitrogen for reduction of the nitrogen fertilization. However, Valladares et al. (2002) reported low fertilizers and rhizobial inoculants for achieving high-quality seedlings. Integrated nutrient supply seems to be essential not only for increasing the crop productivity but also for the maintenance and possibly improvement of soil fertility for sustainable crop productivity (Belay et al. 2002; Shrotriya 2005). Results from various cropping system showed the positive interaction of the integrated use of mineral fertilizers, organic manures, and biofertilizers for maintaining the growth throughout the crop duration (Halder et al. 1990; Ghosh and Das 1998; Ayala and Rao 2002; Wu et al. 2005). Additions of biofertilizers not only help in proliferation of beneficial microbes in soil but also provide residual effect on subsequent crops (Deshwal et al. 2006) and help in decomposition of organic matter (Patersion 2003) and availability of nutrients to plants (Amir et al. 2003; Gudadhe et al. 2005; Cakmakci et al. 2006). Biofertilizers besides improving the physical, chemical and biological properties of soils and resulting into better agricultural environment. Microbial inoculants as a source of biofertilizers have become a hope in most countries as far as economical and environmental view points are concerned (Singhal et al. 2003; Kloepper 2003; Gudadhe et al. 2005; Kumar et al. 2006; Tilak et al. 2006).

## 1.4 PGPR and Oilseed Crop Improvement

Application of biofertilizers, *Azospirillum*, *Phosphobacterium* along with inorganic fertilizers and nutrient sources (NPK) resulted in higher growth parameters, stalk and seed yield in sunflower than during the sole application of nutrient sources (Reddy et al. 2005). Integrated use of *Rhizobium*, PSM, along with recommended dose of inorganic fertilizer showed maximum dry matter, seed oil content and seed yield of soybean (Lanje et al. 2005). Microbial inoculants not only increase the nutritional assimilation but also improve soil properties, such as organic matter content and total N in soil (Wu et al. 2005). Experimental findings revealed the functional and structural soil microbial property influenced by organic and inorganic fertilization (N, P and K) in long-term field experiments. Earlier, Penrose and Glick (2003) obtained bacterial strains that could utilize ACC as sole source of nitrogen from the soil. All of these PGPR strains induced root elongation in canola seedling under the gnotobiotic conditions.

## 1.5 Conclusion

Scanty information is available regarding the nature of biochemical changes occurred due to the application of INM, growth hormones and biofertilizers. The production constraints facing each of the oilseed crops are diverse in nature. However, also some common production constraints exist which are applicable across all the oilseed crops. A multitude of factors such as cultivation in submarginal and marginal lands, poor management of the crop with little or high or no nutrient inputs, ignorance of biofertilizers and INM are responsible for low oilseed crops yields. Therefore, it is essential that these aspects must be studied in depth under diverse soil and climatic conditions across the world on intensive long-term basis research for each agro ecological region. Hence, research focus is required on enhancing fertilizer use efficiency and reduction in use of synthetic chemicals including inorganic fertilizers, pesticides and fungicides. This aspect can be strengthened by following interactive effect of biofertilizers with INM approach for the production of high-quality, safe and affordable edible oil in sustainable manner.

**Acknowledgement** Thanks are due to UCOST (Dehradun), UGC and CSIR (New Delhi) for providing financial support in the form of research project to D.K.M. The authors are thankful to Dr. Piyush Pandey, Associate Professor, Department of Biotechnology, S. B. S. P. G. Institute of Biomedical Sciences and Research, Balawala, Dehradun, Uttarakhand, India, for critical reading of the manuscript.

## References

Abd-El-Ghany AK, Seoud MB, Azab MW, Mahmoud BK, El-Alfy KAA, Abd-El-Gwad MA (1974) Tests with different varieties and strains of sesame for resistance to root rot and wilt diseases. Agric Res Rev 52:75–83

Adesemoye AO, Torbert HA, Kloepper JW (2009) Plant growth-promoting rhizobacteria allow reduced application rates of chemical fertilizers. Plant Microb Interact. doi:10.1007/s00248-009-9531-y

Aeron A, Pandey P, Maheshwari DK (2010) Differential response of sesame under influence of indigenous and non-indigenous rhizosphere competent fluorescent pseudomonads. Curr Sci 99 (2):166–168

Agarwal PC, Dev V, Singh B, Rani I, Khetarpal RK (2006) Seed-borne fungi detected in germplasm of *Sesamum indicum* L. introduced into India during last three decades. Ind J Microbiol 46(2):161–164

Alagawadi AR, Gaur AC (1992) Inoculation of *Azospirillum brasilense* and phosphate solubilizing bacteria on yield of sorghum (*Sorghum bicolor* L. *moench*) in dry land. Trop Agric 69:347–350

Amir HG, Shamsuddin ZH, Halimi MS, Ramlan MF, Marziah (2003) $N_2$ fixation, nutrient accumulation and plant growth promotion by rhizobacteria in association with oil Palm seedlings. Pak J Biol Sci 6(4):1269–1272

Anonymous (2010a) The Financial Express, Wednesday, 10 November

Anonymous (2010b) The Financial Express, Wednesday, 11 November

Arora NK, Kang SC, Maheshwari DK (2001) Isolation of siderophore producing strains of *Rhizobium meliloti* and their biocontrol potential against *Macrophomina phaseolina* that causes charcoal rot of groundnut. Curr Sci 81:673–677

Arora NK, Khare E, Oh JH, Kang SC, Maheshwari DK (2008) Diverse mechanisms adopted by fluorescent *Pseudomonas* PGC2 during the inhibition of *Rhizoctonia solani* and *Phytophthora capsici*. World J Microbiol Biotechnol 24:581–585

Arora NK, Khare E, Singh S, Maheshwari DK (2009) Effect of Al and heavy metals on enzymes of nitrogen metabolism of fast and slow growing rhizobia under explanta conditions. World J Microbiol Biotechnol. doi:10.1007/s11274-009-0237-6

Arshad M, Frankenberger WT Jr (1991) Microbial production of plant hormones. Plant Soil 133:1–8

Asghar HN, Zahir ZA, Arshad M, Khaliq A (2002) Relationship between in vitro production of auxins by rhizobacteria and their growth-promoting activities in *Brassica juncea* L. Biol Fertil Soils 35:231–237

Ayala S, Rao EVSP (2002) Perspective of soil fertility management with a focus on fertilizer use for crop productivity. Curr Sci 82:797–807

Bansode BU, Shelke DK (1991) Integrated weed management in *Sesamum*. J Maha Agric Univ 16:275

Becking JH (1995) Pleomorphic in *Azospirillum*. In: Klingmuller W (ed) *Azospirillum* III genetics physiology ecology. Springer, Heidelberg, pp 243–262

Bedigian D (2004) History and lore of sesame in Southwest Asia. Econ Bot 58:329–353

Belay A, Claassens AS, Wehner FC (2002) Effect of direct nitrogen and potassium and residual phosphorous fertilizers on soil chemical properties, microbial components and maize yield under long-term crop rotation. Biol Fertil Soils 35:420–427

Bhatia S, Dubey RC, Maheshwari DK (2003) Antagonistic effect of fluorescent pseudomonads against *Macrophomina phaseolina* that causes charcoal rot of groundnut. Ind J Expt Biol 41:1442–1446

Bhatia S, Dubey RC, Maheshwari DK (2005) Enhancement of plant growth and suppression of collar rot of sunflower caused by *Sclerotium rolferii* through fluorescent pseudomonads. Ind Phytopathol 58:17–24

Bhatia S, Maheshwari DK, Dubey RC, Arora DS, Bajpai VK, Kang SC (2008) Beneficial effects of fluorescent pseudomonads on seed germination, growth promotion, and suppression of charcoal rot in groundnut (*Arachis hypogea* L). J Microbiol Biotechnol 18(9):1578–1583

Bhatnagar A, Gupta A (1998) Chlorpyriphos, Quinalphos, and Lindane residues in sesame (*Sesamum indicum* L.) seed and oil. Bull Envirn Contam Toxicol 60:569–600

Biswas JC, Ladha JK, Dazzo FB (2000) Rhizobia inoculation improves nutrient uptake and growth of lowland rice. Sci Soc Am J 64:1344–1650

Cakmakci R, Donmez F, Adyin A, Sahin F (2006) Growth promotion of plants by plant growth-promoting rhizobacteria under greenhouse and two different field soil conditions. Soil Biol Biochem 38:1482–1487

Chandra S (2004) Impact of rhizobia and chemical nutrients status on productivity of non-leguminous crop (*Brassica compestris* L. var. local), PhD Thesis, Gurukul, Kangri University, Haridwar (India)

Chandra S, Choure K, Dubey RC, Maheshwari DK (2007) Rhizosphere competent *Mesorhizobium loti* MP6 induced root hair curling, inhibits *Sclerotinia sclerotiorum* and enhance growth of Indian mustard (*Brassica compestris*). Braz J Microbiol 38:24–130

Chattopadhyay C, Kalpna S (2002) Combining viable disease control tools for management of sesame stem root rot caused by *Macrophomina phaseolina* (Tassi) Goid. Ind J Plant Prot 30 (2):132–138

Chauhan DR, Paroda S, Kataria OP, Singh KP (1995) Response of Indian mustard (*Brassica juncea)* to biofertilizers and nitrogen. Ind J Agron 40(1):86–90

Deshmukh MR, Jain HC, Duhoon SS, Goswami U (2002) Integrated nutrient management in sesame (*Sesamum indicum* L.) for Kymore zone of Madhya Pradesh. J Oilseeds Res 19:73–75

Deshwal VK, Dubey RC, Maheshwari DK (2003) Isolation of plant growth promoting *Bradyrhizobium* (Arachis) sp. with biocontrol potential against *Macrophomina phaseolina* causing charcoal root of peanut. Curr Sci 84:443–448

Deshwal VK, Kumar T, Dubey RC, Maheshwari DK (2006) Long term effect of *Pseudomonas aeruginosa* $GRC_1$ on yield of subsequent crops of paddy after mustard seed bacterization. Curr Sci 91(4):423–424

Dey R, Pal KK, Bhatt DM, Chauhan SM (2004) Growth promotion and yield enhancement of peanut (*Arachis hypogaea* L.) by application of plant growth-promoting rhizobacteria. Microbiol Res 159(4):371–394

Dinakaran D, Manoharan N (2001) Identification of resistant sources to root rot of sesame caused by *Macrophomina phaseolina* (Tassi.) Goid. Sesame and Safflower. Newslett 16:68–71

Dubey SK (1998) Response of soybean (*Glycin max*) to biofertilizer with and without nitrogen, phosphorous and potassium on swell-shrink soil. Ind J Agron 43(3):546–549

Epert JM, Digat B (1995) Biocontrol of *Sclerotinia* wilt of sunflower by *Pseudomonas fluorescens* and *Pseudomonas putida* strains. Can J Microbiol 41:685–691

Forchetti G, Masciarelli O, Alemano S, Alvarez D, Abdala G (2007) Endophytic bacteria in sunflower (*Helianthus annuus* L.) isolation, characterization, and production of jasmonates and abscisic acid in culture medium. Appl Microbiol Biotechnol 76:1145–1152

Fravel DR (2005) Commercialization and implementation of biocontrol. Annu Rev Phytopathol 43:337–359

Garcia de Salamone IE, Hynes RK, Nelson LM (2001) Cytokinin production by plant growth-promoting rhizobacteria and selected mutants. Can J Microbiol 47:404–411

Ghosh DC, Das AK (1998) Effect of biofertilizers and growth regulators on growth and productivity of potato (*Solanum tuberosum*). Ind Agric 42:109–113

Glick BR, Liu C, Ghosh S, Dumbroff EB (1997) Early development of canola seedling in the presence of plant growth-promoting rhizobacterium *Pseudomonas putida* GR12-2. Soil Biol Biochem 29:1233–1239

Gricher WJ, Sestak DC, Brewer KD, Besler BA, Stichler R, Charles SDT (2001a) Sesame (*Sesamum indicum*) tolerance and weed control with soil-applied herbicides. Crop Protect 20 (5):389–394

Gricher WJ, Sestak DC, Brewer KD, Besler BA, Stichler RC, Smith DT (2001b) Sesame (*Sesamum indicum*) tolerance and with various post-emergence herbicides. Crop Protect 20(8):389–394

Gudadhe NN, Mankar PS, Dongarkar KP (2005) Effect of biofertilizers on growth and yield of mustard (*Brassica juncea* L). Soils Crops 15(1):160–162

Gupta CP, Sharma A, Dubey RC, Maheshwari DK (1999) *Pseudomonas aeruginosa* as a strong antagonist of *Macrophomina phaseolina* and *Fusarium oxysporum*. Cytobios 99:183–189

Gupta CP, Sharma A, Dubey RC, Maheshwari DK (2001) Effect of metal ions on the growth of *Pseudomonas aeruginosa* and siderophore and protein production. Ind J Exp Biol 39:1318–1321

Gupta CP, Dubey RC, Maheshwari DK (2002) Plant growth enhancement and suppression of *Macrophomina phaseolina* causing charcoal rot of peanut by fluorescent *Pseudomonas*. Biol Fertil Soils 35:399–405

Halder AK, Mishra AK, Bhattacharya P, Chakrabartty PK (1990) Solubilization of rock phosphate by *Rhizobium* and *Bradyrhizobium*. J Gen Appl Microbiol 36:81–92

Hera C (1996) The role of inorganic fertilizers and their management practices (ed. Rodriguez – Barrueco, C.). Fert Environ 131–149

Hibasami T, Fujikawa T, Takeda H, Nishibe S, Satoh TF, Nakashima K (2000) Induction of apoptosisi by *Acanthopanax senticosus* HARMS and its component, sesamin in human stomach cancer KATO III cells. Oncol Rep 7:1213–1216

Hirsch AM, Fang Y, Asad S, Kapulnik Y (1997) The role of phytohormones in plant microbes symbiosis. Plant Soil 194:171–184

Jadhav RS, Desai AJ (1996) Affect of mutational and environmental condition on siderophore production by apnea Rhizobium GNI peanut inset. Ind J Exp Biol 34:436–439

Jeffries P, Gianinazzi S, Perotto S, Turnau K, Barea JM (2003) The contribution of arbuscular mycorrhizal fungi in sustainable maintenance of plant health and soil fertility. Biol Fertil Soils 37:1–16

Jones NP, Sreenivasa MN (1993) Response of sunflower to the inoculation of VA mycorrhiza and/or phosphate solubilizing bacteria in black clayey soil. J Oilseeds Res 10(1):86–92

Joshi KK, Kumar V, Dubey RC, Maheshwari DK (2006) Effect of chemical fertilizer adaptive variants, *Pseudomonas aeruginosa* $GRC_2$ and *Azotobacter chroococcum* $AC_1$ on *Macrophomina phaseolina* causing charcoal rot of *Brassica juncea*. Kor J Env Agric 25(3):228–235

Kantachote D, Naidu R, Singleton I, McClure N, Harch BD (2001) Resistance of microbial populations in DDT-contaminated and uncontaminated soils. Appl Soil Ecol 16:85–90

Katayal V, Gangwar KS, Gangwar B (2003) Long term effect of fertilizer use on yield sustainability and soil fertility in rice-wheat system in sub-tropical India. Fert News 48(7):43–46

Kloepper JW (2003) 6th International PGPR workshop. 81–90. Indian Institute of Spices Research, Calicut, India

Kloepper JW, Lifshitz R, Zablotowich RK (1989) Free-living bacterial inoculum for enhancing crop productivity. Trends Biotechnol 7:39–43

Kumar T, Bajpai VK, Maheshwari DK, Kang SC (2005a) Plant growth promotion and suppression of root disease complex due to *Meloidogyne incognita* and *Fusarium oxysporum* by fluorescent pseudomonads in tomato. Agric Chem Biotechnol 48:79–83

Kumar T, Kang SC, Maheshwari DK (2005b) Nematicidal activity of some fluorescent pseudomonads on cyst forming nematode, *Heterodera cajani* and growth of *Sesamum indicum* var. RT1. Agric Chem Biotechnol 48(4):161–166

Kumar P, Tripathi N, Verma O (2006) Biofertilizer: a boon for agriculture. Agri Update 1(3):45–47

Kumar S, Pandey P, Maheshwari DK (2009) Reduction in dose of chemical fertilizers and growth enhancement of sesame (*Sesamum indicum* L.) with application of rhizospheric competent *Pseudomonas aeruginosa* LES4. Eur J Soil Biol 45:334–340

Lanje PW, Buldeo AN, Zade SR, Gulhane VG (2005) The effect of Rhizobium and phosphorous solubilizers on nodulation, dry matter, seed protein, oil and yield of soybean. J Oilseeds Res 15(1):132–135

Lemcke-Norojarvi M, Kamal-Eldin A, Appelqvist LA, Dimberg MO, Vessby B (2001) Corn and sesame oils increase serum gamma-tocopherol concentrations in healthy Swedish women. J Nutr 131:1195–1201

Lifshitz R, Kloepper JW, Kozlwski M, Cacison J, Tipping EM, Zalestha I (1987) Growth promotion of canola (rape seed) seedling by a strain of *Pseudomonas putida* under gnotobiotic conditions. Can J Microbiol 33:390–395

Maheshwari DK, Kumar S, Kumar B, Pandey P (2011) Co-inoculation of urea and DAP tolerant *Sinorhizobium meliloti* and *Pseudomonas aeruginosa* as integrated approach for growth enhancement of *Brassica Juncea*. Ind J Microbiol 50(4):425–431

Mahmoud SAZ, Ramadan EM, Thabet FM, Khater T (1984) Production of plant growth promoting substances by Rhizosphere microorganisms. Zbl Mikrobiol 139:227–232

Mandal BK, Mandal BB, Murmu NN, Tudu BC, Khanda CM (2003) Growth and yield of sunflower as influenced by dates of sowing and levels of N, P and K. Ind J Plant Physiol 8:103–105

Meena R, Gautam RC (2005) Effect of integrated nutrient management on productivity, nutrient uptake and moisture-use functions of pearlmillet (*Pennisetum glaucum*). Ind J Agron 50(4):305–307

Meena B, Radhajeyalakshmi R, Marimuthu T, Vidhyasekaran P, Velazhahan R (2002) Biological control of groundnut late leaf spot and rust by seed and foliar applications of a powder formulation of *Pseudomonas fluorescens*. Biocont Sci Technol 12:195–204

Meena B, Marimuthu T, Velazhahan R (2006) Role of fluorescent pseudomonads in plant growth promotion and biological control of late leaf spot of groundnut. Acta Phytopathol Entomol Hung 41(3–4):203–212.

Mehta N, Saharan GS, Kaushik CD (1996) Efficiency and economic of fungicidal management of white rust and downey mildew complex in rapeseed and mustard. Ind J Mycol Plant Pathol 26(3):243–247

Mia MAB, Shamsuddin ZH, Mahmood M (2010) Use of plant growth promoting bacteria in banana: a new insight for sustainable banana production. Int J Agric Biol 12(3):459–467

Miyahara Y, Hibasami H, Katsuzaki H, Komiya Imai K (2001) Sesamolin from sesame seed inhibits proliferation by inducing apoptosis in human lymphoid leukemia Molt 4B cells. Int J Mol Med 7:369–371

Mohiuddin MD, Das AK, Ghosh DC (2000) Growth and productivity of wheat as influenced by integrated use of chemical fertilizer, biofertilizers and growth regulator. Ind Plant Physiol 5:334–338

Mordukhova EA, Skvortsova NP, Kochaetko VV, Dubeikovskii AN, Boronin AM (1991) Synthesis of the phytohormone indole-3 acetic acid by rhizosphere bacteria by the genus *Pseudomonas*. Mikrobiologiya 60:494–500

Muthukumarasamy R, Revathi G, Loganathan P (2002) Effect of inorganic N on the population, *in vitro* colonization and morphology of *Acetobcater diazotrophicus* (syn. *Gluconoacetobacter diazotrophicus*). Plant Soil 243:91–102

Negi YK, Garg SK, Kumar J (2005) Cold tolerant fluorescent *Pseudomonas* isolates from Garhwal Himalayas as potential plant growth promoting and biocontrol agents in pea. Curr Sci 89:2151–2156

Noel TC, Sheng C, York CK, Pharis RP, Hynes MF (1996) *Rhizobium leguminasorum* as a plant growth-promoting rhizobacterium: direct growth promotion of canola and lettuce. Can J Microbiol 42:279–283

Okon Y, Labandera-Gonzales CA (1994) Agronomic applications of Azospirillum: an evaluation of 20 years worldwide field inoculation. Soil Biol Biochem 26:1591–1601

Parasuraman P, Mani AK (2003) Integrated nutrient management for groundnut (*Arachis hypogaea*)-horsegram (*Macrotyloma uniflorum*) cropping sequence under rainfed Entisol. Ind J Agro 48(2):82–85

Park M, Kim C, Yang J, Lee H, Shin W, Kim S, Sa T (2005) Isolation and characterization of diazotrophic growth promoting bacteria from rhizosphere of agricultural crops of Korea. Microbial Res 160(2):127–133

Patel SR, Thakur DS (2003) Effect of crop geometry, phosphorous levels and phosphate solubilizing bacteria on growth, yield and oil content of sunflower (*Helianthus annuus*). J Oilseeds Res 20(1):153–154

Patel VA, Vaishnava MV (1987) Assessment of losses in groundnut due to rust and tikka leaf spots in Gujarat. Res J Guj Agric Uni 12:52–53

Patersion E (2003) Importance of rhizodeposition in the coupling of plant and microbial productivity. Eur J Soil Sci 54:741–750

Patten CL, Glick BR (2002) Role of Pseudomonas putida indole acetic acid in development of the host plan root system. Appl Environ Microbiol 68: 3795–3801.

Peng GX, Tan ZY, Wang ET, Reinhold-Hurek B, Chen WF, Chen WX (2002) Identification of isolates from soybean nodules in Xinjiang region as *Sinorhizobium xinjiangense* and genetic differentiations of *S. xinjiangense* from *Sinorhizobium fredii*. Int J Syst Evol Microbiol 52:457–462

Penrose DM, Glick BR (2003) Methods for isolating and characterizing ACC deaminase-containing plant growth-promoting rhizobacteria. Physiol Plant 118:10–15

Planziski J, Innes RW, Rolfe BG (1985) Expression of *Rhizobium trifoli* early nodulation genes on maize and rice plants. J Bacteriol 163:812–815

Raj L, Singh VP, Kumar P, Berwal KK (1997) Stability analysis for reproductive traits in rapeseed *Brassica napus* subsp. *oleifera*\\, indian mustard *B. juncea*\\ and gobhi sarson*B. napus* subsp. *oleifera* var.*annua*\\. Ind J Agric Sci 67(4):168–170

Ram N (1994) Environmental pollution hazards by fertilizers. Ind Far Dig 27(2):14–15

Ramesh P, Hegde DM (2010) Directorate of oilseed research. News Lett 16(2):1–8

Rao SVCK, Soren G (1991) Response of sunflower culture to planting density and nutrient application. Ind J Agron 36:95–98

Reddy PM, Reddiramu Y, Ramakrishna Y (2005) Conjunctive use of biological, organic and inorganic fertilizers in sunflower (*Helianthus annuus* L.). J Oilseeds Res 22(1):59–62

Reed PM, Glick BR (2004) Agricultural and environmental uses of plant growth promoting bacteria. Anton Van Leeuwenhoek 86:1–25

Roy RN, Finck A, Blair GJ, Tandon HLS (2006) Plant nutrition for food security a guide for integrated nutrient management. Food and Agriculture Organization of the United Nations, Rome

Sankar D, Sambandam MR, Rao PKV (2004) Impact of sesame oil on nifedipine in modulating oxidative stress and electrolytes in hypertensive patients. Asia Pac J Clin Nutr 13:107

Saraf M, Sood N (2002) Influence of monocrotophos on growth oxygen uptake and exopolysaccharide production of *Rhizobium* NICM 2771 on chickpea. J Ind Bot Soc 81:154–157

Shanmugam V, Senthil N, Raguchander T, Ramanathan A, Samiyappan R (2002) Interaction of *Pseudomonas fluorescens* and *Rhizobium* for their effect on the peanut root rot. Phytoparasitica 30(2):169–176

Shehata MM, El-khawas SA (2003) Effect of two biofertilizers on growth parameters, yield characters, nitrogenous components, nucleic acids content, minerals, oil content, protein profiles and DNA banding pattern of sunflower (*Helianthus annus* L.cv. vedock) yield. Pak J Biol Sci 6(4):1257–1268

Shrotriya GC (2005) Promoting potash application for balanced fertilizer use. Ind J Fert 1(2):31–38, 41–43

Shukla V (1984) Chemical weed control in sesame. Pesticides 18:13–14

Singh RK, Abraham T (2001) Effect of rhizobium inoculation and phosphorous on growth and yield ofsoybean (*Glycine max*). Agron Digest 1:97–98

Singh R, Sinsinwar BS (2006) Effect of integrated nutrient management on growth, yield, oil content and nutrient uptake of Indian mustard (*Brassica juncea*) in eastern part of Rajasthan. Ind J Agric Sci 76(5):322–324

Singhal V, Pratibha, Sengar RS (2003) Biofertilizer: boon for farmers. Ind Farm (April):11–12

Smiley RW (1981) Non-target effects of pesticides on turf grasses. Plant Disease 65:17–23

Smith RS (1992) Legume inoculant formulation and application. Can J Microbiol 38:485–492

Strzeleowa A (1970) The effect of urea on spheroplast formation in *Rhizobium*. Acta Microbiol Polan 2:23–24

Suja KP, Abraham Thamizh SN, Jayalekshmy A, Arumughan C (2004) Antioxidant efficacy of sesame cake extract in vegetable oil protection. Food Chem 84:393–400

Tein TM, Gaskins MH, Hubbel DH (1979) Plant growth substances browned by *Azospirillum brasitencwe* and their effect on the growth of pearl millet. Appl Environ Microbiol 37:1016–1024

Tilak KVBR, Ranganayaki N, Pal KK, De R, Saxena AK, Shekhar Nautiyal C, Mittal S, Tripathi AK, Johri BN (2006) Diversity of plant growth and soil health supporting bacteria. Curr Sci 89(1):136–150

Triouchi N, Syono K (1990) *Rhizobium* attachment and curling in asparagus, rice and oat plants. Plant Cell Physiol 31:119–127

Valladares F, Salvador PV, Dominguez S, Fernandez M, Penuelas JL, Pugnaire FI (2002) Enhancing the early performance of the leguminous *shrub Retama sphaerocarpa* (L.) biosis: fertilization versus *Rhizobium* inoculation. Plant Soil 240:253–262

Vassilev N, Vassileva M, Mikelaeva I (2006) Simultaneous P-solubilizing and biocontrol activity of microorganism: Potential and future trends. Appl Microbiol Biotechnol 71:137–144

Vyas SP (2003) Efficacy of biofertilizer on *Brassica* genotypes in arid Gujarat. Fert News 48(7):49–51

Wu SC, Cao ZH, Li ZG, Cheung KC, Wong MH (2005) Effects of biofertilizer containing N-fixer, P and K solubilizers and AM fungi on maize growth: a greenhouse trial. Geoderma 125:155–166

Yadav PIP, Matew PB, Sheela KR (2002) Effect of seed soaking with growth promoters on germination and seedling characters of rice (*Oryza sativa*). Agron Digest 2:29–39

# Chapter 2
# Mechanisms Used by Plant Growth-Promoting Bacteria

**Elisa Gamalero and Bernard R. Glick**

## 2.1 Introduction

The world population, currently ~7 billion, continues to increase so that by 2020 it is estimated to reach ~8 billion. There is a real concern regarding our ability to feed all of these people, an endeavor that requires that agricultural productivity continues to increase. Thus, more than ever, obtaining high yields is the main challenge for agriculture. In addition, in recent years both producers and consumers have increasingly focused on the health and quality of foods, as well as on their organoleptic and nutritional properties.

Stimulated by increasing demand, and by the awareness of the environmental and human health damage induced by overuse of pesticides and fertilizers (Avis et al. 2008; Leach and Mumford 2008), worldwide agricultural practice is moving to a more sustainable and environmental friendly approach. As an example, the amount of organically cultivated land in the European Union increased by ~21% per year between 1998 and 2002 and has continued to expand since then. In Italy, the EU Member State with the largest number of agricultural producers and the highest number of hectares devoted to organic agriculture, consumption of organic foods increased by 11% in 2007 alone (http://www.ec.europa.eu/agriculture/organic/eu-policy/data-statistics_it).

In this context, soil microorganisms with beneficial activity on plant growth and health represent an attractive alternative to conventional agricultural (Antoun and Prévost 2005). In recent years, several microbial inoculants have been formulated,

E. Gamalero (✉)
Dipartimento di Scienze dell'Ambiente e della Vita, Università del Piemonte Orientale, Viale Teresa Michel 11, Alessandria 15121, Italy
e-mail: elisa.gamalero@unipmn.it

B.R. Glick
Department of Biology, University of Waterloo, Waterloo, Ontario, Canada N2L 3G1

D.K. Maheshwari (ed.), *Bacteria in Agrobiology: Plant Nutrient Management*,
DOI 10.1007/978-3-642-21061-7_2, 

produced, marketed, and applied successfully by an increasing number of growers (Reed and Glick 2004).

Although all parts of the plant are colonized by microorganisms, the rhizosphere represents the main source of bacteria with plant-beneficial activities. These bacteria are generally defined as plant growth-promoting bacteria (PGPB) (Bashan and Holguin 1998). They typically promote plant growth in two ways: direct stimulation and biocontrol (i.e., suppressive activity against soil-borne diseases) (Glick 1995). Stimulation and protection of different crops by PGPB has been demonstrated many times under controlled conditions and field trials and a large number of papers on this topic are available (reviewed by Reed and Glick 2004). The positive effect of many soil bacteria on plants is mediated by a range of mechanisms including improvement of mineral nutrition, enhancement of plant tolerance to biotic and abiotic stress, modification of root development, as well as suppression of soil-borne diseases (Glick 1995; Glick et al. 1999; Kloepper et al. 1989). The bacterial traits involved in these activities, include nitrogen fixation, phosphate solubilization, iron sequestration, synthesis of phytohormones, modulation of plant ethylene levels, and control of phytopathogenic microorganisms.

This chapter provides an overview of the main mechanisms used by PGPB (Fig. 2.1). In the first section, the mechanisms involved in plant-growth promotion via mineral nutrition improvement are described with special reference to nitrogen fixation, phosphate solubilization, and iron chelation. In the second section, the effects of phytohormones whose levels are modulated by PGPB, auxins, cytokinins, and gibberellins, which are synthesized by PGPB, and the enzyme 1-aminocyclopropane-1-carboxylate (ACC) deaminase, which lowers plant ethylene levels, on plant growth and development are discussed. The ability of some PGPB to inhibit the growth of phytopathogens via competition for nutrients or colonization sites, the synthesis of antibiotics and lytic enzymes, and induced systemic resistance is discussed in the third section. Finally, given the very large body of literature in this area, for the most part, our attention is focused on the more recent literature.

## 2.2 Provision of Nutrients

Plant growth promotion by bacteria can also occur as a consequence of the provision of nutrients that are not sufficiently available in the soil; these nutrients include phosphate, nitrogen, and iron. The main mechanisms involved, as explained below, are the solubilization of phosphate, nitrogen fixation, and iron chelation through siderophores.

### *2.2.1 Phosphate Solubilization*

Although the amount of phosphorus (P) usually in soil is between 400 and 1,200 mg $kg^{-1}$ of soil, the concentration of soluble P in soil is typically ~1 mg $kg^{-1}$

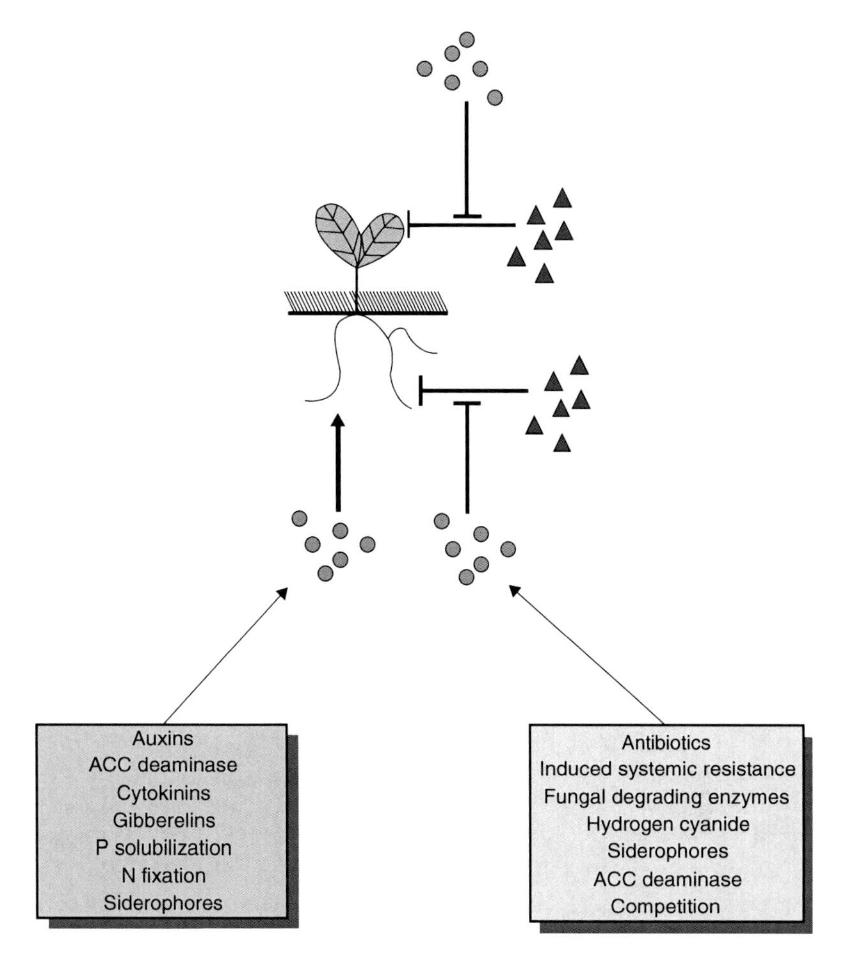


**Fig. 2.1** Facilitation of plant growth by plant growth-promoting bacteria (PGPB). The PGPB (*circles*) may either promote plant growth directly (*arrow*), generally by first interacting with plant roots, or indirectly (⊥) by preventing pathogens (*triangles*) from damaging the plant. Some of the bacterial traits/mechanisms that contribute to direct and indirect plant growth promotion are highlighted. Similar PGPB traits contribute to the biocontrol of root and leaf pathogens

or less (Goldstein 1994). P in soil is present in two main insoluble forms: mineral forms such as apatite, hydroxyapatite, and oxyapatite, and organic forms including inositol phosphate (soil phytate), phosphomonoesters, phosphodiesters, and phosphotriesters (Khan et al. 2007).

Since P is an essential macronutrient for plant growth and has only limited bioavailability, it is considered to be one of the elements that limits plant growth (Feng et al. 2004). To satisfy plants' nutritional requirements, P is usually added to soils as fertilizers synthesized through high-energy-intensive processes (Goldstein et al. 1993). However, plants can use only a small amount of this P since 75–90% of added P is precipitated by metal–cation complexes, and rapidly becomes fixed in soil. Thus, solubilization and mineralization of P by phosphate-solubilizing bacteria (PSB) is one of the most important bacterial physiological traits in soil biogeochemical cycles (Jeffries et al. 2003), as well as in plant growth promotion by PGPB (Rodriguez and Fraga 1999; Richardson 2001).

The major mechanism used by PSB for solubilization of inorganic P is based on the synthesis of low molecular weight organic acids such as gluconic and citric acid (Bnayahu 1991; Rodriguez et al. 2004). These organic acids bind phosphate with their hydroxyl and carboxyl groups thereby chelating cations and also inducing soil acidification, both resulting in the release of soluble phosphate (Kpomblekou and Tabatabai 1994; Bnayahu 1991). Other mechanisms that have been implicated in solubilization of inorganic phosphate are the release of $H^+$ (Illmer and Schinner 1992), the production of chelating substances (Sperber 1958; Duff and Webley 1959) and inorganic acids (Hopkins and Whiting 1916). In addition, exopolysaccharides synthesized by PSB participate indirectly in the solubilization of tricalcium phosphates by binding free P in the medium, affecting the homeostasis of P solubilization (Yi et al. 2008).

The mineralization of organic P occurs through the synthesis of phosphatases, including phosphomonoesterase, phosphodiesterase, and phosphotriesterase, catalyzing the hydrolysis of phosphoric esters (Rodriguez and Fraga 1999). In addition, P solubilization and mineralization can coexist in the same bacterial strain (Tao et al. 2008).

Genera able to solubilize phosphate include *Pseudomonas* (Di Simine et al. 1998; Gulati et al. 2008; Park et al. 2009; Malboobi et al. 2009), *Bacillus* (De Freitas et al. 1997; Toro et al. 1997; Rojas et al. 2001; Sahin et al. 2004), *Rhizobium* (Halder et al. 1991; Abd-Alla 1994; Chabot et al. 1996), *Burkholderia* (Tao et al. 2008; Jiang et al. 2008), *Enterobacter* (Toro et al. 1997; Sharma et al. 2005), and *Streptomyces* (Molla et al. 1984; Mba 1994; Hamdali et al. 2008; Chang and Yang 2009). Recently, the potential of nonstreptomycete actinomycetes to solubilize insoluble phosphates in soil and to promote plant growth has been investigated (El-Tarabily et al. 2008). In particular, a highly rhizosphere competent isolate of *Micromonospora endolithica* able to solubilize considerable amounts of P, to produce acid and alkaline phosphatases as well as several organic acids, was found to be unable to synthesize any other stimulatory compounds (such as auxin, cytokinin, and gibberellin) and yet promoted the growth of beans.

The role of phosphate solubilization in plant growth promotion is often overshadowed by other plant beneficial activities expressed by the PSB. When Poonguzhali et al. (2008) selected ten pseudomonads on the basis of their high phosphate solubilization activity on tricalcium phosphate and inoculated seeds with these strains, which also synthesize indole-3-acetic acid, ACC deaminase, and siderophores, the plants showed increased root elongation and biomass, however, under the conditions employed, P uptake was unaffected. Notwithstanding the difficulty that sometimes exists in pinpointing the contribution of phosphate solubilization activity to plant growth promotion, numerous reports demonstrate direct connections between phosphate solubilization activity and increased P in tissues of plants inoculated with PSB (Rodriguez and Fraga 1999).

Besides the low rhizospheric competence of some PSB strains, specificity for the host plant or soil type could play a role. For example, solubilization of Ca–P complexes is quite prevalent among PSB, whereas the release of P by Fe–P or Al–P is very rare. Thus, release of soluble P is prevalent in calcareous soils and low in alfisols (Gyaneshwar et al. 2002). Frequently, the relatively high PSB density in soil does not correspond to the amount of soluble P present in soil. The efficiency of various PSB also depends upon their physiological status, and the level of P released by phosphate solubilization is considered to be inadequate to induce a substantial increase of biomass. To obviate this problem, plants are often inoculated with PSB at concentrations that are higher than what is normally present in soil.

As a consequence of the heterogeneous results obtained by inoculating plants with PSB, the commercial application of PSB-based biofertilizers has been quite limited. Longer bacterial survival of PSB may be achieved by cell encapsulation inside nontoxic polymers such as alginate that increase the shelf life of the bacteria, protect them against many environmental stresses, and release them to the soil gradually (Bashan and Gonzalez 1999; Bashan et al. 2002). This may be more effective than the application of cell suspensions (Vassileva et al. 2000, 2006a, b; Vassilev and Vassileva 2004) with, e.g., improvement of growth promotion efficacy related to enhanced phosphate solubilization activity in lettuce (*Lactuca sativa*) inoculated with encapsulated but not free-living *Enterobacter* sp. cells (Vassileva et al. 1999).

The highest efficiency in stimulating plant growth was observed when PSB were co-inoculated with bacteria with other physiological capabilities such as N fixation (Rojas et al. 2001; Valverde et al. 2006; Matias et al. 2009), or with mycorrhizal (Ray et al. 1981; Azcón-Aguilar et al. 1986; Toro et al. 1997; Babana and Antoun 2006; Matias et al. 2009) or nonmycorrhizal fungi (Babana and Antoun 2006). Thus, the use of mixed inocula with different plant beneficial activities appears to be a promising strategy. In one set of experiments, increased amounts of both nitrogen fixation and phosphate solubilization were observed in mangrove seedlings treated with a mixture of the nitrogen-fixing *Phyllobacterium* sp. and the PSB *Bacillus licheniformis*, compared to plants inoculated with individual cultures (Rojas et al. 2001). In fact, when the two bacterial species were cocultivated in vitro, they affected one another's metabolism: N fixation increased in *Phyllobacterium* sp., and phosphate solubilization increased in *B. licheniformis*. However, the growth of the coinoculated plants did not differ from that of plants treated with a single bacterium.

Finally, the genetic manipulation of PGPB to obtain expression or over-expression of genes involved in phosphate solubilization is an attractive strategy for improving the efficacy of some bacterial inoculants. With this approach, it may be possible to avoid competition among microorganisms that is often observed when mixed inoculants are employed. Unfortunately, largely for political rather than scientific reasons, the deliberate release of genetically modified organisms to the environment is still controversial in many jurisdictions (Rodriguez et al. 2006).

### 2.2.2 Iron Chelation and Siderophores

Iron is the fourth most abundant element on earth (Ma 2005); however, in aerobic soils, iron is mostly precipitated as hydroxides, oxyhydroxides, and oxides so that the amount of iron available for assimilation by living organisms is very low, ranging from $10^{-7}$ to $10^{-23}$ M at pH 3.5 and 8.5, respectively. Both microbes and plants have a quite high iron requirement (i.e., $10^{-5}$ to $10^{-7}$ and $10^{-4}$ to $10^{-9}$ M, respectively), and this condition is more accentuated in the rhizosphere where plant, bacteria, and fungi compete for iron (Guerinot and Ying 1994; Loper and Buyer 1991). To survive with a limited supply of iron, in bacteria, cellular iron deficiency induces the synthesis of low-molecular weight siderophores, molecules with an extraordinarily high affinity for $Fe^{+3}$ ($K_a$ ranging from $10^{23}$ to $10^{52}$) as well as membrane receptors able to bind the Fe–siderophore complex, thereby allowing iron uptake by microorganisms (Neilands 1981).

Many *Pseudomonas* spp. and related genera produce yellow–green, water soluble, fluorescent pigments collectively called pyoverdines, composed of a quinoleinic chromophore bound together with a peptide and an acyl chain, conferring a characteristic fluorescence to the bacterial colonies (Meyer and Abdallah 1978). About 100 different pyoverdines have been identified (Budzikiewicz 2004; Meyer et al. 2008) and represent about 20% of the microbial siderophores that have been characterized (Boukhalfa and Crumbliss 2002). Pyoverdine-mediated iron uptake confers a competitive advantage on to fluorescent pseudomonads over other microorganisms (Mirleau et al. 2000, 2001). Regulation of pyoverdine synthesis is not only based on iron availability but also on quorum sensing whereby cell-to-cell communication mediated by N-acyl homoserines lactones occurs activating siderophore synthesis (Stintzi et al. 1998).

In plants, active iron uptake occurs mainly through two strategies (Curie and Briat 2003). Strategy I, exploited by dicotyledonous and nongraminaceous monocotyledonous plants, is based on acidification of the rhizosphere by $H^{+}$ excretion, leading to the reduction of $Fe^{+3}$ to $Fe^{+2}$ and its transport inside root cells (Robinson et al. 1999; Marschner 1995; Eide et al. 1996; Vert et al. 2002). Strategy II, used in grasses and graminaceous plants including wheat (*Triticum aestivum*), barley (*Hordeum vulgare*), rice (*Oryza sativa*), and maize (*Z. mays*), relies on the synthesis of $Fe^{+3}$ chelators called phytosiderophores and on the uptake of the Fe–phytosiderophore complex in root cells mediated by specific transporter molecules (Curie et al. 2001; Von Wirén et al. 1994). The iron dynamics in the rhizosphere are under the control of the combined effects of soil properties and plant and microbially produced compounds (Robin et al. 2008; Lemanceau et al. 2009).

Plant iron nutrition can affect the structure of bacterial communities in the rhizosphere. For example, transgenic tobacco that overexpresses ferritin and accumulates more iron than nontransformed tobacco has less bioavailable iron in the rhizosphere (Robin et al. 2006). As a consequence, the composition of the rhizosphere bacterial community differed significantly when compared to nontransformed tobacco lines.

Siderophores are involved both in plant growth promotion and health protection (Robin et al. 2008). The benefits of microbial siderophores have been demonstrated by supplying radiolabeled ferric-siderophores to plants as a sole source of iron (Crowley et al. 1988; Duijff et al. 1994a, b; Walter et al. 1994; Yehuda et al. 1996; Siebner-Freibach et al. 2003; Jin et al. 2006). The role of siderophores in plant nutrition is further supported by the absence of iron-deficiency symptoms (i.e., chlorosis) and by the fairly high iron content in roots of plants grown in nonsterile soils compared to plants grown in sterile systems (Masalha et al. 2000). Thus, mung bean (*Vigna radiata* L. Wilzeck) plants, inoculated with the siderophore-producing *Pseudomonas* strain GRP3 and grown under iron-limiting conditions, showed reduced chlorotic symptoms and an enhanced chlorophyll level compared to uninoculated plants (Sharma et al. 2003). In addition, the Fe–pyoverdine complex synthesized by *Pseudomonas fluorescens* C7 was efficiently taken up by the plant *Arabidopsis thaliana*, leading to an increase of iron content inside plant tissues and to improved plant growth (Vansuyt et al. 2007).

Plant iron nutrition improvement by soil bacteria is even more important when the plant is exposed to an environmental stress such as heavy metal pollution. Metal mobility in soil can be affected by microbial metabolites and especially by siderophores that can bind to magnesium, manganese, chromium (III), gallium (III), cadmium, copper, nickel, arsenic, lead, and zinc and radionuclides, such as plutonium (IV) as well as to iron (Malik 2004; Nair et al. 2007). In addition, by supplying iron to the plants, siderophores may help to alleviate the stresses imposed on plants by high soil levels of heavy metals (Diels et al. 2002; Belimov et al. 2005; Braud et al. 2006). *Kluyvera ascorbata*, a PGPB able to synthesize siderophores was able to protect canola, Indian mustard, canola, and tomato from heavy metal (nickel, lead, and zinc) toxicity (Burd et al. 1998, 2000). The siderophore overproducing mutant SUD165/26 of this bacterium provided even greater protection, as indicated by the enhanced biomass and chlorophyll content in plants cultivated in nickel-contaminated soil (Burd et al. 2000).

When two mutants of strain *Pseudomonas putida* ARB86, one impaired in siderophore synthesis and the other overproducing siderophores were used to inoculate *A. thaliana* plants exposed to nickel, symptoms induced by the metal were relieved to the same extent in plants inoculated with both mutants and wild type suggesting that alleviation of Ni toxicity in this case is siderophore independent (Someya et al. 2007). Similarly, two siderophore-producing bacterial strains reduced Zn uptake by willow (*Salix caprea*) suggesting that bacterial siderophores may bind to heavy metals from soil and inhibit their uptake by plants. On the other hand, enhancement of Zn and Cd uptake in willow inoculated with a *Streptomyces* strain unable to produce siderophores, highlights the importance of other physiological traits for heavy metal accumulation by *S. caprea* (Kuffner et al. 2008). The bottom line for these seemingly contradictory results is that the effect of siderophores in the presence of high concentrations of metals is quite complex, depending on soil composition, metal type and concentration, and the siderophore(s) and plant(s) utilized. Thus, the impact of siderophore in metal-contaminated soils needs to be assessed on a case by case basis.

### 2.2.3 Nitrogen Fixation

Agriculture has become increasingly dependent on chemical sources of nitrogen derived at the expense of petroleum. Besides being costly, the production of chemical nitrogen fertilizers depletes nonrenewable resources and poses human and environmental hazards. To complement and eventually substitute mineral fertilizers with biological nitrogen fixation would represent an economically beneficial and ecologically sound alternative. However, despite nitrogen's abundance in the atmosphere, it must first be reduced to ammonia before it can be metabolized by plants to become an integral component of proteins, nucleic acids, and other biomolecules (Bøckman 1997). The most important microorganisms that are currently used agriculturally to improve the nitrogen content of plants, include a range of *Rhizobia*, each specific for a limited number of plants. Other nitrogen-fixing bacteria, notably *Azospirillum* spp., are also employed as bacterial inoculants; however, it is generally thought that for free-living bacteria, the provision of fixed nitrogen is only a very small part of what the bacterium does for the plant (James and Olivares 1997).

Nitrogen-fixing (diazotrophic) bacteria fix atmospheric nitrogen by means of the enzyme nitrogenase, a two component metalloenzyme composed of (a) dinitrogenase reductase, a dimer of two identical subunits that contains the sites for MgATP binding and hydrolysis, and supplies the reducing power to the dinitrogenase, and (b) the dinitrogenase component that contains a metal cofactor (Dean and Jacobson 1992). Overall, nitrogenase biosynthesis (*nif*) genes include structural genes, genes involved in the activation of the Fe protein, iron molybdenum cofactor biosynthesis, electron donation, several genes of unknown function, and the regulatory genes required for the synthesis and function of the nitrogenase. The *nif* genes may be carried on plasmids as in most *Rhizobium* species or, more commonly, in the chromosome of free-living (Fischer 1994) and associative nitrogen-fixing bacteria (Colnaghi et al. 1997). The *nif* genes from many different diazotrophs are arranged in a single cluster of approximately 20–24 kb with seven separate operons that together encode 20 distinct proteins. All of the *nif* genes are transcribed and translated in a concerted fashion, under the control of the *nifA* and *nifL* genes. NifA protein is a positive regulatory factor which turns on the transcription of all of the *nif* operons (except its own). The DNA-bound NifA protein interacts with transcription initiation factor sigma 54 before transcription from the *nif* promoter is initiated. NifL protein is a negative regulatory factor which, in the presence of either oxygen or high levels of fixed nitrogen, acts as an antagonist of the NifA protein. Because of the complexity of the *nif* system, genetic strategies to improve nitrogen fixation have been elusive.

Since nitrogen fixation requires a large amount of ATP, it would be advantageous if rhizobial carbon resources were directed toward oxidative phosphorylation, which results in the synthesis of ATP, rather than glycogen synthesis, which results in the storage of energy as glycogen. A strain of *R. tropici* with a deletion in the gene for glycogen synthase was constructed (Marroquí et al. 2001). Treatment of bean plants with this mutant strain resulted in a significant increase in both the number of nodules that formed and the plant dry weight in comparison with treatment with the wild-type rhizobial strain.

Oxygen is both inhibitory to nitrogenase and is a negative regulator of *nif* gene expression; however, it is required for *Rhizobia* spp. bacteroid respiration. This difficulty can be resolved by the introduction of leghemoglobin, which binds free oxygen tightly resulting in an increase in nitrogenase activity. Since the globin portion of leghemoglobin is produced by the plant, more efficient strains of *Rhizobium* spp. may be engineered by transforming strains with genes encoding bacterial hemoglobin (Ramírez et al. 1999). Following transformation of *Rhizobium etli* with a plasmid carrying the *Vitreoscilla* sp. (a gram negative bacterium) hemoglobin gene, at low levels of dissolved oxygen in the medium, the rhizobial cells had a two- to threefold higher respiratory rate than the nontransformed strain. In greenhouse experiments, when bean plants were inoculated with either nontransformed or hemoglobin-containing *R. etli* the plants inoculated with the hemoglobin-containing strain had approximately 68% more nitrogenase activity. This difference in nitrogenase activity leads to a 25–30% increase in leaf nitrogen content and a 16% increase in the nitrogen content of the seeds that are produced (Ramírez et al. 1999).

The most common strain of *R. etli* encodes three copies of the nitrogenase reductase (*nifH*) gene, each under the control of a separate promoter. To increase the amount of nitrogenase, the strongest of the three *nifH* promoters (i.e., *PnifHc*) was coupled to the *nifHcDK* operon, which encodes the nitrogenase structural genes (*nifHc* is one of the three *nifH* genes). When the *PnifHc–nifHcDK* construct was introduced into the wild-type strain, the net result was a significant increase in nitrogenase activity, plant dry weight, seed yield, and the nitrogen content of the seeds. This genetic manipulation worked as well or better when the *PnifHc–nifHcDK* construct was introduced into the Sym plasmid from *R. etli* that contains all of the *nif* genes (Peralta et al. 2004). In addition, expression of the *PnifHc–nifHcDK* construct in a poly-β-hydroxybutyrate negative strain of *R. etli* enhanced plant growth to an even greater extent than when this construct was expressed in a wild-type poly-β-hydroxybutyrate positive strain. This is probably because in the poly-β-hydroxybutyrate negative strain there is an increased flux of carbon through the citric acid cycle and hence an increase in the amount of ATP to power nitrogen fixation (Peralta et al. 2004).

An undesirable side reaction of nitrogen fixation is the reduction of $H^+$ to $H_2$ by nitrogenase. ATP is wasted on the production of hydrogen and only 40–60% of the electron flux through the nitrogenase system is transferred to $N_2$, lowering the overall efficiency of nitrogen fixation. Some diazotrophic strains contain hydrogenase that can take up $H_2$ from the atmosphere and convert it into $H^+$ and the presence of a hydrogen uptake system in a symbiotic diazotroph improves its ability to stimulate plant growth by binding and then recycling the hydrogen gas that is formed inside the nodule by the action of nitrogenase. Although it is clearly beneficial to the plant to obtain its nitrogen from a symbiotic diazotroph that has a hydrogen uptake system, this trait is not common in naturally occurring rhizobial strains.

In *Rhizobium leguminosarum*, 18 genes are associated with hydrogenase activity. There are 11 *hup* (hydrogen uptake) genes responsible for the structural components of the hydrogenase, the processing of the enzyme, and electron transport. There are also seven *hyp* (hydrogenase pleitropic) genes that are involved in processing the nickel that is part of the active center of the enzyme. The *hup* promoter is dependent on the NifA protein so that *hup* genes are only expressed within bacteroids. On the other hand, the *hyp* genes are transcriptionally regulated by an FnrN-dependent promoter that is turned on by low levels of oxygen so that the *hyp* genes are expressed both in bacteroids and microaerobically. By modifying the chromosomal DNA of *R. leguminosarum* and exchanging the *hup* promoter for an FnrN-dependent promoter, a derivative of the wild type with an increased level of hydrogenase was created (Ureta et al. 2005). The engineered strain displayed a twofold increase in hydrogenase activity compared to the wild type and no discernible amount of hydrogen gas was produced as a byproduct of nitrogen fixation with the net result that the amount of fixed nitrogen and hence plant productivity was greater.

A small localized rise in plant ethylene that can inhibit subsequent rhizobial infection and nodulation is often produced following the initial stages of *Rhizobia* infection. Some *Rhizobia* strains increase the number of nodules that form on the roots of a host legume by limiting the rise in ethylene either by synthesizing a small molecule called rhizobitoxine (Yuhashi et al. 2000) that chemically inhibits ACC synthase, one of the ethylene biosynthetic enzymes, or by producing ACC deaminase and removing some of the ACC before it can be converted to ethylene (Ma et al. 2002). The result of lowering the level of ethylene is that both the number of nodules and the biomass of the plant is increased by 25–40% (Ma et al. 2003). In the field, approximately 1–10% of rhizobial strains possess ACC deaminase (Duan et al. 2009) thus it is possible to increase the nodulation efficiency of *Rhizobia* strains that lack ACC deaminase by engineering these strains with isolated *Rhizobia* ACC deaminase genes (and regulatory regions). In fact, insertion of an ACC deaminase gene from *R. leguminosarum* bv. *viciae* into the chromosomal DNA of a strain of *Sinorhizobium meliloti* that lacked this enzyme dramatically increased both nodule number and biomass of host alfalfa plants (Ma et al. 2004). Because of political/regulatory considerations, genetically engineered strains of *Rhizobia* may not currently be acceptable for use in the field; however, several commercial inoculant producers are already screening their more recently isolated *Rhizobia* strains for active ACC deaminase.

## 2.3 Modulation of Phytohormone Levels

The phytohormones auxins, cytokinins, gibberellins, and ethylene and abscisic acid (ABA) all play key roles in the regulation of plant growth and development (Salisbury and Ross 1992). When plants encounter suboptimal environmental conditions, the levels of endogenous phytohormones are often insufficient for optimal growth (De Salamone et al. 2005). In this context, some phytohormones or hormone-like substances that stimulate seed and tuber germination, root formation or fruit ripening, are included as a part of commercial plant growth stimulators (Tsakelova et al. 2006).

Many rhizosphere microorganisms produce or modulate phytohormones under in vitro conditions (De Salamone et al. 2005). Consequently, many PGPB with the ability to alter phytohormone levels can affect the plant's hormonal balance. The production of IAA, cytokinins, and gibberellins by PGPB and their effect on plant growth are discussed in Sects. 2.3.1 and 2.3.2. The modulation of ethylene synthesis by ACC deaminase-producing bacteria and their role in supporting plant growth in natural and stressed environment is described in Sect. 2.3.3.

### *2.3.1 IAA*

Besides influencing division, extension, and differentiation of plant cells and tissues, auxins stimulate seed and tuber germination; increase the rate of xylem and root development; control processes of vegetative growth; initiate lateral and adventitious roots; mediate responses to light and gravity, florescence, and fructification of plants; and also affect photosynthesis, pigment formation, biosynthesis of various metabolites, and resistance to stressful conditions (Tsakelova et al. 2006). Although several naturally occurring auxins have been described, indole-3-acetic acid (IAA) is the most studied auxin, and frequently auxin and IAA are considered as interchangeable terms. However, in plants most IAA is generally present as conjugated forms that are mainly implicated in transport, storage, and protection of IAA catabolism (Seidel et al. 2006).

Different auxin concentrations have diverse effects on the physiology of plants with plant responses being a function of the type of plant, the particular tissue involved, and the developmental stage of the plant. The actual range of effective auxin concentrations varies according to plant species and to the sensitivity of the plant tissue to auxin; levels below this range have no effect, whereas higher concentrations inhibit growth (Peck and Kende 1995). For example, Evans et al. (1994) found that only exogenous concentrations between $10^{-10}$ and $10^{-12}$ M stimulated primary root elongation in *A. thaliana* seedlings. Moreover, the endogenous pool of auxin in the plant is affected by soil microorganisms able to synthesize this phytohormone and also the impact of bacterial IAA on plant development ranges from positive to negative effects according to the amount of auxin available to the plant and to the sensitivity of the host plant to the phytohormone. In addition, the level of auxin synthesized by the plant itself may be important in determining whether bacterial IAA will stimulate or suppress plant growth. In plant roots, endogenous IAA may be suboptimal or optimal for growth (Pilet and Saugy 1987) and additional IAA from bacteria could alter the auxin level to either optimal or supraoptimal, resulting in plant growth promotion or inhibition, respectively.

Production of auxin is widespread among soil bacteria (estimated to be ~80% of all soil bacteria). This ability has been detected in a wide range of soil bacteria as well as in streptomycetes, methylobacteria, cyanobacteria, and archaea. Several of these microorganisms, are involved in plant pathogenesis, whereas others are free-living or symbiotic PGPB.

Five of the six pathways for auxin biosynthesis in bacteria rely on tryptophan as the main IAA precursor. These pathways, constitutively expressed or inducible, encoded by genomic or plasmid DNA, have been classified according to their intermediate as indole-3-acetamide, indole-3-pyruvate, tryptamine, tryptophan side-chain oxidase, indole-3-acetonitrile, and tryptophan independent and they have been extensively reviewed (Patten and Glick 1996; Spaepen et al. 2007).

Auxin biosynthesis in bacteria is affected by a number of factors including environmental stress, pH, osmotic and matrix stress, carbon starvation, and the composition of the root exudates. However, due to the diversity of IAA expression and regulation according to the biosynthetic pathways and bacterial species, all of these factors cannot easily be integrated into a comprehensive regulatory scheme of IAA biosynthesis in bacteria (Spaepen et al. 2007). IAA synthesized by bacteria is involved at different levels in plant–microorganism interactions: in particular, plant growth promotion and root nodulation can be affected by IAA.

One of the main effects of bacterial IAA is the enhancement of lateral and adventitious rooting leading to improved mineral and nutrient uptake and root exudation that in turn stimulates bacterial proliferation on the roots (Dobbelaere et al. 1999; Lambrecht et al. 2000; Steenhoudt and Vanderleyden 2000). The role of IAA synthesized by the PGPB *P. putida* GR12-2, which produces relatively low levels of the phytohormone, in the development of the canola roots has been studied following the construction of an IAA-deficient mutant of this strain (Patten and Glick 2002). Seed inoculation with wild-type GR12-2 induced the formation of tap roots that were 35–50% longer than the roots from seeds treated with the IAA-deficient mutant and the roots from uninoculated seeds. Conversely, inoculation of mung bean cuttings with the mutant *aux1* of the same strain, which overproduces IAA, yielded a greater number of shorter roots compared with controls (Mayak et al. 1999). This result was explained by the combined effect of auxin on growth promotion and inhibition of root elongation by ethylene (Jackson 1991). The bacterial IAA incorporated by the plant stimulates the activity of ACC synthase, resulting in increased synthesis of ACC (Jackson 1991), and a rise in ethylene which, in turn, inhibited root elongation (Riov and Yang 1989). Therefore, the production of IAA alone does not account for growth promotion capacity of *P. putida* GR12-2 (Xie et al. 1996).

Most *Rhizobium* species produce IAA (Badenochjones et al. 1983) and several studies have suggested that changes in auxin levels in the host plant are necessary for nodule organogenesis (Mathesius et al. 1998). Treatment of plants with low concentrations (up to $10^{-8}$ M) of exogenous IAA can enhance nodulation on *Medicago* and *Phaseolus vulgaris*, whereas higher concentrations inhibit nodulation (Plazinski and Rolfe 1985; van Noorden et al. 2006). In addition, the amount of IAA in root nodules was higher than in nonnodulated roots (Badenochjones et al. 1983;

Basu and Ghosh 1998; Theunis 2005). On the other hand, mutants of *Bradyrhizobium elkanii* that were defective in IAA synthesis induced fewer nodules on soybean roots than did the wild-type strain (Fukuhara et al. 1994). Furthermore, in nodules induced by low IAA-producing mutants of *Rhizobium* sp. NGR234, the IAA content is lower than in nodules induced by the wild-type strain, supporting the idea that part of the IAA found in nodules is of prokaryotic origin (Theunis 2005).

It has been suggested that PGPB-synthesizing IAA may prevent the deleterious effects of environmental stresses (Lindberg et al. 1985; Frankenberger and Arshad 1995). For example, IAA stimulated lengthening of the root and shoot of wheat seedling exposed to high levels of saline (Egamberdieva 2009). An increased tolerance of *Medicago truncatula* against salt stress was also observed in plants nodulated by the IAA-overproducing strain *S. meliloti* DR-64 (Bianco and Defez 2009); plants inoculated with this mutant accumulated a high amount of proline, and showed enhanced levels of the antioxidant enzymes superoxide dismutase, peroxidase, glutathione reductase, and ascorbate peroxidase compared with plants inoculated with the parental strain.

On the other hand, IAA is a readily biodegradable compound and bacteria able to catabolize IAA have been recovered from various environments, including soil (Gieg et al. 1996) and plant tissues (Libbert and Risch 1969; Strzelczyk et al. 1973; Leveau and Lindow 2005). Degradation of IAA has been reported for strains belonging to *Pseudomonas* (Gieg et al. 1996; Leveau and Lindow 2005), *Arthrobacter* (Mino 1970), *Alcaligenes* (Claus and Kutzner 1983) and *Bradyrhizobium* (Jensen et al. 1995; Jarabo-Lorenzo et al. 1998) genera. Recently, the *iac* locus a cluster of ten genes for the catabolism of IAA that showed some similarity to genes encoding enzymes that catabolized indole or amidated aromatics, was detected in *P. putida* 1290 (Leveau and Gerards 2008). In this regard, degradation of IAA, or its inactivation, provides bacteria with the potential for manipulation of the plant's IAA pool and its related impact on plant physiology and growth.

### 2.3.2 Cytokinins and Gibberellins

Cytokinins are N6-substituted aminopurines that play a key role in a wide range of physiological processes such as plant cell division, interruption of the quiescence of dormant buds, activation of seed germination, promotion of branching, root growth, accumulation of chlorophyll, leaf expansion, and delay of senescence (Salisbury and Ross 1992). In addition, cytokinins regulate the expression of the gene coding for expansin, a protein that induces the loosening of plant cell walls and thereby facilitates turgor-driven plant cell expansion, which affects both the size and the shape of the cells (Downes and Crowell 1998; Downes et al. 2001).

The gene encoding the enzyme responsible for the synthesis of cytokinins was initially characterized in *Agrobacterium tumefaciens* (Nester et al. 1984) and subsequently found in methylotrophic and methanotrophic bacteria (Koenig et al. 2002; Ivanova et al. 2001). Since then, many PGPB including *Azotobacter*,

*Azospirillum*, *Rhizobium*, *Bacillus*, and *Pseudomonas* spp., have been found to produce this hormone (Nieto and Frankenberger 1989; Timmusk et al. 1999; Salamone et al. 2001; Taller and Wong 1989).

Interestingly, unlike the situation in bacteria, the gene encoding isopentenyl-transferase, the cytokinin biosynthesis enzyme, was not definitively identified in plants until 2001 (Kakimoto 2001), putting an end to speculation regarding the supposed inability of plants to produce cytokinins.

Seed inoculation with cytokinin-producing bacteria usually leads to a higher cytokinin content in the plants, with a concomitant influence on plant growth and development (Arkhipova et al. 2005). Various environmental stresses such as drought may also cause plant cytokinin levels to become elevated (Arkhipova et al. 2007), often inducing an increase in plant ethylene levels which in turn inhibits root elongation (Werner et al. 2003).

A positive correlation has been observed in several legume species between the level of cytokinins in plants and the ability of *Rhizobia* to form nodules on the roots of those plants (Yahalom et al. 1990; Hirsch and Fang 1994; Lorteau et al. 2001). In addition, cytokinins are believed to be involved in rhizobial infection and nodule differentiation (Frugier et al. 2008). A strain of *Rhizobium* sp., impaired in the synthesis of the Nod factor (Nod$^-$) and therefore unable to nodulate its legume host, but genetically modified for the production of the cytokinin transzeatin, induced the formation, on *Medicago sativa* roots, of a nodule-like structure which remained uncolonized by *Rhizobia*, suggesting that cytokinins can mimic some of the morphogenetic effects of Nod factors.

Recently, the role that cytokinin receptors play in plant growth stimulation by cytokinin-producing PGPB was elaborated. Plant growth promotion and modification of root architecture with the development of short tap roots and highly branched lateral roots with long root hairs were induced by *Bacillus megaterium* UMCV1 on *A. thaliana*; effects that were all ascribed to cytokinin synthesis, independent of auxin and ethylene signaling (Ortiz-Castro et al. 2008). Since a number of PGPB that synthesize cytokinins stimulate the growth of different crops, it's likely that this plant beneficial activity is mediated by different cytokinin receptor homologs (Ortiz-Castro et al. 2008).

Gibberellins are synthesized by higher plants, fungi, and bacteria; they are diterpenoid acids consisting of isoprene residues (generally with four rings); to date 136 different gibberellins have been identified and characterized (MacMillan 2002). They affect cell division and elongation and are involved in several plant developmental processes, including seed germination, stem elongation, flowering, fruit setting, and delay of senescence in many organs of a range of plant species (MacMillan 2002). Gibberellins have also been implicated in promotion of root growth since they regulate root hair abundance (Bottini et al. 2004). However, in these processes gibberellins interact with other phytohormones and alter the plant's hormonal balance thereby affecting plant growth (Trewavas 2000).

The ability of bacteria to synthesize gibberellins-like substances was first described in *Azospirillum brasilense* (Tien et al. 1979) and *Rhizobium* (Williams and Sicardi de Mallorca 1982); it has since been detected in different bacterial genera that inhabit the plant root system including *Azotobacter*, *Arthrobacter*, *Azospirillum*, *Pseudomonas*, *Bacillus*, *Acinetobacter*, *Flavobacterium*, *Micrococcus*, *Agrobacterium*, *Clostridium*, *Rhizobium*, *Burkholderia*, and *Xanthomonas* (Mitter et al. 2002; Tsakelova et al. 2006; Joo et al. 2009). Plant growth promotion by gibberellin-producing PGPB has been reported by several labs, and this positive effect on plant biomass is frequently associated with an increased content of gibberellins in plant tissues (Atzhorn et al. 1988; Gutierrez-Manero et al. 2001; Joo et al. 2005, 2009; Kang et al. 2009). Modification of the gibberellin concentration in plants is the result of either (a) gibberellin synthesis (Lucangeli and Bottini 1997; Piccoli et al. 1999), (b) deconjugation of glucosylgibberellins (Piccoli et al. 1997), or (c) chemical activation of inactive gibberellins by PGPB (Cassán et al. 2001a, b).

*Azospirillum* spp. is a nitrogen fixing and IAA-producing PGPB that is well known to induce enhancement of plant growth and yield (Okon and Labandera-Gonzalez 1994) under both nonstressed as well as stressful conditions such as drought (Creus et al. 1997). Besides improving N nutrition under some conditions, plant growth promotion activity by *Azospirillum* spp. may also be related to gibberellin synthesis. Moreover, the capability to deconjugate the conjugated form of gibberellins or to activate inactive forms of gibberellins has been implicated in plant growth promotion and reversal of dwarf phenotype in rice and maize, which lack the ability to synthesize gibberellin, by *A. lipoferum* and *A. brasilense* (Lucangeli and Bottini 1997; Cassán et al. 2001a, b).

### 2.3.3 Ethylene

The synthesis of ethylene in all higher plants is based on three enzymes (a) S-adenosyl-L-methionine (SAM) synthetase, which catalyzes the conversion of methionine to SAM (Giovanelli et al. 1980), (b) 1-aminocyclopropane-1-carboxylic acid (ACC) synthase, which mediates the hydrolysis of SAM to ACC and 5′-methylthioadenosine (Kende 1989), and (c) ACC oxidase which metabolizes ACC to ethylene, carbon dioxide, and cyanide (John 1991). Although several phases of plant growth (i.e., fruit ripening, flower senescence, leaf and petal abscission) are regulated by ethylene, this phytohormone is also important for its role in plant responses to biotic and abiotic stresses (Abeles et al. 1992). The term "stress ethylene" (Abeles 1973), describes the increase in ethylene synthesis associated with environmental stresses including extremes of temperature, high light, flooding, drought, the presence of toxic metals and organic pollutants, radiation, wounding, insect predation, high salt, and various pathogens including viruses, bacteria, and fungi (Morgan and Drew 1997).

The increased level of ethylene formed in response to environmental stresses can exacerbate symptoms of stress or it can lead to responses that enhance plant survival under adverse conditions. Thus, stress ethylene has been suggested to both alleviate and exacerbate some of the effects of the stress, depending upon the plant species, its age and the nature of the stress (Van Loon and Glick 2004). This behavior is explained by a two-phase model (Glick et al. 2007). When plants are exposed to stress, they quickly respond with a small peak of ethylene that initiates a protective response by the plant, such as transcription of pathogenesis-related genes and induction of acquired resistance (Ciardi et al. 2000; Van Loon and Glick 2004). If the stress is chronic or intense, a second much larger peak of ethylene occurs, often 1–3 days later. This second ethylene peak induces processes such as senescence, chlorosis, and abscission that may lead to a significant inhibition of plant growth and survival.

In 1978, an enzyme capable of degrading the ethylene precursor, ACC, to ammonia and α-ketobutyrate was isolated from *Pseudomonas* sp. strain ACP (Honma and Shimomura 1978). Further studies demonstrated the presence of ACC deaminase activity in a wide range of soil microorganisms including the fungus *Penicillium citrinum* (Honma 1993), and various bacteria (Jacobson et al. 1994; Glick et al. 1995; Burd et al. 1998; Belimov et al. 2001; Ma et al. 2003; Ghosh et al. 2003; Sessitsch et al. 2005; Blaha et al. 2006; Madhaiyan et al. 2007; Kuffner et al. 2008; Chinnadurai et al. 2009). Bacterial ACC deaminase activity is relatively common. In one study, 12% of isolated *Rhizobium* spp. from various sites in southern and central Saskatchewan possessed this enzyme (Duan et al. 2009). In another study, ACC deaminase activity/genes were found in a wide range of bacterial isolates including *Azospirillum*, *Rhizobium*, *Agrobacterium*, *Achromobacter*, *Burkholderia*, *Ralstonia*, *Pseudomonas*, and *Enterobacter* (Blaha et al. 2006).

In a model described by Glick et al. (1998), PGPB colonize the seed or root of a developing plant and, in response to tryptophan and other small molecules in seed or root exudates (Bayliss et al. 1997; Penrose and Glick 2001), the bacteria synthesize and secrete IAA (Patten and Glick 1996, 2002). This IAA, and the endogenous plant IAA, can either stimulate plant growth or induce the synthesis of ACC synthase, which converts SAM to ACC. A portion of the ACC produced by this reaction is exuded from seeds or plant roots (Bayliss et al. 1997; Penrose and Glick 2001), taken up by the bacteria, and converted by ACC deaminase to ammonia and α-ketobutyrate. As a result of this activity, the amount of ethylene produced by the plant is reduced. Therefore, root colonization by bacteria that synthesize ACC deaminase prevents limits ethylene levels that might otherwise be growth inhibitory (Glick 1995). The main visible effect of seed inoculation with ACC deaminase-producing bacteria, under gnotobiotic conditions, is the enhancement of root elongation (Glick et al. 1995; Hall et al. 1996; Shah et al. 1997).

In addition, other processes such as nodulation of legumes and mycorrhizal establishment in the host plant induce local increases in ethylene content. In this context, ACC deaminase-producing bacteria, lowering the ethylene content in the plants, can increase both nodulation and mycorrhizal colonization in pea (Ma et al. 2003) and cucumber (Gamalero et al. 2008), respectively.

Bacteria that have ACC deaminase facilitate plant growth under a variety of ethylene-producing environmental stresses including flooding (Grichko and Glick 2001; Farwell et al. 2007), pollution by organic toxicants such as polycyclicaromatic hydrocarbons, polycyclic biphenyls, and total petroleum hydrocarbons

(Saleh et al. 2004; Huang et al. 2004a, b; Reed and Glick 2005) and by heavy metals including nickel, lead, zinc, copper, cadmium, cobalt, and arsenic (Burd et al. 1998, 2000; Belimov et al. 2001, 2005; Nie et al. 2002; Glick 2003; Reed and Glick 2005; Farwell et al. 2006; Rodriguez et al. 2008), salinity (Mayak et al. 2004b; Saravanakumar and Samiyappan 2006; Cheng et al. 2007; Gamalero et al. 2009), drought (Mayak et al. 2004a), and phytopathogen attack (Wang et al. 2000; Hao et al. 2007).

## 2.4 Soil-borne Disease Suppression

Plant disease suppression by soil microorganisms is a possible alternative means of reducing the chemical input in agriculture (Compant et al. 2005). Biocontrol of plant pathogenic microorganisms relies on different traits including competition for colonization site or nutrients, production of antibiotics and enzymes, and induction of systemic resistance (ISR) against the pathogens (Raaijmakers et al. 2009).

Competitive colonization of the root system and successful establishment in the zones of the roots that are preferentially colonized by the pathogen is a prerequisite for effective biocontrol (Weller 1988; Raaijmakers et al. 1995). In addition, the synthesis of several antagonistic molecules through quorum sensing is directly linked to the proliferation of the PGPB on the roots. Moreover, PGPB can outcompete some pathogens by degrading organic compounds or sequestering micronutrients (i.e., iron), which are also required for the growth and the development of deleterious microorganisms (Fravel et al. 2003; Lemanceau et al. 1992).

A number of factors such as soil composition, temperature, relative humidity, composition of root exudates, presence of recombinant plasmids as well as the interactions with other soil biota can affect the persistence of a PGPB on the root system making it difficult to predict the behavior of the bacterial strain under natural conditions. Therefore, PGPB that are effective in the laboratory frequently do not show any significant impact on plants in the field (Glick et al. 1999).

The synthesis of antibiotics is the mechanism that is most commonly associated with the ability of a PGPB to suppress pathogen development (Haas and Keel 2003; Whipps 2001; Mazurier et al. 2009). The antibiotics synthesized by PGPB include agrocin 84, agrocin 434, herbicolin, 2,4-diacetylphloroglucinol, oomycin, cyclic lipopeptides, hydrogen cyanide, phenazines, pyoluteorin, and pyrrolnitrin. Although the main target of these antibiotics are the electron transport chain (phenazines, pyrrolnitrin), metalloenzymes such as copper-containing cytochrome c oxidases (hydrogen cyanide), membrane integrity (biosurfactants), or cell membrane and zoospores (2,4-diacetylphloroglucinol, DAPG, biosurfactants) (Haas and Défago 2005; Raaijmakers et al. 2006) their mode of action are still largely unknown.

Other PGPB behave as biocontrol agents by producing enzymes such as chitinase, cellulose, β-1,3 glucanase, protease, or lipase, that induce lysis of fungal cell walls (Chet and Inbar 1994). In particular, chitinase is considered crucial for the biocontrol activity exhibited by PGPB against phytopathogenic fungi such as *Botrytis cinerea* (Frankowski et al. 2001), *Sclerotium rolfsii* (Ordentlich et al. 1988), *Fusarium oxysporum* f.sp. *cucumerinum* (Singh et al. 1999) and *Phytophthora* (Kim et al. 2008) and β-glucanase for the suppression of *R. solani*, and *Pythium ultimum* cell walls (Frankowski et al. 2001; Ordentlich et al. 1988).

Some PGPB can trigger the phenomenon of induced systemic resistance (ISR) which is phenotypically similar to systemic acquired resistance (SAR) which occurs when plants activate their defense mechanism in response to primary infection by a pathogen. ISR involves jasmonate and ethylene signaling within the plant that stimulates the host plant's response to a range of pathogens without requiring direct interaction between the resistance-inducing microorganisms and the pathogen (Bakker et al. 2007). Besides ethylene and jasmonate, other bacterial molecules such as the *O*-antigenic side chain of the bacterial outer membrane protein lipopolysaccharide (Leeman et al. 1995), flagellar fractions (Zipfel et al. 2004), pyoverdine (Maurhofer et al. 1994), DAPG (Iavicoli et al. 2003; Siddiqui and Shoukat 2003), cyclic lipopeptide surfactants (Ongena et al. 2007; Tran et al. 2007) and, in some instances, salicylic acid (van Loon et al. 1998) have been implicated as signals for the induction of systemic resistance.

Most studies of systemic resistance have been carried out using fungal pathogens; however, this approach may also have potential in the control of bacterial pathogens such as *P. syringae* pv. *lachrymans*, the causal agent of bacterial angular leaf spot (Liu et al. 1995). ISR can induce alterations to host physiology leading to an overexpression of plant defensive chemicals including pathogenesis-related proteins such as chitinases, peroxidases, superoxide dismutase phenylalanine ammonia lyase, phytoalexins, and polyphenol oxidase enzymes (Bakker et al. 2007).

Using genetic engineering techniques, it should be possible to create superior biocontrol strains that utilize two or more different mechanisms to protect plants against phytopathogens. For example, genes encoding enzymes such as chitinase may readily be expressed in biocontrol bacteria that produce specific antibiotics. Moreover, the efficacy of any biocontrol strain may be improved by the introduction of an ACC deaminase gene (Wang et al. 2000; Hao et al. 2007).

## 2.5 Conclusions

In the past 10–15 years, there has been an increasing interest in the possibility of utilizing PGPB as adjuncts to agricultural and horticultural practice as well as environmental cleanup. Moreover, with this interest there has been a major effort worldwide to better understand many of the fundamental mechanisms that PGPB use to facilitate plant growth. These basic studies are predicated on the notion that it is necessary to first understand the fundamental genetic and biochemical mechanisms that govern the relationship between PGPB and plants before using them on a massive scale in the environment. With increasing concern about the natural environment and the understanding that the era of the large scale use of chemicals in the environment needs to come to an end, PGPB offer an attractive alternative that contains the possibility of developing more sustainable approaches to agriculture. Finally, it is likely to be much simpler and more efficacious to select or engineer PGPB so that they confer plants with specific desirable traits than to genetically engineer the plants themselves to the same end.

**Acknowledgments** The work from our laboratories that is cited here was supported by funds from the Italian Ministry of Education, University and Research and the Natural Sciences and Engineering Research Council of Canada.

## References

Abd-Alla MH (1994) Use of organic phosphorus by *Rhizobium leguminosarum* biovar *vicea* phosphatases. Biol Fert Soils 18:216–218

Abeles FB (1973) Ethylene in plant biology. Academic, New York, p 302

Abeles FB, Morgan PW, Saltveit ME Jr (1992) Ethylene in plant biology, 2nd edn. Academic, New York

Antoun H, Prévost D (2005) Ecology of plant growth promoting rhizobacteria. In: Siddiqui ZA (ed) PGPR: biocontrol and biofertilization. Springer, Dordrecht, pp 1–38

Arkhipova TN, Veselov SU, Melentiev AI, Martynenko EV, Kudoyarova GR (2005) Ability of bacterium *Bacillus subtilis* to produce cytokinins and to influence the growth and endogenous hormone content of lettuce plants. Plant Soil 272:201–209

Arkhipova TN, Prinsen E, Veselov SU, Martinenko EVA, Melentiev I, Kudoyarova GR (2007) Cytokinin producing bacteria enhance plant growth in drying soil. Plant Soil 292:305–315

Atzhorn R, Crozier A, Wheeler CT, Sandberg G (1988) Production of gibberellins and indole-3-acetic acid by *Rhizobium phaseoli* in relation to nodulation of *Phaseolus vulgaris* roots. Planta 175:532–538

Avis TJ, Gravel V, Antoun H, Tweddell RJ (2008) Multifaceted beneficial effects of rhizosphere microorganisms on plant health and productivity. Soil Biol Biochem 40:1733–1740

Azcón-Aguilar C, Gianinazzi-Pearson V, Fardeau JC, Gianinazzi S (1986) Effect of vesicular-arbuscular mycorrhizal fungi and phosphate-solubilizing bacteria on growth and nutrition of soybean in a neutral-calcareus soil amended with $^{32}$P-$^{45}$Ca-tricalcium phosphate. Plant Soil 96:3–15

Babana AH, Antoun H (2006) Effect of *Tilemsi* phosphate rock-solubilizing microorganisms on phosphorus uptake and yield of field-grown wheat (*Triticum aestivum* L.) in Mali. Plant Soil 287:51–58

Badenochjones J, Rolfe BG, Letham DS (1983) Phytohormones, *Rhizobium* mutants, and nodulation in legumes. 3. Auxin metabolism in effective and ineffective pea root nodules. Plant Physiol 73:347–352

Bakker PAHM, Pieterse CMJ, van Loon LC (2007) Induced systemic resistance by fluorescent *Pseudomonas* spp. Phytopathology 97:239–243

Bashan Y, Gonzalez LE (1999) Long-term survival of the plant growth-promoting bacteria *Azospirillum brasilense* and *Pseudomonas fluorescens* in dry alginate inoculant. Appl Microbiol Biotechnol 51:262–266

Bashan Y, Holguin G (1998) Proposal for the division of plant growth-promoting rhizobacteria into two classifications: biocontrol-PGPB (plant growth-promoting bacteria) and PGPB. Soil Biol Biochem 30:1225–1228

Bashan Y, Hernandez JP, Leyva LA, Bacilio M (2002) Alginate microbeads as inoculant carriers for plant growth-promoting bacteria. Biol Fert Soils 35:359–368

Basu PS, Ghosh AC (1998) Indole acetic acid and its metabolism in root nodules of a monocotyledonous tree *Roystonea regia*. Curr Microbiol 37:137–140
Bayliss C, Bent E, Culham DE, MacLellan S, Clarke AJ, Brown GL, Wood JM (1997) Bacterial genetic loci implicated in the *Pseudomonas putida* GR12-2R3-canola mutualism: identification of an exudate-inducible sugar transporter. Can J Microbiol 43:809–818
Belimov AA, Safronova VI, Sergeyeva TA, Egorova TN, Matveyeva VA, Tsyganov VE, Borisov AY, Tikhonovich IA, Kluge C, Preisfeld A, Dietz KJ, Stepanok VV (2001) Characterization of plant growth promoting rhizobacteria isolated from polluted soils and containing 1-aminocyclopropane-1-carboxylate deaminase. Can J Microbiol 47:642–652
Belimov AA, Hontzeas N, Safronova VI, Demchinskaya SV, Piluzza G, Bullitta S, Glick BR (2005) Cadmium-tolerant plant growth-promoting bacteria associated with the roots of Indian mustard (*Brassica juncea* L. Czern.). Soil Biol Biochem 37:241–250
Bianco C, Defez R (2009) *Medicago truncatula* improves salt tolerance when nodulated by an indole-3-acetic acid-overproducing *Sinorhizobium meliloti* strain. J Exp Bot 60:3097–3107
Blaha D, Prigent-Combaret C, Mirza MS, Moënne-Loccoz Y (2006) Phylogeny of the 1-aminocyclopropane-1-carboxylic acid deaminase-encoding gene *acdS* in phytobeneficial and pathogenic *Proteobacteria* and relation with strain biogeography. FEMS Microbiol Ecol 56:455–470
Bnayahu BY (1991) Root excretions and their environmental effects: influence on availability of phosphorus. In: Waisel Y, Eshel A, Kafkafi U (eds) Plant roots: the hidden half. Dekker, New York, pp 529–557
Bøckman OC (1997) Fertilizers and biological nitrogen fixation as sources of plant nutrients: perspectives for future agriculture. Plant Soil 194:11–14
Bottini R, Cassán F, Piccoli P (2004) Gibberellin production by bacteria and its involvement in plant growth promotion and yield increase. Appl Microbiol Biotechnol 65:497–503
Boukhalfa H, Crumbliss AL (2002) Chemical aspects of siderophore mediated iron transport. Biometals 15:325–339
Braud A, Jézéquel K, Vieille E, Tritter A, Lebeau T (2006) Changes in extractability of Cr and Pb in a polycontaminated soil after bioaugmentation with microbial producers of biosurfactants, organic acids and siderophores. Water Air Soil Pollut 6:261–279
Budzikiewicz H (2004) Siderophores of the Pseudomonadaceae *sensu stricto* (fluorescent and non fluorescent *Pseudomonas* spp.). In: Herz W, Falk H, Kirby GW (eds) Progress in the chemistry of organic natural products. Springer, Vienna, pp 81–237
Burd GI, Dixon DG, Glick BR (1998) A plant growth-promoting bacterium that decreases nickel toxicity in seedlings. Appl Environ Microbiol 64:3663–3668
Burd GI, Dixon DG, Glick BR (2000) Plant growth promoting bacteria that decrease heavy metal toxicity in plants. Can J Microbiol 46:237–245
Cassán F, Bottini R, Schneider G, Piccoli P (2001a) *Azospirillum brasilense* and *Azospirillum lipoferum* hydrolyze conjugates of GA20 and metabolize the resultant aglycones to GA1 in seedlings of rice dwarf mutants. Plant Physiol 125:2053–2058
Cassán F, Lucangeli C, Bottini R, Piccoli P (2001b) *Azospirillum* spp. metabolize [17,17–2H2] Gibberellin A20 to [17,17–2H2] Gibberellin A1 in vivo in dy rice mutant seedlings. Plant Cell Physiol 42:763–767
Chabot R, Antoun H, Cescas MP (1996) Growth promotion of maize and lettuce by phosphate-solubilizing *Rhizobium leguminosarum* biovar *phaseoli*. Plant Soil 184:311–321
Chang CH, Yang SS (2009) Thermo-tolerant phosphate-solubilizing microbes for multi-functional biofertilizer preparation. Bioresour Technol 100:1648–1658
Cheng Z, Park E, Glick BR (2007) 1-Aminocyclopropane-1-carboxylate deaminase from *Pseudomonas putida* UW4 facilitates the growth of canola in the presence of salt. Can J Microbiol 53:912–918
Chet I, Inbar J (1994) Biological control of fungal pathogens. Appl Biochem Biotechnol 48:37–43
Chinnadurai C, Balachandar D, Sundaram SP (2009) Characterization of 1-aminocyclopropane-1-carboxylate deaminase producing methylobacteria from phyllosphere of rice and their role in ethylene regulation. World J Microbiol Biotechnol 25:1403–1411

Ciardi JA, Tieman DM, Lund ST, Jones JB, Stall RE, Klee HJ (2000) Response to *Xanthomonas campestris* pv. *vesicatoria* in tomato involves regulation of ethylene receptor gene expression. Plant Physiol 123:81–92

Claus G, Kutzner HJ (1983) Degradation of indole by *Alcaligenes* spec. Syst Appl Microbiol 4:169–180

Colnaghi R, Green A, Luhong HE, Rudnick P, Kennedy C (1997) Strategies for increased ammonium production in free-living or plant associated nitrogen fixing bacteria. Plant Soil 194:145–154

Compant S, Duffy B, Nowak J, Clement C, Barka EA (2005) Use of plant growth-promoting bacteria for biocontrol of plant diseases: principles, mechanisms of action, and future prospects. Appl Environ Microbiol 71:4951–4959

Creus C, Sueldo R, Barassi C (1997) Shoot growth and water status in *Azospirillum*-inoculated wheat seedlings grown under osmotic and salt stresses. Plant Physiol Biochem 35:939–944

Crowley DE, Reid CPP, Szaniszlo PJ (1988) Utilization of microbial siderophores in iron acquisition by oat. Plant Physiol 87:685–688

Curie C, Briat JF (2003) Iron transport and signalling in plants. Annu Rev Plant Biol 54:183–206

Curie C, Panaviene Z, Loulergue C, Dellaporta SL, Briat JF, Walker EL (2001) Maize yellow stripe 1 encodes a membrane protein directly involved in Fe(III) uptake. Nature 409:346–349

De Freitas JR, Banerjee MR, Germida JJ (1997) Phosphate-solubilizing rhizobacteria enhance the growth and yield but not phosphorus uptake of canola (*Brassica napus* L). Biol Fert Soils 24:358–364

De Salamone IEG, Hynes RK, Nelson LM (2005) Role of cytokinins in plant growth promotion by rhizosphere bacteria. In: Siddiqui ZA (ed) PGPR: biocontrol and biofertilization. Springer, The Netherlands, pp 173–195

Dean DR, Jacobson R (1992) Biochemical genetics of nitrogenase. In: Stacey G, Burris RH, Evans HJ (eds) Biological nitrogen fixation. Chapman and Hall, New York, pp 763–817

Di Simine CD, Sayer JA, Gadd GM (1998) Solubilization of zinc phosphate by a strain of *Pseudomonas fluorescens* isolated from a forest soil. Biol Fert Soils 28:87–94

Diels L, van der Lelie N, Bastiaens L (2002) New developments in treatment of heavy metal contaminated soils. Rev Environ Sci Biotechnol 1:75–82

Dobbelaere S, Croonenborghs A, Thys A, Vande Broek A, Vanderleyden J (1999) Phytostimulatory effect of *Azospirillum brasilense* wild type and mutant strains altered in IAA production on wheat. Plant Soil 212:155–164

Downes BP, Crowell DN (1998) Cytokinin regulates the expression of a soybean b-expansin gene by a posttranscriptional mechanism. Plant Mol Biol 37:437–444

Downes BP, Steinbaker CR, Crowell DN (2001) Expression and processing of a hormonally regulated b-expansin from soybean. Plant Physiol 126:244–252

Duan J, Müller KM, Charles TC, Vesely S, Glick BR (2009) 1-Aminocyclopropane-1-carboxylate (ACC) deaminase genes in *Rhizobia* from southern Saskatchewan. Microb Ecol 57:423–436

Duff RB, Webley DM (1959) 2-Ketogluconic acid as a natural chelator produced by soil bacteria. Chem Ind 1959:1376–1377

Duijff BJ, Bakker PAHM, Schippers B (1994a) Ferric pseudobactin 358 as an iron source for carnation. J Plant Nutr 17:2069–2078

Duijff BJ, De Kogel WJ, Bakker PAHM, Schippers B (1994b) Influence of pseudobactin 358 on the iron nutrition of barley. Soil Biol Biochem 26:1681–1994

Egamberdieva D (2009) Alleviation of salt stress by plant growth regulators and IAA producing bacteria in wheat. Acta Physiol Plant 31:861–864

Eide D, Broderius M, Fett J, Guerinot M (1996) A novel iron regulated metal transporter from plants identified by functional expression in yeast. Proc Natl Acad Sci USA 93:5624–5628

El-Tarabily K, Nassar AH, Sivasithamparam K (2008) Promotion of growth of bean (*Phaseolus vulgaris* L.) in a calcareous soil by a phosphate-solubilizing, rhizosphere-competent isolate of *Micromonospora endolithica*. Appl Soil Ecol 39:161–171

Evans ML, Ishikawa H, Estelle MA (1994) Responses of *Arabidopsis* roots to auxin studied with high temporal resolution: comparison of wild type and auxin-response mutants. Planta 194:215–222

Farwell AJ, Vesely S, Nero V, Rodriguez H, Shah S, Dixon DG, Glick BR (2006) The use of transgenic canola (*Brassica napus*) and plant growth-promoting bacteria to enhance plant biomass at a nickel-contaminated field site. Plant Soil 288:309–318

Farwell AJ, Vesely S, Nero V, Rodriguez H, McCormack K, Shah S, Dixon DG, Glick BR (2007) Tolerance of transgenic canola plants (*Brassica napus*) amended with plant growth-promoting bacteria to flooding stress at a metal-contaminated field site. Environ Pollut 147:540–545

Feng K, Lu HM, Sheng HJ, Wang XL, Mao J (2004) Effect of organic ligands on biological availability of inorganic phosphorus in soils. Pedosphere 14:85–92

Fischer HM (1994) Genetic regulation of nitrogen fixation in rhizobia. Microbiol Rev 58:352–386

Frankenberger JWT, Arshad M (1995) Microbial synthesis of auxins. In: Arshad M, Frankenberger WT (eds) Phytohormones in soils. Dekker, New York, pp 35–71

Frankowski J, Lorito M, Scala F, Schmidt R, Berg G, Bahl H (2001) Purification and properties of two chitinolytic enzymes of *Serratia plymuthica* HRO-C48. Arch Microbiol 176:421–426

Fravel D, Olivain C, Alabouvette C (2003) *Fusarium oxysporum* and its biocontrol. New Phytol 157:493–502

Frugier F, Kosuta S, Murray JD, Crespi M, Szczyglowski K (2008) Cytokinin: secret agent of symbiosis. Trends Plant Sci 13:115–120

Fukuhara H, Minakawa Y, Akao S, Minamisawa K (1994) The involvement of indole-3-acetic acid produced by *Bradyrhizobium elkanii* in nodule formation. Plant Cell Physiol 35:1261–1265

Gamalero E, Berta G, Massa N, Glick BR, Lingua G (2008) Synergistic interactions between the ACC deaminase-producing bacterium *Pseudomonas putida* UW4 and the AM fungus *Gigaspora rosea* positively affect cucumber plant growth. FEMS Microbiol Ecol 64:459–467

Gamalero E, Berta G, Massa N, Glick BR, Lingua G (2009) Interactions between *Pseudomonas putida* UW4 and *Gigaspora rosea* BEG9 and their consequences for the growth of cucumber under salt stress conditions. J Appl Microbiol 108:236–245

Ghosh S, Penterman JN, Little RD, Chavez R, Glick BR (2003) Three newly isolated plant growth-promoting bacilli facilitate the seedling growth of canola, *Brassica campestris*. Plant Physiol Biochem 41:277–281

Gieg LM, Otter A, Fedorak PM (1996) Carbazole degradation by *Pseudomonas* sp LD2: metabolic characteristics and the identification of some metabolites. Environ Sci Technol 30:575–585

Giovanelli J, Mudd SH, Datko AH (1980) Sulfur amino acids in plants. In: Miflin BJ (ed) Amino acids and derivatives, the biochemistry of plants: a comprehensive treatise, vol 5. Academic, New York, pp 453–505

Glick BR (1995) The enhancement of plant growth by free-living bacteria. Can J Microbiol 41:109–117

Glick BR (2003) Phytoremediation: synergistic use of plants and bacteria to clean up the environment. Biotechnol Adv 21:383–393

Glick BR, Karaturovíc D, Newell P (1995) A novel procedure for rapid isolation of plant growth-promoting rhizobacteria. Can J Microbiol 41:533–536

Glick BR, Penrose DM, Li J (1998) A model for the lowering of plant ethylene concentrations by plant growth-promoting bacteria. J Theor Biol 190:63–68

Glick BR, Patten CL, Holguin G, Penrose DM (1999) Biochemical and genetic mechanisms used by plant growth promoting bacteria. Imperial College Press, London

Glick BR, Todorovic B, Czarny J, Cheng Z, Duan J (2007) Promotion of plant growth by bacterial ACC deaminase. Crit Rev Plant Sci 26:227–242

Goldstein AH (1994) Involvement of the quinoprotein glucose dehydrogenase in the solubilization of exogenous phosphates by gram-negative bacteria. In: Torriani-Gorini A, Yagil E, Silver S (eds) Phosphate in microorganisms: cellular and molecular biology. ASM, Washington, DC, pp 197–203

Goldstein AH, Rogers RD, Mead G (1993) Mining by microbe. Biotechnology 11:1250–1254
Grichko VP, Glick BR (2001) Amelioration of flooding stress by ACC deaminase-containing plant growth-promoting bacteria. Plant Physiol Biochem 39:11–17
Guerinot ML, Ying Y (1994) Iron: nutritious, noxious, and not readily available. Plant Physiol 104:815–820
Gulati A, Rahi P, Vyas P (2008) Characterization of phosphate-solubilizing fluorescent pseudomonads from the rhizosphere of seabuckthorn growing in the cold deserts of Himalayas. Curr Microbiol 56:73–79
Gutierrez-Manero FJ, Ramos-Solano B, Probanza A, Mehouachi J, Tadeo FR, Talon M (2001) The plant-growth-promoting rhizobacteria *Bacillus pumilis* and *Bacillus licheniformis* produce high amounts of physiologically active gibberellins. Physiol Plant 111:206–211
Gyaneshwar P, Naresh Kumar G, Parekh LJ, Poole PS (2002) Role of soil microorganisms in improving P nutrition of plants. Plant Soil 245:83–93
Haas D, Défago G (2005) Biological control of soil-borne pathogens by fluorescent pseudomonads. Nat Rev Microbiol 3:307–319
Haas D, Keel C (2003) Regulation of antibiotic production in root colonizing *Pseudomonas* spp. and relevance for biological control of plant disease. Annu Rev Phytopathol 41:117–153
Halder AK, Misra AK, Chakrabarty PK (1991) Solubilization of inorganic phosphates by *Bradyrhizobium*. Indian J Exp Biol 29:28–31
Hall JA, Peirson D, Ghosh S, Glick BR (1996) Root elongation in various agronomic crops by the plant growth promoting rhizobacterium *Pseudomonas putida* GR 12-2. Isr J Plant Sci 44:37–42
Hamdali H, Hafidi M, Virolle MJ, Ouhdouch Y (2008) Rock phosphate-solubilizing *Actinomycetes*: screening for plant growth-promoting activities. World J Microbiol Biotechnol 24:2565–2575
Hao Y, Charles TC, Glick BR (2007) ACC deaminase from plant growth promoting bacteria affects crown gall development. Can J Microbiol 53:1291–1299
Hirsch AM, Fang Y (1994) Plant hormones and nodulation: what's the connection? Plant Mol Biol 26:5–9
Honma M (1993) Stereospecific reaction of 1-aminocyclopropane-1-carboxylate deaminase. In: Pech JC, Latché A, Balagué C (eds) Cellular and molecular aspects of the plant hormone ethylene. Kluwer, Dordrecht, pp 111–116
Honma M, Shimomura T (1978) Metabolism of 1-aminocyclopropane-1-carboxylic acid. Agric Biol Chem 42:1825–1831
Hopkins CG, Whiting AL (1916) Soil bacteria and phosphates. III. Agric Exp Stn Bull 190:395–406
Huang XD, El-Alawi Y, Penrose DM, Glick BR, Greenberg BM (2004a) Responses of three grass species to creosote during phytoremediation. Environ Pollut 130:453–463
Huang XD, El-Alawi Y, Penrose DM, Glick BR, Greenberg BM (2004b) A multi-process phytoremediation system for removal of polycyclic aromatic hydrocarbons from contaminated soils. Environ Pollut 130:465–476
Iavicoli A, Boutet E, Buchala A, Métraux JP (2003) Induced systemic resistance in *Arabidopsis thaliana* in response to root inoculation with *Pseudomonas fluorescens* CHA0. Mol Plant Microbe Interact 16:851–858
Illmer P, Schinner F (1992) Solubilization of inorganic phosphates by microorganisms isolated from forest soil. Soil Biol Biochem 24:389–395
Ivanova EG, Doronina NV, Ya T (2001) Aerobic methylobacteria are capable of synthesizing auxins. Mikrobiologiya 70:452–458
Jackson MB (1991) Ethylene in root growth and development. In: Mattoo AK, Suttle JC (eds) The plant hormone ethylene. CRC, Boca Raton, FL, pp 169–181
Jacobson CB, Pasternak JJ, Glick BR (1994) Partial purification and characterization of 1-aminocyclopropane-1-carboxylate deaminase from the plant growth promoting rhizobacterium *Pseudomonas putida* GR12-2. Can J Microbiol 40:1019–1025
James EK, Olivares FL (1997) Infection of sugar cane and other graminaceous plants by endophytic diazotrophs. Crit Rev Plant Sci 17:77–119

Jarabo-Lorenzo A, Perez-Galdona R, Vega-Hernandez M, Trujillo J, Leon-Barrios M (1998) Indole-3-acetic acid catabolism by bacteria belonging to the *Bradyrhizobium* genus. In: Elmerich C, Kondorosi A, Newton WE (eds) Biological nitrogen fixation for the 21st century. Kluwer, Dordrecht, p 484

Jeffries P, Gianinazzi S, Perotto S, Turnau K (2003) The contribution of arbuscular mycorrhizal fungi in sustainable maintenance of plant health and soil fertility. Biol Fert Soils 37:1–16

Jensen JB, Egsgaard H, Vanonckelen H, Jochimsen BU (1995) Catabolism of indole-3-acetic-acid and 4-chloroindole-3-acetic and 5-chloroindole-3-acetic acid in *Bradyrhizobium japonicum*. J Bacteriol 177:5762–5766

Jiang CY, Sheng XF, Qian M, Wang QY (2008) Isolation and characterization of a heavy metal-resistant *Burkholderia* sp. from heavy metal-contaminated paddy field soil and its potential in promoting plant growth and heavy metal accumulation in metal-polluted soil. Chemosphere 72:157–164

Jin CW, He YF, Tang CX, Wu P, Zheng SJ (2006) Mechanisms of microbially enhanced Fe acquisition in red clover (*Trifolium pratense* L.). Plant Cell Environ 29:888–897

John P (1991) How plant molecular biologists revealed a surprising relationship between two enzymes, which took an enzyme out of a membrane where it was not located, and put it into the soluble phase where it could be studied. Plant Mol Biol Rep 9:192–194

Joo GJ, Kim YM, Kim JT, Rhee IK, Kim JH, Lee IJ (2005) Gibberellins-producing rhizobacteria increase endogenous gibberellins content and promote growth of red peppers. J Microbiol 43:510–515

Joo GJ, Kang SM, Hamayun M, Kim SK, Na CI, Shin DH, Lee IJ (2009) *Burkholderia* sp. KCTC 11096BP as a newly isolated gibberellin producing bacterium. J Microbiol 47:167–171

Kakimoto T (2001) Plant cytokinin biosynthetic enzymes as dimethylallyl diphosphate: ATP/ADP isopentenyltransferases. Plant Cell Physiol 42:677–685

Kang SM, Joo GJ, Hamayun M, Na CI, Shin DH, Kim HY, Hong JK, Lee IJ (2009) Gibberellin production and phosphate solubilization by newly isolated strain of *Acinetobacter calcoaceticus* and its effect on plant growth. Biotechnol Lett 31:277–281

Kende H (1989) Enzymes of ethylene biosynthesis. Plant Physiol 91:1–4

Khan MS, Zaidi A, Wani PA (2007) Role of phosphate-solubilizing microorganisms in sustainable agriculture – a review. Agron Sustain Dev 27:29–43

Kim YC, Jung H, Kim KY, Park SK (2008) An effective biocontrol bioformulation against *Phytophthora* blight of pepper using growth mixtures of combined chitinolytic bacteria under different field conditions. Eur J Plant Pathol 120:373–382

Kloepper JW, Lifshitz R, Zablotowitz RM (1989) Free living bacteria inocula for enhancing crop productivity. Trends Biotechnol 7:39–43

Koenig RL, Morris RO, Polacco JC (2002) tRNA is the source of low-level trans-zeatin production in *Methylobacterium* spp. J Bacteriol 184:1832–1842

Kpomblekou K, Tabatabai MA (1994) Effect of organic acids on release of phosphorus from phosphate rocks. Soil Sci 158:442–453

Kuffner M, Puschenreiter M, Wieshammer G, Gorfer M, Sessitsch A (2008) Rhizosphere bacteria affect growth and metal uptake of heavy metal accumulating willows. Plant Soil 304:35–44

Lambrecht M, Okon Y, Vande Broek A, Vanderleyden J (2000) Indole-3-acetic acid: a reciprocal signalling molecule in bacteria-plant interactions. Trends Microbiol 8:298–300

Leach AW, Mumford JD (2008) Pesticide environmental accounting: a method for assessing the external costs of individual pesticide applications. Environ Pollut 151:139–147

Leeman M, van Pelt JA, Denouden FM, Heinsbroek M, Bakker PAHM, Schippers B (1995) Induction of systemic resistance against *Fusarium* wilt of radish by lipopolysaccharides of *Pseudomonas fluorescens*. Phytopathology 85:1021–1027

Lemanceau P, Bakker PAHM, de Kogel WJ, Alabouvette C, Schippers B (1992) Effect of pseudobactin 358 production by *Pseudomonas putida* WCS358 on suppression of *Fusarium* wilt of carnation by nonpathogenic *Fusarium oxysporum* Fo47. Appl Environ Microbiol 58:2978–2982

Lemanceau P, Bauer P, Kraemer S, Briat JF (2009) Iron dynamics in the rhizosphere as a case study for analyzing interactions between soils, plants and microbes. Plant Soil 321:513–535
Leveau JHJ, Gerards S (2008) Discovery of a bacterial gene cluster for catabolism of the plant hormone indole 3-acetic acid. FEMS Microbiol Ecol 65:238–250
Leveau JHJ, Lindow SE (2005) Utilization of the plant hormone indole-3-acetic acid for growth by *Pseudomonas putida* strain 1290. Appl Environ Microbiol 71:2365–2371
Libbert E, Risch H (1969) Interactions between plants and epiphytic bacteria regarding their auxin metabolism. V. Isolation and identification of IAA-producing and -destroying bacteria from pea plants. Physiol Plant 22:51–58
Lindberg T, Granhall U, Tomenius H (1985) Infectivity and acetylene reduction of diazotrophic rhizosphere bacteria in wheat (*Triticum aestivum*) seedlings under gnotobiotic conditions. Biol Fert Soils 1:123–129
Liu L, Kloepper JW, Tuzun S (1995) Induction of systemic resistance in cucumber against bacterial angular leaf spot by plant growth-promoting rhizobacteria. Phytopathology 85:843–847
Loper JE, Buyer JS (1991) Siderophores in microbial interactions on plant surfaces. Mol Plant Microbe Interact 4:5–13
Lorteau MA, Ferguson BJ, Guinel FC (2001) Effects of cytokinin on ethylene production and nodulation in pea (*Pisum sativum* cv. *sparkle*). Physiol Plant 112:421–428
Lucangeli C, Bottini R (1997) Effects of *Azospirillum* spp. on endogenous gibberellins content and growth of maize (*Zea mays* L.) treated with uniconazole. Symbiosis 23:63–72
Ma JF (2005) Plant root responses to three abundant soil minerals: silicon, aluminum and iron. Crit Rev Plant Sci 24:267–281
Ma W, Penrose DM, Glick BR (2002) Strategies used by rhizobia to lower plant ethylene levels and increase nodulation. Can J Microbiol 48:947–954
Ma W, Guinel FC, Glick BR (2003) The *Rhizobium leguminosarum* bv. *viciae* ACC deaminase protein promotes the nodulation of pea plants. Appl Environ Microbiol 69:4396–4402
Ma W, Charles TC, Glick BR (2004) Expression of an exogenous 1-aminocyclopropane-1-carboxylate deaminase gene in *Sinorhizobium meliloti* increases its ability to nodulate alfalfa. Appl Environ Microbiol 70:5891–5897
MacMillan J (2002) Occurrence of gibberellins in vascular plants, fungi, and bacteria. J Plant Growth Regul 20:387–442
Madhaiyan M, Poonguzhali S, Sa T (2007) Characterization of 1-aminocyclopropane-1-carboxylate (ACC) deaminase containing *Methylobacterium oryzae* and interactions with auxins and ACC regulation of ethylene in canola (*Brassica campestris*). Planta 226:867–876
Malboobi MA, Behbahani M, Madani H, Owlia P, Deljou A, Yakhchali B, Moradi M, Hassanabadi H (2009) Performance evaluation of potent phosphate solubilizing bacteria in potato rhizosphere. World J Microbiol Biotechnol 25:1479–1484
Malik A (2004) Metal bioremediation through growing cells. Environ Int 30:261–278
Marroquí S, Zorreguieta A, Santamaría C, Temprano F, Soberón M, Megías M, Downie JA (2001) Enhanced symbiotic performance by *Rhizobium tropici* glycogen synthase mutants. J Bacteriol 183:854–864
Marschner H (1995) Mineral nutrition of higher plants, 2nd edn. Academic, London
Masalha J, Kosegarten H, Elmaci O, Mengel K (2000) The central role of microbial activity for iron acquisition in maize and sunflower. Biol Fert Soils 30:433–439
Mathesius U, Schlaman HRM, Spaink HP, Sautter C, Rolfe BG, Djordjevic MA (1998) Auxin transport inhibition precedes root nodule formation in white clover roots and is regulated by flavonoids and derivatives of chitin oligosaccharides. Plant J 14:23–34
Matias SR, Pagano MC, Carvalho-Muzzi F, Oliveira CA, Almeida-Carneiro A, Horta SN, Scotti MR (2009) Effect of rhizobia, mycorrhizal fungi and phosphate-solubilizing microorganisms in the rhizosphere of native plants used to recover an iron ore area in Brazil. Eur J Soil Biol 45:259–266

Maurhofer M, Hase C, Meuwly P, Metraux JP, Défago G (1994) Induction of systemic resistance of tobacco to tobacco necrosis virus by the root-colonizing *Pseudomonas fluorescens* strain CHA0: influence of the *gacA* gene and of pyoverdine production. Phytopathology 84:139–146

Mayak S, Tirosh T, Glick BR (1999) Effect of wild type and mutant plant growth-promoting rhizobacteria on the rooting of mung been cuttings. J Plant Growth Regul 18:49–53

Mayak S, Tirosh T, Glick BR (2004a) Plant growth-promoting bacteria that confer resistance to water stress in tomato and pepper. Plant Sci 166:525–530

Mayak S, Tirosh T, Glick BR (2004b) Plant growth-promoting bacteria confer resistance in tomato plants to salt stress. Plant Physiol Biochem 42:565–572

Mazurier S, Corberand T, Lemanceau P, Raaijmakers JM (2009) Phenazine antibiotics produced by fluorescent pseudomonads contribute to natural soil suppressiveness to *Fusarium* wilt. ISME J 3:977–991

Mba CC (1994) Rock phosphate solubilizing and cellulolytic actinomycetes isolates of earthworm casts. Environ Manage 18:257–261

Meyer JM, Abdallah MA (1978) The fluorescent pigment of *Pseudomonas fluorescens*: biosynthesis, purification and physico-chemical properties. J Gen Microbiol 107:319–328

Meyer JM, Gruffaz C, Raharinosy V, Bezverbnaya I, Schäfer M, Budzikiewicz H (2008) Siderotyping of fluorescent *Pseudomonas*: molecular mass determination by mass spectrometry as a powerful pyoverdine siderotyping method. Biometals 21:259–271

Mino Y (1970) Studies on destruction of indole-3-acetic acid by a species of *Arthrobacter*. IV. Decomposition products. Plant Cell Physiol 11:129–138

Mirleau P, Delorme S, Philippot L, Meyer JM, Mazurier S, Lemanceau P (2000) Fitness in soil and rhizosphere of *Pseudomonas fluorescens* C7R12 compared with a C7R12 mutant affected in pyoverdine synthesis and uptake. FEMS Microbiol Ecol 34:35–44

Mirleau P, Philippot L, Corberand T, Lemanceau P (2001) Involvement of nitrate reductase and pyoverdine in competitiveness of *Pseudomonas fluorescens* strain C7R12 in soil. Appl Environ Microbiol 67:2627–2635

Mitter N, Srivastava AC, Renu AS, Sarbhoy AK, Agarwal DK (2002) Characterization of gibberellin producing strains of *Fusarium moniliforme* based on DNA polymorphism. Mycopathologia 153:187–193

Molla MAZ, Chowdhury AA, Islam A, Hoque S (1984) Microbial mineralization of organic phosphate in soil. Plant Soil 78:393–399

Morgan PW, Drew CD (1997) Ethylene and plant responses to stress. Physiol Plant 100:620–630

Nair A, Juwarkar AA, Singh SK (2007) Production and characterization of siderophores and its application in arsenic removal from contaminated soil. Water Air Soil Pollut 180:199–212

Neilands JB (1981) Iron adsorption and transport in microorganisms. Annu Rev Nutr 1:27–46

Nester EW, Gordon MP, Amasino RM, Yanofsky MF (1984) Crown gall: a molecular and physiological analysis. Annu Rev Plant Physiol 35:387–413

Nie L, Shah S, Rashid A, Burd GI, Dixon DG, Glick BR (2002) Phytoremediation of arsenate contaminated soil by transgenic canola and the plant growth-promoting bacterium *Enterobacter cloacae* CAL2. Plant Physiol Biochem 40:355–361

Nieto KF, Frankenberger WT Jr (1989) Biosynthesis of cytokinins by *Azotobacter chroococcum*. Soil Biol Biochem 21:967–972

Okon Y, Labandera-Gonzalez C (1994) Agronomic applications of *Azospirillum*: an evaluation of 20 years of worldwide field inoculation. Soil Biol Biochem 26:1591–1601

Ongena M, Jourdan E, Adam A, Paquot M, Brans A, Joris B, Arpigny JL, Thonart P (2007) Surfactin and fengycin lipopeptides of *Bacillus subtilis* as elicitors of induced systemic resistance in plants. Environ Microbiol 9:1084–1090

Ordentlich A, Elad Y, Chet I (1988) The role of chitinase of *Serratia marcescens* in biocontrol of *Sclerotium rolfsii*. Phytopathology 78:84–88

Ortiz-Castro R, Valencia-Cantero E, Lopez-Bucio J (2008) Plant growth promotion by *Bacillus megaterium* involves cytokinin signaling. Plant Signal Behav 3:263–265

Park KH, Lee CY, Son HJ (2009) Mechanism of insoluble phosphate solubilization by *Pseudomonas fluorescens* RAF15 isolated from ginseng rhizosphere and its plant growth-promoting activities. Lett Appl Microbiol 49:222–228

Patten C, Glick BR (1996) Bacterial biosynthesis of indole-3-acetic acid. Can J Microbiol 42:207–220

Patten CL, Glick BR (2002) The role of bacterial indoleacetic acid in the development of the host plant root system. Appl Environ Microbiol 68:3795–3801

Peck SC, Kende H (1995) Sequential induction of the ethylene biosynthetic enzymes by indole-3-acetic acid in etiolated peas. Plant Mol Biol 28:293–301

Penrose DM, Glick BR (2001) Levels of ACC and related compounds in exudate and extracts of canola seeds treated with ACC deaminase containing plant growth-promoting bacteria. Can J Microbiol 47:368–372

Peralta H, Mora Y, Salazar E, Encarnación S, Palacios R, Mora J (2004) Engineering the *nifH* promoter region and abolishing poly-β-hydroxybutyrate accumulation in *Rhizobium etli* enhance nitrogen fixation in symbiosis with *Phaseolus vulgaris*. Appl Environ Microbiol 70:3272–3281

Piccoli P, Lucangeli C, Schneider G, Bottini R (1997) Hydrolisis of 17,17-[$2H_2$]-gibberellin A20-glucoside and 17,17-[$2H_2$]-gibberellin A20-glucosyl esther by *Azospirillum lipoferum* cultured in nitrogen-free biotin-based chemically defined medium. Plant Growth Regul 23:179–182

Piccoli P, Masciarelli O, Bottini R (1999) Gibberellin production by *Azospirillum lipoferum* cultured in chemically defined medium as affected by water status and oxygen availability. Symbiosis 27:135–146

Pilet PE, Saugy M (1987) Effect on root growth of endogenous and applied IAA and ABA. Plant Physiol 83:33–38

Plazinski J, Rolfe BG (1985) Influence of *Azospirillum* strains on the nodulation of clovers by *Rhizobium* strains. Appl Environ Microbiol 49:984–989

Poonguzhali S, Munusamy M, Sa T (2008) Isolation and identification of phosphate solubilizing bacteria from Chinese cabbage and their effect on growth and phosphorus utilization of plants. J Microbiol Biotechnol 18:773–777

Raaijmakers JM, Leeman M, Van Oorschot MMP, Van der Sluis I, Schippers B, Bakker PAHM (1995) Dose-response relationships in biological control of fusarium wilt of radish by *Pseudomonas* spp. Phytopathology 85:1075–1081

Raaijmakers JM, de Bruijn I, de Kock MJD (2006) Cyclic lipopeptide production by plant-associated *Pseudomonas* species: diversity, activity, biosynthesis and regulation. Mol Plant Microb Interact 19:699–710

Raaijmakers JM, Paulitz TC, Steinberg C, Alabouvette C, Moënne-Loccoz Y (2009) The rhizosphere: a playground and battlefield for soilborne pathogens and beneficial microorganisms. Plant Soil 321:341–361

Ramírez M, Valderrama B, Arredondo-Peter R, Soberon M, Mora J, Hernández G (1999) *Rhizobium etli* genetically engineered for the heterologous expression of *Vitreoscilla* sp. hemoglobin effects on free-living symbiosis. Mol Plant Microbe Interact 12:1008–1015

Ray J, Bagyaraj DJ, Manjunath A (1981) Influence of soil inoculation with versicular arbuscular mycorrhizal (VAM) and a phosphate dissolving bacteria on plant growth and $^{32}P$ uptake. Soil Biol Biochem 13:105–108

Reed MLE, Glick BR (2004) Applications of free living plant growth-promoting rhizobacteria. Antonie Van Leeuwenhoek 86:1–25

Reed MLE, Glick BR (2005) Growth of canola (*Brassica napus*) in the presence of plant growth-promoting bacteria and either copper or polycyclic aromatic hydrocarbons. Can J Microbiol 51:1061–1069

Richardson A (2001) Prospects for using soil microorganisms to improve the acquisition of phosphorus by plants. Aust J Plant Physiol 28:897–907

Riov J, Yang SF (1989) Ethylene and auxin-ethylene interaction in adventitious root formation in mung bean (*Vigna radiata*) cuttings. J Plant Growth Regul 8:131–141

Robin A, Mougel C, Siblot S, Vansuyt G, Mazurier S, Lemanceau P (2006) Effect of ferritin overexpression in tobacco on the structure of bacterial and pseudomonad communities associated with the roots. FEMS Microbiol Ecol 58:492–502
Robin A, Vansuyt G, Hinsinger P, Meyer JM, Briat JF, Lemanceau P (2008) Iron dynamics in the rhizosphere: consequences for plant health and nutrition. Adv Agron 99:183–225
Robinson NJ, Procter CM, Connolly EL, Guerinot ML (1999) A ferricchelate reductase for iron uptake from soils. Nature 397:694–697
Rodriguez H, Fraga R (1999) Phosphate solubilizing bacteria and their role in plant growth promotion. Biotechnol Adv 17:319–339
Rodriguez H, Gonzalez T, Goire I, Bashan Y (2004) Gluconic acid production and phosphate solubilization by the plant growth-promoting bacterium *Azospirillum* spp. Naturwissenschaften 91:552–555
Rodriguez H, Fraga R, Gonzalez T, Bashan Y (2006) Genetics of phosphate solubilization and its potential applications for improving plant growth-promoting bacteria. Plant Soil 287:15–21
Rodriguez H, Vessely S, Shah S, Glick BR (2008) Effect of a nickel-tolerant ACC deaminase-producing *Pseudomonas* strain on growth of nontransformed and transgenic canola plants. Curr Microbiol 57:170–174
Rojas A, Holguin G, Glick BR, Bashan Y (2001) Synergism between *Phyllobacterium* sp. ($N_2$-fixer) and *Bacillus licheniformis* (P-solubilizer), both from a semiarid mangrove rhizosphere. FEMS Microbiol Ecol 35:181–187
Sahin F, Çakmakçi R, Kantar F (2004) Sugar beet and barley yields in relation to inoculation with $N_2$-fixing and phosphate solubilizing bacteria. Plant Soil 265:123–129
Salamone GIE, Hynes RK, Nelson LM (2001) Cytokinin production by plant growth promoting rhizobacteria and selected mutants. Can J Microbiol 47:404–411
Saleh S, Huang X-D, Greenberg BM, Glick BR (2004) Phytoremediation of persistent organic contaminants in the environment. In: Singh A, Ward O (eds) Applied bioremediation and phytoremediation, vol 1, Soil biology. Springer, Berlin, pp 115–134
Salisbury FB, Ross CW (1992) Plant physiology. Wadsworth, Belmont, CA
Saravanakumar D, Samiyappan R (2006) ACC deaminase from *Pseudomonas fluorescens* mediated saline resistance in groundnut (*Arachis hypogea*) plants. J Appl Microbiol 102:1283–1292
Seidel C, Walz A, Park S, Cohen JD, Ludwig-Müller J (2006) Indole-3-acetic acid protein conjugates: novel players in auxin homeostasis. Plant Biol 8:340–345
Sessitsch A, Coenye T, Sturz AV, Vandamme P, Ait Barka E, Salles JF, Van Elsas JD, Faure D, Reiter B, Glick BR, Wang-Pruski G, Nowak J (2005) *Burkholderia phytofirmans* sp. nov., a novel plant-associated bacterium with plant-beneficial properties. Int J Syst Evol Microbiol 55:1187–1192
Shah S, Li J, Moffatt BA, Glick BR (1997) ACC deaminase genes from plant growth promoting rhizobacteria. In: Kobayashi K, Hemma Y, Kodema F, Kondo N, Akino S, Ogoshi A (eds) Plant growth-promoting rhizobacteria. Present status and future prospects. Organization for Economic Cooperation and Development, Paris, pp 320–324
Sharma A, Johria BN, Sharma AK, Glick BR (2003) Plant growth-promoting bacterium *Pseudomonas* sp. strain GRP3 influences iron acquisition in mung bean (*Vigna radiata* L. Wilzeck). Soil Biol Biochem 35:887–894
Sharma V, Kumar V, Archana G, Kumar GN (2005) Substrate specificity of glucose dehydrogenase (GDH) of *Enterobacter asburiae* PSI3 and rock phosphate solubilization with GDH substrates as C sources. Can J Microbiol 51:477–482
Siddiqui IA, Shoukat SS (2003) Suppression of root-knot disease by *Pseudomonas fluorescens* CHA0 in tomato: importance of bacterial secondary metabolite, 2,4-diacetylphloroglucinol. Soil Biol Biochem 35:1615–1623
Siebner-Freibach H, Hadar Y, Chen Y (2003) Siderophores sorbed on Ca-montmorillonite as an iron source for plants. Plant Soil 251:115–124
Singh PP, Shin YC, Park CS, Chung YR (1999) Biological control of *Fusarium* wilt of cucumber by chitinolytic bacteria. Phytopathology 89:92–99

Someya N, Sato Y, Yamaguchi I, Hamamoto H, Ichiman Y, Akutsu K, Sawada H, Tsuchiya K (2007) Alleviation of nickel toxicity in plants by a rhizobacterium strain is not dependent on its siderophore production. Commun Soil Sci Plant Anal 38:1155–1162

Spaepen S, Vanderleyden J, Remans R (2007) Indole-3-acetic acid in microbial and microorganism-plant signaling. FEMS Microbiol Rev 31:425–448

Sperber JI (1958) The incidence of apatite-solubilizing organisms in the rhizosphere and soil. Aust J Aric Res 9:778–781

Steenhoudt O, Vanderleyden J (2000) *Azospirillum*, a free-living nitrogen-fixing bacterium closely associated with grasses: genetic, biochemical and ecological aspects. FEMS Microbiol Rev 24:487–506

Stintzi A, Evans K, Meyer JM, Poole K (1998) Quorum-sensing and siderophore biosynthesis in *Pseudomonas aeruginosa*: lasR/lasI mutants exhibit reduced pyoverdine biosynthesis. FEMS Microbiol Lett 166:341–345

Strzelczyk E, Kampert M, Dahm H (1973) Production and decomposition of indoleacetic-acid (IAA) by microorganisms isolated from root zone of 2 crop plants. Acta Microbiol Pol B 5:71–79

Taller BJ, Wong TY (1989) Cytokinins in *Azotobacter vinelandii* culture medium. Appl Environ Microbiol 55:266–267

Tao GC, Tian SJ, Cai MY, Xie GH (2008) Phosphate-solubilizing and -mineralizing abilities of bacteria isolated from soils. Pedosphere 18:515–523

Theunis M (2005) IAA biosynthesis in rhizobia and its potential role in symbiosis. Ph.D. thesis, Universiteit Antwerpen

Tien T, Gaskin M, Hubbel D (1979) Plant growth substances produced by *Azospirillum brasilense* and their effect on the growth of pearl millet (*Pennisetum americanum* L.). Appl Environ Microbiol 37:1016–1024

Timmusk S, Nicander B, Granhall U, Tillberg E (1999) Cytokinin production by *Paenobacillus polymyxa*. Soil Biol Biochem 31:1847–1852

Toro M, Azcon R, Barea J (1997) Improvement of arbuscular mycorrhiza development by inoculation of soil with phosphate solubilizing rhizobacteria to improve rock phosphate bioavailability ($^{32}$P) and nutrient cycling. Appl Environ Microbiol 63:4408–4412

Tran H, Ficke A, Asiimwe T, Höfte M, Raaijmakers JM (2007) Role of the cyclic lipopeptide massetolide A in biological control of *Phytophthora infestans* and in colonization of tomato plants by *Pseudomonas fluorescens*. New Phytol 175:731–742

Trewavas AJ (2000) Signal perception and transduction. In: Buchannan B, Gruisem W, Jones R (eds) Biochemistry and molecular biology of plants, American Society of Plant Physiology, Rockville, MD, pp 530–587

Tsakelova EA, Klimova SY, Cherdyntseva TA, Netrusov AI (2006) Microbial producers of plant growth stimulators and their practical use: a review. Appl Biochem Microbiol 42:117–126

Ureta AC, Imperial J, Ruiz-Argüeso T, Palacios JM (2005) *Rhizobium leguminosarum* biovar *viciae* symbiotic hydrogenase activity and processing are limited by the level of nickel in agricultural soils. Appl Environ Microbiol 71:7603–7606

Valverde A, Burgos A, Fiscella T, Rivas R, Velazquez E, Rodrìguez-Barrueco C, Cervantes E, Chamber M, Igual JM (2006) Differential effects of coinoculations with *Pseudomonas jessenii* PS06 (a phosphate-solubilizing bacterium) and *Mesorhizobium ciceri* C-2/2 strains on the growth and seed yield of chickpea under greenhouse and field conditions. Plant Soil 287:43–50

Van Loon LC, Glick BR (2004) Increased plant fitness by rhizobacteria. In: Sandermann H (ed) Molecular ecotoxicology of plants, ecological studies, vol 170. Springer, Berlin, pp 177–205

Van Loon LC, Bakker PAHM, Pieterse CMJ (1998) Systemic resistance induced by rhizosphere bacteria. Annu Rev Phytopathol 36:453–483

van Noorden GE, Ross JJ, Reid JB, Rolfe BG, Mathesius U (2006) Defective long distance auxin transport regulation in the *Medicago truncatula* super numerary nodules mutant. Plant Physiol 140:1494–1506

Vansuyt G, Robin A, Briat JF, Curie C, Lemanceau P (2007) Iron acquisition from Fe-pyoverdine by *Arabidopsis thaliana*. Mol Plant Microbe Interact 4:441–447

Vassilev N, Vassileva M (2004) Multifunctional properties of a plant growth promoting bacterium entrapped in k-carrageenan. In: Pedraz JL, Orive G, Poncelet D (eds) XII international workshop on bioencapsulation, Vitoria, Univ Pais Vasco, 24–26 September 2004, pp 162–166

Vassilev N, Medina A, Vassileva M (2006a) Microbial solubilization of rock phosphate on media containing agro-industrial wastes and effect of the resulting products on plant growth and P uptake. Plant Soil 287:77–84

Vassilev N, Vassileva M, Nikolaeva I (2006b) Simultaneous P-solubilizing and biocontrol activity of microorganisms: potentials and future trends. Appl Microbiol Biotechnol 71:137–144

Vassileva M, Azcon R, Barea JM, Vassilev N (1999) Effect of encapsulated cells of *Enterobacter* sp. on plant growth and phosphate uptake. Bioresour Technol 67:229–232

Vassileva M, Azcon R, Barea JM, Vassilev N (2000) Rock phosphate solubilization by free and encapsulated cells of *Yarrowia lipolytica*. Process Biochem 35:693–697

Vert G, Grotz N, Dedaldechamp F, Gaymard F, Guerinot ML, Briat JF, Curie C (2002) IRT1, an *Arabidopsis* transporter essential for iron uptake from the soil and for plant growth. Plant Cell 14:1223–1233

Von Wirén N, Mori S, Marschner H, Römheld V (1994) Iron inefficiency in maize mutant ys1 (*Zea mays* L. cv yellow-stripe) is caused by a defect in uptake of iron phytosiderophores. Plant Physiol 106:71–77

Walter A, Römheld V, Marschner H, Crowley DE (1994) Iron nutrition of cucumber and maize: effect of *Pseudomonas putida* YC3 and its siderophore. Soil Biol Biochem 26:1023–1031

Wang C, Knill E, Glick BR, Defago G (2000) Effect of transferring 1-aminocyclopropane-1-carboxylic acid (ACC) deaminase genes into *Pseudomonas fluorescens* strain CHA0 and its *gacA* derivative CHA96 on their growth-promoting and disease-suppressive capacities. Can J Microbiol 46:898–907

Weller DM (1988) Biological control of soilborne pathogens in the rhizosphere with bacteria. Annu Rev Phytopathol 26:379–407

Werner T, Motyka V, Laucou V, Smets R, Onckelen HV, Schmülling TH (2003) Cytokinin-deficient transgenic *Arabidopsis* plants show multiple developmental alterations indicating opposite functions of cytokinins in the regulation of shoot and root meristem activity. Plant Cell 15:2532–2550

Whipps JM (2001) Microbial interactions and biocontrol in the rhizosphere. J Exp Bot 52:487–511

Williams PM, Sicardi de Mallorca M (1982) Abscisic acid and gibberellin-like substances in roots and root nodules of *Glycine max*. Plant Soil 65:19–26

Xie H, Pasternak JJ, Glick BR (1996) Isolation and characterization of mutants of the plant growth-promoting rhizobacterium *Pseudomonas putida* GR12-2 that overproduce indoleacetic acid. Curr Microbiol 32:67–71

Yahalom E, Okon Y, Dovrat A (1990) Possible mode of action of *Azospirillum brasilense* strain Cd on the root morphology and nodule formation in burr medic (*Medicago polymorpha*). Can J Microbiol 36:10–14

Yehuda Z, Shenker M, Romheld V, Marschner H, Hadar Y, Chen Y (1996) The role of ligand exchange in the uptake of iron from microbial siderophores by gramineous plants. Plant Physiol 112:1273–1280

Yi Y, Huang W, Ge Y (2008) Exopolysaccharide: a novel important factor in the microbial dissolution of tricalcium phosphate. World J Microbiol Biotechnol 24:1059–1065

Yuhashi KI, Ichikawa N, Ezura H, Akao S, Minakawa Y, Nukui N, Yasuta T, Minamisawa K (2000) Rhizobitoxine production by *Bradyrhizobium elkanii* enhances nodulation and competitiveness on *Macroptilium atropurpureum*. Appl Environ Microbiol 66:2658–2663

Zipfel C, Robatzek S, Navarro L, Oakeley EJ, Jones JDG, Felix G, Boller T (2004) Bacterial disease resistance in *Arabidopsis* through flagellin perception. Nature 428:764–767

# Chapter 3
# Microbial Zinc Solubilization and Their Role on Plants

**V.S. Saravanan, M. Rohini Kumar, and T.M. Sa**

## 3.1 Introduction

Zinc (Zn) can best be regarded as a double-edged sword in living systems, serving as a nutrient in minute concentrations yet causing toxicity problems at higher levels. In natural environments, toxicity by Zn is a less common phenomenon compared to deficiency, which is more often encountered in living organisms, especially in plants, animals and humans. Zinc deficiency in humans leads to various abnormalities in growth, immunity and brain development. A survey has pointed out that in developing countries; 2.9 billion children under the age of five are deficient in Zn nutrients. Deficiency of Zn among members of a community, including animals and humans, is a manifestation of Zn deficiency in the crops that supply the community. Although no permanent solution is available at present, several approaches are being followed by scientists around the world to address this problem. These include, crop breeding approaches that increase plant tolerance to low levels of Zn without compromising productivity, even inducing secretion of higher amounts of plant root exudates. The type and quantity of exudates can influence Zn uptake by plant and consequently the supplements of Zn in the food commonly consumed by the people. Crop production-based approaches such as regular application of micronutrient fertilizers and adoption of crop rotation (Frossard et al. 2000) are also

V.S. Saravanan
Department of Microbiology, Indira Gandhi College of Arts and Science, Pondicherry 605007, India

M. Rohini Kumar
Division of Industrial Biotechnology, School of Bio Sciences and Technology, Vellore Institute of Technology, Vellore 632014, India

T.M. Sa (✉)
Department of Agricultural Chemistry, Chungbuk National University, Cheongju, Chungbuk 361-763, South Korea
e-mail: tomsa@chungbuk.ac.kr

D.K. Maheshwari (ed.), *Bacteria in Agrobiology: Plant Nutrient Management*,
DOI 10.1007/978-3-642-21061-7_3, 

being considered. In India, however, it is a known fact that, although total Zn concentration in soil is adequate, the quantity that is readily available to plants is insufficient to meet the demand of the growing crop (Singh 1991, 2001). This is where the role of bacteria that are able to solubilize insoluble Zn compounds and increase their availability in the soil solution, similar to that of P nutrition (Saravanan et al. 2007a). This chapter focuses on the properties and importance of Zn in living systems; their deficiency and toxicity; microbial solubilization of insoluble Zn compounds; and the benefits that crops could harness through increased amounts of solubilized Zn available for plant uptake.

## 3.2 Importance of Zn in the Living Systems

The cation $Zn^{2+}$ can perform several functions in the living system since it has a similar size to that of other cations useful in biological systems, such as $Mn^{2+}$, $Fe^{2+}$, $Co^{2+}$, $Ni^{2+}$, $Cu^{2+}$ and $Zn^{2+}$ (all have ionic diameter between 138 and 160 pm). Cells do not usually differentiate between toxic and physiologically required metals during the uptake. Many metal cations are transported across the cell membrane by an unspecific and fast, chemiosmotic driven uptake system like the CorA membrane integral protein, which is part of the MIT (metal inorganic transport) family. These are constitutively expressed, unspecific, open-gate system transporters that pump the cation inside the cell when the extracellular concentration of the metal increases. This can eventually lead to metal toxicity (Nies 1999).

Inside the cell, Zn has an essential role in the proper functioning of enzymes. Zinc performs a pivotal role in more than 300 metalloenzymes (Vallee 1991) and is the only element found in all the six enzyme classes (oxidoreductases, transferases, hydrolases, lyases, isomerases and ligases) as designated by the International Union of Biochemistry and Molecular Biology (IUBMB). Some of these important enzymes include carbonic anhydrase and carboxy peptidase, which are involved in $CO_2$ regulation and digestion of proteins, respectively. Zinc is also found in the DNA-binding proteins that play an important role in DNA transcription. TFIIIA was the first transcription factor identified as a Zn protein among the functionally and structurally diverse proteins associated with transcription. In some proteins, Zn organizes the domain and makes them capable of molecular recognition by nucleic acids, proteins and lipids (Laity et al. 2001). A proteome analysis of all the three forms of life (bacteria, archaea and eukaryotes) revealed that 4–10% of cellular proteins are basically Zn-binding proteins (Andrieni et al. 2006).

Interestingly, the importance of Zn in the living system was first identified by Raulin in 1869 in the bread mold, *Aspergillus niger*. In plants, more than 90% of Zn is present in soluble forms. It plays major roles in carbohydrate metabolism, through photosynthesis; in sucrose and starch formation; protein metabolism; membrane integrity; auxin metabolism; and reproduction (Alloway 2008).

## 3.3 Zinc and Evolutionary Significance

During the process of evolution, biological recruitment of elements was influenced by their chemical state and availability (Frausto Da Silva and Williams 1991). It was deduced that during early evolution, metals such as Zn and Cu would have been locked up in insoluble sulfides. For Fe, Co and Ni, it was suggested that at a predicted pH, close to the solubility product of the precipitates, small metal–sulfate clusters may have existed in the sea water solution. It is thus believed that biology may have taken advantage of the sulfide clusters of these three metals; however, Cu, being so insoluble, may have been totally excluded. The appearance of di oxygen may have led to the removal of sulfides from the sea, resulting in an increased availability of metal ions such as Zn and Cu. This was evident from findings that carbonic anhydrase, a Zn-containing enzyme, was one of the ancient enzymes found distributed in all prokaryotes, including archaea. It was found that this enzyme also played an important role in the autotrophic evolution of life.

The first organisms that evolved in the autotrophic life were the cyanobacteria, which commanded a major evolutionary transition in the toxicity and use of these metal ions. The cyanobacteria have since undergone rapid evolutionary radiation (Giovannoni et al. 1988). Thus, during the time when the availability of many metal ions was relatively changing, new species of cyanobacteria were rapidly evolving. Exposure to elevated concentrations of metals, such as Cu and Zn, would have initially favored the evolution of resistance determinants, including the Zn and Cu efflux systems. However, exploitation of newly available ions eventually became a determinant for subsequent evolution.

## 3.4 Zn Deficiency and Toxicity: The Two Poles in the Soil System

Zn deficiency is a problem commonly encountered in plants and animals, including humans. Deficiency symptoms in microbes, however, are not well documented. In plants, deficiency problems arise due to insufficient Zn available in the soil solution. Zn concentration varies according to the soil's geographical location and the type of crop cultivated. When the concentration of available Zn in soils falls below 0.5 mg Zn/kg dry soil, deficiency symptoms arise (Alloway 2001). The concentration of Zn in the soil solution is usually very low (0.002–0.196 mg/L) despite a total Zn content of around 50–100 mg/L (Srivastava and Gupta 1996) in most soils. Another way of assessing the deficiency is analyzing the leaf Zn levels. If the leaf Zn level is below 20 mg Zn/kg of the plant tissue, deficiency symptoms may become apparent (Alloway 2008).

Zinc toxicity symptoms are exhibited in both plant and microbial populations. Toxicity results from Zn pollution of soils due to atmospheric deposition from nearby smelting industries, Zn mines' wastes; flooding of alluvial soils with Zn polluted river water and sediments; excessive application of Zn-rich materials including Zn-rich sewage sludge; or from industrial waste waters. In general, most plant species have high tolerance potential to excessive Zn concentration in soils. In a few cases, however, toxicity symptoms, such as chlorosis in new leaves and depressed plant growth, can be observed. Toxicity results when the Zn concentration range is from 100 to 500 mg/kg of soil (Macnicol and Beckett 1985).

High soil Zn concentration affects microbial diversity, with up to 25% decrease in their population (Moffett et al. 2003) in agricultural soils based on accounting the OTU (operational taxonomic unit) numbers. In general, diazotrophs are sensitive to Zn toxicity. In particular, *Rhizobium leguminosarum* and *Bradyrhizobium* were found to be more sensitive to Zn ions, apart from $Cu^{2+}$ (Biro et al. 1995). It was further shown that concentrations of up to 25 mmol Zn/kg do not affect resident nitrifying communities in long-term contaminated soils. However, the addition of the same total concentration in the form of $ZnCl_2$ to uncontaminated soil potentially eliminated nitrifying populations, showing that long-term exposure may lead to tolerance development among the nitrifying communities in the soil (Mertens et al. 2006).

## 3.5 Bacterial Mechanisms of Zn Resistance

Zn enters the cell along with other essential metal ions through specific and non-specific transport systems. The toxic effects of Zn on bacteria are mainly by inhibition of the respiratory electron transport chain (Beard et al. 1995). Microbes have evolved several mechanisms for Zn resistance and detoxification, including (a) binding of the metal to the outer membrane (b) efflux by antiport system (c) efflux by P type ATPase (d) Zn-binding proteins (Choudhury and Srivastava 2001) and (e) complexation by organic acids (Appanna and Whitmore 1995) (Fig. 3.1).

### *3.5.1 Zinc Transport*

Zn is one the important cations that are available in trace quantities in environment, which needs to be transported inside the cells to meet out their nutritional needs. Conversely, the metal cation causes toxicity when bacteria were dwelling in highly Zn-contaminated areas. So under both the situations, the transporter proteins embedded in the cell membrane play an important role in the uptake or efflux of Zn.

Zn uptake into the cell involves two distinct phases, one with rapid initial energy-independent process and the latter one is the metabolism-dependant intracellular cation uptake process. This kind of biphasic Zn uptake was identified in *Neoosmospora vasinfecta*, *Chlorella vulgaris* and *Candidda utilils* (Bucheder and Broda 1974). However, in bacteria, Zn uptake is mainly a plasmid coded mechanism, especially in *Cupriavidus eutrophs* CH34 (syn. *Ralstonia eutrophus* or *Alcaligenes eutrophus*). It found to possess two plasmids, of which one codes for Zn resistance, and this plasmid houses the *Czc* gene that encodes for an ABC

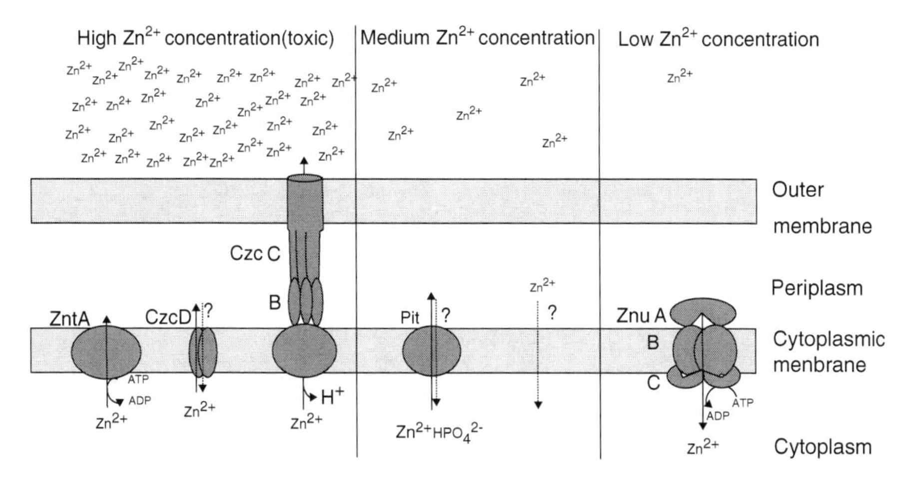


**Fig. 3.1** Diagram showing the different mechanisms of Zn resistance in bacteria. (**a**) extracellular accumulation; (**b**) binding on the outer membrane; (**c**) active efflux via an efflux pump p type ATPase or proton antiporter; and (**d**) sequestration by periplasmic or cytoplasmic proteins or other ligands, like polyphosphate granules (adopted from Choudhury and Srivastava 2001)

protein complex that functions as a Zn efflux system (Nies 1992). In this bacterium, other novel kinds of Zn resistance were also identified (Collard et al. 1993). Possession of the plasmids allows it to grow in a minimal medium with a Zn concentration of 12 μM Zn in the form of $ZnSO_4$ (Mergeay et al. 1985). The plasmid encoding Zn resistance is capable of self-transmissible during homologus mating. In addition to Zn, this bacterium is also resistant and flourishes to grow in millimolar levels in Ni, Co and Cd amended media. This strain was recently subjected to a complete genome analysis; it serves to be a model organism for heavy metal detoxification and for biotechnological applications (Rozyeki and Nies 2009) (Fig. 3.2, Table 3.2).

## 3.6 Zn Uptake Systems

Bacterial growth in an environment is always subjected to fluctuation of environmental conditions and the nutrient resources available in their vicinity. Bacteria are deprived of Zn under the conditions of Zn deficiency; transport systems play an important role in their uptake and help in fulfilling the Zn requirements of an organism.

### *3.6.1 ABC Transporters*

These are the transport systems that function under Zn-deficient condition in an organism; they help to maintain Zn homeostasis in the organism. This kind of

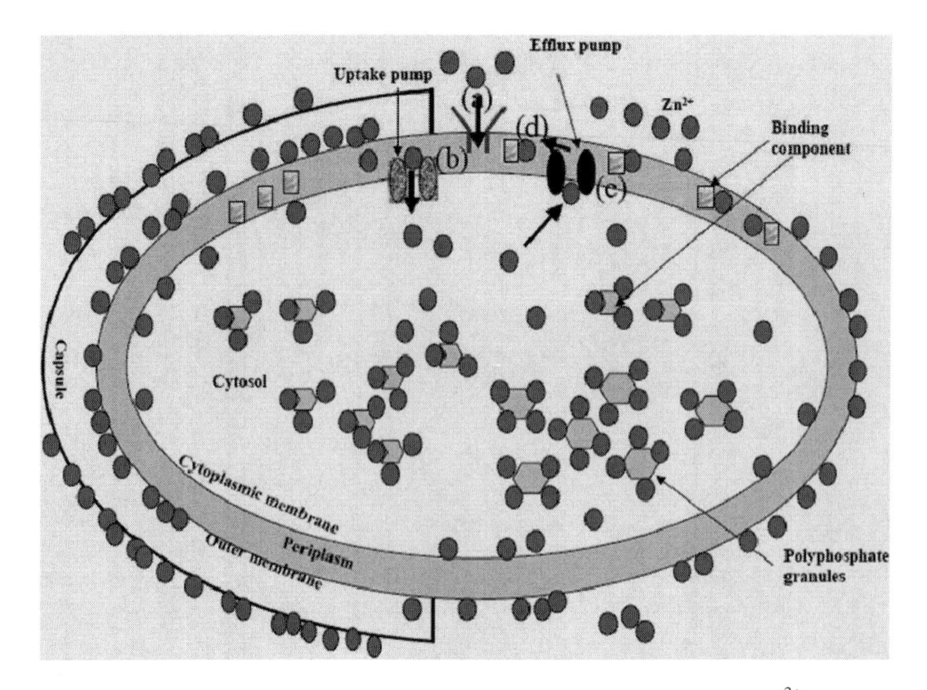


**Fig. 3.2** Various kinds of transport systems involved in uptake and efflux of $Zn^{2+}$ in Gram-negative bacteria. On the basis of the availability of $Zn^{2+}$ concentration in the medium, different types of $Zn^{2+}$ transporters are synthesized. At limiting $Zn^{2+}$ concentrations, binding protein-dependent ABC transporters is induced (*right*). The Pit-like proteins might act as co-transporters and satisfy the $Zn^{2+}$ demands of the cell under $Zn^{2+}$-depleted conditions (*middle*). At toxic levels of Zn, exporters of the *CzcABC*-like transporters that belong to RND protein family seem to be very efficient in protecting the cells against toxic $Zn^{2+}$ concentrations. Also, *CzcD*-like cation facilitators and P-type ATPases such as *ZntA* protect the cells against high $Zn^{2+}$ concentrations (*left*) (adopted from Hantke 2001)

system was characterized in both Gram-negative and Gram-positive bacteria. It was demonstrated in *Escherichia coli* (Patzer and Hantke 1998) and also in *Streptococcus pneumoniae*. Interestingly, in *S. pneumoniae* two distinct proteins (Znu A and Psa A) which belong to ABC family of proteins were also observed (Dintilhac et al. 1997).

## *3.6.2 Low Affinity Zn Uptake Systems*

A less information is available at present about the low affinity uptake systems; they are operating under rich, non-toxic levels of Zn. They co-transport $Zn^{2+}$ with phosphate via pit inorganic phosphate transport system in *E. coli* (Beard et al. 2000). In *Bacillus subtilis*, the metal–citrate uptake protein Cit M has been shown some broad specificity and can transport $Zn^{2+}$ (Krom et al. 2000).

In *Amycolatopsis orientalis*, ethylene diamine disuccinic acid (EDDS) is produced under $Zn^{2+}$ limiting conditions; this complexes Zn and other divalent cations. EDDS is an isomer of the common chelator EDTA. It is a well-known fact that under iron-limiting conditions siderophore is produced by an organism, and it is involved in binding $Fe^{3+}$ ions and uptake. Thus EDDS function as zincophore under $Zn^{2+}$ limited conditions and satisfy the $Zn^{2+}$ demands of an organism (Zwider et al. 1997).

## 3.7 Zn Efflux Systems

These are the transport systems functioning under high concentration of Zn ions present in the surrounding environment. As previously mentioned, owing to a similar size of metal ions $Zn^{2+}$ ions may not be differentiated by cells, and they unspecifically bind to the transporter proteins and taken inside the cell. Thus the efflux system protects the cytoplasm from high toxic levels of $Zn^{2+}$ and maintains Zn homeostasis (Hantke 2001).

### *3.7.1 Cation Diffusion Facilitators*

They are coded by the gene *Czc* gene cluster found on the large plasmid and determines the high metal resistance by encoding two systems for the export of $Zn^{2+}$, $Co^{2+}$ and $Cd^{2+}$ (Anton et al. 1999). This transport system was identified in *Cupriavidus metallidurans* and *Staphylococcus aureus*.

### *3.7.2 RND Type Exporters*

These efflux systems basically belong to a widely distributed RND (resistance, nodulation, division) protein family (Tseng et al. 1999). Under this category the Czc protein seems to play an important role in *C. metallidurans* against high $Zn^{2+}$ concentrations (Anton et al. 1999). The presence of other proteins such as Czc (cation–proton anti-porter protein located in cytoplasmic membrane), Czc B (protein in periplasm is a part for the acriflavin pump, and it is distally related to Tol C protein) and Czc C (protein present in outer membrane) forms a flexible transport system that helps in exporting Zn across the cytoplasm and outer membranes, they also protect the periplasmic space, and collectively they form the CzcABC protein complex.

### 3.7.3 P-type ATPases

They refer to phosphorylated aspartate enzyme intermediates, which form large family of cation-transporting membrane proteins found in the eukaryotes, bacteria and archaea. These types of transporters are encoded by the genes *CadA* and *CadC* in *S. aureus* plasmid pI258 (Yoon and Silver 1991). The same transport system was further identified in *Stenotrophomonas maltophila* and *E. coli* (Alonoso et al. 2000).

## 3.8 Properties of Zn and Its Compounds

Zinc is a relatively soft, bluish-white metal that belongs to group "IIb" in the Periodic Table. Like Fe and Cu, it is considered a heavy metal, with density above 5 g/cm$^3$, and it occurs exclusively as divalent cation $Zn^{2+}$ with completely filled "d" orbitals. The $Zn^{2+}$ cation cannot readily undergo redox changes under simple biological conditions. This metal reacts readily with inorganic (oxides, phosphates, sulfates), as well as organic compounds. It has a fast ligand exchange capacity and possesses amphoteric properties, with intermediate polarizability. Table 3.1 shows the list of the physical properties of the compounds; understanding their physical nature is much helpful in conducting studies related to Zn solubilization by bacteria (Table 3.2).

**Table 3.1** Zinc Uptake and efflux systems adapted under Zn deficient and toxic conditions

| SI. No. | Name of the genes and the bacterium | Protein involved in transport and their function |
|---|---|---|
| Zn uptake system | | |
| 1. | *Znu ABC* (*Streptococcus pneumoniae*) | ABC type, transports $Zn^{2+}$ |
| 2. | *Adc A* (*S. pneumoniae*) | ABC type, transports $Zn^{2+}$ |
| 3. | *Ycd H* (*Bacillus subtilis*) | ABC type, transports $Zn^{2+}$ |
| 4. | *Znu A* (*Haemophilus ducreyi*) | ABC type, transports $Zn^{2+}$ |
| 5. | *Pzp1*(*Haemophilus influenzae*) | ABC type, transports $Zn^{2+}$ |
| 6. | *Znu A* (*E. coli* and *Thermotaga maritima*) | ABC type, transports $Zn^{2+}$ |
| 7. | *Pit A* (*E. coli*) | Low affinity, $Zn^{2+}$–phosphate co-transport system |
| 8. | *Cit M* (*Bacillus subtilis*) | Low affinity, $Zn^{2+}$–citrate co-transport system |
| Zn efflux system | | |
| | *Czc ABC* (*Cupriavidus metallidurans*) | ABC type, belongs to RND protein family, cation/proton antiporter for $Cd^{2+}$, $Co^{2+}$ and $Zn^{2+}$ |
| 9. | *Czc D (Cupriavidus metallidurans)* | |
| 10. | *Cad AC* (*Staphylococcus aureus*) | P type ATPase of $Cd^{2+}$>$Zn^{2+}$ |
| 11. | *Znt A* (*Escherichia coli*) | P type ATPase of $Cd^{2+}$>$Zn^{2+}$ |

Source: Hantke (2002)

**Table 3.2** Chemical and physical properties of zinc and zinc compounds

| | Zinc | Zinc chloride | Zinc sulfate | Zinc sulfide | Zinc oxide |
|---|---|---|---|---|---|
| Synonym(s) | Zinc dust; zinc powder | Butter of zinc | Acid zinc salt; zincum sulfuricum; sulfuric acid, zinc | Wurtzite (alpha); sphalerite (beta) | Calaminec; zincitec |
| Formula | Zn | $ZnCl_2$ | $ZnSO_4$ | ZnS | ZnO |
| CAS registry | 7440-66-6 | 7646-85-7 | 7733-02-0 | 1314-98-3 | 1314-13-2 |
| NIOSH RTECS | ZG8600000 | ZH1400000 | ZH5260000 | No data | ZH4810000 |
| Molecular weight | 65.38 | 136.29 | 161.44 | 97.44 | 81.38 |
| Color | Bluish-white | White granules | Colorless | Colorless | White-yellow powder |
| Boiling point (°C) | 908 | 732 | No data | No data | No data |
| Melting point (°C) | 419.5 | 290 | 600 (decomposes) | About 1,700 | Decomposes at 100 |
| Density ($g/cm^3$) @25°C | 7.14 | 2.907 | 3.54 | 4.1 | 5.607 (20°C) |
| Water solubility (mg/L) | Insoluble | $4.36 \times 10^6$ (25°C) | Soluble | No data | Poorly soluble |
| Vapor pressure | 1 mm Hg (487°C) | 1 mm Hg (428°C) | No data | No data | No data |

*CAS* chemical abstracts services, *NIOSH* National Institute for Occupational Safety and Health, *RTECS* Registry of Toxic Effects. (Adopted from Barceloux 1999)

## 3.9 Microbial Solubilization of Insoluble Zn Compounds

Zn is an element that occurs in soil systems in several forms. The possible mineral forms of Zn in soils include zincite (ZnO), smithsonite ($ZnCO_3$), sphalerite (ZnFe)S, zinc silicates ($ZnSiO_3$), willemite ($ZnSiO_4$) and zinc sulfide (ZnS), which are formed only under reduced conditions. In general, Zn solubility increases with decrease in pH and its activity declines upon precipitation as hydroxide, phosphate, carbonate and silicate at slightly acid to alkaline pH (Baruah and Barthakur 1999; Srivastava and Gupta 1996). Microbial metabolites could have an effect on the solubilization of these insoluble materials. Till date, most of the studies conducted in this regard are related to bacteria and fungi. Both autotrophs and heterotrophs were also found to solubilize Zn through metabolic processes. Microbial Zn solubilization was previously focused on autotrophic bacteria mediated solubilization, particularly by *Thiobacillus ferroxidans* mainly in relation to leaching of metal ores (White et al. 1997). These studies can be grouped into three broad topics: Zn solubilization associated phytoextraction, Zn solubilization associated nutrient enhancement for crop system and Zn mineral weathering by fungi.

## 3.10 Zn Solubilization Associated Phytoextraction

Recently, the microbe-mediated phytoextraction of Zn and other toxic substances gained momentum (Whiting et al. 2001; Coles et al. 2001; Fasim et al. 2002). In this approach, the microbial inocula were added into the rhizosphere of the plants, such as *Thalaspi* sp., *Calminaria* sp. and Indian mustard, which are commonly used for phytoextraction. This may also be referred as rhizoremediation approach for extracting metals.

A few researches in this line have identified a few bacteria that can solubilize Zn; these include *Microbacterium saperdae*, *Pseudomonas monteilli*, *Enterobacter cancerogens* (Whiting et al. 2001), *Pseudomonas fluorescens* (Di Simine et al. 1999) and *Pseudomonas aeruginosa* (Fasim et al. 2002). The efficient ZnO solubilizng strains *Microbacterium saperdae*, *P. monteilli* and *Enterobacter cancerogens* were isolated from the rhizosphere of Zn hyper-accumulating plant *Thlaspi caerulescens* that have Zn concentration greater than 10,000 μg/g of plant tissue when growing in natural habitats that have high concentrations of Zn. When these strains were inoculated into the rhizosphere of the germinating seeds of Thlaspi plants, the bacteria increased the water soluble Zn portion in the soil system both in the rhizosphere of a Zn hyperaccumulator (*T. caerulescens*) and non-Zn accumulator (*Thlaspi arvense*). However, only the Zn hyper-accumulating plant *T. caerulescens* effectively absorbed the Zn and transported to the roots and shoots, and the increased water soluble Zn concentrations available to both *Thlaspi* species under bacterial inoculation were 22–67% higher compared to control (Whiting et al. 2001). *T. caerulescens* were also found to house certain fungi that are capable of Zn solubilization, thus the root endophyte, soil isolates and root surface associated fungi solubilized both ZnO and $Zn_3(PO_4)_2$. Strains belonging to *Abisidia cylindrospora* (root surface), *A. glauca* (root surface and rhizosphere soil) and *A. spinosa* (root surface and rhizosphere soil) efficiently solubilized both the forms of Zn; apart from these a few root endophytic members of *Penicillium* (*P. aurantiogriseum, P. brevicompactum* and *P. simplicissium*) also solubilized both the forms of Zn (Coles et al. 2001).

## 3.11 Zn Solubilization Associated Nutrient Enhancement for Plant System

In the approach of Zn solubilization mediated nutrient enhancement for crop plants, the bacterial strains are inoculated into soils containing high levels of total Zn but are generally unavailable to the plant system. The bacteria work by releasing Zn from its unavailable complexed form, making it available to the plant system. Nutrient enhancement by solubilization is of considerable importance since soils in most parts of the globe exhibit Zn deficiency (Alloway 2001), This is particularly true in countries like India and Pakistan where 70% of soil is deficient in Zn (Alloway 2001; Singh 2001). Solubilization of fixed Zn near the root zones can alleviate Zn deficiency and help maintain the critical level of Zn (2 μg/kg of soil) necessary for crop nutrition.

Kucey (1998) discovered that certain *Penicillium* sp. inoculated in soil enhanced Fe, Cu and Zn content inside plants. Meyer and Linderman (1986) reported that co-inoculation of plant growth-promoting rhizobacteria like *Pseudomonas putida* and arbuscular mycorhizae (AM) increased the concentration of Fe, Cu, Al, An, Co and Ni in shoot tissues. They further hypothesized that *P. putida* may be capable of producing 2-ketogluconic acid that renders native micronutrients available for plant uptake through the mycorrhizal association. It was also found that AM influence the Zn uptake of *Pinus radiata* and *Araucaria* sp. (Sharma and Srivastava 1991; Bowen et al. 1974). The biocontrol fungus, *Trichoderma harzianum* Rifai 1295–22, showed the ability to solubilize the metallic form, but the mechanism behind this action remains unclear (Altomare et al. 1999). Metallic Zn solubilization was later reported however in the Acetobacteraceae member, *Gluconacetobacter diaztorophicus*, which involved the production of gluconic acid and its derivatives in the culture broth (Saravanan et al. 2007b).

The term Zn solubilizing bacteria was coined to refer to certain bacteria that are efficient in transforming insoluble Zn compounds into soluble forms. A pioneering study in this area described solubilization by garden and paddy soil isolated *Bacillus* sp. and *P. fluorescens* of zinc oxide (ZnO), zinc carbonate ($ZnCO_3$) and the zinc ore sphalerite (ZnS) under in vitro conditions (Saravanan et al. 2004). These bacteria were neither tested for solubilization of the essential but limited soil macronutrient, phosphorus or of other micronutrients, nor were their mechanism

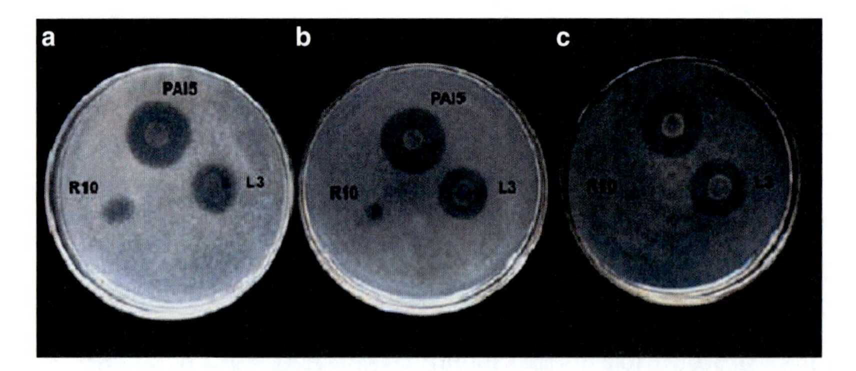


**Fig. 3.3** Solubilization halos produced by *Gluconacetobacter diazotrophicus* strains (Pal 5, R10 and L3), on LGI minimal medium supplemented with 0.1%w/v of (**a**) ZnO, (**b**) $ZnCO_3$ and (**c**) $Zn_3(PO_4)_2$ (Saravanan et al. 2007a)

of solubilization elucidated. Identifying them only as Zn solubilizers, therefore, is practically inaccurate. However, a few field studies showed the success of these bacteria in turmeric and provided evidence for the ability of the bacterial inoculants also alter other macro- and micronutrients (Senthil Kumar et al. 2004).

Several rhizosphere bacteria exhibit potential influence on Zn availability in the plant system. Among these is the metal-tolerant *Burkholderia cepaciae,* which was isolated from the rhizosphere of metal hyper-accumulating plants. After re-inoculation, it was able to significantly enhanced plant growth and uptake of metals (Zn and Cd); favor translocation of these metals from roots to shoots; and enhance organic acid (tartaric acid) secretion in the Cd treated rhizosphere (Li et al. 2007). Another study demonstrated the ability of the soil bacteria *Azotobacter chroococcum*, *Bacillus megaterium* and *Bacillus edaphicus* to increase the bioavailability of Zn and Pb in the soil system (Wu et al. 2006). Recently, solubilization of insoluble Zn compounds was successfully demonstrated in *Gluconacetobacter diazotrophicus*, a plant growth-promoting bacterium associated with sugarcane. The key player in the solubilization process was identified as 5 keto gluconic acid, a derivative acid that efficiently solubilizes Zn (Saravanan et al. 2007a, 2008). Further studies reinforced the molecular mechanism of this solubilization phenomenon, providing evidence for gluconic acid production by glucose dehydrogenase and furthermore identification of the genes responsible for this property (Intorne et al. 2009) (Fig. 3.3).

## 3.12 Fungal Zn Solubilization

Interestingly, the importance of Zn in the nutrition of microorganisms was first recognized in *A. niger*, which was not able to grow in the absence of Zn (Raulin 1869). A few fungal genera showed capacity for solubilization, including *A. niger*,

*Penicillium simplicissimum* (Franz et al. 1993), *Abisidia cylindrospora*, *A. glauca* and *A. spinosa* (Coles et al. 2001), all of which were found to solubilize both zinc oxide and zinc phosphate.

The ZnO addition seems to cause a buffer action, thus a minimal medium supplemented with ZnO filter dust allowed an increased production of citric acid by *P. simplicissimum*; this action was achieved even in the absence of an extracellular buffer, and in the presence of amino acids and urea which were used as nitrogen source, the acid production was parallel to the biomass production and occurred in a pH between 4 and 7; in contrast, metals like Cu and Fe have no effects on the citric acid excretion, and the buffer (ZnO) stimulated production of citric acid is mainly due to metabolic energy and an energized plasma membrane (Franz et al. 1993). However, the citric acid production was reduced due to the presence of ammonium and trace elements in the broth.

Among the mycorrhizae, ericoid mycorrhizae showed potential for solubilization of insoluble Zn compounds. These include *Paxillus involutus*, *Hymenoscyphus ericae Oidiodendron maius*, *Suillus luteus* strains 21 and 22, *S. bovinus* LSt8 and *Beauveria caledonica* (Fomina et al. 2005a, b; Schöll et al. 2006; Gadd 2007). Certain ectomycorrhizae possess distinct mechanisms that enable them to survive in both Zn limited and excess conditions. The ericoid mycorrhizae were isolated from unpolluted natural soils and found to produce abundant quantity of organic acids like fumarate, citrate and malate which improved recovery of micronutrients for both the plant and mycorrhizae. However, in polluted environments with high concentrations of available and potentially toxic Zn, isolated ericoid mycorrhizae seem to exhibit unknown mechanisms of reducing bioavailability of soluble Zn by converting them to insoluble forms. These mycorrhizae therefore have a dual mechanism for surviving in both environments (Martino et al. 2003). In most of the ectomycorrhizal strains studied, acidification of the medium was the foremost factor affecting solubilization. Furthermore, their ability to grow and tolerate the presence of metals supplemented to the media enhanced solubilization (Fomina et al. 2005a). Some ectomycorrhizal fungi also occur as endophytic and entomopathogenic strains like *Beauveria caledonica,* which readily solubilize chemical forms such as zinc phosphate compared to mineral forms such as pyromorphite. Their mechanism for metal mobilization is by acidolysis and complexolyis. Oxalic acid production helped in solubilization, with this increasing in the presence of toxic minerals (Fomina et al. 2004).

Another study documented the increased dissolution of zinc phosphate and Zn accumulation by inoculated Zn-resistant mycorrhiza under low phosphorus conditions, showing the influence of other nutrients on the solubilization of Zn minerals in plants (Fomina et al. 2006). The mechanism of reduction of bioavailability and toxicity of the metals was partly understood from studies on *B. caledonica*. From these studies, it was noted that the previously known dissolution agent, oxalic acid, was primarily responsible for transforming zinc phosphate into zinc oxalate under enhanced excretion. Thick hydrated mucilaginous sheaths that covered *B. caledonica* hyphae provided a microenvironment for chemical reactions, crystal deposition and growth (Fomina et al. 2005b).

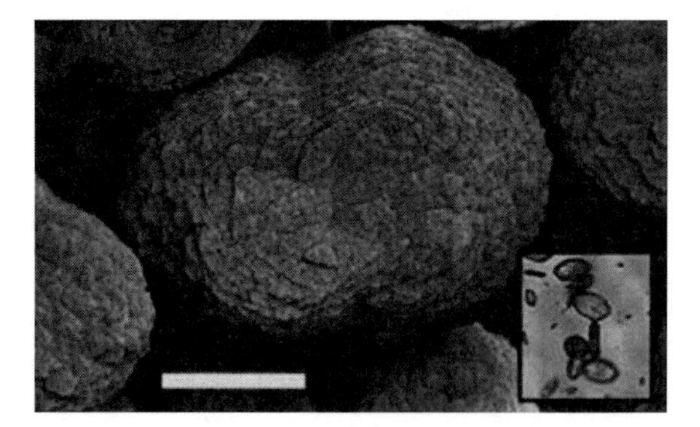

**Fig. 3.4** Scanning electron microscopic image and the inset showing the light microscopic image of zinc oxalate dehydrate formed by transforming zinc phosphate mediated by oxalic acid produced by the ectomycorrhizal fungus *Beauveria caledonica*, scale bars = 100 μm (Fomina et al. 2005b)

The role of arbuscular mycorrhiza (AM) in Zn nutrition is not only as a solubilizer but also for mobilizing available Zn, generally in soils of low Zn availability. However, under high Zn concentrations in the environment, they also "protect" against toxic accumulation of Zn in plant tissues (Cavagnaro 2008). The roles of fungal isolates in the mobilization of Zn and various toxic minerals and their physiology and survival have been well reviewed (Gadd 2007) (Fig. 3.4).

## 3.13 Future Thrust

Screening for Zn-solubilizing potential of bacteria has a two-dimensional application: one involves its use as a bacterial agent for rhizoremediation in the areas of high metal contamination where they pose toxicity problems to plants. This sort of phytoextraction of metals is rapidly gaining momentum. The presence of bacteria in the rhizosphere region plays a major role in the nutrient transformation even in certain zincophilic plants such as *Thlaspi caeulescens*.

On the other hand, some potential strains were tested as nutrient enhancers. Mineral content of plants can be maintained at an optimum level through this process, which in turn offers other benefits like disease and pest resistance (Graham and Webb 1991). In this area, considerable research focus is needed to understand whether they solubilize other minerals including P, apart from Zn. In solubilizing toxic elements, their tolerance potential toward a toxic ion, their mechanism of solubilization, their survival in rhizosphere and field level solubilization and their enhancement of these minerals inside the plant system need to be determined.

**Acknowledgements** This work was supported by the research grant of Chungbuk National University 2010. VSS was supported by PONSHE, Puducherry, and MR likes to thank UGC for providing Rajiv Gandhi National Fellowship (RGNF) and Management of VIT University.

## References


Alloway BJ (2001) Zinc the vital micronutrient for healthy, high-value crops. International Zinc Association, Brussels, Belgium http://www.interzinc.org/pdf/ZP_Zinc_TheVitalMicronutrient.pdf

Alloway BJ (2008) Zinc in soils and crop nutrition, 2nd ed. IZA and IFA, Brussels, p 139 http://www.zinc-crops.org/Documents/Zn_in_Soils_and_Crop_Nutrition_2008.pdf

Alonoso A, Sanchez P, Martinez JL (2000) *Stenotrophomonas maltophila* D457R contains acluster of genes from gram positive bacteria involved in antibiotic and heavy metal resistance. Antimicrob Agents Chemother 44:1778–1782

Altomare C, Norvell WA, Bjorkman T, Harman GE (1999) Solubilization of phosphates and micronutrients by the plant-growth promoting and biocontrol fungus *Trichoderma harzianum* Rifai 1295–22. Appl Environ Microbiol 65:2926–2933

Andrieni C, Bansi L, Bertini I, Rosato A (2006) Zinc through the three domains of life. J Proteome Res 5:3173–3178

Anton A, Grosse C, Reissman J, Pribyl T, Nies DH (1999) Czcd is a heavy metal ion transporter involved in regulation of heavy metal resistance in *Ralstonia* sp. strain CH34. J Bacteriol 181:6876–6881

Appanna VD, Whitmore L (1995) Biotransformation of zinc by *Pseudomonas fluorescens*. Microbios 82:149–1551

Barceloux DG (1999) Zinc. Clin Toxicol 37:279–292

Baruah TC, Barthakur HP (1999) A textbook of soil analysis. Vikas Publishers, New Delhi, p 334

Beard SJ, Hughes MN, Poole RK (1995) Inhibition of the cytochrome bd-terminated NADH oxidase system in *Escherichia coli* K-12 by divalent metal cations. FEMS Microbiol Lett 131:205–210

Beard SJ, Hashim R, Wu G, Binet HR, Hughes MN, Poole RK (2000) Evidence for transport of zinc (II) ions via the pit inorganic phosphate transport system in *Escherichia coli*. FEMS Microbiol Lett 184:231–235

Biro S, Bayoumi HE, Balazsy S, Kecskes M (1995) Metal sensitivity of some symbiotic nitrogen fixing bacteria and *Pseudomonas* strains. Acta Biol Hung 46:9–16

Bowen GD, Skinner MF, Bevege DI (1974) Zinc uptake by mycorrhizal and uninfected roots of *Pinus radiata* and *Auraucaria* sp. Soil Biol Biochem 6:141–144

Bucheder F, Broda E (1974) Energy dependent zinc transport in *Escherichia coli*. Eur J Biochem 45:555–559

Cavagnaro TR (2008) The role of arbuscular mycorrhizas in improving plant zinc nutrition under low soil zinc concentrations: a review. Plant Soil 304:315–325

Choudhury R, Srivastava S (2001) Zinc resistance mechanisms in bacteria. Curr Sci 81:768–775

Coles KE, David JC, Fisher PJ, Lappin-Scott HM, Macnair MR (2001) Solubilization of zinc compounds by fungi associated with the hyperaccumulator *Thlaspi caerulescens*. Bot J Scotl 51:237–247

Collard JM, Provoost A, Taghavi S, Mergeay M (1993) A new type of *Alcaligenes eutrophus* CH34 zinc resistance generated by mutation affecting regulation of their *cnR*, cobalt-nickel resistance system. J Bacteriol 175:779–784

Di Simine CD, Sayer JA, Gadd GM (1999) Solubilization of zinc phosphate by a strain of *Pseudomonas fluorescens* isolated from a forest soil. Biol Fertil Soils 28:87–94

Dintilhac A, Alloing G, Granadel C, Claverys JP (1997) Competence and virulence of *Streptococcus pneumoniae Adc A* and *Psa A* mutants exhibit a requirement for Zn and Mn resulting from inactivation of putative ABC metal permeases. Mol Microbiol 25:727–739

Fasim F, Ahmed N, Parsons R, Gadd GM (2002) Solubilization of zinc salts by a bacterium isolated from the air environment of a tannery. FEMS Microbiol Lett 213:1–6

Fomina M, Alexander IJ, Hillier S, Gadd GM (2004) Zinc phosphate and pyromorphite solubilization by soil plant-symbiotic fungi. Geomicrobiol J 21:351–366

Fomina M, Alexander IJ, Colpaert JV, Gadd GM (2005a) Solubilization of toxic metal minerals and metal tolerance of mycorrhizal fungi. Soil Biol Biochem 37:851–866

Fomina M, Hillier S, Charnock JM, Melville KI, Alexander J, Gadd GM (2005b) Role of oxalic acid overexcretion in transformations of toxic metal minerals by *Beauveria caledonica*. Appl Environ Microbiol 71:371–381

Fomina M, Charnock JM, Hillier S, Alexander IJ, Gadd GM (2006) Zinc phosphate transformations by the *Paxillus involutus*/pine ectomycorrhizal association. Microb Ecol 52:322–333

Franz A, Burgstaller W, Muller B, Schinner F (1993) Influence of medium components and metabolic inhibitors on citric acid production by *Penicillium simplicissimum*. J Gen Microbiol 139:2101–2107

Frau´sto da Silva JJR and Williams RJP (1991) Zinc: Lewis acid catalysis and regulation. In: The Biological Chemistry of the Elements- The inorganic chemistry of life. Clarendon Press, Oxford, pp 315–338

Frossard E, Bucher M, Machler F, Mozafar A, Hurrell R (2000) Potential for increasing the content and bioavailability of Fe, Zn and Ca in plants for human nutrition. J Sci Food Agric 80: 861–879

Gadd GM (2007) Geomycology: biogeochemical transformations of rocks, minerals, metals and radionuclides by fungi, bioweathering and bioremediation. Mycol Res 111:3–49

Giovannoni SJ, Turner S, Olsen GJ, Barns S, Lane DJ, Pace NR (1988) Evolutionary relationships among cyanobacteria and green chloroplasts. J Bacteriol 170:3584–3592

Graham DR and Webb MJ (1991) Micronutrients and disease resistance and tolerance in plants. In: Mortvedt JJ, Cox FR, Shuman LM and Welch R.M (ed) Micronutrients in Agriculture (second ed), Soil Science Society of America, Inc., Madison, WI, USA , pp. 329–370

Hantke K (2001) Bacterial zinc transporters and regulators. Biometals 14:239–249

Hantke K (2002) Bacterial Zn transport. In: Winkelmann G (ed) Microbial transport systems. Wiley-VCH, Weinheim, pp 313–324

Intorne AC, Marcos VV, de Oliveira LML, da Silva JF, Olivares FL, de Souza Filho GA (2009) Identification and characterization of *Gluconacetobacter diazotrophicus* mutants defective in the solubilization of phosphorus and zinc. Arch Microbiol 191:477–483

Krom BP, Warner JB, Konings WN, Lolkema JS (2000) Complementary metal ion specificity of metal-citrate transporters Cit M an Cit H of *Bacillus subtilis*. J Bacteriol 182:6374–6381

Kucey RMN (1998) Effect of *Penicillium bilaji* on the solubility and uptake of P and micronutrients from soil by wheat. Can J Soil Sci 68:261–270

Laity JH, Lee BM, Wright PE (2001) Zinc finger proteins: new insights into structural and functional diversity. Curr Opin Struct Biol 11:39–46

Li WC, Ye ZH, Wong MH (2007) Effects of bacteria on enhanced metal uptake of the Cd/Zn-hyperaccumulating plant, *Sedum alfredii*. J Exp Bot 58:4173–4178

Macnicol RG, Beckett PHT (1985) Critical tissue concentrations of potentially toxic elements. Plant Soil 85:107–130

Martino E, Perotto S, Parsons R, Gadd GM (2003) Solubilization of insoluble inorganic zinc compounds by ericoid mycorrhizal fungi derived from heavy metal polluted sites. Soil Biol Biochem 35:133–141

Mergeay M, Nies D, Schlgel HG, Gerits J, Charles PG, Gijsegem FV (1985) *Alcaligenes eutrophus* CH34 is a facultative chemolithotroph with plasmid bound resistance to heavy metals. J Bacteriol 162:328–334

Mertens J, Springael D, De Troyer I, Cheyns K, Wattiau P, Smolders E (2006) Long-term exposure to elevated zinc concentrations induced structural changes and zinc tolerance of the nitrifying community in soil. Environ Microbiol 8:2170–2178

Meyer JR, Linderman RG (1986) Response of subterranean clover to dual inoculation with vesicular-arbuscular mycorrhizal fungi and a plant growth promoting bacterium *Pseudomonas putida*. Soil Biol Biochem 18:185–190

Moffett BF, Nicholson FA, Uwakwe NC, Chambers BJ, Harris JA, Hill TCJ (2003) Zinc contamination decreases the bacterial diversity of agricultural soil. FEMS Microbiol Ecol 43:13–19
Nies DH (1992) *CzcR* and *CzcD* gene products affecting regulation of resistance to Co, Zn and Cd (Czc system) in Alcaligenes *eutrophus*. J Bacteriol 174:8102–8110
Nies DH (1999) Microbial heavy-metal resistance. Appl Microbiol Biotechnol 51:730–750
Patzer SI, Hantke K (1998) The Znu ABC high affinity zinc uptake system and its regulator Zur in *Escherichia coli*. Mol Microbiol 28:1199–1210
Raulin J (1869) Etudes Chimique sur la vegetation. Ann Sci Nat 11:93–299
Rozyeki TV, Nies DH (2009) *Cupriavidus metallidurans*: evolution of a metal resistant bacterium. Anton Leeuw Int J Gen 96:115–139
Saravanan VS, Subramoniam SR, Raj SA (2004) Assessing *in vitro* solubilization potential of different zinc solubilizing bacterial (ZSB) isolates. Braz J Microbiol 34:121–125
Saravanan VS, Madhaiyan M, Thangaraju M (2007a) Solubilization of zinc compounds by the diazotrophic, plant growth promoting bacterium *Gluconacetobacter diazotrophicus*. Chemosphere 66:1794–1798
Saravanan VS, Osborne J, Madhaiyan M, Mathew L, Chung J, Ahn K, Sa T (2007b) Zinc metal solubilization by *Gluconacetobacter diazotrophicus* and induction of pleomorphic cells. J Microbiol Biotechnol 17:1477–1488
Saravanan VS, Madhaiyan M, Osborne J, Thangaraju M, Sa TM (2008) Ecological occurrence of *Gluconacetobacter diazotrophicus* and nitrogen-fixing *Acetobacteraceae* members: Their possible role in plant growth promotion. Microb Ecol 55:130–140
Schöll LV, Smits MM, Hoffland E (2006) Ectomycorrhizal weathering of the soil minerals muscovite and hornblende. New Phytol 171:805–814
Senthil Kumar PS, Aruna Geetha S, Savithri P, Jagadeeswaran R, Ragunath KP (2004) Effect of Zn enriched organic manures and zinc solubilizer application on the yield, curcumin content and nutrient status of soil under turmeric cultivation. J Appl Hortic 6:82–86
Sharma AK, Srivastava PC (1991) Effect of vesicular – arbuscular mycorrhiza and zinc application on dry matter and zinc uptake of green gram (*Vigna radiata* L. Wilczek). Biol Fertil Soils 11:52–56
Singh MV (1991) Proceedings of workshop on rice–wheat cropping system. Project Directorate Cropping System Research, Modipuram pp 133–144
Singh MV (2001) Evaluation of current micronutrient stocks in different agro ecological zones of India for sustainable crop production. Fert News 46:25–28
Srivastava PC, Gupta UC (1996) Trace elements in crop production. Oxford & IBH Publishers, New Delhi, p 356
Tseng TT, Gratwick KS, Kollman J, Park D, Nies DH, Goffeau A, Saier MHJ (1999) The RND permease super family: an ancient, ubiquitous and diverse family that includes human disease and development proteins. J Mol Microbiol Biotechnol 1:107–125
Vallee BL (1991) Introduction to metallothionein. Methods Enzymol 205:3–7
White C, Sayer JA, Gadd GM (1997) Microbial solubilization and immobilization of toxic metals: key biogeochemical processes for treatment of contamination. FEMS Microbiol Rev 20:503–516
Whiting SN, de Souza MP, Terry N (2001) Rhizosphere bacteria mobilize Zn for hyperaccumulation by *Thalaspi caerulescens*. Environ Sci Technol 35:3144–3150
Wu SC, Luo YM, Cheung KC, Wong MH (2006) Influence of bacteria on Pb and Zn speciation, mobility and bioavailability in soil: A laboratory study. Environ Pollut 144:765–773
Yoon KP, Silver S (1991) A second gene in the *Staphyloccus aureus Cad A* cadmium resistance determinant of plasmid pI 258. J Bacteriol 173:7636–7642
Zwider N, Theobald U, Zahner H, Fiedler HP (1997) Optimization of fermentation conditions for the production of ethylene diamine-disuccinic acid by *Amycolatopsis orientalis*. J Ind Microbial Biotechnol 19:280–285

# Chapter 4
# Effectiveness of Phosphate Solubilizing Microorganism in Increasing Plant Phosphate Uptake and Growth in Tropical Soils

**Nelson Walter Osorio**

## 4.1 Soil Phosphate Sorption

Phosphate fixation is a serious problem for plant productivity in agricultural soils, particularly in highly weathered soils (Trolove et al. 2003; Sanchez and Uehara 1980). Sanchez and Logan (1992) estimated that the soils that exhibit high phosphate fixation capacity occupy 1,018 million ha in the tropics. In tropical America there are 659 million ha affected, 210 in Africa, and 199 in Asia. The term phosphate fixation is used to describe reactions that remove available phosphate from the soil solution into the soil solid phase (Barber 1995). There are two types of reactions (a) phosphate sorption on the surface of soil minerals and (b) phosphate precipitation by free $Al^{3+}$ and $Fe^{3+}$ in the soil solution (Havlin et al. 1999).

Phosphate sorption can occur in two ways: *nonspecific sorption* that consists of electrostatic attraction of phosphate ions by positive charges on the surface of a soil mineral; the *specific sorption* takes place when phosphate ions are exchanged by hydroxyl (–OH) groups of soil minerals forming a strong bond with the mineral (Bohn et al. 1985). The strength and extension of phosphate sorption vary widely among soils and follows the next order: andisols > ultisols > oxisols > mollisols, vertisols > histosols. Soil phosphate sorption capacity can be easily measured by isotherms (Fox and Kamprath 1970). To this purpose, an aliquot of 3 g of soil (dry basis) is transferred into centrifuge tubes. Then, 30 mL of 0.01 M $CaCl_2$ with graded amount of soluble phosphate and two drops of toluene are added into each tube. Then, the centrifuge tubes are shaken in a reciprocal shaker for 7 days. After that, the tubes are centrifuged at 4,000 rpm for 15 min and the supernatant is filtered through a filter paper. Soluble phosphate concentration is measured by the molybdate-blue method (Murphy and Riley 1962). The $P_{0.2}$-value is an index of the soil

N.W. Osorio (✉)
Facultad de Ciencias, Universidad Nacional de Colombia, Calle 59 A No. 63-20, Medellin, Colombia
e-mail: nwosorio@unal.edu.co

D.K. Maheshwari (ed.), *Bacteria in Agrobiology: Plant Nutrient Management*,
DOI 10.1007/978-3-642-21061-7_4, © Springer-Verlag Berlin Heidelberg 2011

phosphate sorption capacity, it means the amount of phosphorus required to achieve a soil solution phosphorus concentration of 0.2 mg $L^{-1}$, which is considered optimal for most plant crops (Fox 1979).

## 4.2 Phosphate Solubilizing Microorganisms

Many soil microorganisms can solubilize inorganic soil phosphate compounds, reversing the process of phosphate fixation (Gyaneshwar et al. 2002; Rao 1992). Soil bacteria capable of plant growth promotion such as the genera *Pseudomonas*, *Enterobacter*, and *Bacillus* are particularly active as phosphate solubilizers (Canbolat et al. 2006; Pandey et al. 2006; Xavier and Germida 2003; Kim et al. 1998a, b; Barea et al. 1975). Soil fungi especially those of the genus *Penicillium* and *Aspergillus* also have been demonstrated to be effective phosphate solubilizing microorganisms (PSMs) (Reddy et al. 2002; Whitelaw 2000; Kang et al. 2008). Kucey and Leggett (1989) found in Mollisols of Canada that 0.5 and 0.1% of the total population of bacteria and fungi, respectively, exhibited the ability to solubilize insoluble Pi compounds.

## 4.3 Mechanisms of Microbial Phosphate Solubilization

Several mechanisms have been proposed to explain the microbial solubilization of phosphate compounds. The mechanisms consist of (a) release of organic acids produced during organic carbon decomposition (Kang et al. 2002; Hameeda et al. 2006; Bar-Yosef et al. 1999); (b) excretion of protons due to $NH_4^+$ assimilation by microorganisms (Whitelaw 2000; Ilmer et al. 1995; Abd-Alla 1994; Asea et al. 1988; Roos and Luckner 1984; Kucey 1983); (c) formation of complexes between organic acids/anions with cations ($Al^{3+}$, $Fe^{3+}$, $Ca^{2+}$) (Welch et al. 2002); and (d) phosphate desorption from soil minerals (Osorio 2008). Bacteria of the genera *Nitrosomonas* and *Thiobacillus* species can also dissolve phosphate compounds by producing nitric and sulfuric acids (Azam and Memom 1996). Equally, phosphate compounds may be solubilized by carbonic acid formed as a result of organic matter decomposition (Memon 1996). In addition, Lopez-Bucio et al. (2007) reported that *Bacillus megaterium* (a known PSM) promoted plant growth and stimulated root branching of *Arabidopsis thaliana*, which can increase plant phosphate uptake. Increase in soil phosphate availability may be caused by several reactions involving microorganisms that produce organic acids (Stevenson 1986). These substances can replace or compete with phosphate ions for sorption sites.

Kim et al. (1997) found that the production of acidity was a major mechanism in the solubilization of hydroxyapatite by *Enterobacter agglomerans* under in vitro conditions. For comparison, Kim and coworkers employed citric acid, oxalic acid, lactic acid, and HCl at the same pH produced by *E. agglomerans*. They found that at pH 4.0–4.1 (and a shaking time of 48–50 h) there were no significant differences among phosphate solubilization produced by this bacterium and that produced by citric acid, oxalic, and HCl. However, lactic acid exhibited a lower capacity for solubilizing hydroxyapatite.

The fungus *Aspergillus niger* produced organic acids but other PSM species did not produce detectable amounts of the organic acids. Under in vitro conditions, the pH of the growth medium decreased as a result of acid production by PSMs. Osorio and Habte (2001) found an inverse relation between culture medium pH and Pi released from rock phosphate (RP) by PSMs (bacteria and fungi) isolated from Hawaiian soils. The microbial solubilization of rock phosphate was associated with the ability of microorganisms to depress the pH of the growth medium by producing organic acids.

Some of the organic acids (or their respective anions) commonly associated with microbial solubilization of phosphate are gluconic acid (Di-Simine et al. 1998; Bar-Yosef et al. 1999), oxalic acid (Osorio 2008), citric acid (Kim et al. 1997; Kucey and Leggett 1989), lactic acid, tartaric acid, and aspartic acid (Venkateswardu et al. 1984). These acids are products of microbial metabolism, in some cases by oxidative respiration or by fermentation of carbonaceous substrates (e.g., glucose) (Trolove et al. 2003; Jones et al. 2003; Gyaneshwar et al. 2002; Prescott et al. 1999; Mathews et al. 1999; Atlas and Bartha 1997). The reactions of phosphate solubilization are believed to occur in the rhizosphere where carbonaceous compounds are released and where the solubilized phosphate may be taken up by the root or mycorrhizal system. Lynch and Ho (2005) showed that in wheat plants up to 33–40% of the total carbon fixed by photosynthesis could be excreted into the rhizosphere; Amos and Walters (2006) estimated a value of 29% for maize. Many rhizosphere microorganisms are heterotrophs and might use these carbonaceous substrates to produce organic acids. Recently, Hameeda et al. (2006) found that the type of carbon source affected the effectiveness of rock phosphate solubilizing bacteria. For *Serratia marcescens* and *Pseudomonas* sp. the more favorable carbon source for rock phosphate solubilization followed the order, glucose > galactose > xylose > mannose = maltose > cellobiose > arabinose. No solubilization of RP was detected with the last carbon source of this series. The bacteria were capable of solubilizing rock phosphate using different kinds of composted crop residues (rice, pigeon pea, and a grass). Reyes et al. (2006) also compared the effect of the carbon source on rock phosphate solubilization and found that *Penicillium* sp. and *Azotobacter* sp. were more effective if the medium contained sucrose than dextrose.

When PSMs were inoculated in neutral or alkaline soils, the production of acids decreased rhizosphere pH, favoring the solubility of soil native calcium-phosphate and added rock phosphate (Kim et al. 1998a). These results have commonly been found in temperate-zone soils of Europe and North America (Kucey and Leggett 1989; Kucey 1983, 1987, 1988) and other countries, e.g., Egypt (Omar 1998) where calcareous soils are abundant.

Welch et al. (2002) found that organic acid/anions produced by microorganisms were capable of dissolving apatite by forming a complex with Ca either in solution and/or directly at the mineral surface. To illustrate this reaction, let us see dissolution of a hydroxyapatite under acidic conditions. If proton activity increases (low pH) the reaction proceeds to right (as written):

$$Ca_5(PO_4)_3OH + 7H^+ \leftrightarrow 3H_2PO_4^{2-} + 5Ca^{2+} + H_2O \quad K = 10^{14.5}$$

It represents a change in free energy ($\Delta G$) equal to $-82.72$ kJ mol$^{-1}$.

Moreover, if in the products there is also a reduction in the activity of $Ca^{2+}$, produced by the formation of an organic complex with $Ca^{2+}$ (e.g., $Ca^{2+}$-citrate and $Ca^{2+}$-oxalate), the reaction will be more favorable thermodynamically ($10^{38.8}$ and $10^{31.7}$, respectively).

In the presence of citrate:

$$Ca_5(PO_4)_3OH + 7H^+ + 5\,\text{citrate} \leftrightarrow 3H_2PO_4^{2-} + 5\,\text{citrate} - Ca^{2+} + H_2O$$
$$K = 10^{37.9}$$

It represents a change in free energy ($\Delta G$) equal to $-216.21$ kJ mol$^{-1}$.

In the presence of oxalate:

$$Ca_5(PO_4)_3OH + 7H^+ + 5\,\text{citrate} \leftrightarrow 3H_2PO_4{}^{2-} + 5\,\text{citrate} - Ca^{2+} + H_2O$$
$$K = 10^{31.7}$$

It represents a change in free energy ($\Delta G$) equals to $-180.73$ kJ mol$^{-1}$.

Thus, the microbial solubilization of soil phosphate seems to be associated with the presence of calcium phosphates. In fact, most of the research on microbial solubilization has been done with solubilizers of tricalcium phosphate ($Ca_3PO_4$) (Pikovskaia 1948; Sperber 1957, 1958; Louw and Webley 1959; Agnihorti 1970; Paul and Rao 1971; Banik and Dey 1981a, b, c) or rock phosphate (Kim et al. 1998b; Osorio and Habte 2001). However, researchers on PSMs no longer recommend PSM isolation using culture medium with $Ca_3PO_4$ because it can supply free phosphate. Some authors reported that in vitro microbial solubilization not only occurred with calcium phosphate but also with Al and Fe phosphates. However, the solubilization was higher with calcium phosphates (Ilmer et al. 1995; Banik and Dey 1983; Rose 1957).

Solubilization of Al and Fe phosphates can be easily observed under in vitro conditions where no soil minerals interfere with phosphate solubility. Ilmer et al. (1995) found that *A. niger*, *Penicillium simplicissium*, *Pseudomonas aurantiogriseum*, and *Pseudomonas* sp. were effective in solubilizing $AlPO_4$ under in vitro conditions via organic acid production or proton excretion due to $NH_4^+$ assimilation. Aluminum and Fe ions in solution may be chelated by organic anions (e.g., oxalate and citrate) (Kang et al. 2002; Bolan et al. 1994), favoring the dissolution of Al and Fe phosphates. Whether organic acids released by PSMs can solubilize phosphate from Al and Fe phosphate under acidic soil conditions must be studied.

On the other hand, organic anions produced by PSMs also can compete with phosphate for phosphate sorption sites on the surface of soil minerals. He and Zhu

(1997, 1998) suggested that phosphate sorbed on the surfaces of some minerals was displaced when a culture medium was inoculated with soil samples containing microorganisms (unidentified) that presumably excreted organic acids. Osorio (2008) further reported that a PSM was capable of phosphate desorption from minerals and soil samples, but this was controlled by the phosphate desorption (higher phosphate desorption at low $P_{0.2}$ value). For minerals, the magnitude on which phosphate desorbed was in the order: montmorillonite > kaolinite > goethite > allophane (null desorption). Consequently, for soils the order was: mollisol > oxisol > ultisol > andisol. The amount of phosphate desorbed by the PSM was higher when the minerals or soils had higher levels of sorbed phosphate; this is when the saturation of phosphate sorption sites was higher.

## 4.4 Effects of PSMs on Plant Phosphate Uptake

During the 1950s–1960s, inoculation with *B. megaterium* var. phosphaticum (phosphobacterin) in Russian soils (mainly Mollisols) was the best known use of PSMs (Stevenson 1986; Kucey and Leggett 1989). At that time, the mechanisms of phosphate solubilization were not fully understood, but the mineralization of organic phosphate was proposed as the major mechanism. Trials carried out in many locations demonstrated little consistency in plant response; apparently other factors such as liming and/or organic material addition affected the effectiveness of phosphobacterin. The lack of response to phosphobacterin in many locations, pointed to possible intensified organic matter decomposition, and the poor understanding of the mechanisms of phosphate solubilization carried out by this microorganism discouraged its use. Since then, the research was oriented toward the study of microbial solubilization of inorganic phosphate compounds in the soil (Kucey and Leggett 1989).

Inoculation with PSMs has produced positive results on crop yield, plant growth, and phosphate uptake of several plant species (Kucey and Leggett 1989). For instance, some effective fungal PSMs are *A. niger* (Omar 1998; Rosendahl 1942) and *Penicillium bilaii* (Kucey 1983, 1987, 1988; Asea et al. 1988; Kucey and Leggett 1989; Gleddie 1993). Whitelaw (2000) and Kucey and Leggett (1989) reviewed the literature in the subject matter and showed several reports of increase in plant growth and phosphate uptake.

However, the effectiveness of PSMs in enhancing plant phosphate uptake has been questioned by some authors (Bolan 1991; Tinker 1980) because (a) organic substances required for these microorganisms are scarce in nonrhizospheric sites, (b) antagonism and competition by other microorganisms in the rhizosphere can reduce the effectiveness of PSMs, and (c) low translocation of solubilized phosphate through the soil because it can be refixed by soil components. This latter point is more important in soils with a high phosphate fixation capacity as discussed already.

## 4.5 Dual Inoculation with PSMs and AMF

Coinoculation with rhizosphere PSMs and AMF of soils with high phosphate fixation capacity may overcome the limitations mentioned on the effectiveness of PSMs in enhancing plant phosphate uptake. First, mycorrhizal plants can release higher amounts of carbonaceous substances into the rhizosphere (Linderman 1988; Rambelli 1973) than nonmycorrhizal plants. Rhizosphere PSMs can use these carbon substrates for their metabolic processes, which are responsible for organic acid production in the rhizosphere and/or proton excretion (Azcon and Barea 1996). Second, the extensive mycorrhizal hyphae network formed around roots can efficiently take up phosphate released by PSMs, thus minimizing its refixation. As long as PSMs grow in the rhizosphere (or mycorrhizosphere), there is a great opportunity to satisfy their carbon requirement and deliver phosphate into the soil solution.

Barea et al. (2002) reported that the combined inoculation with phosphate-solubilizing-rhizobacteria, mycorrhizal fungi, and *Rhizobium* increased the phosphate uptake in several legumes fertilized with rock phosphate. Some reports on the beneficial effects on dual inoculation with AMF and PSMs are presented in Table 4.1.

In tomato, beneficial results were found with *E. agglomerans* and *Glomus etunicatum* (Kim et al. 1998a) (Table 4.2). Synergistic effects have been also found in sunflower (*Helianthus annuus*) with the triple inoculation of two PSMs (*Azotobacter chroococcum* and *Penicillium glaucum*) and the AMF *G. fasciculatum* (Gururaj and

**Table 4.1** Effect of PSM inoculation on plant phosphate uptake of mycorrhiza-free and mycorrhized plants grown on temperate soils

| | | | Increase of plant P uptake due to PSM inoculation (%) | | |
|---|---|---|---|---|---|
| Soil type/plant | P added | PSMs | −AMF | +AMF | References |
| | | *Azospirillum* | 33 | 0–33 | |
| Calcareous mixed | | *Penicillium* | 33 | 0–44 | |
| with sand, pH 6.7 | | Unidentified | 33 | 22–38 | Toro et al. |
| (plant: kudzu) | RP | *Pseudomonas* | 33 | 33–50 | (1996) |
| Vertic Epiaqualf | | | 54 (35 days)[a] | 124 | |
| mixed with sand | | | 27 (55 days) | 27 | |
| and vermiculite, | | | | | |
| pH 5.9 (plant: | | | | | Kim et al. |
| tomato) | RP | *Enterobacter agglomerans* | 8 (75 days) | 11 | (1998a) |
| | | *Bacillus* | 26 | 52 | Singh and |
| | | *Cladosporium* | 47 | 69 | Kapoor |
| | None | *Bacillus* + *Cladosporium* | 73 | 98 | (1999) |
| | | *Bacillus* | – | 248 | |
| Sandy soil pH 7.6 | | *Cladosporium* | – | 301 | |
| (plant: wheat) | RP | *Bacillus* + *Cladosporium* | 51 | 344 | |

*RP* rock phosphate, [a]Days after planting

**Table 4.2** Effects of *E. agglomerans* (PSM) and *G. etunicatum* (AMF) inoculation on tomato plant growth and phosphate uptake 75 days after inoculation

| Treatment | Shoot dry weight (g/plant) | Root dry weight (g/plant) | Shoot phosphate content (mg/plant) |
|---|---|---|---|
| Control | 42.2 (100)[a] | 4.3 (100) | 116.6 (100) |
| PSM | 48.5 (115) | 5.1 (118) | 125.3 (107) |
| AMF | 47.6 (113) | 5.6 (130) | 120.9 (104) |
| PSM + AMF | 54.6 (129) | 6.8 (158) | 134.4 (115) |
| LSD ($P \leq 0.05$) | 1.96 | 0.5 | 9.8 |

[a]In parenthesis percentage of the control. Source: Kim et al. (1998a)

Mallikarjunaiah 1995). Similar effects were found in cotton with the inoculation of *Pseudomonas striata* and *Azospirillum* sp. (PSMs) and *G. fasciculatum* (AMF) (Prathibha et al. 1995). In rice, favorable effects were also reported with *P. striata* (PSM), *Bacillus polymyxa* (PSM), and *G. fasciculatum* (AMF) (Mohod et al. 1991). In chili (*Capsicum annuum*) synergistic effects were reported with two AMF, *G. fasciculatum* or *G. macrocarpum*, and a PSM *P. striata* (Sreenivasa and Krishnaraj 1992). Moreover, positive results have been obtained in wheat with several combinations that include *P. striata* (PSM) and *G. fasciculatum* (AMF), *P. putida*, *P. aeruginosa* and *P. fluorescens* (PSMs) with *G. clarum* (AMF), *P. striata* and *Agrobacterium radiobacter* (PSMs) combined with two AMF, *G. fasciculatum* and *Gigaspora margarita* (Gaur et al. 1990).

Kopler et al. (1988) indicated that more legume nodulation was obtained with concurrent inoculation of *Rhizobium* and *Pseudomonas* spp. (PSMs). Sturz et al. (1997) found that nodulation by *Rhizobium leguminosarum* b.v. trifolii of red clover (*Trifolium pratense*) was promoted when it was coinoculated with the PSM *Bacillus insolitus*, *B. brevis*, or *Agrobacterium rhizogenes*. Similar results were obtained with the inoculation of *G. mosseae* (AMF) and *Azorhizobium caulinodans* (PSMs) in *Sesbania rostrata* (Rahman and Parsons 1997). In soybean, the combination of *Bradyrhizobium japonicum* ($N_2$ fixer) with *P. fluorescens* (PSMs) and *G. mosseae* (AMF) has given equally good results (Shabayey et al. 1996). Such results are likely to be due to a higher plant phosphate uptake promoted by the combined action of PSM and AMF, which may satisfy the high phosphate requirements of the $N_2$-fixing process (Azcon and Barea 1996; Young et al. 1990).

Peix et al. (2001) found that the $N_2$-fixing bacterium *Mesorhizobium mediterraneum* was able to solubilize $Ca_3(PO_4)_2$ under in vitro and soil conditions. Inoculation with *M. mediterraneum* of seeds of the chickpea (*Cicer arietinum*) and barley (*Hordeum vulgare*) planted in a Calcic Rhodoxeralf significantly increased plant growth and total nitrogen and phosphate content in both plants. Further increase was observed when *M. mediterraneum* and $Ca_3PO_4$ were concurrently applied. Benefits of this bacterium were not only due to the symbiotic $N_2$ fixation when associated with chickpea but also due to the enhancement of plant phosphate uptake of both plants.

Apparently, there is a certain degree of specificity among PSM, AMF, and phosphate source. Toro et al. (1996) studied the combined effect of AMF (*Glomus* spp.) and eight PSMs (bacteria) on plant growth and phosphate nutrition of the tropical legume kudzu (*Pueraria phaseoloides*). The PSMs were isolated from an Oxisol and were characterized by their ability to solubilize rock phosphate, Al, and Fe phosphate compounds. In general, PSM inoculation of the kudzu–*Rhizobium*–AMF association increased plant growth, yield, and nutritional status. However, this interaction was not observed in all combinations of AMF + PSM. For instance, the three PSMs *Azospirillum* sp., *Bacillus* sp., and *Enterobacter* sp. had a higher effect when they were coinoculated with *G. mosseae*. In contrast, *Pseudomonas* sp. and an unidentified isolate had a better performance when they were combined with *G. fasciculatum*. On the other hand, Fe phosphate solubilizers were more effective if they were alone, whereas Al-phosphate and rock phosphate solubilizers performed better when they were concurrently inoculated with AMF. Reasons for these differences may be due to interactions between the microorganisms such as a more effective stimulation of rapid mycorrhizal colonization, and enhancing the length, distribution, and/or survival of external fungal mycelium. Mycorrhizal fungi might differ in the amount and type of hyphal exudates released into the mycorrhizosphere. In addition, a high capacity to solubilize phosphate might stimulate plant growth and favor mycorrhizal activity.

Kucey (1987) inoculated a Mollisol (pH 7.2) of Canada with *P. bilaii*, in which either mycorrhizal or nonmycorrhizal wheat or beans were grown. In the case of wheat, mycorrhizal inoculation alone increased significant phosphate uptake (30%), but *P. bilaii* alone did not do so. However, *P. bilaii* increased phosphate uptake of mycorrhizal wheat by 10% in the unfertilized soil, but not in the soil fertilized with rock phosphate. In the case of bean, mycorrhizal inoculation alone did not increase plant phosphate uptake, but *P. bilaii* alone was able to significantly increase it by 31%. Dual inoculation did not increase phosphate uptake beyond the level obtained with *P. bilaii*. In other words, there was not synergism between both microorganisms on bean phosphate uptake in unfertilized and rock phosphate-fertilized soil. Mollisols usually exhibit a low phosphate fixation capacity. For this reason, it is not surprising that inoculation with this PSM alone increased bean phosphate uptake. Differences between wheat and bean could be due to different types and amounts of root exudates that would stimulate acid production in the rhizosphere by PSMs.

Barea et al. (1975) found that inoculation with PSMs (*Pseudomonas* + *Agrobacterium*) did not increase plant phosphate uptake of mycorrhiza-free and mycorrhizal lavender (*Lavandula spica* L. cv. Vera) and maize (*Zea mays* L.) plants grown in an unfertilized red mediterranean soil (pH 7.5, 0.01 M $CaCl_2$–P: 0.021 mg $L^{-1}$). In contrast, PSM inoculation significantly increased plant phosphate uptake of mycorrhizal maize (24%) and mycorrhiza-free lavender (42%), but not of mycorrhizal lavender in a gray-meridional soil (pH 7.6, 0.01 M $CaCl_2$–P: 0.008 mg $L^{-1}$). Maize plants achieved 75 and 95% of their maximum yield at P concentrations of 0.008 and 0.025 mg $L^{-1}$ (Fox 1979), indicating that maize plants (and perhaps lavender too) satisfied most of their phosphate requirements in the red mediterranean soil.

Currently, *P. bilaii* is commercially available in North America under the name of Provide$^{TM}$, which has been successfully tested to enhance plant phosphate uptake of some crop plants (Whitelaw 2000). Whitelaw et al. (1997) inoculated an acidic phosphate-deficient soil of Australia (pH 4.6) with *Penicillium radicum* in combination with several levels of $KH_2PO_4$ (added P: 0–20 kg $ha^{-1}$). The inoculation with this PSM increased wheat phosphate uptake by 8% in the unfertilized soil. When the fungus was inoculated in combination with phosphate fertilization, plant P increased between 2 and 28%; the increase was highest when the rate of added P was 15 kg $ha^{-1}$.

Young et al. (1990) found that inoculation with either PSM or AMF significantly increased peanut (*Arachis hypogea*) production in two subtropical–tropical acidic soils of Taiwan. Inoculation with either AMF or PSMs in unfertilized soils was as effective as the addition of rock phosphate alone. Inoculation with AMF or PSMs in rock phosphate-fertilized soils did not increase peanut yield above that obtained with AMF or PSM inoculation in unfertilized soils. Unfortunately, dual inoculation of AMF and PSMs was not evaluated.

In addition, Young et al. (1990) found that the responses to single or mixed inoculations with PSM and/or AMF had variable effects on plant growth of leucaena grown in three soils of Taiwan. Inoculation with PSMs was not as effective as AMF inoculation in enhancing plant growth in the soil with the lowest available phosphate level (Hinshe soil). In the Wunfun soil (also with a low soil available phosphate level), PSM inoculation was ineffective in increasing plant growth unlike AMF. In the alkaline soil containing the highest soil available phosphate (and presumably rich in calcium-phosphates), PSM inoculation alone significantly increased plant growth (40%) above the AMF inoculation effect, which did not increase growth.

Effectiveness of PSM inoculation alone to enhance plant phosphate uptake in subtropical and tropical acidic soils is relatively low and variable. The increases recorded were 8% (Whitelaw et al. 1997), 13% (Osorio and Habte 2001), and 24–25% (Young et al. 1990) compared with those reported in less-weathered soils (mostly Mollisols) of the temperate zone, where soil phosphate-fixation capacity is low. By contrast, the effectiveness of PSM inoculation in enhancing plant phosphate uptake of mycorrhizal plants grown in tropical or subtropical soils can be relatively higher compared to data reported in temperate soils.

Mycorrhizal association in combination with PSMs is often needed to obtain improvements in plant phosphate uptake in highly weathered soils, in contrast to results obtained in less-weathered soils. In these less-weathered soils that normally exhibit low soil phosphate sorption, the inoculation of PSMs alone has been enough to increase plant phosphate uptake of nonmycorrhizal plants (Peix et al. 2001; Omar 1998; Gleddie 1993; Kucey and Leggett 1989; Asea et al. 1988; Kucey 1983, 1987, 1988). Most of the soils used by these authors were mollisols, calcareous soils, or sandy soils, which are characterized by a low phosphate sorption capacity and relatively high soil Ca–phosphate content (Cross and Schlesinger 1995). Therefore, the freshly released phosphate by PSMs can remain longer in the soil solution until its absorption by the roots.

For instance, Toro et al. (1998) found that the PSM *Enterobacter* sp. alone was as effective as the mycorrhizal fungus *G. mosseae* alone in increasing twofold the plant phosphate uptake of alfalfa grown in a calcareous soil of Spain. Duponnois et al. (2006) found that the inoculation alone with the fungus *Arthrobotrys oligospora* increased the phosphate uptake and shoot dry weight of *Acacia holoserica* grown in a sandy soil of Senegal by 56% and 46%. The increase in plant phosphate uptake and growth were even higher when rock phosphate was added with the PSMs (74 and 103%, respectively). The addition of rock phosphate alone did not increase significantly plant phosphate uptake and growth. Similar results were reported by Wakelin et al. (2004a, b) on wheat with the inoculation of *P. radicum* in some sandy soils of Australia with neutral to alkaline soil reactivity. In these soils, Wakelin and coworkers observed increases in plant growth between 34 and 76%. Even higher was the effect of *Penicillium thomii* on plant phosphate uptake of mint (*Mentha piperita*) grown in a soilless medium (vermiculite–perlite) fertilized with rock phosphate (Cabello et al. 2005). In that experiment, the inoculation with the fungus increased plant phosphate content by more than threefold compared to control plants (uninoculated and unfertilized). The rock phosphate alone was ineffective. This significant increase in plant phosphate uptake is understandable given the very low phosphate sorption on these kind of substrates.

## 4.6 In Vitro Test to Evaluate Effectiveness of PSMs

A method to quantify PSM effectiveness in dissolving rock phosphate was developed by Osorio and Habte (2001). This consists of preparing a soluble P-free-medium without agar (a rock phosphate is used as the unique source of P) and transfer 75 mL of this into a 250-mL Erlenmeyer flask. Then, the flasks and their contents are autoclaved (120°C, 0.1 MPa, 30 min). Later, a selected microbe (bacterium or fungus) is transferred from a Petri dish with a sterile loop. The flask also can be inoculated with 1 mL of a fungal or bacterial suspension. A control uninoculated is included for comparison. The flasks are shaken continuously in an orbital shaker at 100 rpm, 28°C for 7 days.

After the incubation, pH is measured directly. An aliquot of 20 mL is transferred into a centrifuged tube and centrifuged at 5,000 rpm for 15 min. Later, the supernatant is filtered through a filter paper (2 μm) and then by a Milipore membrane (0.45 μm). Soluble phosphate concentration is determined by the molybdate-blue method (Murphy and Riley 1962). Figure 4.1 illustrates the results of an in vitro test for 32 microbes isolated from Hawaiian soils. At the end of the incubation, the uninoculated control had a medium pH 7.0, whereas the inoculated flask had lower pHs. Significant increases in solution phosphorus concentration are detected when the pH decreases below 4.7. An attempt to classify microbes in categories of effectiveness to dissolve rock phosphate as a function of final pH is as follows: ineffective when final pH is >4.7, moderately effective when pH is 4.0–4.7. It is also possible to determine phosphate concentration in the microbial

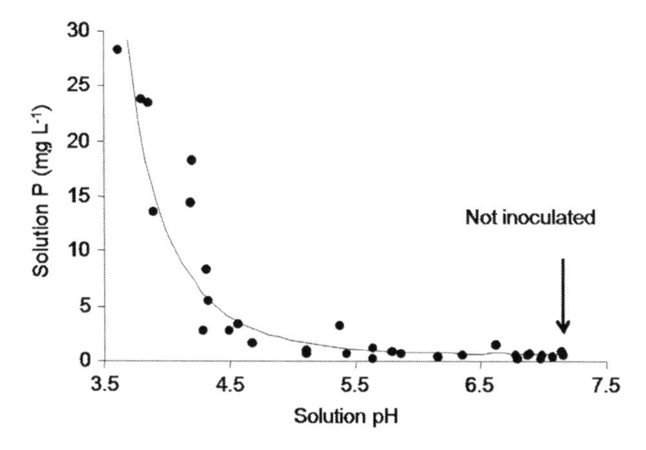


**Fig. 4.1** Relationship between pH and concentration of phosphate in liquid medium inoculated or not inoculated with phosphate solubilizing microorganisms (Source: Osorio and Habte 2001)

biomass that remains at the bottom of the centrifuge tube; special care must be taken to separate the microbial biomass from the particles of rock phosphate undissolved.

## 4.7 In Vivo Test with Plants

In order to evaluate the effectiveness of PSMs in enhancing plant phosphate uptake and growth, a simple method has been developed. First, a sample of an autoclaved and phosphate-deficient soil is transferred into plastic pots (ca. 1 L). Then, the soil sample is amended or unamended with rock phosphate and other nutrients (based on soil test). Right before planting, the soil is inoculated or not with a PSM and a mycorrhizal fungus. In the hole of planting, author used 5–10 mL of a microbial suspension containing $10^5$–$10^6$ UFC $mL^{-1}$ of a given PSM; 20–25 g of a crude mycorrhizal inoculum containing 40 infective propagules per gram is good enough. *Leucaena leucocephala* is our preferred indicator plant given its high phosphate requirement and its inability to absorb it in phosphate-deficient soils. After a short period of growth (60–90 days), plants are harvested and shoot dry matter and phosphate uptake is measured (Table 4.3).

In these experiments, PSM inoculation (alone) significantly increased plant growth of leucaena by 71% in the Mollisol (low phosphate sorption) and by 25% in the Oxisol de Wahiawa (HI, USA); in the other Oxisols (Quilichao and Carimagua, Colombia, medium phosphate sorption) and in the Andisol, PSM alone was ineffective in increasing plant growth. On the other hand, when the PSM was concomitantly inoculated with the AMF (AMF + PSM), the effect was significantly ($P \leq 0.05$) higher than with the AMF inoculation alone (AMF), except in the Andisol given its very high phosphate sorption capacity. In the Oxisols, the increase

**Table 4.3** Shoot dry weight (g) of leucaena inoculated or not with a PSM and AMF in soils that differed in their phosphate sorption capacity

| Soil | Soil phosphate sorption category | Control | PSM | AMF | AMF + PSM |
|---|---|---|---|---|---|
| Mollisol – Neira ($P_{0.2} = 45$) | Low | 0.7d | 1.2c | 1.4b | 1.5a |
| Oxisol – Quilichao ($P_{0.2} = 328$) | Medium | 0.3c | 0.3c | 2.0b | 2.6a |
| Oxisol – Carimagua ($P_{0.2} = 417$) | Medium | 0.3c | 0.3c | 0.8b | 1.0a |
| Oxisol – Wahiawa ($P_{0.2} = 450$) | Medium | 0.4d | 0.5c | 1.0b | 1.4a |
| Andisol – La Selva ($P_{0.2} = 2{,}222$) | Very high | 0.3b | 0.3b | 0.5a | 0.5a |

Source: Osorio (2008) and Londoño (2010)

ranged from 25 to 40%. In the Mollisol, the increase was only 7%; in this case, it is likely that leucaena plants already have enough phosphate absorbed by the micorrhizal association.

## 4.8 Conclusions

The effectiveness of PSMs in increasing plant phosphate uptake and growth is controlled by the phosphate sorption capacity. In soils with low phosphate sorption ($P_{0.2} < 100$), PSM inoculation alone can increase plant performance. In soils with medium and high phosphate sorption ($100 < P_{0.2} < 500$; $500 < P_{0.2} < 1{,}000$) PSMs alone are less effective or even ineffective, their effectiveness depends on the presence of mycorrhizal association. In soils with very high phosphate sorption ($P_{0.2} > 1{,}000$), PSMs seem to be ineffective even in the presence of AMF.

In soils conducive for PSMs, the ability to solubilize phosphate might have some practical implications such as lowering soil phosphate fertilizer requirement and enhancing the residual effect of soluble phosphate fertilizers. Their effectiveness in increasing soil solution phosphate by dissolving rock phosphate, native soil phosphate, and/or desorbing sorbed phosphate can play an important role in the alleviation of phosphate deficiency in soils, particularly in the tropics where the high phosphate sorption capacity of soils constrains plant productivity. Moreover, the role that PSMs play in soil phosphate availability ought to be an important consideration in the development of biotechnological approaches to the managements of soils.

## References

Abd-Alla MH (1994) Use of organic phosphorus by *Rhizobium leguminosarum* biovar. viceae phosphatases. Biol Fert Soils 18:216–218

Agnihorti VP (1970) Solubilization of insoluble phosphates by some soil fungi isolated from nursery seedbeds. Can J Microbiol 16:877–880

Amos B, Walters DT (2006) Maize root biomass and net rhizodeposited carbon: an analysis of the literature. Soil Sci Soc Am J 70:1489–1503
Asea PEA, Kucey RMN, Stewart JWB (1988) Inorganic phosphate solubilisation by 2 *Penicillium* species in solution culture and soil. Soil Biol Biochem 20:459–464
Atlas R, Bartha R (1997) Microbial ecology. Addison Wesley Longman, New York
Azam F, Memom GH (1996) Soil organisms. In: Bashir E, Bantel R (eds) Soil science. National Book Foundation, Islamabad, pp 200–232
Azcon C, Barea JM (1996) Interactions of arbuscular mycorrhiza with rhizosphere microorganisms. In: Guerrero E (ed) Mycorrhiza. Biological soil resource, FEN, Bogota, pp 47–68
Banik S, Dey BK (1981a) Phosphate solubilizing microorganisms of a lateritic soil. I. Solubilization of inorganic and production of organic acids by microorganisms isolated in sucrose calcium phosphate agar plates. Zentralblatt Bakteriol. Parasitenkunde infectionskarankheiten, hygiene. 2. Naturwiss Miikrobiol Landwirtsch 136:476–486
Banik S, Dey BK (1981b) Phosphate solubilizing microorganisms of a lateritic soil. II. Effect of some tricalcium phosphate-solubilizing microorganisms on available phosphorus content of the soil. Zentralblatt Bakteriol. Parasitenkunde infectionskarankheiten, hygiene. 2. Naturwiss Miikrobiol Landwirtsch 136:487–492
Banik S, Dey BK (1981c) Phosphate solubilizing microorganisms of a lateritic soil. III. Effect of inoculation of some tricalcium phosphate-solubilizing microorganisms on available phosphorus content of rhizosphere soils of rice (*Oryza sativa* L. cv. IR-20). Zentralblatt Bakteriol. Parasitenkunde infectionskarankheiten, hygiene. 2. Naturwiss Miikrobiology Landwirtsch 136:493–501
Banik S, Dey BK (1983) Phosphate solubilizing potentiality of the microorganisms capable of utilizing aluminium phosphate as a sole phosphate source. Zbl Mikrobiol 138:17–23
Barber SA (1995) Soil nutrient bioavailability. A mechanistic approach. Wiley, New York
Barea JM, Azcon R, Hayman DS (1975) Possible synergistic interactions between endogone and phosphate-solubilizing bacteria in low-phosphate soils. In: Mosse B, Tinker PB (eds) Endomycorrhizas. Academic, London, pp 409–417
Barea JM, Toro M, Orozco M, Campos E, Azcon R (2002) The Application of isotopic ($^{32}P$ and $^{15}N$) dilution technique to evaluate the interactive effect of phosphate-solubilizing-rhizobacteria, mycorrhizal fungi and *Rhizobium* to improve the agronomic efficiency of rock phosphate form legume crops. Nutr Cycl Agroecosyst 63:35–42
Bar-Yosef B, Rogers RD, Wolfram JH, Richman E (1999) *Pseudomonas cepacia*-mediated rock phosphate solubilization in kaolinite and montmorillonite suspensions. Soil Sci Soc Am J 63:1703–1708
Bohn H, McNeal BL, O'connor G (1985) Soil chemistry. Wiley, New York
Bolan NS (1991) A critical review on the role of mycorrhizal fungi in the uptake of phosphorus by plants. Plant Soil 134:189–207
Bolan NS, Naidu R, Mahimairaja S, Baskaran S (1994) Influence of low-molecular-weight organic acids on the solubilization of phosphates. Biol Fert Soils 18:311–319
Cabello M, Irrazabal G, Bucsinszky AM, Saparrat M, Schalamuk S (2005) Effect of an arbuscular mycorrhizal fungus, *Glomus mosseae*, and a rock-phosphate-solubilizing fungus, *Penicillium thomii*, on *Mentha piperita* growth in a soilless medium. J Basic Microbiol 45:182–289
Canbolat MC, Bilen S, Cakmakci R, Sahin F, Aydin A (2006) Effect of plant growth-promoting bacteria and soil compaction on barley seedling growth, nutrient uptake, soil properties, and rhizosphere microflora. Biol Fert Soils 42:350–357
Cross AF, Schlesinger WH (1995) A literature review and evaluation of the Hedley fractionation: application to the biogeochemical cycle of soil phosphorus in natural ecosystems. Geoderma 64:197–214
Di-Simine CD, Sayer JA, Gadd GM (1998) Solubilization of zinc phosphate by a strain of *Pseudomonas fluorescens* isolated from a forest soil. Biol Fert Soils 28:87–94
Duponnois R, Kisa M, Plenchette C (2006) Phosphate-solubilizing potential of the nematophagous fungus *Arthrobotrys oligospora*. J Plant Nutr Soil Sci 169:280–282

Fox RL (1979) Comparative responses of field grown crops to phosphate concentrations in soil solutions. In: Munsell H, Staples R (eds) Stress physiology in crop plants. Wiley, New York, pp 81–106

Fox RL, Kamprath E (1970) Phosphate sorption isotherms for evaluating phosphorus requirements of soils. Soil Sci Soc Am Proc 34:902–907

Gaur A, Rana J, Jalali B, Chand H (1990) Role of VA mycorrhizae, phosphate solubilizing bacteria and their interactions on growth and uptake of nutrients by wheat crops. In: Trends in mycorrhizal research. Proceedings of the national conference on mycorrhizae, Hisar, India, pp 105–106

Gleddie SC (1993) Response of pea and lentil to inoculation with the phosphate-solubilizing fungus *Penicillium bilaii* (provide). In: Proceedings of the soils and crops workshops, Saskatoon, Saskatchewan, pp 47–52

Gururaj R, Mallikarjunaiah R (1995) Interactions among *Azotobacter chroococcum*, *Penicillium glaucum* and *Glomus fasciculatum* and their effect on the growth and yield of sunflower. Helia 18(23):73–84

Gyaneshwar P, Kumar GN, Parekh LJ, Poole PS (2002) Role of soil microorganisms in improving P nutrition of plants. Plant Soil 245:83–93

Hameeda B, Kumar YH, Rupela OP, Kumar GN, Reddy G (2006) Effect of carbon substrates on rock phosphate solubilization by bacteria from compost and macrofauna. Curr Microbiol 53:298–302

Havlin J, Beaton J, Tisdale SL, Nelson W (1999) Soil fertility and fertilizers. An introduction to nutrient management. Prentice Hall, Upper Saddle River, NJ

He ZL, Zhu J (1997) Transformation and bioavailability of specifically sorbed phosphate on variable-charge mineral soils. Biol Fert Soils 25:175–181

He ZL, Zhu J (1998) Microbial utilization and transformation of phosphate adsorbed by variable charge minerals. Soil Biol Biochem 30:917–923

Ilmer P, Barbato A, Schinner F (1995) Solubilization of hardly-soluble $AlPO_4$ with P-solubilizing microorganisms. Soil Biol Biochem 27:265–270

Jones DL, Dennis PG, Owen AG, van Hees PAW (2003) Organic acid behavior in soils – misconceptions and knowledge gaps. Plant Soil 248:31–41

Kang SC, Chul GH, Lee TG, Maheshwari DK (2002) Solubilization of insoluble inorganic phosphates by a soil inhabiting, fungus, *Fomitopsis* spp. PS 102. Curr Sci 25:439–442

Kang SC, Pandey P, Khillon R, Maheshwari DK (2008) Process of rock phosphate solubilization by *Aspergillus* spp. PS 104 in soil amended medium. J Environ Biol 29(5):743–746

Kim KY, McDonald GA, Jordan D (1997) Solubilization of hydroxyapatite by *Enterobacter agglomerans* and cloned *Escherichia coli* in culture medium. Biol Fert Soils 24:347–352

Kim KY, McDonald GA, Jordan D (1998a) Effect of phosphate solubilizing bacteria and vesicular-arbuscular mycorrhizae on tomato growth and soil microbial activity. Biol Fert Soils 26:79–87

Kim KY, Jordan D, McDonald GA (1998b) *Enterobacter agglomerans*, phosphate solubilizing bacteria and microbal activity in soil. Effect of carbon sources. Soil Biol Biochem 30:995–1003

Kopler J, Lifshitz R, Schroth M (1988) *Pseudomonas* inoculants to benefit plant production. ISI Atl Sci Anim Plant Sci 1:60–64

Kucey RMN (1983) Phosphate solubilising bacteria and fungi in various cultivated and virgin Alberta soils. Can J Soil Sci 63:671–678

Kucey RMN (1987) Increased phosphorus uptake by wheat and field beans inoculated with a phosphorus solubilising *Penicillium bilaii* strain and with vesicular-asbuscular mycorrhizal fungi. Appl Environ Microbiol 53:2699–2703

Kucey RMN (1988) Effect of *Penicillium bilaii* on the solubility and uptake of P and micronutrients from soil by wheat. Can J Soil Sci 68:261–270

Kucey RMN, Leggett ME (1989) Microbial mediated increases in plant available phosphorus. Adv Agron 42:199–228

Linderman RG (1988) Mycorrhizal interaction with the rhizosphere microflora: the mycorhizosphere effect. Phytopathology 78:366–371

Londoño A (2010) Efecto de la inoculación con un hongo micorrizal y un hongo solubilizador de fósforo en la absorción de fosfato y crecimiento de leucaena en un oxisol. Tesis M.Sc. Ciencias Agrarias, Universidad Nacional de Colombia, Medellín, p 58

Lopez-Bucio J, Campo-Cuevas JC, Hernandez-Calderon E, Velásquez-Becerra C, Farias-Rodriguez R, Macias-Rodriguez LI, Valencia-Cantero E (2007) *Bacillus megatherium* rhizobacteria promote growth and alter root-system architecture through an auxin- and ethylen-independing signaling mechanism in *Arabidopsis thaliana*. Mol Plant Microbe Interact 20:207–217

Louw HA, Webley DM (1959) The solubilization of insoluble phosphates. V. The action of some organic acids on iron and aluminium phosphates. NZ J Sci 2:215–218

Lynch JP, Ho MD (2005) Rhizoeconomics: carbon costs of phosphorus acquisition. Plant Soil 269:45–56

Mathews CK, Van Holde KE, Ahern KG (1999) Biochemistry. Benjamin Cummings, San Francisco, CA

Memon KS (1996) Soil and fertilizer phosphorus. In: Bashir E, Bantel R (eds) Soil science. National Book Foundation, Islamabad, pp 291–314

Mohod S, Gupta DN, Chavan AS (1991) Effects of P solubilizing organisms on yield and N uptake by rice. J Maharashtra Agric Univ 16:229–231

Murphy J, Riley JP (1962) A modified single solution method for the determination of phosphate in natural waters. Anal Chim Acta 27:31–35

Omar SA (1998) The role of rock-phosphate-solubilizing fungi and vesicular-arbuscular-mycorrhiza (VAM) in growth of wheat plants fertilized with rock phosphate. World J Microbiol Biotechnol 14:211–218

Osorio NW (2008) Effectiveness of microbial solubilization of phosphate in enhancing plant phosphate uptake in tropical soils and assessment of the mechanisms of solubilization. Ph.D. dissertation, University of Hawaii, Honolulu

Osorio NW, Habte M (2001) Synergistic influence of an arbuscular mycorrhizal fungus and P solubilizing fungus on growth and plant P uptake of *Leucaena leucocephala* in an Oxisol. Arid Land Res Manag 15:263–274

Pandey A, Trivedi P, Kumar B, Palni LMS (2006) Characterization of a phosphate solubilizing microorganism and antagonistic strain of *Pseudomonas putida* (B0) isolated from a sub-alpine location in the Indian Central Himalaya. Curr Microbiol 53:102–107

Paul NB, Rao WVBS (1971) Phosphate-dissolving bacteria in the rhizosphere of some cultivated legumes. Plant Soil 35:127–132

Peix A, Rivas-Boyero AA, Mateos PF, Rodriguez-Barrueco C, Martinez-Molina E, Velasquez E (2001) Growth promotion of chickpea and barley by a phosphate solubilizing strain of *Mesorrhizobium mediterraneum* under growth chamber conditions. Soil Biol Biochem 33:103–110

Pikovskaia RI (1948) Mobilization of phosphates in soil in connection with the vital activities of some microbial species. Mikrobiologiia 17:362–370

Prathibha CK, Alagawadi A, Sreenivasa M (1995) Establishment of inoculated organisms in rhizosphere and their influence on nutrient uptake and yield of cotton. J Agric Sci 8:22–27

Prescott L, Harley J, Klein DA (1999) Microbiology. McGraw-Hill, Boston, MA

Rahman MK, Parsons JW (1997) Effects of inoculation with *Glomus mosseae, Azorhizobium caulinodans* and rock phosphate on the growth of and nitrogen and phosphorus accumulation in *Sesbania rostrata*. Biol Fert Soils 25:47–52

Rambelli A (1973) The rhizosphere of mycorrhyzae. In: Marks GC, Kozlowski TT (eds) Ectomycorrhyzae, their ecology and physiology. Academic, London, pp 299–343

Rao S (1992) Biofertilizers in agriculture. AA Balkema, Rotterdam

Reddy MS, Kumar S, Babita K, Reddy MS (2002) Biosolubilization of poorly soluble rock phosphates by *Aspergillus tubigensis* and *Aspergillus niger*. Bioresour Technol 84:187–189

Reyes I, Valery A, Valduz Z (2006) Phosphate-solubilizing microrganisms isolated from rhizospheric and bulk soils of colonizer plants at an abandoned rock phosphate mine. Plant Soil 287:69–75

Roos W, Luckner M (1984) Relationships between proton extrusion and fluxes of ammonium ions and organic acids in *Penicillium cyclopium*. J Gen Microbiol 130:1007–1014

Rose RE (1957) Techniques of determining the effect of microorganisms on insoluble inorganic phosphates. NZ J Sci Technol 38:773–780
Rosendahl RO (1942) The effect of mycorrhizal and non-mycorrhizal fungi on the availability of difficulty soluble potassium and phosphorus. Soil Sci Soc Am Proc 7:477–479
Sanchez P, Logan T (1992) Myths and science about the chemistry and fertility of soils in the tropics. In: Lal R, Sanchez P (eds) Myths and science of soils of the tropics. Soil Science Society of America, Madison, WI, pp 35–46
Sanchez P, Uehara G (1980) Management considerations for acid soils with high phosphorus fixation capacity. In: Khasawneh FE (ed) The role of phosphorus in agriculture. Soil Science Society of America, Madison, WI, pp 471–514
Shabayey VP, Smolin VY, Mudrik VA (1996) Nitrogen fixation and $CO_2$ exchange in soybeans inoculated with mixed cultures of different microorganisms. Biol Fert Soils 23:425–430
Singh S, Kapoor KK (1999) Inoculation with phosphate-solubilizing microorganisms and a vesicular-arbuscular mycorrhizal fungus improves dry matter yield and nutrient uptake by wheat grown in a sandy soil. Biol Fert Soils 28:139–144
Sperber JI (1957) Solution of mineral phosphates by soil bacteria. Nature 180:994–994
Sperber JI (1958) Solution of apatite by soil microorganisms producing organic acids. Aust J Agric Res 9:782–787
Sreenivasa M, Krishnaraj M (1992) Synergistic interaction between VA mycorrhizal fungi and a phosphate solubilizing bacterium in chili. ZBL Mikrobiol 147:126–130
Stevenson FJ (1986) Cycles of soil. Wiley, New York
Sturz AV, Christie BR, Matheson BG, Nowak J (1997) Biodiversity of endophytic bacteria which colonize red clover nodules, roots, stems and foliage and their influence on host growth. Biol Fert Soils 25:13–19
Tinker PB (1980) Role of rhizosphere microorganisms in phosphorus uptake by plants. In: Khasawneh FE, Sample EC, Kamprath EJ (eds) The role of phosphorus in agriculture. Soil Science Society of America, Madison, WI, pp 617–654
Toro M, Azcon R, Herrera R (1996) Effects on yield and nutrition of mycorrhizal and nodulated *Pueraria phaseolides* exerted by P-solubilizing rhizobacteria. Biol Fert Soils 21:23–29
Toro M, Azcon R, Barea JM (1998) The use of isotopic dilution techniques to evaluate the interactive effects of rhizobium genotypes, mycorrhizal fungi, phosphate-solubilizing rhizobacteria and rock phosphate on nitrogen and phosphorus acquisition by *Medicago sativa*. New Phytol 138:265–273
Trolove SN, Hedley MJ, Kirk GJD, Bolan NS, Loganathan P (2003) Progress in selected areas of rhizosphere research on P acquisition. Aust J Soil Res 41:471–499
Venkateswardu B, Rao AV, Raina P (1984) Evaluation of phosphorus solubilization by microorganisms isolated from aridisols. J Indian Soc Soil Sci 32:273–277
Wakelin SA, Warren RA, Ryder MH (2004a) Effect of soil properties on growth promotion of wheat by *Penicillium radicum*. Aust J Soil Res 42:897–904
Wakelin SA, Warren RA, Harvey PR, Ryder MH (2004b) Phosphate solubilization by *Penicillium* spp. closely associated with wheat roots. Biol Fert Soils 40:36–43
Welch S, Taunton AE, Banfiled JF (2002) Effect of microorganisms and microbial metabolites on apatite dissolution. Geomicrobiol J 19:343–367
Whitelaw MA (2000) Growth promotion of plants inoculated with phosphate-solubilizing fungi. Adv Agron 69:99–151
Whitelaw MA, Harden TJ, Bender GL (1997) Plant growth promotion of wheat inoculated with *Penicillium radicum* sp. nov. Aust J Soil Res 35:291–300
Xavier LJC, Germida JJ (2003) Bacteria associated with *Glomus clarum* spores influence mycorrhizal activity. Biol Fert Soils 35:471–478
Young CC, Chen CL, Chao CC (1990) Effect of *Rhizobium*, vesicular-arbuscular mycorrhiza, and phosphate solubilizing bacteria on yield and mineral phosphorus uptake of crops in subtropical-tropical. In: 14th international congress of soil science. Transactions, vol. III, International Society of Soil Science, Kyoto, Japan, pp 55–60

# Chapter 5
# Sulfur-oxidizing Bacteria: A Novel Bioinoculant for Sulfur Nutrition and Crop Production

**R. Anandham, P. Indira Gandhi, M. SenthilKumar, R. Sridar, P. Nalayini, and Tong-Min Sa**

## 5.1 Introduction

Sulfur (S) is increasingly being recognized as the fourth major plant nutrient after nitrogen, phosphorus, and potassium. It is a constituent of amino acids cysteine and methionine, which act as a precursor for the synthesis of all other compounds containing reduced sulfur (Scherer 2001). Sulfur is essential for both plant and animal life. Although the element is required by plants in amounts comparable to phosphorus, the first field case of sulfur deficiency was reported only in 1933. In wetland rice, sulfur deficiency was first reported in 1938. During the last 10 years,

R. Anandham
Department of Agricultural Microbiology, Agricultural College and Research Institute, Tamil Nadu Agricultural University, Madurai, Tamil Nadu 625 104, India

Department of Agricultural Chemistry, Chungbuk National University, Cheongju, Chungbuk 361-763, Republic of Korea

P.I. Gandhi
Krishi Vigyan Kendra, Tamil Nadu Agricultural University, Vriddhachalam, Tamil Nadu 606 001, India

Department of Agricultural Chemistry, Chungbuk National University, Cheongju, Chungbuk 361-763, Republic of Korea

M. SenthilKumar • R. Sridar
Department of Agricultural Microbiology, Centre for Plant and Molecular Biology, Tamil Nadu Agricultural University, Coimbatore, Tamil Nadu 641 003, India

P. Nalayini
Division of Crop Production, Central Institute for Cotton Research, Regional Station, Indian Council of Agricultural Research, Coimbatore, Tamil Nadu 641 003, India

T.-M. Sa (✉)
Department of Agricultural Chemistry, Chungbuk National University, Cheongju, Chungbuk 361-763, Republic of Korea
e-mail: tomsa@chungbuk.ac.kr

D.K. Maheshwari (ed.), *Bacteria in Agrobiology: Plant Nutrient Management*,
DOI 10.1007/978-3-642-21061-7_5, 

sulfur deficiency has been recognized as an important growth-limiting factor for both dry land crops and wetland rice. Sulfur deficiency generally occurs in Andosols, Vertisols, Alfisols, Ultisols, and Oxisols of the tropics. In Asia, sulfur deficiency of wetland rice has been reported in Bangladesh, Burma, India, Indonesia, Japan, the Philippines, and Sri Lanka. Responses to sulfur have been reported for 23 crops in 40 tropical countries. This occurrence of sulfur deficiencies has been accentuated by the increase in use of low sulfur fertilizers, decrease in use of organic manures, intensive cropping, and reduced atmospheric deposition (Islam and Ponnamperuma 1982). To alleviate sulfur deficiency, sulfur fertilizers are invariably added to soils, usually in a reduced form, such as elemental sulfur. Use of S oxidizers enhances the rate of natural oxidation of S and speeds up the production of sulfates, and makes them available to plants at their critical stages, consequently resulting in an increased plant yield (Wainright 1984; Grayston and Germida 1991; Scherer 2001; Anandham et al. 2007a, 2008a; Stamford et al. 2007a, b, 2008a, b). In this review, forms of sulfur and sulfur fertilizers and physiological role of sulfur in plants and interaction of sulfur with other elements in soil are discussed. Further, insight was provided in the area of ecological niches for isolation of sulfur-oxidizing bacteria and their role in sulfur oxidation in soil and sulfur nutrition to crop plants.

## 5.2 Soil Sulfur

Of 105 elements that make up the earth's crust, sulfur concentration was measured to be about 0.03%; the 17th most abundant element. It occurs in soil mainly as elemental sulfur, mineral sulfides, sulfates, and hydrogen sulfides, and organic sulfur compounds. Sulfur in soils principally originates from the pyrite ($FeS_2$) of primary minerals. During weathering and soil formation, the S from pyrite is oxidized to several forms of different states, which are intimately connected with soil's reduction–oxidation potential. Oxidation and reduction reactions of S occur rather easily, accounting for the diversity of reactions that S undergoes in the soil. Total S is usually related to the following soil properties (1) organic matter content, (2) clay percentage and type of clay minerals, (3) content of active iron and aluminum oxides, and (4) soil texture.

Organic S content of soil increases with the level of humus content. Particularly, peat and marsh soils are rich in organic S. In the south pacific, total S tends to be less in soil with low organic matter content (Morrison et al. 1987). The highly weathered and leached tropical soils are typically low in organic matter and S (Stevenson 1986). In the soils of south China, total S is positively correlated with clay percentage (Liu 1986). Total S is higher when kaolinite (1:1) type clay minerals are dominant relative to montmorillonite (2:1) type clays (Morrison et al. 1987). S retention in soils is favored by the presence of iron and aluminum oxides (Shin 1987). In the soils of India, total S is higher in fine textured soils than in coarse textured soils, partly because of their greater organic matter content (Tandon 1991).

## 5.3 Fractions of S in Soil

The total content of S in soils is usually only a weak indicator of availability of S to plants. Plants mainly absorb S in the form of sulfates. However, soil S occurs in several forms and, therefore, the availability of S to plants depends on the transformations of S in the soil and the interactions between the different S fractions. Figure 5.1 provides a rough estimate of the distribution of different fractions of S in soils (Kleinhenz 1999).

### *5.3.1 Organic S*

The distribution of organic S within a soil profile follows the pattern of organic matter and decreases with depth. Soil organic S can be divided into three fractions: C-bonded S, non C-bonded S, and the soil biomass (Stevenson 1986). The greater part of total organic S in soils of humid and semiarid regions is present as C-bonded S. Separation of this fraction is difficult since these S-containing compounds undergo extensive transformations. Although it is anticipated that S present in

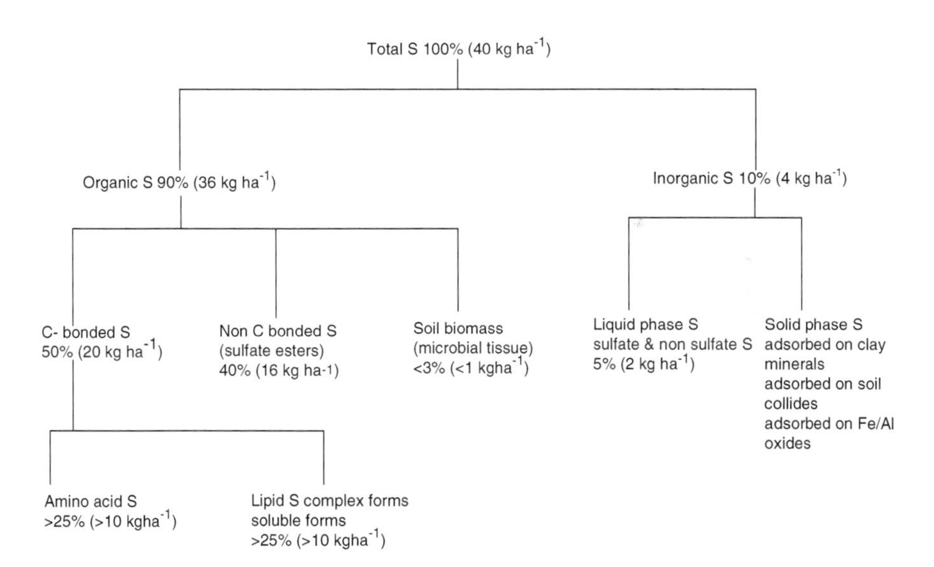


**Fig. 5.1** Distribution of sulfur fractions in soil (Kleinhenz 1999)

amino acid accounts for only a minor part of the carbon bonded S, trace quantities of this fraction can be identified in soil whereas for the other forms (lipid S, complex forms, and soluble forms) it is usually possible only to demonstrate their occurrence. The S-containing amino acids include cysteine, cystine, and methionine. However, only a small fraction of the C-bonded S can be accounted for these known compounds. The greater part occurs as lipid-S (sulfolipids), complex S-forms and soluble S-forms. These forms occur as numerous products resulting from other complex compounds and only exist in soil as intermediate products between synthesis and destruction by microorganisms. Non C-bonded S is assumed to occur as unknown ester sulfates, such as sulfated polysaccharides. Only a small part of the organic S resides in soil biomass. Nevertheless, the fraction of S in microbial tissue is extremely labile and responsible for the turnover of S with consequences for the availability to plants (Kleinhenz 1999).

### *5.3.2 Inorganic S*

Inorganic S occurs in soil largely as sulfate. S is available to plants as sulfate in the liquid phase of the soil. Under anaerobic conditions, S is present in reduced forms. A major fraction of the S in calcareous and saline soils occurs as gypsum ($CaSO_4 \cdot 2H_2O$). In arid regions, high amount salts such as $CaSO_4$, $MgSO_4$, and $Na_2SO_4$ can accumulate. Solid phase S comprises a $SO_4^{2-}$ retained in an adsorbed form. Sulfate can be adsorbed to clay minerals and active Fe and Al oxides. Sorption is due primarily to an ion exchange by charges on clay minerals and oxides and increases with decreasing soil pH (Table 5.1) (Kleinhenz 1999).

**Table 5.1** Important forms of inorganic soil sulfur (Kleinhenz 1999)

| Oxidation state | Sulfur form | Molecular formula |
|---|---|---|
| −2 | Sulfide | $S^{2-}$ |
| −2 | Hydrogen sulfide | $H_2S$ |
| −2, 0 | Pyrite | $FeS_2$ |
| 0 | Elemental sulfur | $S^0$ |
| −1, +5 | Thiosulfate | $S_2O_3^{2-}$ |
| +4 | Sulfite | $SO_3^{2-}$ |
| +6 | Sulfate | $SO_4^{2-}$ |
| +6 | Sulfuric acid | $H_2SO_4$ |
| +3 | Hyposulfite (dithionite) | $S_2O_4^{2-}$ |
| +4 | Dithionate | $S_2O_6^{2-}$ |
| −2, +6 | Trithionate | $S_3O_6^{2-}$ |
| −2, +6 | Tetrathionate | $S_4O_6^{2-}$ |
| −2, +6 | Pentathionate | $S_5O_6^{2-}$ |

## 5.4 Physiological Role of Sulfur in Plants

A review of the physiological role of all sulfur compounds is hardly to be found. The role of sulfur can be considered in two ways; directly in compounds and indirectly by affecting synthesis of compounds or other nutrients. Sulfate is the main sulfur source for crop plants and is activated by forming APS (adenosine-5′-phosphosulfate) and then active sulfate, PAPS (3′-phosphoadenosine-5′-phosphosulfate). Cysteine is first formed and becomes the precursor of methionine and cysteine sulfur amino acids (Giovanelli et al. 1980). These amino acids are building blocks of proteins and keep specific structures of each protein through the formation of disulfide bonds (–S–S–). Enzymes are proteins, thus S–S bridges contribute to the conformation of enzyme protein, which is a prerequisite for enzyme activity. Plant sulfolipid (diacylphoquinivosylglycerol), a ubiquitous constituent of the leaves of higher plants is associated with the chloroplast, especially in the lamellar membranes, suggesting a close association with photosystem I activity. Sulfolipid roles may be summarized for boundary lipid, chlorophyll binding and morphology in chloroplast, and for electron transport and over all process in photosynthesis (Harwood 1980). These strongly suggest that sulfur is involved in chlorophyll synthesis. Ferredoxin is classified into iron–sulfur enzymes, which have no group but characteristically contain equal numbers of iron and sulfur atoms in a special labile form decomposed by acid. Iron–sulfur enzymes function in electron transferring reaction in plant, animal and bacterial cells. The nitrogenase complex catalyzes nitrogen fixation. The reducing equivalents of NADPH are transferred to the iron–sulfur protein ferredoxin. Ferredoxin is the immediate donor of reducing equivalents for reduction of nitrogen. The role of sulfur in TCA cycle and glycolysis is also proved.

Sulfur contributes to lipid synthesis in two ways, one as a constituent in sulfolipid and the other is to help lipid synthesis. A number of other sulfur-containing lipids such as a ceramide sulphonic acid, a sterol sulfur ester, phosphatidylsulfocholine, and chlorosulfolipids are found in simpler organisms (Harwood 1980). S was identified as important compounds in the synthesis of flavonoids and carotenoids. Also, S was identified as important components in the synthesis of polyamines. Biochemical functions of naturally occurring polyamines are the regulation of physical and chemical properties of membrane, nucleic acid structure–function regulation, modulation of enzyme activities, and regulation of molecular synthesis. Physiological functions of polyamines are regulation of growth and differentiation, retardation of senescence and hormone-mediated responses (Slocum et al. 1984; Smith 1985). Glucosinolates have the general formula (I) and can be degraded by myrosinase to give D-glucose sulfate and isothiocyanates (II), organic cyanide (III) plus sulfur, of thiocyanate (IV), and other minor products such as nitriles. Glucosinolates are nonvolatile and can therefore be recognized by insects only on contact, whereas isothiocyanates are generally volatile and can induce olfactory responses (Larsen 1981).

Isothiocyanates also have antibacterial and antifungal properties. *Brassica* species inhibit the invasion of mildew pathogen *Peronospora parasitica* by allylisothiocyanate formed enzymatically from sinigrin following tissue damage (Bell 1981). There is a correlation between isothiocyanate content and disease resistance in cultivated *Brassica oleracea*. Thus allylisothiocyanate content was 630 in a resistant variety and between 21 and 450 μg g dry weight$^{-1}$ in susceptible varieties. Allylisothiocyanate is not only a repellent but also can actually be toxic to insects. When the larvae of the black swallowtail butterfly are fed celery, one of its normal food plants into which a 0.1% solution (based on leaf fresh weight) of sinigrin is infiltrated, the mortality is 100%. This indicated that sinigrin in cabbage plants has a clear defensive function against insects not normally feeding on *Brassicaceae* because it contains about 0.1% of sinigrin (Harborne 1982). Proline and glycine betaine are the two most widespread compounds in adaptation to the saline condition. Certain plants may replace these with two sulfonium derivatives, S-methylsulfonium propanoic acid and S-dimethylsulfonium pentanoic acid, which have a role of cytoplasmic osmoticum in higher plants. The effect of sulfur on other mineral nutrient uptake is not fully understood. In pea plants the lower level of S nutrition decreased the N, Si, Fe, Ca, Mg, S, and crude ash content but increased the P and Al contents of both shoot and root tissues (Bugakova et al. 1981). Between S and Zn application of one of the two nutrients depressed the concentration of the other nutrient in shoots. S–Zn interaction perhaps occurred both at absorption sites and within plants (Shukla and Prasad 1979).

Allylisothiocyanate is the acrid, sharp-tasting principle of mustard and man takes it in small amounts as table condiments. The isothiocyanate is also the major flavor principle of cabbage and other vegetable crucifers (Harborne 1982). The local horse radish types (wild and cultivated) were generally more pungent than the imported ones. Rising sulfur rates increased root allylisothiocyanate content and augmented root pungency. The garlic plants showed significant positive correlation ($r = 0.723, n = 15, p = 0.01$) between allylsulfide content and the available sulfur content in soils in Korea (Lee and Ham 1986). Allylthiocyanate value of seed oil of Raya (*Brassica juncea*) is increased by sulfur application (Singh and Singh 1977).

## 5.5 Crop Response to Sulfur

Legumes usually require almost equal amounts of phosphorus and sulfur. When P and S are present below the critical level in soil, the plant growth and qualities of produce are affected adversely (Dubey and Mishra 1970). Dhillion and Dev (1978) indicated that soybean is quite responsive to sulfur application and it has a high sulfur requirement due to higher quantities of protein and S-containing amino acids (methionine, cysteine, and cystine) thus vital for protein synthesis. Application of elemental sulfur increased the dry matter, nodule biomass, pod yield, shoot length; reduced the chlorosis of groundnut (Hago and Salama 1987); and increased the oil content by 5% over control (Dimkee et al. 1997). Higher levels of sulfur favorably increased the transformation of carbohydrates in oil that increased the oil content and yield by hydrolyzing more glucosides (Rathore and Manohar 1989).

Application of 50 kg S $ha^{-1}$ increased the oil content by 3.2% over control in sesame (Chaplot et al. 1991). Sulfur application increased the seed yield and oil content of mustard (Mohan and Sharma 1992). Application of sulfur at the rate of 30 kg $ha^{-1}$ increased the leaf area, dry matter, number of pods and 100 seed weight in soybean (Singh and Singh 1995). Application of 30 kg S $ha^{-1}$ increased the wheat yield by 2.19 q $ha^{-1}$ (Raghuwanshi et al. 1997) and increased the uptake of macronutrients like N, P, and K. Application of 30 kg $ha^{-1}$ of sulfur significantly increased the yield contributing characters and yield of blackgram (Singh and Aggarwal 1998). Legha and Giri (1999) reported that application of sulfur at 30 kg $ha^{-1}$ significantly increased the number of seeds and seed weight per capitulum resulting in significantly higher seed yield in sunflower. Riley et al. (2000) reported that application of micronised sulfur increased wheat yield and sulfur uptake by 36 and 164%, respectively. Histuda et al. (2005) observed early plant growth promotion of rice, maize, field bean, wheat, cotton, sorghum, and sunflower due to sulfur application.

## 5.6 Deficiency Symptoms

Sulfur, as a constituent of nitrate reductase is involved in the conversion of nitrate into organic nitrogen. Sulfur deficiency consequently interferes with nitrogen metabolism, which explains why sulfur deficiency resembles nitrogen deficiency in many crops. However, the symptoms usually are not dramatic and are not localized on the older leaves. Lack of sulfur appears as light green coloring of the whole plant. Legumes, especially alfalfa, have a high sulfur requirement, so deficiencies usually appear on these crops first. Sulfur deficiency in corn sometimes mimics the other deficiencies such as manganese or magnesium in that it causes interveinal chlorosis; the upper leaves tend to be stripped, with the veins remaining a darker green than the area between the veins. Sulfur containing fertilizers are normally recommended to alleviate sulfur deficiency of the crops (Table 5.2) (Kleinhenz 1999).

## 5.7 Interaction Between S and Other Nutrients in Soil

Availability of anions such as $SO_4^{2-}$ in soil solution depends on the existence of equivalent amounts of counter cation such as $Ca^{2+}$, $Mg^{2+}$, $Na,^{+}$ and $K^{+}$. Therefore, availability of sulfate also depends on the concentration of cations in the soil solution. In the oxidation of reduced S species to $SO_4^{2-}$ hydrogen ions are produced that can release other cations by exchange from soil colloids. These cations may play a role in balancing $SO_4^{2-}$ in soil solution. However, cations like $Ca^{2+}$ may obscure availability of $SO_4^{2-}$ by the formation of insoluble species such as $CaSO_4$. The addition of $CaCO_3$ can lead to an increase in soluble $SO_4^{2-}$.

**Table 5.2** Fertilizer sources of sulfur

| Name of fertilizer | Chemical formula | Analysis $N–P_2O_5–K_2O$ (%) | Sulfur (%) |
|---|---|---|---|
| Highly soluble | | | |
| Ammonium sulfate | $(NH_4)_2SO_4$ | 21–0–0 | 24 |
| Ammonium thiosulfate (60% aqueous solution) | $(NH_4)_2S_2O_3$ | 12–0–0 | 26 |
| Magnesium sulfate | $MgSO_4·7H_2O$ (Epsom salt) | 0–0–0 | 14 |
| Superphosphate | $Ca(H_2PO_4)_2 + CaSO_4$ | 0–20–0 | 14 |
| Potassium magnesium sulfate | $K_2SO_4·2MgSO_4$ | 0–0–22 | 23 |
| Potassium sulfate | $K_2SO_4$ | 0–0–50 | 18 |
| Least soluble | | | |
| Calcium sulfate | $CaSO_4·2H_2O$ | 0–0–0 | 17 |
| Insoluble | | | |
| Elemental sulfur | S | 0–0–0 | 88–98 |

This may be due to the release of adsorbed $SO_4^{2-}$ by the increase in soil pH (Stevenson 1986).

Several authors have studied interactions between sulfate and phosphorus. This interrelationship may be due to competition for anion adsorption sites in soil. The adsorption strength of $PO_4^{3-}$ is expected to be higher than for $SO_4^{2-}$. Dressings of fertilizer P may therefore result in desorption of $SO_4^{2-}$ since $PO_4^{3-}$ substitutes sulfate on the adsorption sites (Abd-Elfattah et al. 1990; Pasricha and Aulakh 1990). This may increase availability of S to plants but makes S at the same time more vulnerable to leaching. Heavy application of sulfur-free P fertilizer can result in heavy leaching of S into subsoil (Vonuexkull 1986). Adetuni (1992) studied some soils and warned of S deficiencies following the application of phosphate.

Some authors have mentioned the effect of lime (CaO) on availability of $SO_4^{2-}$. The principle effect of lime is to react with $H_2O$ to form calcium hydroxide, which neutralizes the effect of lime in the soil solution. Adsorbed $SO_4^{2-}$ is released to the soil liquid phase and therefore made available. This may be desired to neutralize acid soil. However, in high rainfall tropical climate released sulfate will be subjected to rapid loss through leaching (Adetuni 1992). Application of elemental sulfur to alkaline soil usually increases the availability of other nutrients. This is due to the acidifying effect of oxidation of S to sulfuric acid. By lowering the pH, S can increase availability of P on high pH, calcareous soil (Vonuexkull 1986). Availability of micronutrients such as Fe, Zn, and Mn is augmented upon the acidifying effect of elemental sulfur in high pH soils (Abdel-Samad et al. 1990; Zhu and Alva 1993). The effect of S oxidation on the availability of these nutrients is to lower oxidation-reduction potential of the soil and increases their solubility by reducing them (e.g., $Fe^{3+}$ is reduced to soluble $Fe^{2+}$). Sulfur has synergistic a relationship with nitrogen, phosphorus, potassium, magnesium and zinc and antagonistic interaction with boron and molybdenum (Tiwari 1997).

## 5.8 Sulfur-oxidizing Bacteria

Microorganisms contribute to the biogeochemical cycling of sulfur (Kelly et al. 1997; Friedrich et al. 2005). Various microorganisms (prokaryotes, green sulfur bacteria) have been described which utilize sulfur compounds obligately or facultatively as electron donors and oxidize them to sulfate. Prokaryotes oxidize sulfur belonging to *Alpha-*, *Beta-*, and *Gammaproteobacteria* subclass. Prokaryotes can be divided into various physiological groups such as chemolithotrophs, photolithotrophs, mixotrophs, photoheterotrophs, and heterotrophs. Each of these groups includes some geomicrobially important organisms. Chemolithotrophs (chemosynthetic autotrophs) includes members of both the bacteria and the Archaea. They are the microorganisms that derive energy for doing metabolic work from the oxidation of inorganic compounds and that assimilate carbon as $CO_2$, $HCO_3^-$, or $CO_3^{2-}$ (Wood 1988). Photolithotrophs (photosynthetic autotrophs) include a variety of the bacteria but no known Archaea. They are the microorganisms that derive the energy for doing metabolic work by converting radiant energy from the sun into chemical energy and that assimilate carbon as $CO_2$, $HCO_3^-$ or $CO_3^{2-}$ (photosynthesis). Some of these microbes are anoxygenic (do not produce oxygen from photosynthesis), whereas others are oxygenic (produce oxygen from photosynthesis). Photoheterotrophs include mostly bacteria but also few Archaea (extreme halophilies). They derive all or a part of their energy from sunlight but derive carbon by assimilating organic carbon. Heterotrophs include members of both the bacteria and Archaea. They derive energy from the oxidation of organic compounds and most or all of their carbon from the assimilation of organic compounds. They may respire (oxidize their energy source) anaerobically or they ferment energy source by disproportionation. Mixotrophs include some members of bacteria or Archaea. They may derive energy simultaneously from the oxidation of reduced carbon compounds and oxidizible inorganic compounds; either they may derive their carbon simultaneously from organic carbon and $CO_2$ or they may derive energy totally from the oxidation of an inorganic compound but their carbon from organic compounds (Tables 5.3 and 5.4).

Mixotrophic life might be the preferred metabolic traits of sulfur-oxidizing bacteria. Since low concentrations of sulfur compounds may limit the growth, usage of organic carbon for biomass synthesis, or even co-oxidation of sulfur compounds together with organic substrates might ensure better survival and growth of sulfur oxidizing bacteria in various environments (Graff and Stubner 2003). In a mixotrophic medium with organic carbon, protein yields were higher with added thiosulfate. It is indicated that bacteria gains metabolically useful energy from thiosulfate oxidation that in turn facilitates greater heterotrophic carbon assimilation (Padden et al. 1998). Increased cellular yields were observed in marine heterotrophic thiosulfate oxidizing bacterial strains 12W and 16B, *Acidiphilium acidophilum*, *Catenococcus thiocyclus,* and *Citreicella thiooxidans* (Tuttle 1980; Pronk et al. 1990; Sorokin et al. 1996, 2005a) in mixotrophic medium. Fujimura and Kuraishi (1980) and Sorokin et al. (2005b) observed the oxidation of

**Table 5.3** Some of aerobic sulfur oxidizing bacteria (Ehrlich 2002)

| Autotrophic | Mixotrophic | Heterotrophic |
|---|---|---|
| *Thiobacillus thioparus* | *Thiobacillus intermedius* | *Thiomonas perometabolis* (*Thiobacillus perometabolis*) |
| *Halothiobacillus neapolitanus* (*Thiobacillus neapolitanus*) | *Paracoccus versutus* (*Thiobacillus versutus*[a]) | *Beggiatoa* spp. |
| *Thermithiobacillus tepidarius* (*Thiobacillus tepidarius*) | *Thiobacillus organoparus* | |
| *Acidithiobacillus caldus* (*Thiobacillus caldus*) | *Pseudomonas* spp. | |
| *Thiobacillus denitrificans*[b] | | |
| *Starkeya novella* (*Thiobacillus novella*[c]) | | |
| *Thiobacillus thermophilic*[c] | | |
| *Sulfobacillus thermosulfidooxidans*[c] | | |
| *Acidithiobacillus thiooxidans* (*Thiobacillus thiooxidans*) | | |
| *Acidithiobacillus albertensis* (*Thiobacillus albertis*) | | |
| *Acidithiobacillus ferrooxidans* (*Thiobacillus ferrooxidans*) | | |
| *Beggiatoa alba* | | |
| *Sulfobacillus acidocaldarius* | | |
| *Acidianus brierleyi*[d] | | |
| *Thermothrix thiopara*[c] | | |

Names in parenthesis are old name of respective bacteria
[a]Can also grow autotrophically and heterotrophically
[b]Facultative anaerobe[c]Facultative autotroph[d]Archea

**Table 5.4** Some of anaerobic sulfur oxidizing bacteria (Ehrlich 2002)

| Photolithotrophs | Chemolithotrophs |
|---|---|
| *Chromatium* spp. | *Thermothrix thiopara*[a,b] |
| *Chlorobium* spp. | *Thiobacillus denitrificans*[b] |
| *Ectothiorhodospira* spp. | |
| *Rhodopseudomonas* spp.[a] | |
| *Chloroflexus aurantiacus*[b] | |
| *Oscillatoria* sp.[b] | |
| *Lyngbya* spp.[b] | |
| *Aphanothece*[b] | |
| *Microcoleus* spp.[b] | |
| *Phormidium* spp.[b] | |

[a]Facultative autotroph
[b]Facultative anaerobe

thiosulfate and enhanced growth in *Starkeya novella* and *Thioclava pacifica* grown in mixotrophic medium containing glucose/glutamate/acetate and thiosulfate. Although several researchers reported that bacteria grown in medium containing glucose or organic matter with thiosulfate, in which organic matter affected the thiosulfate oxidation in *T. intermedius* and *Thiovirga sulfuroxydans* (London and Rittenberg 1966; Ito et al. 2005), on the other hand, thiosulfate inhibited the uptake of glucose by affecting the enzymes of glucose metabolic pathway, i.e., Entner–Doudoroff pathway and pentose phosphate pathway in *T. intermedius*, *S. novella,* and *Paracoccus versutus* (previously *Thiobacillus* A2) (Romano et al. 1975; Matin et al. 1980; Wood and Kelly 1980). Recently, Anandham et al. (2007c, 2009a, c) identified mixotrophic metabolism with thiosulfate/succinate or acetate in *Methylobacterium oryzae*, *Methylobacterium fujisawaense*, *Methylobacterium goesingense,* and *Burkholderia kururiensis* subsp. *thiooxydans*.

## 5.9 Isolation and Characterization of Sulfur-oxidizing Bacteria

Beijerinck (1904) isolated thiosulfate-oxidizing organisms (*Thiobacillus thioparus*) from fresh water canal mud and salt water. He also reported that *T. thioparus* produced a thick pellicle consisting of sulfur enclosing bacterial bodies. *T. thioparus* was isolated from soil, ditchwater, sewage, and sea water and its oxidation of sulfur and sulfide has been studied by various authors (Joffe 1922; Brown 1923). Waksman and Joffe (1922) isolated an organism, which was able to oxidize sulfur very rapidly to sulfuric acid. It was a small colorless, nonthread-forming organism using primarily elemental sulfur as a source of energy, not accumulating any sulfur within or outside its cells. Starkey (1935) isolated sulfur-oxidizing bacteria from black clay loams. He also reported that the characteristics of growth were diverse; in some cases, the medium was turbid. In others, sulfur precipitated on the liquid surface or upon the wall of the flasks and in still other cultures, thin membranes, free from sulfur, appeared on the surface. Combinations of these effects were also noted. As growth progressed the reactions of the media became somewhat acidic in all cases, although some variation among the cultures was noticed. Transfers were made to fresh media, in which the thiosulfate was again rapidly oxidized.

Alkali soils are poor in sulfur-oxidizing bacteria. By appropriate enrichment, two bacterial strains resembling *A. thiooxidans* and *S. novella* have been isolated from alkali soils (Rupela and Tauro 1973). From the Galapagos hydrothermal vent, obligate heterotrophic sulfur oxidizers were repeatedly isolated that presumably oxidized thiosulfate either to sulfate (acid producing *Thiobacillus* like) or to polythionates (base producing *Pseudomonas*) (Ruby et al. 1981). Kelly and Harrison (1988) isolated 'S'-oxidizing bacteria from its natural habitat by the use of mineral media containing elemental sulfur or thiosulfate as an energy source. The use of media at different pH assists differential selection of the neutrophilic and acidophilic species. The use of acid ferrous sulfate medium frequently selects for *Acidithiobacillus ferrooxidans* and the use of anaerobic thiosulfate medium (pH 7) supplemented with nitrate will be electively successful in selection for *T. denitrificans*, *Acidiphilium acidophilum* (previously *Thiobacillus acidophilus*) originally isolated as a commensal of *A. ferrooxidans*. Dave and Upadhyay (1993) used an improved medium using gelrite as an operating gel for isolation and enumeration of acidophilic chemolithotrophic *Thiobacillus* strains. Dark brown circular colonies, which were robust and well differentiated, developed on this medium within 48h. *Sulfolobus* and *Thermus*-like strains and thiosulfate oxidizing microorganisms were isolated from spring waters. Most suitable conditions for thiosulfate oxidation were pH around 4–6, temperature of 40°C and yeast extract concentration of 60 mg 100 $ml^{-1}$. Shooner and Tyagi (1995) isolated a moderate thermophilic organism *Thiobacillus thermosulfatus* from sewage sludge. Takano et al. (1997) isolated *A. thiooxidans* from Yugama Crater Lake of Japan and drew the influence of *A. thiooxidans* on the budget of sulfate. Vlasceanu et al. (1997) isolated *Thiobacillus* sp. from a floating microbial mat located in Movile cave. Thiosulfate oxidizing base producing *Pseudomonas stutzeri*, *Pseudoalteromonas* and *Halomonas deleya* were isolated from marine and hydrothermal vents of New England (Teske et al. 2000). A novel chemolithoheterotrophic member of *Betaproteobacteria* was isolated from a hot spring of Portugal later classified as *Tepidimonas ignava,* which can oxidize the thiosulfate to sulfate (Moreira et al. 2000). Oligonucleotide probes were developed to identify *Thiobacillus* and *Acidiphilium* from environmental samples (Peccia et al. 2000). Ryu et al. (2003) isolated *Acidithiobacillus* from sewage sludge using modified Waksman medium. Ito et al. (2004) isolated aerobic chemolithotrophic sulfur-oxidizing bacterium from wastewater biofilms.

The existence of chemolithotrophic sulfur-oxidizing bacteria capable of growth in an extremely alkaline and saline environment has not been recognized until recently. Extensive studies of saline, alkaline (soda) lakes located in Central Asia, Africa, and North America have now revealed the presence, at relatively high numbers, of a new branch of obligately autotrophs at double extreme environments. All of the isolated strains have the potential to grow optimally at around pH 10 in media strongly buffered with sodium carbonate/bicarbonate but cannot grow at pH $< 7.5$ and $Na^+ < 0.2$ M. The majority of the isolate fell into two distinct groups with differing phylogeny and physiology that have been described as two new genera of *Gammaproteobacteria* viz., *Thioalkalimicrobium* and *Thioalkalivibrio*. The third genus, *Thioalkalispira* contains a single obligate microaerophilic species *Thioalkalispira microaerophila* (Sorokin and Kuenen 2005). Sulfur-oxidizing bacteria such as *Bosea thiooxidans*, *Pseudaminobacter salicylatoxidans*, *Paracoccus bengalensis*, *Mesorhizobium thiogangeticum*, *Tetrathiobacter kashmirensis*, *Paracoccus pantotrophus,* and *Paracoccus thiocyanatus* were isolated from rhizosphere and bulk soils of agricultural fields of India (Das et al. 1996; Deb et al. 2004; Ghosh and Roy 2006a, b, 2007; Ghosh et al. 2005, 2006). Presence of *T. thioparus* was reported in Canadian and Scottish agricultural soils (Germida et al. 1985; Chapman 1990). Also, thiosulfate-oxidizing *Xanthobacter tagetidis* and *Methylobacterium thiocyanatum* were isolated from rhizosphere of marigold and Persian onion, respectively (Padden et al. 1997; Wood et al. 1998). Recently, Anandham et al. (2005, 2007a, 2008a, b, 2009b, c, 2010) have isolated several obligate and facultative chemolithotrophic thiosulfate-oxidizing bacteria in rhizosphere soils and documented their ubiquitous presence in rhizosphere of crop plants of Korea (Yim et al. 2008).

## 5.10 Role of Sulfur Bacteria in Sulfur Nutrition

Most of the sulfur in soil environments (>95% of total sulfur) is bound to organic molecules, and therefore not directly available to plants. Use of sulfur-oxidizing bacteria enhances the rate of natural oxidation of sulfur and speeds up the production of sulfates. This process makes sulfur more available to plants at their critical stages and consequently results in an increased plant yield. Recently the use of sulfur-oxidizing bacteria as plant growth promoting bacteria is gaining momentum.

### *5.10.1 Sulfur Oxidation in Soils*

The soil-incubation experiments were performed to assess the sulfur oxidation in Korean soils. The physicochemical properties of the selected clay, silty clay, and sandy loam soils were as follows: sand 21.7, 2.2, 73.4%; silt 32.4, 46.60, 13.6%; clay 45.9, 51.20, 13.0%; pH 7.3, 6.0, 6.3; EC 327, 206, 133 dS $m^{-1}$; organic matter 3.4, 1.63, 1.8%; total nitrogen 0.24, 0.03, 0.09%, total phosphorous 132.2, 284.0, 295.7 mg $kg^{-1}$; total $SO_4$–S 18, 20, 12 μg $g^{-1}$, respectively. In the clay and sandy loam soils, inoculation of *Dyella ginsengisoli* accumulated the maximum sulfate–sulfur (1,927 and 2,527 μg $SO_4$–S $g^{-1}$ soil, respectively) when compared to the other strains on day 30. Meanwhile, in the silty clay soil, *Microbacterium phyllosphaerae* registered the highest sulfate–sulfur content, followed by *D. ginsengisoli*. The sulfur oxidation rate was higher in the sandy loam, followed by the clay soil, however, the sulfur oxidation activity of the thiosulfate-oxidizing bacteria was severely inhibited in the silty clay soil (Fig. 5.2). The sulfur oxidation and sulfate accumulation in the tested soils were accompanied by a decrease in the soil pH (Fig. 5.2) (Anandham and Sa unpublished data). The effect of sulfur inoculated with *Acidithiobacillus* oxidative bacteria was more efficient than gypsum in soil pH reduction. Gypsum or sulfur applied individually was not satisfactory for soil reclamation. Exchangeable cations were leached in highest amounts after 15 days of incubation (Stamford et al. 2002, 2007a, 2008a).

### *5.10.2 Bioacidulation of Rock Phosphate*

Phosphorus and S interact on growth of a variety of legumes when they are grown in soils deficient in both nutrients. In the recent years the possibility of practical use of rock phosphate (RP) as fertilizer has received significant interest (Anandham et al.

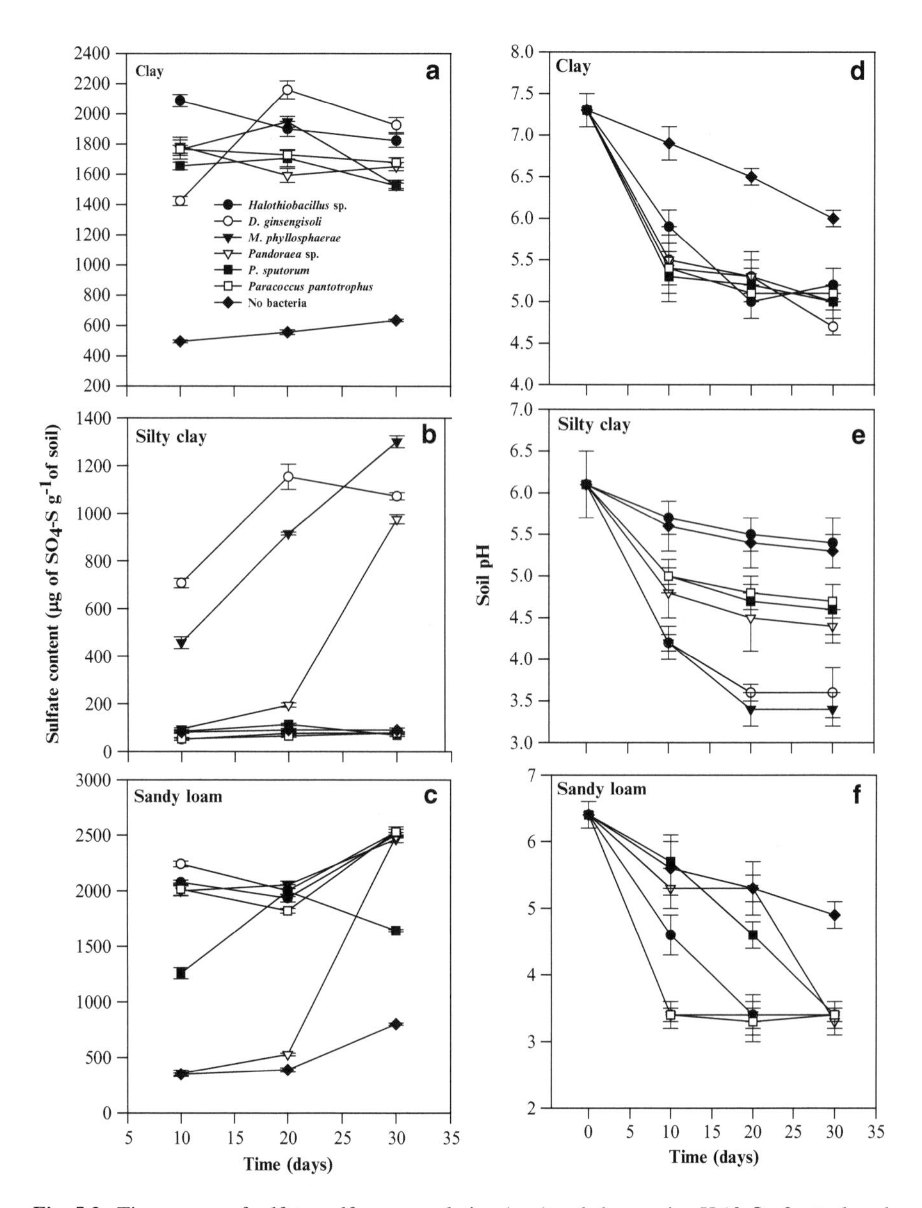


**Fig. 5.2** Time course of sulfate–sulfur accumulation (**a–c**) and changes in pH (**d–f**) of autoclaved soil amended with elemental sulfur and inoculated with thiosulfate oxidizing bacteria. *Error bars* are standard deviation of treatment means and significantly different at $P \leq 0.05$ when compared by LSD. Treatment means are the average of three replications (Anandham and Tong Min Sa unpublished data)

2007b). RP is theoretically the cheapest P fertilizer but most phosphate rock deposits found in the world are classified as low reactive therefore, its direct application is not always effective without previous treatment. Microbiological

**Table 5.5** Changes in water extractable-P of RP incubated with thiosulfate and inoculated with thiosulfate oxidizing bacteria (Anandham et al. 2008a)

| | Water extractable-P (μg P g $RP^{-1}$) | | |
|---|---|---|---|
| | Day of incubation | | |
| Bacterial strains/incubation conditions | 15 | 30 | 45 |
| No glucose | | | |
| *Halothiobacillus* sp. (ATSB2)[a] | 125.90 ± 5f | 149.25 ± 10cd | 1,166 ± 66a |
| *D. ginsengisoli* (ATSB10)[b] | 85.43 ± 3g | 133.26 ± 3d | 950.96 ± 25h |
| *M. phyllosphaerae* (ATSB31)[b] | 58.45 ± 4h | 149.25 ± 7c | 1,076 ± 75def |
| *Pandoraea* sp. (ATSB30)[b] | 134.89 ± 7f | 197.23 ± 7cd | 1,147 ± 47bc |
| *P. sputorum* (ATSB28)[b] | 53.96 ± 2h | 149.25 ± 7cd | 1,087 ± 85cdef |
| 0.5% Glucose | | | |
| *Halothiobacillus* sp. (ATSB2) | 175.91 ± 10c | 355.22 ± 5b | 1,038 ± 35fg |
| *D. ginsengisoli* (ATSB10) | 218.55 ± 9a | 242.81 ± 4c | 808.10 ± 20i |
| *M. phyllosphaerae* (ATSB31) | 157.37 ± 7de | 170.58 ± 5cd | 1,061 ± 40ef |
| *Pandoraea* sp. (ATSB30) | 191.90 ± 5b | 454.14 ± 10b | 1,055 ± 45fg |
| *P. sputorum* (ATSB28) | 184.35 ± 4c | 293.18 ± 3b | 1,219 ± 25a |
| 1.0% Glucose | | | |
| *Halothiobacillus* sp. (ATSB2) | 154.58 ± 10cd | 571.04 ± 20a | 1,135 ± 35bcd |
| *D. ginsengisoli* (ATSB10) | 154.58 ± 4d | 224.82 ± 10cd | 1,048 ± 40fg |
| *M. phyllosphaerae* (ATSB31) | 165.25 ± 5cd | 188.85 ± 3cd | 986.14 ± 30ag |
| *Pandoraea* sp. (ATSB30) | 138.59 ± 8cd | 152.88 ± 7cd | 1,094 ± 27cdef |
| *P. sputorum* (ATSB28) | 154.58 ± 7cd | 188.85 ± 10cd | 1,128 ± 43bcde |
| LSD ($P \leq 0.05$) | 13.70 | 101.85 | 70.12 |

Each value represents mean ± SD of four replications per treatment. The data are statistically analyzed using DMRT. In the same column significant differences according to LSD ($P \leq 0.05$) levels are indicated by different letter(s)

[a]Obligate chemolithoautotroph

[b]Facultative chemolithoautotroph

approaches have been proposed to improve the agronomic value of RP. Bacterial inoculation significantly enhanced the RP solubilization. On day 15, maximum water extractable-P was noted on RP inoculated with *D. ginsengisoli* amended with 0.5% glucose. On day 45 the highest water extractable-P (1,219 μg P g $RP^{-1}$) was recorded in RP inoculated with *P. sputorum* amended with 0.5% glucose which was on par with RP inoculated with *Halothiobacillus* sp. unamended with glucose (Table 5.5). Similarly on day 45, RP inoculated with *Pandoraea* sp. amended with 1.0% glucose registered the highest bicarbonate extractable-P (Table 5.6). RP solubilization was always accompanied with reduction in pH of the incubation mixture. The lowest pH (3.6) was observed on day 45 in *D. ginsengisoli* and *Halothiobacillus* sp.-inoculated treatments. Bicarbonate extractable-P was significantly higher than the water extractable-P. Glucose amendment significantly enhanced the water and bicarbonate-extractable-P. Irrespective of incubation time and amendment of the glucose, *Halothiobacillus* sp. inoculation significantly enhanced the water and bicarbonate extractable-P (541 and 568 μg P g $RP^{-1}$) (Anandham et al. 2008a). Species of *Acidithiobacillus* react with elemental sulfur-producing sulfuric acid and may increase available P in soil by promoting higher

**Table 5.6** Changes in bicarbonate extractable-P of RP incubated with thiosulfate and inoculated with thiosulfate oxidizing bacteria (Anandham et al. 2008a)

| | Bicarbonate extractable-P (μg P g $RP^{-1}$) | | |
|---|---|---|---|
| | Day of incubation | | |
| Bacterial strains/incubation condition | 15 | 30 | 45 |
| No glucose | | | |
| *Halothiobacillus* sp. (ATSB2) | 213.22 ± 5c | 229.32 ± 9f | 1,203 ± 25c |
| *D. ginsengisoli* (ATSB10) | 139.39 ± 6i | 186.57 ± 3hi | 1,045 ± 35f |
| *M. phyllosphaerae* (ATSB31) | 181.24 ± 2de | 206.83 ± 5g | 1,064 ± 27f |
| *Pandoraea* sp. (ATSB30) | 166.37 ± 6g | 181.24 ± 2i | 1,145 ± 23d |
| *P. sputorum* (ATSB28) | 157.37 ± 4h | 191.9 ± 4h | 1,321 ± 20b |
| 0.5% Glucose | | | |
| *Halothiobacillus* sp. (ATSB2) | 223.88 ± 3b | 341.73 ± 20b | 998.93 ± 13g |
| *D. ginsengisoli* (ATSB10) | 181.24 ± 6de | 184.35 ± 4hi | 995.74 ± 17g |
| *M. phyllosphaerae* (ATSB31) | 184.35 ± 4d | 191.90 ± 2h | 1,113 ± 32de |
| *Pandoraea* sp. (ATSB30) | 135.97 ± 4i | 170.58 ± 5j | 1,106 ± 19e |
| *P. sputorum* (ATSB28) | 207.89 ± 7c | 310.32 ± 4c | 1,038 ± 23f |
| 1.0% Glucose | | | |
| *Halothiobacillus* sp. (ATSB2) | 175.91 ± 2ef | 395.68 ± 20a | 1,326 ± 12b |
| *D. ginsengisoli* (ATSB10) | 170.58 ± 5gf | 247.30 ± 12e | 1,001 ± 19g |
| *M. phyllosphaerae* (ATSB31) | 165.25 ± 9g | 337.23 ± 6b | 1,213 ± 25c |
| *Pandoraea* sp. (ATSB30) | 165.25 ± 5g | 232.01 ± 8f | 1,374 ± 22a |
| *P. sputorum* (ATSB28) | 281.47 ± 10a | 298.51 ± 14d | 1,223 ± 15c |
| LSD ($P \leq 0.05$) | 5.82 | 10.07 | 35.19 |

Each value represents mean ± SD of four replications per treatment. The data are statistically analyzed using DMRT. In the same column significant differences according to LSD ($P \leq 0.05$) levels are indicated by different letter(s)

solubility of phosphate rocks, furnishing phosphorus to the symbiotic process and to plant growth (Stamford et al. 2003a, b, 2007b). Few studies have reported the enhancement of sulfur availability, rock phosphate solubilization and plant growth promotion abilities of sulfur oxidizing bacteria of canola, groundnut and yam bean (Grayston and Germida 1991; Anandham et al. 2007a; Stamford et al. 2007b; Ghani et al. 1994).

### 5.10.3 Plant Growth Promotion

Combined application of *Thiobacillus*, elemental sulfur, and phosphobacteria increased the yield in maize (Kabesh et al. 1989). Pathiratna et al. (1989) reported that pelleted apatite mixed with S and *Thiobacillus* increased the shoot dry matter and yield. In a pot culture study, inoculation of sulfur-oxidizing bacteria significantly increased the root and shoots lengths, leaf width, and pod dry weight of canola (Grayston and Germida 1991). Application of biologically activated sulfur with *Thiobacillus* and *Aspergillus awamori* increased the yield and grain filling in

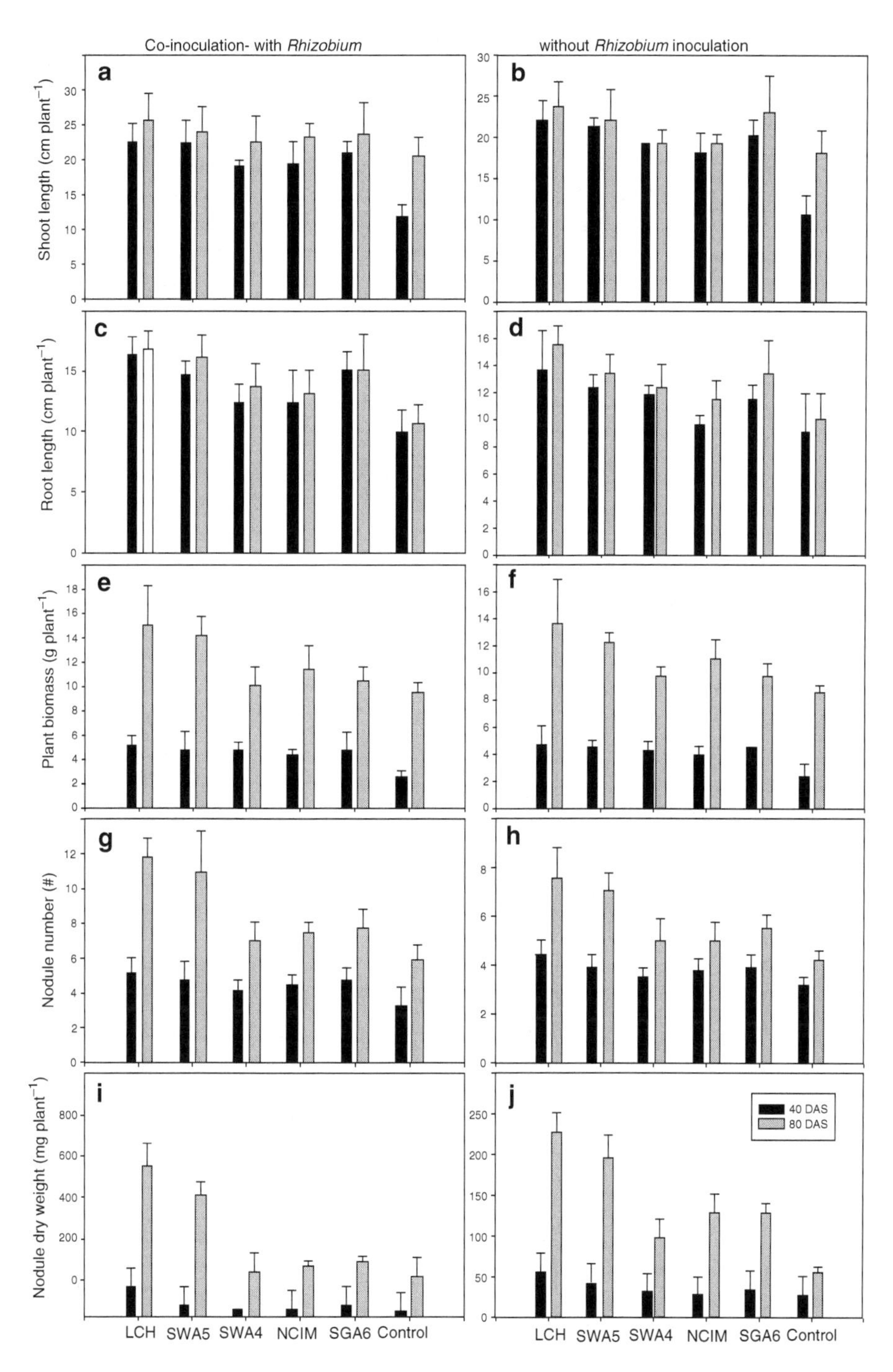


**Fig. 5.3** Effects of co-inoculation of *Thiobacillus* sp. and *Rhizobium* in field trial. *Thiobacillus* sp. inoculated in the form of pellets ($2.5 \times 10^7$ cfu g$^{-1}$ pellet) at 60 g ha$^{-1}$. (**a** and **b**) shoot length,

rice (Muralidharan and Jose 1993). Application of sulfur (300 kg ha$^{-1}$) inoculated with *Thiobacillus* increased the dry matter production of Leucena and decreased the soil salinity, whereas increased concentration of sulfur (600 kg ha$^{-1}$) produced negative results (Stamford et al. 2002). Similarly, application of sulfur with co-inoculation of *Thiobacillus* with *Bradyrhizobium* significantly enhanced the yield of cowpea and yam bean (Stamford et al. 2003a). El-Tarabily et al. (2006) reported that sulfur-oxidizing bacteria isolated from calcareous sandy soils significantly increased the yield and nutrient uptake of maize. Inoculation of *A. thiooxidans*, *T. thioparus*, *Scolecobasidium constrictum*, *Myrothecium* sp., and *Aspergillus terreus* increased the yield 5–40% over control in wheat. In a field trial, *Thiobacillus* sp. strain (LCH) pellets applied at 60 kg ha$^{-1}$ and co-inoculated with *Rhizobium* enhanced the groundnut shoot, root lengths and plant biomass by 43, 27, and 76%, respectively over uninoculated control (Fig. 5.3). In a field trial, *Thiobacillus* sp. applied at 60 kg ha$^{-1}$ significantly increased the groundnut nodule number and nodule dry weight by 32 and 43%, pod yield (2%) and oil content (3%) over pellets applied at 20 kg ha$^{-1}$(Table 5.7) (Anandham et al. 2007a). Similar results were also observed in pot experiment (Table 5.8, Fig. 5.3) (Anandham et al. 2007a). A significant effect on sugarcane stalk dry matter yield from phosphate and potash rocks was observed with the application of sulfur and *Acidithiobacillus* (Stamford et al. 2008a), available P and K and exchangeable Ca and Mg increased with *Acidithiobacillus* application compared to mineral fertilizers and P and K rocks alone. Further, it was concluded that biofertilizers produced from phosphate or potassium rock mixed with sulfur inoculated with *Acidithiobacillus* may be used as an alternative to soluble fertilizers for the fertilization of sugarcane grown in soils with low available P and K (Stamford et al. 2008a). The effectiveness of biofertilizers, produced through mixing powdered rocks and elemental sulfur inoculated with *Acidithiobacillus*, were compared with P and K-soluble fertilizer (triple superphosphate and potassium chloride) and the improved availability of P and K in soil. Further, these biofertilizers increased the cowpea shoot biomass and accumulation of total N, P, and K in shoots, suggesting that they could be important in P and K fertilization, especially in soils with neutral pH due to the acidification effect. In contrast it is necessary to take care when using the biofertilizers in acidic soils (Stamford et al. 2008b). In a pot culture study, the inoculation of *D. ginsengisoli* and *M. phyllosphaerae* significantly enhanced the maize root length (73 and 67%, respectively), shoot length (27 and 31%), and shoot biomass (58 and 45%), along with the nutrient uptake of P, K, S, Mn, Ca, Cu, and Na when compared to the uninoculated control (Anandham and Ton Min Sa unpublished data).

---

**Fig. 5.3** (continued) (**c** and **d**) root length, (**e** and **f**) plant biomass, (**g** and **h**) nodule number (figure drawn for square root of $x + 0.5$ transformed values) and (**i** and **j**) nodule dry weight with and without *Rhizobium* co-inoculation, respectively. Error bars are standard error of treatment means and significantly different at $P \leq 0.05$ when compared by critical difference. Treatment means are the average of three replications (Anandham et al. 2007a)

**Table 5.7** The main effects of inoculation of *Rhizobium*, *Thiobacillus* sp. and *Thiobacillus* sp. pellets at three levels on plant growth, nodulation, yield and oil content of groundnut under field conditions (Anandham et al. 2007a)

| | Shoot length (cm) | | Root length (cm) | | Plant biomass (g plant$^{-1}$) | | Nodule number (# plant$^{-1}$) | | Nodule dry weight (mg plant$^{-1}$) | | | |
|---|---|---|---|---|---|---|---|---|---|---|---|---|
| Treatment | 40 DAS | 80 DAS | 40 DAS | 80 DAS | 40 DAS | 80 DAS | 40 DAS | 80 DAS | 40 DAS | 80 DAS | Yield (kg ha$^{-1}$) | Oil content (%) |
| *Rhizobium* (R) | | | | | | | | | | | | |
| R0 | 17.27 | 19.38 | 10.50 | 11.60 | 3.71 | 10.23 | 12.92 (3.65) | 27.53 (5.19) | 29.0 | 116.0 | 1,830.7 | 46.96 |
| R1 | 18.25 | 22.28 | 12.56 | 12.86 | 3.89 | 10.98 | 16.57 (4.10) | 57.18 (7.40) | 43.0 | 278.0 | 1,876.5 | 47.83 |
| CD (0.05) | 0.79 | 2.06 | 1.20 | 0.41 | 0.07 | 0.08 | 0.09 | 0.33 | 9.0 | 40.0 | 10.5 | 0.08 |
| *Thiobacillus* sp. strains | | | | | | | | | | | | |
| LCH | 19.52 | 22.36 | 13.14 | 13.72 | 4.17 | 12.75 | 17.44 (4.20) | 59.17 (7.41) | 55.0 | 278.0 | 1,929.4 | 49.57 |
| SWA5 | 20.11 | 21.34 | 12.57 | 12.20 | 4.22 | 11.78 | 16.41 (4.09) | 65.22 (7.87) | 39.0 | 302.0 | 1,921.1 | 50.00 |
| SWA4 | 18.30 | 20.21 | 11.52 | 12.35 | 3.99 | 9.68 | 13.33 (3.70) | 32.36 (5.36) | 31.0 | 145.0 | 1,847.8 | 47.68 |
| NCIM2426[a] | 18.00 | 20.42 | 10.06 | 11.63 | 3.79 | 10.83 | 15.57 (15.6) | 35.91 (5.88) | 31.0 | 166.0 | 1,847.8 | 45.57 |
| SGA6 | 19.32 | 21.62 | 12.57 | 13.37 | 4.19 | 9.69 | 16.13 (4.07) | 35.38 (5.91) | 34.0 | 173.0 | 1,847.2 | 46.89 |
| None | 11.34 | 19.05 | 9.34 | 10.08 | 2.44 | 8.88 | 9.60 (3.18) | 26.09 (5.08) | 24.0 | 119.0 | 1,728.3 | 44.65 |
| CD (0.05) | 2.50 | 1.00 | 0.080 | 1.12 | 0.20 | 0.40 | 0.10 | 0.34 | 5.0 | 80.0 | 12.4 | 3.13 |
| *Thiobacillus* pellet inoculum (kg ha$^{-1}$) | | | | | | | | | | | | |
| 20 | 16.75 | 20.19 | 16.75 | 11.60 | 3.55 | 10.22 | 13.19 (3.69) | 37.66 (6.07) | 27.0 | 172.0 | 1,833.9 | 46.86 |
| 40 | 17.43 | 20.38 | 17.43 | 12.00 | 3.69 | 10.40 | 14.41 (3.84) | 38.52 (6.04) | 31.0 | 174.0 | 1,852.8 | 47.28 |
| 60 | 19.04 | 21.99 | 19.04 | 13.14 | 4.15 | 11.19 | 16.64 (4.10) | 50.89 (6.78) | 48.0 | 246.0 | 1,874.2 | 48.04 |
| CD (0.05) | 0.60 | 0.68 | 0.56 | 0.52 | 0.14 | 0.28 | 0.07 | 0.19 | 4.0 | 10.0 | 8.7 | 0.64 |

Values in parenthesis are square root of $x + 0.5$ transformed *R0* Without *Rhizobium*, *R1* With *Rhizobium*, *DAS* Days after sowing, *CD* Critical difference[a]*T. thiooxidans*

**Table 5.8** Effect of interactions of *Rhizobium*, *Thiobacillus* sp. and *Thiobacillus* sp. pellets at three levels on yield and oil content of groundnut (Anandham et al. 2007a)

| | Pot experiment | | | Field experiment | | |
|---|---|---|---|---|---|---|
| *Rhizobium* (R) × *Thiobacillus* | Pellets[a] (g pot$^{-1}$) | Yield (g plant$^{-1}$) | Oil content (%) | Pellets[a] (kg ha$^{-1}$) | Yield (kg ha$^{-1}$) | Oil content (%) |
| | 2 | 5.83 | 47.06 | 20 | 1,833.3 | 46.73 |
| | 4 | 6.86 | 47.53 | 40 | 1,896.7 | 49.96 |
| R0 × LCH | 6 | 7.27 | 50.06 | 60 | 1,940.0 | 52.16 |
| | 2 | 5.38 | 46.72 | 20 | 1,870.0 | 48.37 |
| | 4 | 5.58 | 47.95 | 40 | 1,880.0 | 49.29 |
| R0 × SWA5 | 6 | 6.27 | 48.17 | 60 | 1,916.7 | 51.33 |
| | 2 | 3.50 | 44.42 | 20 | 1,826.7 | 45.12 |
| | 4 | 5.33 | 45.93 | 40 | 1,840.0 | 46.93 |
| R0 × SWA4 | 6 | 5.67 | 47.46 | 60 | 1,843.3 | 48.43 |
| | 2 | 5.30 | 44.34 | 20 | 1,786.7 | 43.30 |
| | 4 | 5.57 | 44.60 | 40 | 1,823.3 | 45.13 |
| R0 × NCIM2426[b] | 6 | 5.73 | 45.06 | 60 | 1,850.0 | 46.47 |
| | 2 | 4.50 | 44.13 | 20 | 1,830.0 | 45.21 |
| | 4 | 4.53 | 44.33 | 40 | 1,830.0 | 46.20 |
| R0 × SGA6 | 6 | 5.20 | 45.67 | 60 | 1,843.3 | 47.91 |
| Absolute control (no | 2 | 3.18 | 43.20 | 20 | 1,680.0 | 43.77 |
| *Rhizobium* and no | 4 | 3.27 | 43.23 | 40 | 1,700.0 | 44.73 |
| *Thiobacillus*) | 6 | 3.25 | 43.28 | 60 | 1,700.1 | 44.11 |
| | 2 | 5.96 | 48.06 | 20 | 1,900.0 | 47.64 |
| | 4 | 7.10 | 48.56 | 40 | 1,950.0 | 48.37 |
| R1 × LCH | 6 | 8.03 | 50.47 | 60 | 2,006.7 | 52.58 |
| | 2 | 5.55 | 48.97 | 20 | 1,886.7 | 49.41 |
| | 4 | 5.68 | 48.97 | 40 | 1,976.7 | 49.27 |
| R1 × SWA5 | 6 | 6.40 | 49.98 | 60 | 1,996.7 | 52.33 |
| | 2 | 3.70 | 46.63 | 20 | 1,843.3 | 48.17 |
| | 4 | 5.43 | 46.93 | 40 | 1,860.0 | 48.33 |
| R1 × SWA4 | 6 | 5.83 | 47.17 | 60 | 1,873.3 | 48.60 |
| | 2 | 5.47 | 46.01 | 20 | 1,856.7 | 45.17 |
| | 4 | 5.74 | 46.30 | 40 | 1,870.0 | 46.17 |
| R1 × NCIM2426[b] | 6 | 5.86 | 46.73 | 60 | 1,900.0 | 47.17 |
| | 2 | 4.67 | 45.33 | 20 | 1,850.0 | 46.39 |
| | 4 | 4.73 | 45.57 | 40 | 1,860.0 | 47.20 |
| R1 × SGA6 | 6 | 5.40 | 47.64 | 60 | 1,870.0 | 48.26 |
| | 2 | 3.30 | 43.40 | 20 | 1,756.7 | 45.06 |
| R1 (*Rhizobium* and no | 4 | 3.33 | 43.43 | 40 | 1,760.0 | 45.13 |
| *Thiobacillus*) | 6 | 3.37 | 43.50 | 60 | 1,760.0 | 45.13 |
| I × D × CD (0.05) | | 0.31 | 1.26 | | 21.4 | 1.56 |
| I × R × CD (0.05) | | NS | NS | | 21.4 | NS |
| I × D × R × CD (0.05) | | NS | NS | | 30.3 | NS |

*R0* Without *Rhizobium*, *R1* With *Rhizobium*, *I Thiobacillus* sp., *D* Level of *Thiobacillus* sp. pellets, *R Rhizobium*, *CD* Critical difference. *Thiobacillus* pellet inoculum *T. thiooxidans*

### 5.10.4 Other Plant Growth Promoting Traits

Sulfur-oxidizing bacteria inhibited the growth of canola pathogen *Rhizoctonia solani* and *Leptosphaeria maculans* (Grayston and Germida 1991). Thiosulfate-oxidizing bacteria isolated from rhizosphere of crop plants possessed different traits related to plant growth promotion (Table 5.9). The ACC-deaminase activity of thiosulfate-oxidizing bacterial isolates ranged from 0.44 in *P. sputorum* ATSB28 to

**Table 5.9** Plant growth promoting traits of thiosulfate oxidizing bacteria (Anandham et al. 2008a)

| Bacterial strains | Growth in Nfb medium[a] | Soluble P ($\mu g\ ml^{-1}$) | Indole production (IAA $\mu g\ ml^{-1}$) | Salicylic acid ($\mu g\ ml^{-1}$) | β-1-3 Glucanase (μg glucose released $min^{-1}$ mg $protein^{-1}$) | Siderophore |
|---|---|---|---|---|---|---|
| *Halothiobacillus* sp. (ATSB2) | ND | ND | ND | ND | ND | ND |
| *Dyella ginsengisoli* (ATSB10) | – | 163.0 ± 10 | – | – | 118.3 ± 12 | – |
| *Burkholderia kururiensis* (ATSB13) | + | 96.2 ± 6 | – | – | – | – |
| *Burkholderia sordidicola* (ATSB16) | + | – | – | – | 178.0 ± 13 | – |
| *Leifsonia shinshuensis* (ATSB20) | + | 137.0 ± 12 | 3.6 ± 0.4 | – | 9.4 ± 1.0 | – |
| *Leifsonia shinshuensis* (ATSB24) | + | 153.1 ± 9 | 5.3 ± 0.2 | – | 4.41 ± 1 | – |
| *Alcaligenes* sp. (ATSB22) | + | – | – | 1.53 ± 0.4 | 4.33 ± 0.8 | – |
| *Pandoraea pnomenusa* (ATSB25) | + | – | – | 1.87 ± 0.4 | 140.3 ± 11 | – |
| *Pandoraea pnomenusa* (ATSB26) | + | – | – | – | 135.9 ± 18 | – |
| *Pandoraea pnomenusa* (ATSB27) | + | – | 3.9 ± 0.5 | – | 170.7 ± 14 | – |
| *Pandoraea sputorum* (ATSB28) | + | – | – | – | 83.3 ± 3 | + |
| *Pandoraea sputorum* (ATSB29) | + | – | – | – | 413.2 ± 27 | + |
| *Pandoraea* sp. (ATSB30) | + | – | – | – | 218.2 ± 20 | – |
| *Pandoraea* sp. (ATSB32) | + | – | – | 2.20 ± 0.1 | 292.0 ± 19 | – |
| *Microbacterium phyllosphaerae* (ATSB31) | + | 181.2 ± 12 | – | – | 6.8 ± 1 | – |

Values are the means of three replications of three experiments ± SD

*ND* Not determined, + Growth, – No growth or absence of particular traits

[a]Bacterial strains showing growth in Nfb medium did not yield positive amplification product with *nif*H specific primers in PCR

53.6 nM of α-keto butyrate formed $min^{-1}$ mg of $protein^{-1}$ in *B. kururiensis*. All the tested isolates listed in Table 5.9 increased the primary root length of canola seedlings between 56 and 166% as compared with controls (Anandham et al. 2008a).

## 5.11 Conclusion

Most geomicrobiologically important microorganisms that oxidize reduced forms of sulfur in relatively large quantities are prokaryotes. They include representatives of the domains Bacteria and Archaea. They comprise aerobes, facultative organisms, and anaerobes and are mostly obligate or facultative autotrophs or mixotrophs. Application of sulfur inoculated with sulfur-oxidizing bacteria proved beneficial in crop production; however, studies are very limited. Till date, none of the studies have addressed the survivability of sulfur oxidizing bacteria in different formulations. In the course of sulfur oxidation, sulfuric acid is produced which could be utilized for partial acidulation of rock phosphate and ultimately leading to an increase in its agronomic efficiency. Thiosulfate-oxidizing bacteria also possessed multiple plant growth promoting characteristics including ACC deaminase activity, which acts as a sink for stress ethylene and increases the plant root length. Two major biochemical pathways for sulfur oxidation have been identified in sulfur oxidizers. The first is the "S4 intermediate" pathway (S4I), which includes the formation and oxidation of tetrathionate or trithionate or polythionate and sulfur from thiosulfate, whereas the second is the "paracoccus sulfur oxidation" (PSO) pathway that directly oxidizes thiosulfate into sulfate (Kelly et al. 1997; Friedrich et al. 2001). In future, sulfur-oxidizing bacteria including various biochemical pathways for sulfur oxidation need to be addressed in different soils in various agroclimatic zones.

**Acknowledgment** This study was supported by the Tamil Nadu Agricultural University, India and Korea Research Foundation.

## References

Abd-Elfattah A, Hilal MH, El-Hahhasha KM, Bakry MD (1990) Amendment of alkaline clay soil by elemental sulfur and its effect on the response of garlic to phosphorous and nitrogen. In: Proceedings Middle East sulfur symposium, Cairo, 12–16 Feb, pp 295–313

Abdel-Samad S, Ismail H, El-Mashhadi HM, Radwan SA (1990) Effects of the interaction between leaching process with sulfur and peat on growth and uptake of nutrients by Barley grown in saline soil. In: Proceedings Middle East sulfur symposium, Cairo, 12–16 Feb, pp 325–337

Adetuni MT (1992) Effect of lime and phosphorus application on sulfate adsorption capacity of south-western Nigerian soils. Indian J Agr Sci 62:150–152

Anandham R, Sridar R, Nalayini P, Poonguzhali S, Madhaiyan M, Indira Gandhi P, Choi KH, Sa TM (2005) Isolation of sulfur oxidizing bacteria from different ecological niches. Korean J Soil Sci Fert 38:180–187

Anandham R, Sridar R, Nalayini P, Poonguzhali S, Madhaiyan M, Sa TM (2007a) Potential for plant growth promotion in groundnut (*Arachis hypogaea* L.) cv. ALR-2 by co-inoculation of sulfur oxidizing bacteria and *Rhizobium*. Microbiol Res 162:139–153

Anandham R, Choi KH, Indira Gandhi P, Yim WJ, Park SJ, Kim KA, Madhaiyan M, Sa TM (2007b) Evaluation of shelf life and rock phosphate solubilization of *Burkholderia* sp. in nutrient amended clay, rice bran and rock phosphate-based granular formulation. World J Microbiol Biotechnol 23:1121–1129

Anandham R, Indira Gandhi P, Madhaiyan M, Kim KA, Yim WJ, Saravanan VS, Chung JB, Sa TM (2007c) Thiosulfate oxidation and mixotrophic growth of *Methylobacterium oryzae*. Can J Microbiol 53:869–876

Anandham R, Indira Gandhi P, Madhaiyan M, Sa TM (2008a) Potential plant growth promoting traits and bioacidulation of rock phosphate by thiosulfate oxidizing bacteria isolated from crop plants. J Basic Microbiol 48:439–447

Anandham R, Indira Gandhi P, Madhaiyan M, Ryu HY, Jee HJ, Sa TM (2008b) Chemolithoautotrophic oxidation of thiosulfate and phylogenetic distribution of sulfur oxidation gene (*soxB*) in rhizobacteria isolated from crop plants. Res Microbiol 159:579–589

Anandham R, Indira Gandhi P, Madhaiyan M, Chung JB, Ryu KY, Jee HJ, Sa TM (2009a) Thiosulfate oxidation, mixotrophic growth of *Methylobacterium goesingense* and *Methylobacterium fujisawaense*. J Microbiol Biotechnol 19:17–22

Anandham R, Indira Gandhi P, Kwon SW, Sa TM, Jee HJ (2009b) Taxonomic characterization of facultative chemolithoautotrophic strains ATSB16 isolated from rhizosphere soils. In: International workshop on microbial sulfur metabolism, Tomar, Portugal, 15–18 Mar, p 151

Anandham R, Indira Gandhi P, Kwon SW, Sa TM, Kim YK, Jee HJ (2009c) Mixotrophic metabolism in *Burkholderia kururiensis* subsp. *thiooxydans* subsp. nov., a facultative chemolithoautotrophic thiosulfate oxidizing bacterium isolated from rhizosphere soil and proposal for classfication of the type strain of *Burkholderia kururiensis* as *Burkholderia kururiensis* subsp. *thiooxydans* subsp. nov. Arch Microbiol 191:885–894

Anandham R, Indira Gandhi P, Kwon SW, Sa TM, Jeon CO, Kim YK, Jee HJ (2010) *Pandoraea thiooxydans* sp. nov., a facultatively chemolithotrophic, thiosulfate-oxidizing bacterium isolated from rhizosphere soils of seasame (*Sesamum indicum* L.). Int J Syst Evol Microbiol 60:21–26

Beijerinck MW (1904) Phenomenes de reduction proguits parles microbes. Arch Sci Exactes et Nat Haarlem Ser 2:131–157

Bell EA (1981) The physiological role(s) of secondary (natural) products. In: Conn EE (ed) The biochemistry of plant secondary plant product. Academic, New York, pp 1–19

Brown HD (1923) Sulfofication in pure and mixed cultures with special reference to sulfate production, hydrogen ion concentration and nitrification. J Am Soc Agron 15:350–382

Bugakova AN, Knorre AF, Lepesheva TM (1981) The effect of sulfur nutrition on the content of certain mineral elements in pea plants. Fiziologiya Rastenii 13:43–46

Chaplot PC, Jain GL, Bansal KN (1991) Effect of phosphorous and sulfur on the oil yield uptake of N, P and S in various seasons. Indian J Trop Agric 9:190–193

Chapman SJ (1990) *Thiobacillus* population in some agricultural soils. Soil Biol Biochem 22:479–482

Das SK, Mishra AK, Tindall BJ, Rainey FA, Stackebrandt E (1996) Oxidation of thiosulfate by a new bacterium *Bosea thiooxidans* gen. nov., sp. nov. analysis of phylogeny based on chemotaxonomy and 16S ribosomal DNA sequencing. Int J Syst Evol Microbiol 46:981–987

Dave SR, Upadhyay NM (1993) Thiosulfate oxidizing organisms from thermal spring. Indian J Microbiol 33:241–244

Deb C, Stackebrandt E, Pradella S, Saha A, Roy P (2004) Phylogenetically diverse new sulfur chemolithotrophs of α-*Proteobacteria* isolated form Indian soils. Curr Microbiol 48:452–455

Dhillion NS, Dev G (1978) Effect of elemental sulphur application on the soybean (*Glycine max* L. Merrill). J Indian Soc Soil Sci 26:55–57
Dimkee SK, Dwivedi N, Hariram K (1997) Effect of sulfur and phosphorous nutrition on yield attributes of groundnut (*Arachis hypogaea* L). Indian J Agron 38:327–328
Dubey SD, Mishra PH (1970) Effect of sulphur deficiency on growth, yield and quality of some of the important leguminous crops. J Indian Soc Soil Sci 4:375–378
Ehrlich HL (2002) Geomicrobiology. Dekker, New York
El-Tarabily KA, Soaud AA, Saleh ME, Matsumoto S (2006) Isolation and characterization of sulfur-oxidising bacteria, including strains of *Rhizobium*, from calcareous sandy soils and their effect on nutrient uptake and growth of maize (*Zea mays*). Aust J Agric Res 57:101–111
Friedrich CG, Rother D, Bardischewsky F, Quentmeier A, Fischer J (2001) Oxidation of inorganic sulfur compounds by bacteria: emergence of a common mechanism? Appl Environ Microbiol 67:2873–2882
Friedrich CG, Bardischewsky F, Rother D, Quentmeier A, Fischer J (2005) Prokaryotic sulfur oxidation. Curr Opin Microbiol 8:253–259
Fujimura YK, Kuraishi H (1980) Characterization of *Thiobacillus novellus* and its thiosulfate oxidation. J Gen Appl Microbiol 26:357–367
Germida JJ, Lawrence JR, Gupta VSSR (1985) Microbial oxidation of sulfur in Saskatchewan soils. In: Terry JW (ed) Proceedings of the international sulfur 84 conference. The Sulfur Development Institute of Canada, Calgary, pp 703–710
Ghani A, Rajan SSS, Lee A (1994) Enhancement of phosphate rock solubility through biological processes. Soil Biol Biochem 26:127–136
Ghosh W, Roy P (2006a) *Mesorhizobium thiogangeticum* sp. nov., novel sulfur-oxidizing chemolithoautotroph from the rhizosphere soil of an Indian tropical leguminous plant. Int J Syst Evol Microbiol 56:91–97
Ghosh W, Roy P (2006b) Ubiquitous presence and activity of sulfur-oxidizing lithoautotrophic microorganisms in the rhizospheres of tropical plants. Curr Sci 91:159–161
Ghosh W, Roy P (2007) Chemolithoautotrophic oxidation of thiosulfate, tetrathionate and thiocyanate by a novel rhizobacterium belonging to the genus *Paracoccus*. FEMS Microbiol Lett 270:124–131
Ghosh W, Bagchi A, Mandal S, Dam B, Roy P (2005) *Tetrathiobacter kashmirensis* gen. nov., sp. nov., a novel mesophilic, neutrophilic, tetrathionate-oxidizing, facultatively chemolithotrophic *betaproteobacterium* isolated from soil from a temperate orchard in Jammu and Kashmir, India. Int J Syst Evol Microbiol 55:1779–1787
Ghosh W, Mandal S, Roy P (2006) *Paracoccus bengalensis* sp. nov., a novel sulfur-oxidizing chemolithoautotroph from the rhizospheric soil of an Indian tropical leguminous plant. Syst Appl Microbiol 29:396–403
Giovanelli J, Mudd SH, and Datko AH (1980) Sulfur amino acids in plants. In: Miflin BJ and PJ Lea (ed) The Biochemistry of Plants, Vol 5. Academic Press, New York, pp. 453–506
Graff A, Stubner S (2003) Isolation and molecular characterization of thiosulfate oxidizing bacteria from an Italian rice field soil. Syst Appl Microbiol 26:445–452
Grayston SJ, Germida JJ (1991) Sulfur oxidizing bacteria as plant growth promoting rhizobacteria for canola. Can J Microbiol 37:521–529
Hago TM, Salama MA (1987) The effect of elemental sulfur on shoot dry-weight, nodulation and pod yield on groundnut under irrigation. Exp Agr 23:93–97
Harborne JB (1982) Introduction to ecological biochemistry. Academic, New York, p 278
Harwood JL (1980) Sulfolipids. In: Stumpt PK (ed) The biochemistry of plants. Lipids structure and function. Academic, New York, pp 301–320
Histuda K, Yamada M, Klepker D (2005) Sulfur requirement of eight crops at early stages of growth. Agron J 97:155–159
Islam MM, Ponnamperuma FN (1982) Soil and plant tests for available sulfur in wetland rice soils. Plant Soil 68:97–113

Ito T, Sugita K, Okabe S (2004) Isolation characterization and in situ detection of a novel chemolithotrophic sulfur oxidizing bacterium in wastewater biofilm growing under microaerophilic conditions. Appl Environ Microbiol 70:3122–3129

Ito T, Sugita K, Yumoto I, Nodasaka Y, Okabe S (2005) *Thivirga sulfuroxydans* gen. nov., a chemolithoautrophic sulfur-oxidizing bacterium isolated from a microaerobic waste-water biofilm. Int J Syst Evol Microbiol 55:1059–1064

Joffee JS (1922) Biochemical oxidation of sulfur and its significance to agriculture. NJ Agric Exp Sta Bull 374:82–90

Kabesh MO, Behairy TG, Saber MSM (1989) Utilization of biofertilizers in field crop production. Effect of elemental sulfur application in the presence and absence of two biofertilizers on growth and yield of maize. Egypt J Agron 14:95–102

Kelly DP, Harrison AP (1988) Genus *Thiobacillus beijerink*. In: Staley GT, Pfenning N, Holt JG (eds) Bergey's manual of systematic bacteriology. Williams and Wilkinson, Baltimore, pp 1842–1871

Kelly DP, Shergill JK, Lu WP, Wood AP (1997) Oxidative metabolism of inorganic sulfur compounds by bacteria. Antonie Van Leeuwenhoek 71:95–107

Kleinhenz V (1999) Sulfur and chloride in the soil plant system. K+S Group, Kassel International Potash Institute, Basel

Larsen PO (1981) Glucosinolates. In: Conn EE (ed) The biochemistry of plants secondary products. Academic, New York, pp 502–525

Lee JS, Ham SH (1986) An investigation on allyl sulfide contents in Korean local garlic cultivars. Hort Abstr 4:42–43

Legha PK, Giri G (1999) Influence of nitrogen and sulfur on growth, yield and oil content of sunflower (*Helianthus annus*) grown in spring season. Indian J Agron 44:408–412

Liu Z (1986) Preliminary study of soil sulfur and sulfur fertilizer efficiency in China. In: Sulfur in agricultural soils. Proceedings of international symposium, Dhaka, 20–22 Apr, pp 371–388

London J, Rittenberg SC (1966) Effects of organic matter on the growth of *Thiobacillus intermedius*. J Bacteriol 91:1062–1069

Matin A, Schleiss M, Perez RC (1980) Regulation of glucose transport and metabolism in *Thiobacillus novellus*. J Bacteriol 142:639–644

Mohan K, Sharma HC (1992) Effect of nitrogen and sulfur on growth, yield attributes, seed and oil yield of Indian mustard (*Brassica*) in seed. Indian J Agron 37:748–754

Moreira C, Rainey FA, Nobre MF, Da Silva MT, Da Costa MS (2000) *Tepidomonas ignava* gen. nov., sp., a new chemolithotrophic and thermophilic member of β-*Proteobacteria*. Int J Syst Evol Microbiol 50:735–742

Morrison RJ, Naidu R, Singh U (1987) Sulfur fertilizer requirements of Papua New Guinea and the South Pacific. In: Proceedings of the symposium on fertilizer sulfur requirements and sources in developing countries of Asia and the Pacific, Bangkok, 26–30 Jan. Fertilizer Advisory, Development and Information Network for Asia and the Pacific (FADINAP), Bangkok, pp 57–66

Muralidharan P, Jose AI (1993) Effect of application of magnesium and sulfur on the growth, yield and uptake in rice. J Trop Agr 31:24–28

Padden N, Rainey FA, Kelly DP, Wood AP (1997) *Xanthobacter tagetidis* sp. nov., an organism associated with *Tagetes* species and able to grow on substituted thiophenes. Int J Syst Bacteriol 47:394–401

Padden AN, Kelly DP, Wood AP (1998) Chemolithoautotrophy and mixotrophy in the thiophene-2-carboxylic acid-utilizing *Xanthobacter tagetidis*. Arch Microbiol 169:249–256

Pasricha NS, Aulakh MS (1990) Effect of phosphorus–sulfur interrelationship on their availability from fertilizer and soil to soybean (*Glycine max*) and linseed (*Linum usitatissimum* L.). In: Proceedings of the Middle East symposium, Cairo, 12–16 Feb, pp 277–279

Pathiratna LSS, Waidyanatha US, Peries OS (1989) The effect of apatite and elemental sulfur mixtures on growth and P content of *Centrosema pubescens*. Fertil Res 21:37–43

Peccia J, Merchand EA, Silverstein J, Hernandez M (2000) Development and application of small sub-unit rRNA probes for assessment of selected *Thiobacillus* species and members of the genus *Acidophilium*. Appl Environ Microbiol 66:3065–3072

Pronk JT, Meulenberg R, Hazeu W, Bos P, Kuenen JG (1990) Oxidation of reduced inorganic sulfur compounds by acidophilic thiobacilli. FEMS Microbiol Rev 75:293–306
Raghuwanshi RKS, Sinha NK, Agarwal SK (1997) Effect of sulfur and zinc in Soy bean (*Glycin max*), wheat (*Triticum aestivum*) cropping sequence. Indian J Agron 42:29–32
Rathore PS, Manohar SS (1989) Effect of sulfur and nitrogen on quality parameters of mustard. Fmg Syst 5:29–32
Riley NG, Zhao FJ, McGrath SP (2000) Availability of different forms of sulphur fertilizers to wheat and oilseed rape. Plant Soil 222:139–147
Romano AH, Van Vranken NJ, Preisand P, Brustolon M (1975) Regulation of the *Thiobacillus intermedius* glucose uptake system by thiosulfate. J Bacteriol 121:577–582
Ruby EG, Wirsen CO, Jannasch HW (1981) Chemolithotrophic sulfur oxidizing bacteria from the Galapagos Rift hydrothermal vents. Appl Environ Microbiol 42:317–327
Rupela OP, Tauro P (1973) Isolation and characterization of *Thiobacillus* from alkali soils. Soil Biol Biochem 5:891–897
Ryu HW, Moon HS, Lee EY, Cho KS, Choi H (2003) Leaching characteristics of heavy metals from sewage sludge by *Acidithiobacillus thiooxidans* MET. J Environ Qual 32:751–759
Scherer HW (2001) Sulphur in crop production. Eur J Agron 14:81–111
Shin JS (1987) Sulfur in Korean agriculture. In: Proceedings of the symposium on fertilizer sulfur requirements and sources in developing countries of Asia and the Pacific, Bangkok, 26–30 Jan. Fertilizer Advisory, Development and Information Network for Asia and the Pacific (FADINAP), Bangkok, pp 76–82
Shooner R, Tyagi RD (1995) Microbial ecology of simultaneous thermophilic microbial leaching and digestion of sewage sludge. Can J Microbiol 41:1071–1080
Shukla UC, Prasad KG (1979) Sulfur zinc interaction in groundnut. J Indian Soc Soil Sci 27:60–64
Singh YP, Aggarwal RL (1998) Effect of sulfur and leaves on yield, nutrient uptake and quality of black gram (*Phaseolus mungo*). Indian J Agron 43:448–452
Singh M, Singh N (1977) Effect of sulfur and selenium on sulfur containing amino acids and quality in Raya (*Brassica juncea* Coss) in normal and sodic soil. Indian J Plant Physiol 20:56–62
Singh D, Singh V (1995) Effect of potassium and sulfur on growth characters, yield attributes and yield of soybean (*Glycine max*). Indian J Agron 40:223–227
Slocum RD, Kaur-Sawhney R, Galston AW (1984) The physiology and biochemistry of polyamines in plants. Arch Biochem Biophys 235:283–303
Smith TA (1985) Polyamines. Annu Rev Plant Physiol 36:117–143
Sorokin DY, Kuenen JG (2005) Haloalkaliphilic sulfur oxidizing bacteria in soda lakes. FEMS Microbiol Rev 29:685–702
Sorokin DY, Robertson LA, Kuenen JG (1996) Sulfur cycling in *Catenococcus thiocyclus*. FEMS Microbiol Ecol 19:117–125
Sorokin DY, Tourova TP, Muyzer G (2005a) *Citreicella thiooxidans* gen. nov., sp. nov., a novel lithoheterotrophic sulfur-oxidizing bacterium from the black sea. Syst Appl Microbiol 28:679–687
Sorokin DY, Tourova TP, Spiridonova EM, Rainey FA, Muyzer G (2005b) *Thioclava pacifica* gen. nov. sp. nov., a novel facultatively autotrophic, marine, sulfur-oxidizing bacteria from a near-shore sulfidic hydrothermal area. Int J Syst Evol Microbiol 55:1069–1075
Stamford NP, Silva AJN, Freitas ADS, Filho A (2002) Effect of sulphur inoculated with *Thiobacillus* on soil salinity and growth of tropical tree legumes. Bioresour Technol 81:53–59
Stamford NP, Freitas ADS, Ferraz DS, Montenegro A, Santos CERS (2003a) Nitrogen fixation and growth of cowpea (*Vigna unguiculata*) and yam bean (*Pachyrhizus erosus*) in sodic soil as affected by gypsum and sulphur inoculated with *Thiobacillus* and rhizobial inoculation. Trop Grasslands 37:11–19
Stamford NP, Santos PR, Moura AMMF, Santos CERS, Freitas ADS (2003b) Biofertilizers with natural phosphate sulfur and *Acidithiobacillus* in a soil with low available-P. Sci Agric 607:63–773

Stamford NP, Ribeiro MR, Cunha KPV, Freitas ADS, Santos CERS, Dias SHL (2007a) Effectiveness of sulfur with *Acidithiobacillus* and gypsum in chemical attributes of a Brazilian sodic soil. World J Microbiol Biotechnol 23:1433–1439

Stamford NP, Santos PR, Santos CERS, Freitas ADS, Dias SHL, Lira MA Jr (2007b) Agronomic effectiveness of biofertilizers with phosphate rock sulfur and *Acidithiobacillus* for Yam bean grown on Brazilian tableland acidic soil. Bioresour Technol 98:1311–1318

Stamford NP, Santos CERS, Silva S Jr, Lira MA Jr, Figueiredo MVB (2008a) Effect of rhizobia and rock biofertilizers with *Acidithiobacillus* on cowpea nodulation and nutrients uptake in a tableland soil. World J Microbiol Biotechnol 24:1857–1875

Stamford NP, Lima RA, Lira MA Jr, Santos CERS (2008b) Effectiveness of phosphate and potash rocks with *Acidithiobacillus* on sugarcane yield and their effects on soil chemical attributes. World J Microbiol Biotechnol 24:2061–2066

Starkey RL (1935) Isolation of some bacteria which oxidize thiosulfate. Soil Sci 39:197–219

Stevenson FJ (1986) Cycles of soil. Wiley, New York

Takano B, Koshida M, Fujiwara Y, Sugimori K, Takayanagi S (1997) Influence of sulfur oxidizing bacteria on the budget of sulfate in Yuma Crater Lake, Kusatsu-Shirane volcano Japan. Biochemistry 38:227–253

Tandon HLS (1991) Sulfur research and agricultural production in India. The Sulfur Institute, Washington, DC

Teske A, Brinkhoff T, Muyzer G, Moser DO, Rethmeier J, Jannasch HW (2000) Diversity of thiosulfate oxidizing bacteria from marine sediments and hydrothermal vents. Appl Environ Microbiol 66:3125–3133

Tiwari KM (1997) Sulfur in balanced fertilization in northern India. In: Proceedings of the TSI/PM/IFA. Symposium on sulfur in balanced fertilization, New Delhi, 13–14 Feb, pp SI-1/1–SI-1/15

Tuttle JH (1980) Organic carbon utilization by resting cells of thiosulfate oxidizing marine heterotrophs. Appl Environ Microbiol 40:516–521

Vlasceanu L, Popa R, Kinkle B (1997) Characterization of *Thiobacillus thioparus* and its distribution in chemoautotrophically based ground water ecosystem. Appl Environ Microbiol 63:3123–3127

Vonuexkull HR (1986) Sulfur interaction with other plant nutrients. In: Sulfur in agricultural soils. Proceedings of international symposium, Dhaka, 20–22 Apr, pp 212–242

Wainright M (1984) Sulfur oxidation in soils. Adv Agron 37:349–396

Waksman SA, Joffe JS (1922) The chemistry of the oxidation of sulfur by microorganisms to sulfuric acid and transformation of insoluble phosphates into soluble forms. J Biol Chem 50:35–45

Wood AP (1988) Chemolithotrophy. In: Anthony C (ed) Bacterial energy transduction. Academic, London, pp 183–230

Wood AP, Kelly DP (1980) Regulation of glucose catabolism in *Thiobacillus* A2 grown in the chemostat under dual limitation by succinate and glucose. Arch Microbiol 128:91–97

Wood AP, Kelly DP, McDonald IR, Jordan SL, Morgan TD, Khan S, Murrell JC, Borodina E (1998) A novel pink-pigmented facultative methylotroph, *Methylobacterium thiocyanatum* sp. nov., capable of growth of thiocyanate or cyanate as sole nitrogen sources. Arch Microbiol 169:148–158

Yim WJ, Anandham R, Indira Gandhi P, Hong IS, Islam MR, Trivedi P, Madhaiyan M, Han GH, Sa TM (2008) Ubiquitous presence and activity of thiosulfate oxidizing bacteria in rhizosphere of economically important crop plants of Korea. Korean J Soil Sci Fert 41:9–17

Zhu B, Alva AK (1993) Trace metal and cation transport in a sandy soil with various amendments. Soil Sci Soc Am J 57:723–727

# Chapter 6
# Role of Siderophores in Crop Improvement

**Anjana Desai and G. Archana**

## 6.1 Introduction

Soils are known to be oligotrophic environments where growth of heterotrophic microbial population is limited by the scarce sources of readily available organic and micronutrients. Most microorganisms therefore face a constant battle for acquiring available resources which provide them with a driving force for developing innovation and diversification. Present chapter highlights on the strategies developed by rhizospheric microorganisms to combat with the competition for one of such limiting resources, iron, which is usually required in high concentration for growth and survival of most of the soil microorganisms.

### *6.1.1 Occurrence of Iron in Nature*

The importance of iron in our civilization is perhaps best summarized by the quotation from Rudyard Kipling cited by Philip Aisen in his article on physico-chemical aspects of iron metabolism:

> Gold is for the mistress – silver for the maid –!
> Copper for the craftsman cunning at his trade.
> "Good!" said the Baron, sitting in his hall,
> "But Iron – Cold Iron – is master of them all."

Certainly, it is due to its biological abundance in the Earth's crust that the practical value of this element was first recognized, particularly compared with elements such as gold, silver, and copper. Iron (Fe) is the fourth most abundant element on Earth's crust, following oxygen, silicon, and aluminum, mostly found in

A. Desai (✉) • G. Archana
Department of Microbiology, M.S. University of Baroda, Vadodara 390002, India
e-mail: desai_aj@yahoo.com

D.K. Maheshwari (ed.), *Bacteria in Agrobiology: Plant Nutrient Management*,
DOI 10.1007/978-3-642-21061-7_6, © Springer-Verlag Berlin Heidelberg 2011

the form of ferro-magnesium silicates. Most of the iron in the soil is found in silicate minerals or iron oxides and hydroxides, forms that are not readily utilizable by microorganisms and plants or not in bioavailable form. The bioavailable form of iron can be defined as the portion of the total iron that can be easily assimilated by living organisms. Iron can exist in aqueous solution in two oxidation states: $Fe^{2+}$ and $Fe^{+3}$ and the solution chemistry of iron is essentially dominated by the hydrolysis and polymerization of aqueous Fe(III) to insoluble and potentially inaccessible ferric hydroxides and oxyhydroxides. The biological importance of iron is a result of its electronic structure, which is capable of reversible changes in oxidation state of over a wide range of redox potential. Iron is 1 of the 12 minerals essential to plant growth and development. It is a microelement, as plants need it in relatively small quantities, as opposed to the macroelements like nitrogen, phosphorous, and potassium, which are required and consumed in larger amounts. Iron is rarely absent from soils, at least in dry climates, but alkaline conditions, which prevent its dissolution in the soil water, render it unavailable to the plants' roots.

## 6.2 Iron Requirement by Plants and Soil Microorganisms

Iron requirements of life forms vary, e.g., $10^{-9}$ M is required for optimal growth of plants, while that required for the optimal growth of microbes is in the range of $10^{-7}$ to $10^{-5}$ M (Raymond et al. 2003), both of which are far greater than the biological availability which is $10^{-17}$ M at physiological pH 7.0. Therefore, there is always an "iron stressed" condition prevalent in the soil. Iron is a vital element required by virtually all living organisms, including bacteria, with the exception of only a few, including *Streptococcus sanguis*, some *Lactobacillus* species, and *Borrelia burgdorferi* (Guiseppi and Fridovich 1982; Archibald 1983; Posey and Gherardini 2000).

### 6.2.1 *Significance of Iron in Cell Metabolism*

Iron either alone or incorporated into iron–sulfur clusters or in heme serves as the catalytic center of enzymes for redox reactions. Owing to its redox potential ranging from −300 to 1,700 mV (Andrew et al. 2003), iron is an all-round prosthetic component for incorporation into proteins that act as biocatalysts or electron carriers. Thus its role is established in central to cellular processes such as electron transport, activation of oxygen, peroxide reduction, amino acid and nucleoside synthesis, DNA synthesis, and photosynthesis.

### 6.2.2 *Sources of Iron*

Insoluble Fe III is not directly assimilable, but it is the major form of iron in aerobic and neutral pH environments. Under anaerobic or reducing conditions, Fe II is the major form of iron. This highly soluble form of iron diffuses freely through the outer membrane (OM) porins of Gram-negative bacteria.

#### 6.2.2.1 Transferrin and Lactoferrin

Transferrin (Tf) is found in serum and lactoferrin in lymph and mucosal secretions. Both proteins exhibit extremely high-affinity constants for Fe III iron (Ka = $10^{20}$ $M^{-1}$) and much lower constants for Fe II (Ka = $10^{3}$ $M^{-1}$). Many bacterial species including *Neisseria meningitidis* and *Neisseria gonorrhoeae* have Tf and/or lactoferrin iron-uptake systems (Cornelissen 2003).

#### 6.2.2.2 Ferritins

Ferritins are cytoplasmic proteins that store iron, making it available in case of iron shortage, and protect the cells from the toxic effects of iron accumulation.

#### 6.2.2.3 Heme Sources

Many hemoproteins as mentioned below serve as the source of iron for different cellular processes in varied systems.

##### Heme

It is the prosthetic group associated with many enzymes. Because of high toxicity it is scarcely found free. However in laboratory conditions, heme is an iron source for many bacterial species. It is also a protoporphyrin source for several bacterial species such as *Enterococcus faecalis*, *Lactococcus lactis*, and *Haemophilus influenzae* which are unable to synthesize the tetrapyrrole ring.

##### Hepatoglobin–Hemoglobin

Hepatoglobin is a steric tetrameric glycoprotein. The hepatoglobin–hemoglobin complex is an iron source for several bacteria (Morton et al. 1999).

##### Hemopexin

Hemopexin is a 60-Kda glycoprotein. The heme–hemopexin complex is transported to the liver and apo-hemoplexin is recycled after heme is discharged intracellularly. Several bacteria have been reported to bind and/or use heme–hemopexin.

Albumin

Albumin is another heme carrier. In laboratory conditions, it is a good heme source for bacteria.

#### 6.2.2.4 Other Hemoproteins

In plants and animals there are many other hemoproteins in which heme is not covalently linked to the apoproteins. These include myoglobin, leghemoglobin, catalase, and type b cytochrome and most are satisfactory heme sources for bacteria in laboratory media.

#### 6.2.2.5 Siderophores and Other Indirect Iron Sources

Compounds synthesized and released by bacteria into the extracellular medium to scavenge iron from various sources fall under the category of indirect iron sources as against the direct contact between bacteria and exogenous iron or heme sources. Two main groups are discussed below.

Hemophores

These are found only in Gram-negative bacteria. They are specialized extracellular proteins that acquire heme from diverse sources and bring it to a specific OM receptor.

Siderophores: Types and Chemistry

Microorganisms have developed various mechanisms in order to overcome the low bioavailability of iron, involving synthesis and secretion of low molecular weight iron specific chelators known as "siderophores," which are iron carriers with a very high affinity, $K_{aff} > 10^{30}\ M^{-1}$, for ferric iron and whose biosynthesis is carefully regulated by ferric iron (Byers and Araceneaux 1998; Boukhalfa and Crumbliss 2002; Raymond and Dertz 2004; Dhungana and Crumbliss 2005). The first siderophore to be isolated was ferrichrome from the culture filtrate of the smut fungus *Ustilago sphaerogena* (Neilands 1952).

Different organisms produce different types of siderophores, and they are categorized on the basis of the iron chelating functional group they possess. Siderophores chelate ferric iron with high affinity typically by virtue of these chelating groups which are generally oxygenated and bind iron to form complexes with great thermodynamic stability. Figures 6.1–6.3 depict the structural formulas for siderophores from various bacteria. There are four main types of iron-coordinating functional groups in siderophores.

### Hydroxamate Type

Siderophores which possess hydroxamic acids as the functional group (Fig. 6.1a), e.g., anguibactin, ferrichrome, rhodotorulic acid (RA) (Fig. 6.1b, c).

*Catecholate type*: Hydroxyls of catechol rings bind to iron (Fig. 6.2a), e.g., enterobactin, anguibactin, and acinetobactin. The three catecholate side chains in *Escherichia coli* siderophore enterobactin make it a complete hexadentate ligand which can bind to ferric ion, accounting for the estimated *Kd* (dissociation constant) of $10^{-52}$ M that makes enterobactin such a good scavenger for iron (Neilands 1981, 1995) (Fig. 6.2b).

Ferrioxamine type: It is represented by ferrioxamine B (FOB) the mesylate salt of which is marketed by Ciba-Geigy as Desferal, the major drug used in chelation therapy of secondary iron overload in man (Fig. 6.3).

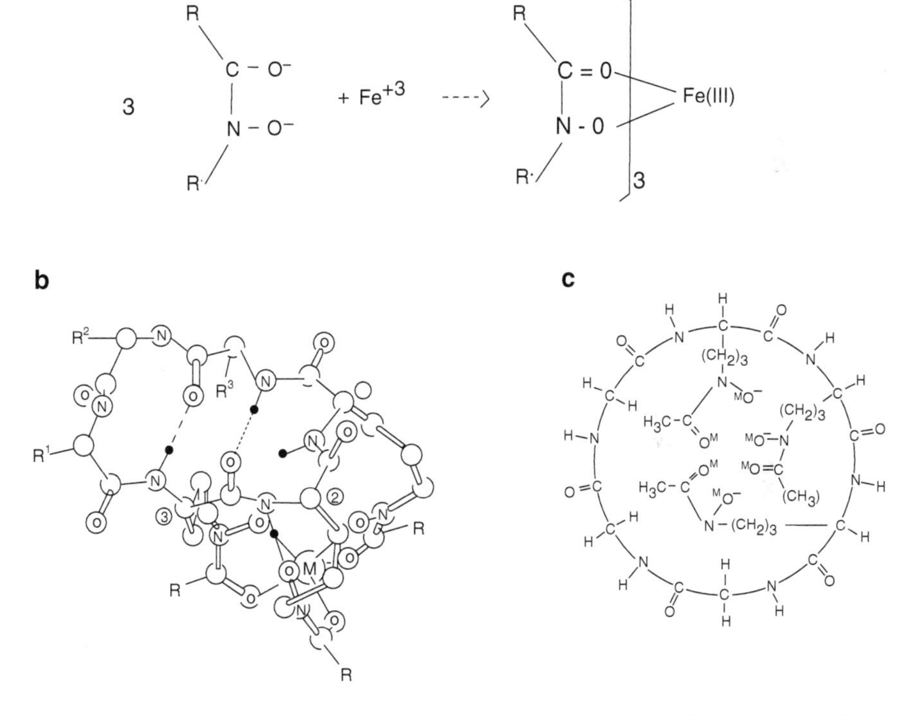


**Fig. 6.1** (**a**) Hydroxamic acid functional group, (**b**) Ferrichrome, the prototype of the hydroxamate type of siderophores (Neilands 1981) and (**c**) Desferrichrome (Byres and Arceneaux 1977)

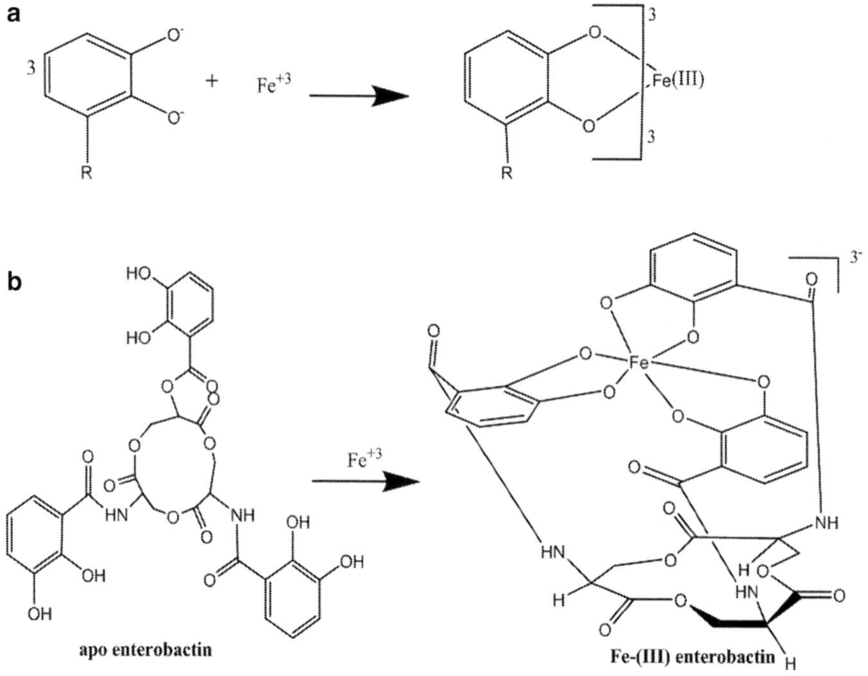

Fig. 6.2 (a) Hydroxyl groups of catecholate interacting with iron, (b) spatial structures of enterobactin without and with $Fe^{3+}$. The groups coordinating with iron are six oxygens from the three diphenolic groups. The complex requires a single molecule of enterobactin

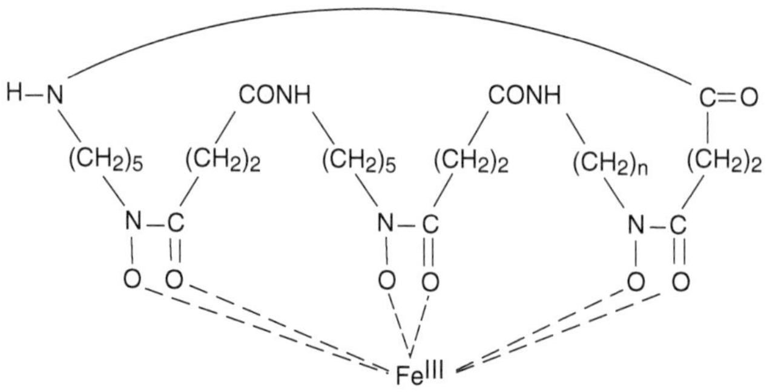

Fig. 6.3 Structure of ferrioxamine family of siderophores. Desferrioxamine or Desferal (Neilands 1977)

## Nitrogen Atoms of Five Membered Thiazoline and Oxazoline Rings

They result from enzymatic cyclization of cysteinyl, seryl, or threonyl side chains, respectively, and can also coordinate $Fe^{+3}$. This type of coordination is a common feature in the pyochelin, yersiniabactin, vibriobactin, anguibactin, and acinetobactin siderophores.

Citrate Type

Carboxylate ion participates in iron binding, the example of which is citrate siderophore produced by *Bradyrhizobium japonicum* strains (Guerinot et al. 1990).

Complex Siderophores

In complex siderophores, different iron-chelating functional groups are combined in the same siderophore, such as mycobactin and anguibactin.

Pyoverdines

Several *Pseudomonas* sp. produce and excrete fluorescent yellow-green siderophores, named pyoverdines or pseudobactins (Budzikiewicz 1993; Meyer 2000), which are composed of a conserved dihydroxyquiniline chromophore and a variable peptide chain. The diversity in the peptide chain is so large that pyoverdines of different strains can be easily differentiated by isoelectric focusing and this approach termed "siderotyping" can be used to distinguish different *Pseudomonas* sp.

Apart from these, there are also a few exceptional siderophores which cannot be classified into any of the above groups, e.g., rhizobactin produced by *Sinorhizobium meliloti* DM4, which possess ethylenediamine as the chelating ligand (Smith et al. 1985). Rhizobactin 1021 produced by *S. meliloti* 2011 is a citrate-based dihydroxamate siderophore that has a core structure identical to that of schizokinin from *Bacillus megaterium* (Persmark et al. 1993).

Phytosiderophores

Mugineic acid (MA) is a phytosiderophore (PS) excreted by roots of graminaceous plants for $Fe^{+3}$ uptake. MA is closely related to its biochemical precursor, nicotinamine, and to a number of other compounds that also have been identified as PS in graminaceous plants: 3-hydroxymugineic acid, 2′-deoxymugineic acid, avenic acid, and distichonic acid (Kraemer et al. 2006; Takagi 1976).

### *6.2.3 Siderophore Estimation from Soils*

One of the problems in elucidating the role of siderophores in microbial competition and plant growth has been the inability to isolate and purify these compounds from soil. Typically, siderophore concentrations in soil are estimated by performing a bioassay with indicator organisms that grow only if a specific ferric siderophore or a general class of ferric siderophores (i.e., ferric siderophores with catechol or hydroxamate groups) provides an exogenous source of iron. By using bioassays, concentrations of bacterial siderophores in soil have been estimated to be between 0.04 and 0.3 nmol/g of soil (Nelson et al. 1988), and the concentration of fungal siderophores may range between 30 and 240 mg/kg or between 0.09 and 0.75 nmol/g of soil (Akers 1983; Powell et al. 1980). By employing monoclonal antibodies, the pyoverdin concentration was estimated to be between 0.2 and 0.5 nmol/g of root (Buyer et al. 1993). All of these assays require sampling large quantities of soil and preclude localization of siderophore production in specific microsites or analysis of their dynamic rates of production over time. *Pseudomonas* species harboring a transcriptional fusion of an iron-regulated promoter of pyoverdine siderophore locus to a promoterless ice nucleation reporter gene (*inaZ*) encoding an outer membrane protein (InaZ) catalyzing ice formation have been developed (Loper and Lindow 1994; Duijff et al. 1994). By using the regression between ice nucleation activity and pyoverdin production determined in vitro in *Pseudomonas fluorescens* Pf-5 (pvd-inaZ), the maximum possible pyoverdin accumulation by this strain in the rhizosphere was estimated to be 0.5 and 0.8 nmol/g of root for lupine and barley, respectively (Marschner and Crowley 1997). This is very low compared to the pyoverdin production measured in vitro under extreme Fe stress which is 300 mM within 48 h (Meyer and Abdallah 1978).

## 6.3 Iron Availability in Soils

Iron, element 26, in the periodic table, has two properties at physiological pH and in aqueous medium which make this element of utmost importance in biological system. Organisms to be able to obtain iron from their environment have evolved a strategy to synthesize and secrete iron chelating compounds called, siderophores which are powerful complexants of Fe(III), thereby making it accessible to microbial cells for its assimilation.

### *6.3.1 Measurement of Bioavailable Iron Content in Soil*

Total soil iron content is usually estimated following chemical extraction procedures, but such measurements which assess the iron content of the bulk soil do not provide any idea about the iron available to bacteria in microhabitats. It is realized that the availability of iron in bulk soil may differ from that in microenvironments such as the rhizosphere due to influences of iron acquisition systems of plants, oxygen depletion, pH differences, and other factors different from bulk soil. The use of reporter genes linked to iron responsive promoters to generate biosensors has proven useful in measuring bioavailability of iron at small spatial scales. Using *Pseudomonas* strains carrying the pvd-inaZ reporter, it was found that *P. fluorescens* Pf-5 encountered an iron-limited environment immediately after it was inoculated onto bean roots planted in agricultural field soils but not after 2 days (Loper and Henkels 1997). Studies were conducted in a soil low in available Fe, planted with an iron-efficient dicot, lupine (*Lupinus albus* L.), or with an iron-efficient grass, barley (*Hordeum vulgare* L.), to examine the induction of the iron stress response in *P. fluorescens* Pf-5 (pvd-inaZ) in different root zones as a function of time (Marschner and Crowley 1997). Ice nucleation activity levels were similar in all root zones examined and were only marginally higher in barley than in lupine. It was concluded that, on average, cells sensed $Fe^{3+}$ concentrations that were intermediate between iron replete and iron-deplete culture media indicating that *P. fluorescens* Pf-5 (pvd-inaZ) is only mildly iron stressed in the rhizosphere. An ice nucleation activity similar to that in the rhizosphere was measured in the in vitro experiment at 25–50 mM $FeCl_3$ suggesting this to be the Fe levels sensed by the bacterium in the rhizosphere. Joyner and Lindow (2000) reported the ferric iron availability to cells of *Pseudomonas syringae* by quantifying the fluorescence intensity of single cell harboring a plasmid-borne transcriptional fusion of an iron-regulated promoter of pyoverdine siderophore receptor with green fluorescent protein (GFP) as the reporter. By analyzing fluorescence intensity of individual bacterial cells on plant surfaces, the relative iron concentration sensed by cells washed from aseptic roots displayed a right-hand skewed distribution, indicating that some cells (estimated as 10%) sensed much less $Fe^{3+}$ than others. Their results suggest that there is substantial heterogeneity of iron bioavailability to cells of *P. syringae* on plants, with a subset of cells experiencing low iron availability. Although *Pseudomonas* species produce a large variety of siderophores and can utilize many heterologous siderophores, yet this organism is mildly Fe stressed in the rhizosphere, thus it is likely that other organisms with poor abilities to utilize and produce siderophores will be strongly limited by iron.

### 6.3.2 Mechanisms of Iron Acquisition by Plants in Soil

Although the total iron (Fe) content in soils usually far exceeds plants requirement for Fe, its bioavailability in the soil, especially in calcareous soils, is often severely limited. Fe-efficient plant species and genotypes have evolved different strategies to ameliorate Fe uptake. Three mechanisms generally adopted by life forms for increasing iron availability in rhizosphere are schematically presented in Fig. 6.4.

Proton extrusion and secretion of organic acids by plants and microorganisms contribute to 100-fold greater acidification in rhizosphere than bulk soil. In addition, root and microbial respiration also contribute to soil acidification by way of elevating $pCO_2$ and dissociation of carbonic acid (Darrah 1993; Hinsinger et al. 2003; Nye 1981). In plants, mainly two strategies have been identified as per Romheld and Marschner (1986). Strategy I, identified in nongraminaceous monocots and dicots, involves response to Fe-deficiency stress typically by inducing ferric chelate reductase and Fe(II) transporter in roots system, by acidifying the

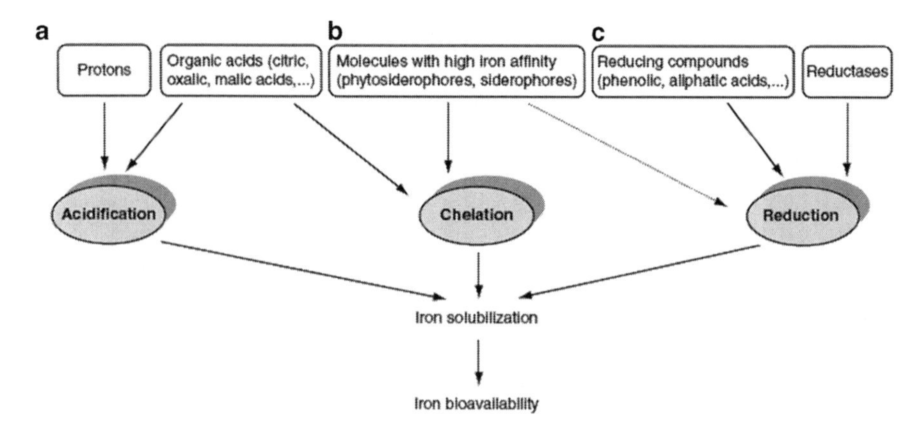


**Fig. 6.4** Schematic representation of mechanisms affecting iron availability in the rhizosphere. The mechanisms include (**a**) acidification through proton extrusion, (**b**) chelation through secretion of complexing molecules like siderophores, phenolics, carboxylic acids, etc., and (**c**) reduction through secretion of compounds having reducing property or reductase activity

rhizosphere medium and exuding organic compounds such as phenolics (Hell and Stephen 2003; Morrissey and Guerinot 2009). Strategy II, in graminaceous monocots, involves a response to Fe-deficiency stress by releasing PS and inducing a specific plasmalemma $Fe^{3+}$-PS transporter in roots system (Hell and Stephen 2003; Morrissey and Guerinot 2009). Release of PS has been reported to be dependent on the diurnal rhythm in wheat (Zhang et al. 1991).

Recent lines of evidence, however, have shown that the Fe-deficiency-induced responses mentioned earlier are insufficient for plants to overcome Fe deficiency under Fe-limited conditions. Masalha et al. (2000) for the first time demonstrated that sunflower and maize grown in sterile soil showed poor growth and lower tissue Fe concentration compared with non-sterile soil-grown plants. Similarly, Rroco et al. (2003) and Jin et al. (2006) reported that Fe acquisition and growth of rape (*Brassica napus*) and red clover (*Trifolium pretense*) were significantly reduced when the plants were grown in sterile soil, but normal growth could be restored by adding Fe-EDDHA to the sterile soil or spraying EDTA-Fe to the leaves. It therefore appeared that soil microbial activity plays an important role in favoring plant Fe uptake. However, the actual mechanism by which soil microbes contribute to plant Fe acquisition still remains to be investigated in detail.

Yang and Crowley (2000) found that the microbial community in barley (a strategy II plant) rhizosphere varied with the plant's Fe nutritional status, and it was proposed to be due to the changes in chemical composition of root exudates. Robin et al. (2006, 2007) showed reduced availability of iron in the rhizosphere and shifts in the microbial community in a transgenic tobacco (a strategy I plant) over-accumulating iron. Previous reports demonstrate that incubation of solution of a calcareous soil on an agar plate containing phenolic root exudates from Fe-deficient red clover supported growth of only those microorganisms which could secrete siderophores under Fe-deficient conditions (Jin et al. 2006, 2008). Generally, siderophores produced by soil microbes are seen as one of the microbial functions most supportive of Fe acquisition by plants, because microbial siderophores have a high affinity for chelating Fe(III), and the resultant chelates have been proven to be an efficient bioavailable Fe source for plants (Chen et al. 1998; Lemanceau et al. 2009).

To eliminate the confounding effects of microorganisms and to examine the direct utilization of microbial siderophores as iron sources by higher plants, a hydroponic cultural system and methodology was developed to grow plants with axenic roots. Using this system the efficacy of the microbial siderophore FOB, compared to the synthetic iron chelate Fe EDTA, and the PS of barley as an iron source for alleviating iron stress in the model dicot cucumber was evaluated. It was concluded that the FOB and iron were taken up by the axenic roots of cucumber in a highly efficient manner, most likely as the iron-siderophore complex, and at rates that could be significant to dicot nutrition. The results also suggested that cucumber may transport FOB through the transpiration stream to upper parts of plants, where the iron would be reductively released from the siderophore for shoot nutrition (Wang et al. 2006). Earlier reports from our laboratory also indicated the involvement of cowpea *Rhizobium* siderophore in the iron nutrition of peanut plants. An iron-inefficient variety of peanut plant, when grown hydroponically with the catechole siderophore of peanut *Rhizobium* isolate showed increased growth and chlorophyll content compared with plants grown with Fe alone. Similar observations were made with desferrioxamine B in the same system (Jadhav et al. 1994).

## 6.3.3 Siderophore Production: The Most Common Mechanism for Iron Acquisition by Microorganisms

Siderophores as mentioned earlier share common features in being low molecular weight, high affinity ferric iron ($Fe^{+3}$) chelators, secreted by the organisms in conditions of iron deficiency. However, with respect to the structural diversity they possess, a less accurate definition which can be given to them is they are small peptidic molecules, readily assembled by short, dedicated metabolic pathways and which contain side chains and functional groups that make them high-affinity ligands for coordination of ferric ions (Crosa 1989; Neilands 1981, 1995). Utilization of Fe-siderophore by microorganisms is generally a receptor-dependent process (Jurkevitch et al. 1992) and the ferri-siderophore complexes formed are internalized by interaction with ligand specific high-affinity transport proteins displayed at the cell surface termed outer membrane receptors (OM).

### 6.3.3.1 Outer Membrane Siderophore Receptors

The three-dimensional atomic structure of three ton B dependent siderophore transporters from *E. coli*, Fec A (Fig. 6.5a), FhuA (Fig. 6.5b), and FepC (Fig. 6.5c)

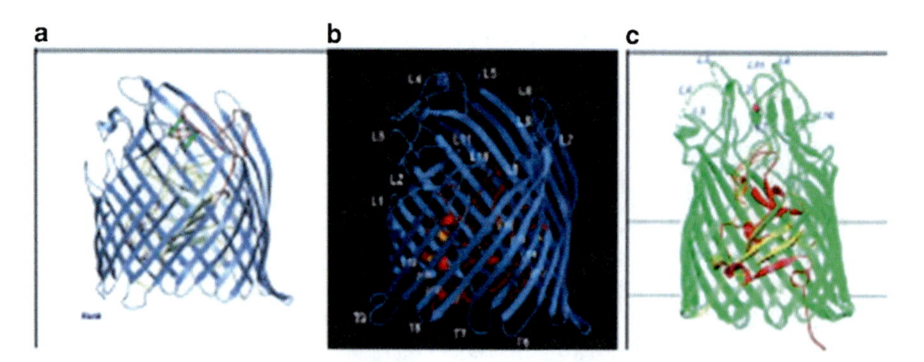


**Fig. 6.5** Crystal structures of (**a**) FecA (ferric-citrate receptor), (**b**) FhuA (ferrichrome receptor), and (**c**) FepA (ferric-enterobactin receptor)

transporting ferric citrate (Ferguson et al. 2002), ferri-ferrichrome (Ferguson et al. 1998), and ferri-entrobactin (Buchanan et al. 1999), respectively, has been determined. FpvA, which transports ferric pyoverdin (Cobessi et al. 2005) has also been worked to some extent. The crystal structures show remarkable similarities, each composed of a β-barrel, consisting of 22 antiparallel β-strands and a globular domain, which is referred to as a "cork" or "plug" because of the way it is oriented into the β-barrel domain (Endriss and Braun 2004). Mutational and structural approaches used to elucidate the binding pocket for iron-loaded siderophores with respect to their specificity and affinity led to no conclusive information. The fundamental questions concerning almost every aspect of the siderophore transport system thus remains to be resolved (Feraldo-Gometz and Sansom 2003).

Most of the ferri-siderophore receptors are multifunctional. For example, FhuA also serves as a receptor for the antibiotic albomycin, colicin M, phages T1, F5, and ϕ 80, FepA for colicins B and D (Braun et al. 1998).

The OM receptors are very specific to the ferri-siderophore it identifies, as given in Table 6.1. The ferri-siderophore complexes of ferrichrome, coprogen, aerobactin, entrobactin, salmochelin, dihydroxybenzoic acid, and citrate bind specifically and without any receptor cross-reactivity to their respective outer membrane proteins (OMP). It can be concluded from the data that the OM receptor is highly specific for a particular ferri-siderophore complex, while the cytoplasmic and periplasmic membrane machinery show relatively relaxed specificity.

#### 6.3.3.2 Iron Siderophore Affinity

The formation constants of the ferric siderophore complexes is of particular interest, since these constants define the ability of the siderophore to extract iron from other biological ferric complexes like lactoferrin, transferrin (animal host), phytoferritins (plant host), and other siderophore complexes produced by other

**Table 6.1** Iron transport systems of *E. coli*

| Substrate | Outer membrane protein | Periplasmic protein | Cytoplasmic membrane proteins |
|---|---|---|---|
| Enterobactin | FepA | FepB | FepD, FepG, FepC |
| Salmochelin | IroN | FepB | FepD, FepG, FepC |
| Catecholates | Cir | FepB | FepD, FepG, FepC |
| Catecholates | Fiu | FepB | FepD, FepG, FepC |
| Ferrichrome | FhuA | FhuD | FhuB, FhuC |
| Aerobactin | IutA | FhuD | FhuB, FhuC |
| Coprogen | FhuE | FhuD | FhuB,FhuC |
| Rhodotorulic acid | | | |
| Citrate | FecA | FecB | FecC, FecD, FecE |
| Heme | ChuA | ChuT | ChuU, ChuV |
| Yersiniabactin | FyuA | NI | YbtP, YbtP |
| $Fe^{+2}$ | | | FeoB |

Braun (2005)

organisms in the soil. It is due to these affinity differences that complex interactions occur in the rhizosphere with respect to iron nutrition.

The affinity of the siderophore is a result of its structure, the amino acid composition, and the way they are arranged with respect to the iron chelating group present. Given below are the affinity constants or the formation constants (at pH 7.0) of ferric-siderophores commonly found in soil (Guerinot 1994).

Ferrichrome, produced by numerous fungi – $10^{25}$
Desferrioxamine B, produced by actinomycetes – $10^{27}$
Enterobactin – $10^{52}$ produced by *E. coli*
Pseudobactin (pyoverdin), produced by pseudomonads – $10^{25}$
Mugineic acid, a phytosiderophore – $10^{17}$

#### 6.3.3.3 Significance of Iron-Siderophore Affinity in Niche Colonization

Plant deleterious organisms possessing siderophores with comparatively high affinity, succeed in colonizing plant rhizosphere and hence cause plant disease. Similarly animal pathogens having siderophores with high affinity for iron can successfully sequester iron from host iron storage proteins-transferrin and lactoferrin and establish pathogenesis. Based on this concept it is obvious that if plant growth promoting (PGP) organisms (free-living or symbiotic) used as bio-inocula produce siderophore of significant high affinity, it will not only succeed to establish itself, but will also be able to inhibit the growth of pathogenic organisms, and hence can act as a biocontrol agent (Chakraborty and Purkayastha 1984; Ehteshamul-Haque et al. 1992).

#### 6.3.3.4 Role of Differential Affinities of Siderophores for Iron Under Iron Limiting Conditions: A Case Study

Root exudates selectively influence the growth of bacteria and fungi that colonize the rhizosphere by altering the chemistry of soil in the vicinity of plant roots and thereby influence the plant growth positively or negatively. The organism's high-affinity iron uptake systems may be of use in the competition among soil microbes for access to available iron and may impart survival advantage. Members of the *Enterobacteriaceae* are known to efficiently colonize the rhizosphere (Halda-Alija 2003). Siderophore production in rhizospheric enterics has not been well addressed but they are reported to produce catecholates, namely entrochelin and enterobactin (O'Brien and Gibson 1970; Perry and San Clemente 1979) with a few exception such as *Enterobacter cloacae* where aerobactin, a hydroxamate siderophore, is a major siderophore produced along with enterochelin (Van Tiel-Mankvald et al. 1982). A systematic study conducted in our laboratory revealed that most of the rhizospheric isolates that belonged to the enteric group were producers of catecholate siderophores. Interaction of rhizospheric organisms with respect to siderophore cross-utilization is important from the point of view of their survivability and functioning under natural conditions. Diverse patterns of siderophore cross-utilization were observed amongst the ground nut isolates from the same field (Joshi et al. 2006a). Based on the number of siderophores being cross-utilized, the isolates were grouped into high (G9) and low (G6) siderophore cross-utilizers. The growth inhibition of G9 in the presence of G6 siderophore, which it failed to utilize, pointed towards the fact that both siderophores had different affinities for iron. Siderophore affinity measurement typically involves its complete purification, stoichiometry determination, and then determination of affinity by the method of competitive displacement of another iron chelator (Cornish and Page 1998; Reid et al. 1993). Cornish and Page (1998) have directly compared the ability of different siderophores of *Azotobacter* to bind $Fe^{3+}$ in a theoretical iron-siderophore system at pH 7.4. The strategy developed by our group to measure relative affinities requires partially purified siderophores and makes use of the rationale that a siderophore with higher affinity for iron can chelate iron from Fe-CAS-HDTMA complex at higher rate as compared to siderophore having lower affinity for iron. The measurement of relative affinities of the siderophores by a simple CAS assay decolorization principle revealed that the affinity of G6 siderophore was three orders of magnitude higher than that of G9 siderophore (Table 6.2) (Joshi et al. 2006b). Differential affinities of siderophores for iron not only enable microbes to quench iron from the soil but also allow mobilization of iron from weaker associated ferric-siderophore complexes from other species. Therefore, an organism producing a siderophore having a higher affinity for iron can have an ecological advantage for growth and survival as opposed to the organisms producing siderophores with weaker affinity. A simplified model based on the concept of differential affinity of siderophores for iron and its impact on bacterial competition in a nich is postulated and presented recently by Hibbing et al. (2010).

**Table 6.2** Pseudo-first-order rate constant for the formation of the $Fe^{+3}$-siderophore complex as indicated by slope of the plot for different siderophores

| Siderophore | Concentration of siderophore (μM) | Slope of the plots obtained | Mean ± SD of the slopes |
|---|---|---|---|
| | 46 | $1.37 \times 10^{-2}$ | |
| | 7.6 | $1.46 \times 10^{-2}$ | |
| G6 | 10.7 | $1.41 \times 10^{-2}$ | $1.41 \times 10^{-2} \pm 4.5 \times 10^{-4}$ |
| | 7.6 | $4.13 \times 10^{-5}$ | |
| | 10.8 | $4.16 \times 10^{-5}$ | |
| G9 | 15.4 | $3.46 \times 10^{-5}$ | $3.91 \times 10^{-5} \pm 3.97 \times 10^{-6}$ |
| | 10 | $1.5 \times 10^{-4}$ | |
| Desferrioxamine | 20 | $1.73 \times 10^{-4}$ | |
| B | 30 | $1.52 \times 10^{-4}$ | $1.58 \times 10^{-4} \pm 1.23 \times 10^{-5}$ |
| | 15 | $1.75 \times 10^{-4}$ | |
| | 20 | $1.89 \times 10^{-4}$ | |
| Citrate | 30 | $1.84 \times 10^{-4}$ | $1.82 \times 10^{-4} \pm 7.27 \times 10^{-6}$ |

G6 and G9 are siderophores from bacterial isolates (Joshi et al. 2006a, b)

#### 6.3.3.5 Utilization of Heterologous Siderophores-Impact on Competitive Survival

Precedents are known for the exploitation of the chelator synthesized by one organism being utilized by another (Neilands 1982). Ferrichrome is produced by fungi but *E. coli* can take up its ferri-hydroxamate derivative for use as a source of iron. Ferrichrome derivatives can also be used as source of iron by higher plants. Iron is one of the major limiting nutrients which affects the growth and propagation of rhizobia and other microorganisms in the soil. The problem becomes more pronounced with rhizobia because nitrogen fixation involves iron-rich proteins like nitrogenase, leg-hemoglobin and cytochromes with nitrogenase and leg-hemoglobin constituting up to 12 and 30% of the total protein in the bacteroids (Verma and Long 1983). Some free living rhizobia and bradyrhizobia not only produce and import their own siderophores but also are benefited by the utilization of heterologous siderophores present in the soil. *B. japonicum* 61A152 ( Bj61A152), a citrate producer (Guerinot et al. 1990), is able to utilize iron bound to hydroxamate-type siderophores like ferrichrome (FC) and RA, produced by soil fungi (Plessner et al. 1993). Pseudomonads are known for their rhizospheric stability, which is reflected by their diverse iron uptake systems. About 32 putative siderophore receptors in *Peudomonas aeruginosa* (Dean and Poole 1993; Ankenbaur and Quan 1994), 29 in *P. putida*, 27 in *P. fluorescens*, and 23 in *P. syringae* (Cornelis and Matthijs 2002) are reported. In addition to these, TonB-dependent receptors of *Pseudomonas* sp. exhibit a high degree of similarity between them, suggesting a possible redundancy of siderophore-mediated uptake systems. It has been shown that a mutant devoid of *pfeA*, the gene encoding receptor for enterobactin in *P. aeruginosa*, could still grow under extreme iron limiting conditions in the presence of enterobactin, suggesting the presence of an alternative, although less efficient, uptake system for this siderophore (Dean and Poole 1993). *P. aeruginosa fpv*A mutant can also grow in the presence of an alternative uptake system for pyoverdin, albeit after a longer lag phase (Cornelis and Matthijs 2002). Due to the presence of varied type of siderophore receptors, they are capable of utilizing a large number of heterologous siderophores, which are postulated to be the main reason for their rhizospheric competence for iron.

The significance of iron acquisition strategies in relation to nitrogen fixation efficiency is not yet clearly understood, but mutations in genes responsible for siderophore production or its uptake in nodulating bacteria had led to defective symbiotic association with the host plant. It has been reported that *fegA* (a gene encoding OM receptor protein for ferrichrome uptake) mutants of Bj61A152 form fix- nodules on soybean (Benson et al. 2005). In addition to this, the efficient performance of rhizobial inoculant strains depends upon their ability of competitive survival in presence of indigenous soil bacteria in rhizosphere, leading to increased nodulation of the host plant. In most cases, strains which are poor survivors under soil conditions do not bring about enhanced legume productivity since majority of nodules formed are not by the inoculated strains, but by indigenous strains in the soil (Hemantaranjan and Garg 1986; O'Hara et al. 1988; Miller and May 1991; Strecter 1994). Construction of genetically engineered inoculum strains of *Rhizobium* with an enhanced ability to survive and greater competitiveness for nodule occupancy may be a valuable approach to address the problem (Archana 2010).

#### 6.3.3.6 Distribution of Siderophore Receptors Among Rhizobiales

There are number of reports of Iron Regulated Outer Membrane Proteins (IROMPs) in rhizobia, correlating with the production and release of specific siderophores (Jadhav and Desai 1994; Reigh and O'Conell 1993; Patel et al. 1994). Not many OM siderophore receptors are reported in rhizobia.

##### The Rhizobial Ferrichrome OM Receptor

The rhizobial ferrichrome OM receptor is the hydroxamate-type siderophore receptor FegA of *B. japonicum* 61A152 (Levier and Guerinot 1996). The *fegA* gene is organized in an operon with *fegB* which probably encodes an inner membrane protein (Benson et al. 2005). Mutant analysis revealed that both genes are required for utilization of ferri-ferrichrome complex and *fegAB* double mutant, but not *fegB* mutant, fails to establish normal symbiosis (Benson et al. 2005).

##### The *FhuA* of *Rhizobium leguminosarum*

It specifies the OM receptor that works in association with *FhuCDB* (inner membrane proteins) for uptake of vicibactin. FhuC is an ABC transporter ATPase, Fhu B is permease, and Fhu D is the periplasmic siderophore binding protein which brings the ferri-vicibactin complex to the inner membrane machinery for its transport from periplasm to the cytoplasm (Stevens et al. 1999).

RhtA

It is the OM receptor responsible for rhizobactin uptake in *S. meliloti,* where a specialized single permease RhtX is responsible for its transport from periplasm to cytoplasm (O'Cuiv et al. 2004). Overall iron uptake genes are not widely studied in rhizobia. Using BLASTp tool, homologs of FegA were searched in the whole genome databases and was found that a homolog 42% identical with Feg A is present in *S. meliloti* (SMc01611), and that with 84% identity to Feg A in *B. japonicum* (strain110 BIJ 4920).

Uptake of Other Iron Sources by Rhizobia

*B. japonicum* can utilize iron bound to heme by the heme uptake system (Hmu) which consists of a TonB-dependent heme receptor (HmuR), a periplasmic heme-binding protein (HmuT), an ABC transporter (HmuUV), and a Ton system (ExbBD and TonB) (Nienaber et al. 2001). Heme is also utilized by *R. leguminosarum* and *S. meliloti*. ShmR is implicated as OM heme receptor in *S. meliloti* (Amerelle et al. 2008).

#### 6.3.3.7 Dependence of Rhizospheric Stability and Colonization of Organism on Iron Acquisition System

The high affinity iron uptake systems may be of use in the competition among soil microbes for access to available iron and may enhance the survival of the free-living forms. Evidence has been presented to indicate that soil competition among rhizospheric pseudomonads may occur at the level of uptake system of the ferri-complexes, specific for the individual *Pseudomonas* sp. siderophores (Buyer and Leong 1986). A comparable competition may also occur among rhizobial strains. *R. meliloti* DM4 and *R. meliloti* 1021 produce different siderophores, each stimulating growth of the source organisms, but antagonizes the growth of the others. It could, therefore, be said that possession of uptake system for the siderophores produced by majority of soil organisms, and hence predominantly present in the soil can have positive implications for growth and survival of receptor possessing organisms. If this organism is a PGP organism like *Rhizobium* or *Pseudomonas*, the growth and propagation of the organism allows it to successfully colonize the rhizoshere and hence contribute to the plant growth. Utilization of foreign or heterologous siderophores is considered to be an important mechanism to attain iron sufficiency. Florescent pseudomonads are known to efficiently colonize various ecological niches and thus known for rhizospheric stability, which is largely attributed to their possession of diverse and sophisticated iron uptake systems. Due to the presence of varied type of siderophore receptors, they are capable of utilizing a large number of heterologous siderophores, which are postulated to be the main reason for their rhizospheric competence for iron.

The plant rhizosphere is a dynamic environment where severe competition amongst microorganisms occurs for various limiting nutrients, one of which is iron. Studies pertaining to siderophore production and siderophore mediated iron nutrition through utilization of homologous and heterologous iron siderophore complexes in rhizobia are of significance since many iron rich proteins play significant role in the process of nitrogen fixation (Verma and Long 1983; Crowley et al. 1987). Hydroxamate type siderophores are main amongst the type of siderophores found in soil (Powell et al. 1980; Crowley et al. 1987) most of it being ferrichrome type, and present in nanomolar concentration (Powell et al. 1983). Our previous observations showed that nodule bacteria are poor producers as well as utilizers of hydroxamate type siderophores (Khan et al. 2006; Joshi et al. 2008). *Mesorhizobium* sp. GN 25, a ground nut isolate, was nonutilizer of ferrichrome and RA which was thought to be a negative fitness factor in iron limiting soil environments. As reported by Jurkevitch et al. (1992), majority of soil bacteria are capable of utilizing hydroxamates and therefore, it was thought that *Mesorhizobium* sp GN25 when used as a biofertilizer would be at a competitive disadvantage while residing free in soil. Successful application of *B. japonicum* 61A152 as a bioinoculant on soybean cultivars has been implicated to its capability of utilizing heterologous siderophores produced by soil fungi (Benson et al. 2005). Genome wide search for *fegA* homologs in rhizobia using BLASTp revealed its absence in genome of *Mesorhizobium loti*. It was therefore hypothesized that introduction of *B. japonicum* 61A152 *fegA* into ferrichrome nonutilizer *Mesorhizobium* sp. GN 25 could make it more competitive with regard to its iron acquisition property. The *Mesorhizobium* sp. GN25pFJ4, engineered with *fegA* displayed ferrichrome utilization ability in addition to the expression of a 79-kDa protein on the OM of the engineered strain as against the wild type strain (Joshi et al. 2008). Increased survival of the engineered strain in comparison to parent strain in presence of *Ustilago maydis*, a ferrichrome producing fungus, reinstated the hypothesis that expression of *fegA* enabled the organism to utilize iron bound to ferrichrome leading to increased growth of the engineered strain. Peanut plant inoculation with GN25pFJ4 led to marked increase in plant growth parameters. Nodule occupancy on peanut plants inoculated with this engineered strain reached to 61% which increased to 66% when co-inoculated with *U. maydis* as compared with 50% shown by the inoculation of parent strain GN25 either alone or in combination with *U. maydis*. A finding similar to ours has also been reported by Brickman and Armstrong (1999), where expression of only *fauA* gene, encoding receptor for alcaligin siderophore, imparts alcaligin utilization to a *P. aeruginosa* strain deficient in alcaligin production. A report from our lab (Rajendran et al. 2007) has shown that introduction of *fhuA* gene alone brings about ferrichrome utilization in rhizobial strains deficient of ferrichrome production as well as utilization.

Further studies from our lab with biofertiizer strains also substantiated the above observation (Khan 2010). *Cajanus cajan* and *Vigna radiata* biofertilizer strains procured from Indian Agriculture Research Institute, New Delhi were found to produce and cross-utilize mainly catecholate type siderophores similar to the native nodule bacteria isolated from locally collected plants. All the rhizobial bioinoculant strains tested, as well as the laboratory isolate ST1, lacked the ability to utilize enterobactin, pyoverdine, and ferrichrome siderophores. *pfeA* and *fpvA* receptor genes for enterobactin and pyoverdine uptake, respectively, from *P. aeruginosa* and *fegAB* receptor gene for ferrichrome uptake from *B. japonicum* 61A152 were cloned and expressed in the rhizobial strains. The resultant transformants were able to utilize enterobactin, pyoverdine, and ferrichrome, indicating the successful expression of *pfeA, fpvA,* and *fegAB* into these rhizobial strains. The ability to utilize enterobactin, pyoverdine, and ferrichrome conferred upon the strains significant growth stimulation in the presence of pure enterobactin, pyoverdine, and ferrichrome as well as when co-inoculated with the enterobactin producing *E. coli* AN102, pyoverdine producing *P. aeruginosa* 7NSK2-562, and ferrichrome producing *U. maydis* under iron limiting laboratory conditions. SDS-PAGE analysis of OMP revealed the presence of an approximately ~80 kDa protein on the OM of all the transformed rhizobia which was absent in the parent, implying the expression of the pfeA, fpvA, and fegAB proteins. The plants treated with the transformed rhizobial strains showed a better plant health with respect to increase in shoot weight, nodule number, and chlorophyll content of leaves as compared to plants inoculated with the parent strain under both autoclaved and un-autoclaved soil conditions. The highest increase in all the growth parameters was observed in plants inoculated with transformants having *feA*, encoding enterobactin receptor which has the highest affinity for $Fe^{+3}$ ($K = 10^{52}\ M^{-1}$) followed by those having fpvA and fegAB constructs encoding pyoverdin and ferrichrome receptors with affinity for Fe in the decreasing order (Fig. 6.6). The studies thus support the hypothesis that presence of specific siderophore receptor gene would lead to increase in iron acquisition and hence increase in survival competence of the organism in the rhizosphere leading to stimulation in plant growth (Khan 2010).

#### 6.3.3.8 Mechanism of Transport of Ferri-siderophores from Outside the Cell to Its Interior

Extensive research has been done to study the transport of ferri-siderophore uptake in Gram-negative bacteria (Braun et al. 1998; van der Helm 1998). Ferri-siderophore complexes are actively transported across the OM in Gram-negative bacteria through specific OM receptor proteins, the genes of which are expressed under iron deficient conditions (Earhart and McIntosh 1977). The transport of ferri-siderophore complex through OM receptor is energy dependent and is mediated through energy transducing complex comprising of proteins, Ton B, Exb B, and Exb D which couple the electrochemical gradient across the cytoplasmic membrane to a highly specific receptor promoting transport of the iron complex across the OM (Fig. 6.7)

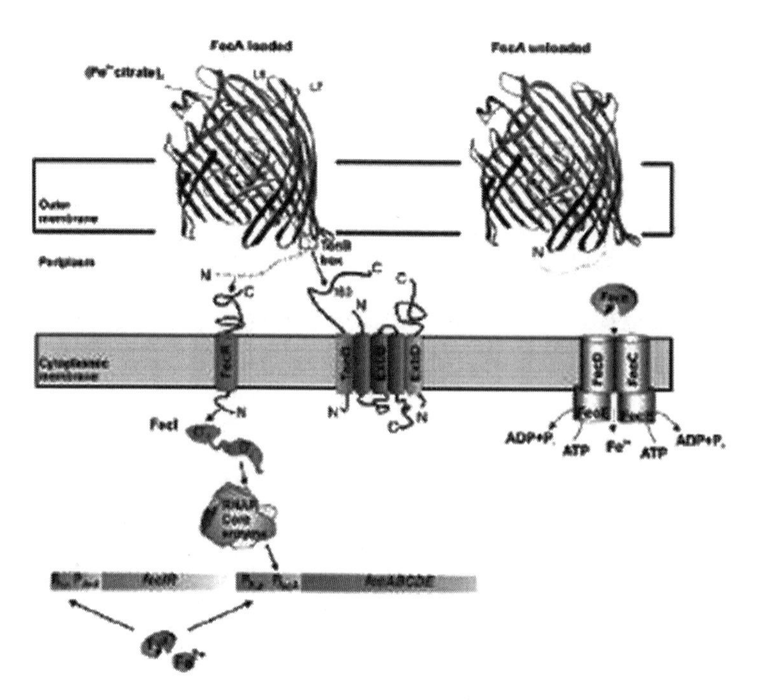


**Fig. 6.6** The ferric citrate transport and regulatory system. The involvement of TonB, ExbB, and ExbD in the signaling pathway from FecA to FecI and transport of iron through the periplasmic FecB protein and the ABC transporter FecCDE proteins

## 6.4 Regulation of Siderophore Uptake Systems

Even though the reactivity of iron atoms makes it useful in many different biological applications, considerable side chain reactions are known to occur. Through Fenton-type chemistry, iron catalyzes the production of toxic hydroxyl radicals from hydrogen peroxide which are generated from the spontaneous combinations of superoxide anions created by oxidative metabolism in the cell (Touti 2000). Oxygen radicals and peroxides are highly destructive, damaging lipids, proteins, and nucleic acid in the cell. However, a simple method to reduce radical formation by iron is to limit the availability of the iron atom itself by sensing the iron levels and limit its uptake.

Different types of ferric uptake regulatory system are reported in bacteria. In majority of Gram-negative organisms, "Fur" (ferric uptake regulator) is considered to be the key regulator for expression of genes involved in iron uptake (Hantke 1981). Fur is a transcriptional repressor of more than 90 different genes involved in iron uptake (Wexler et al. 2003). In some Gram-positive bacteria, iron responsive gene regulation is mediated by members of Dtx R family, identified in *Corynebacterium diphtheria* as a regulator of iron-dependent diphtheria toxin. This protein also uses ferrous iron as a co-repressor (Qian et al. 2002) like Fur, but shows no sequence homology to it (Wexler et al. 2003). One of the members of Fur super family is Irr (iron responsive regulator), which is restricted to a few $\alpha$-proteobacteria including

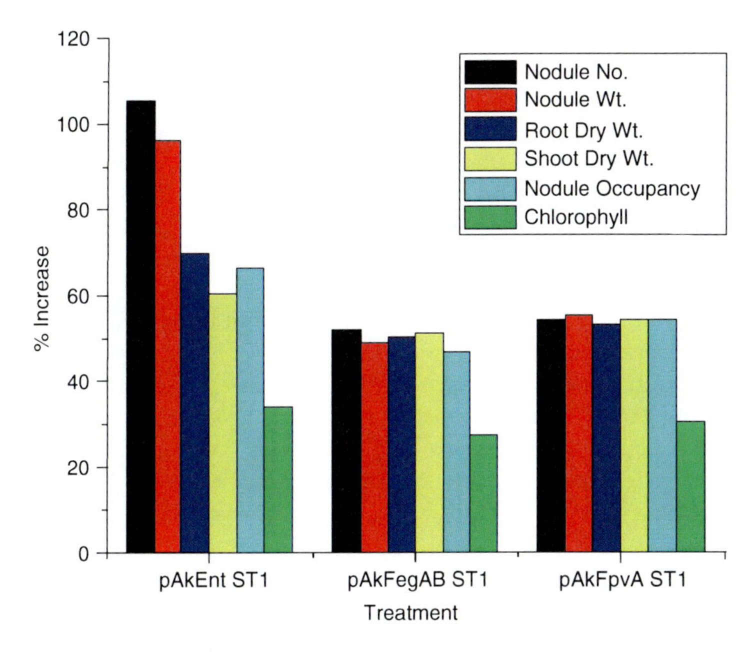


**Fig. 6.7** Comparison of effect inoculation of pAKEnt, pAKFpvA, and pAKFegAB transformed *Rhizobium* sp ST1 on various growth parameters of groundnut plant in unautoclaved soil condition. pAKEntST1, pAKFegABST1, and pAKFpvST1 denote *Rhizobium* sp ST1 constructs having *feA, fegAb,* and *fpvA* genes on a plasmid. The % increase in all the parameters is in comparison with those plants inoculated with parent *Rhizobium* sp. ST1

rhizobia, *Agrobacterium, Brucella*, and *Rhodopseudomonas palustris.* Regulation of iron responsive genes in *R. leguminosarum* and *S. meliloti* is not mediated by Fur but rather by a dissimilar RirA (rhizobial iron regulator) protein, a member of Rtz family of regulators (Todd et al. 2002).

Rice plants utilize the iron chelators known as mugineic acid family phytosiderophores (MAs) to acquire iron from the rhizosphere. Synthesis of MAs and uptake of MA-chelated iron are strongly induced under conditions of iron deficiency. In Fe-deficient barley roots, the expression of the genes corresponding to all of the steps in the biosynthesis of MAs from methionine is induced, as confirmed by northern and microarray analyses (Higuchi et al. 1999; Takahashi et al. 1999; Nakanishi et al. 2000; Negishi et al. 2002)

## 6.5 Mechanisms Other than Siderophore Production to Acquire Iron

In addition to producing one or multiple siderophore acquisition systems, bacteria have evolved other mechanisms for obtaining scarce iron from a wide range of environments. Many Gram-negative bacteria possess a system for transporting ferrous iron, termed the Feo system, which is typically only expressed under anaerobic conditions (Kammler et al. 1993). In some bacteria this system is essential, particularly in organisms that colonize the stomach or intestine, such as *Helicobacter pylori* and *Salmonella enterica* (Wandersman and Delepelaire 2004). Another putative mechanism for iron acquisition is the Sit ABCD system found in *Salmonella* and *Shigella* species and in enteroinvasive *E. coli*. Depending on the organism, this system is thought to transport ferric or ferrous iron (Runyan-Janecky et al. 2003). It is not certain that this system transports iron, but studies have shown that it is partially regulated by Fur and available iron in the environment (Janakiraman and Slauch 2000). Additionally, it has been shown that induction of the *sit* genes can improve growth of mutants deficient in iron transport (Runyan-Janecky et al. 2003). Similar observations were made in microarray study conducted in rice by Kobayashi et al. (2005) which clearly demonstrated that all the genes involved in the synthesis of MAs are induced in Fe-deficient rice roots. These include all the genes participating in the steps from *S* adenosyl-methionine to MAs which have already been shown by northern analysis to be up-regulated by Fe deficiency in roots (Higuchi et al. 2001; Inoue et al. 2003, 2004; Nozoye et al. 2004).

Zhang et al. (2009) reported a previously unidentified mechanism to increase iron assimilation in *Arabidopsis* mediated by the PGP *Bacillus subtilis* GB03. The studies revealed that GB03 transcriptionally up-regulates Fe-deficiency-induced transcription factor 1(*FIT1*), which is necessary for GB03-induction of ferric reductase *FR02* and the iron transporter *IRT 1*. Gb03 also caused acidification of rhizosphere by enhancing root proton release and by direct bacterial acidification thereby facilitating iron mobility. The results thus demonstrate the potential of microbes to control iron acquisition in plants by way of integration of microbial signaling in photosynthetic regulation.

## 6.6 Role of Siderophores Produced by Rhizosphere Colonizing and Nodulating Bacteria

Plant growth promoting rhizobacteria (PGPR) enhance plant growth by direct and indirect mechanisms. One such mechanism, which falls under indirect one, is production of siderophores by rhizospheric microorganisms. The control of phytopathogens is brought about by production of siderophores by these organisms which chelate iron in the rhizosphere thereby making it unavailable to the phytopathogens.

### *6.6.1 Role of* Pseudomonas *Siderophores in Plant Growth Promotion*

The importance of *Pseudomonas* siderophores in plant growth promotion was realized through the pioneering work of Kloepper et al. (1980) who searched for the mechanism of disease suppressiveness of certain soils which were not conducive to disease even though the phytopathogens were present. A specific strain of *Pseudomonas* (B10) isolated from suppressive soils or its purified siderophore, pseudobactin (pyoverdine), when applied to soils conducive to *Fusarium* wilt or take all disease caused by *Gaeumannomyces graminis* converted them to disease suppressive soils. Also, amendment of exogenous iron (III) to disease-suppressive soils converted them to conductive soils by repressing siderophore production by the *Pseudomonas* species. Since then many bacteria are reported to be effective in biocontrol of plant diseases due to their ability to outcompete pathogenic bacteria or fungi for iron by producing a superior iron-chelating complex with a higher affinity for iron, by producing siderophores in greater abundance, or by utilization of multiple siderophores (Raaijmakers et al. 1995; Loper and Buyer 1991). Siderophore production by fluorescent *Pseudomonas* spp. has been suggested or demonstrated to be involved in the suppression of *G. graminis* var. *tritici* (Kloepper et al. 1980), *F. oxysporum* (Elad and Baker 1985), *Pythium* spp. (Becker and Cook 1988; Loper 1988). Antagonistic activity against plant pathogens leads to an improvement in plant health (Loper and Buyer 1991) and that against deleterious microorganisms to an enhanced plant growth (Becker and Cook 1988; Schippers et al. 1987). Siderophores are also implicated in the induction of defense reactions in plants. *P. fluorescens* CHA0 was shown to induce systemic resistance of tobacco, whereas its pvd$^-$ mutant was less efficient than the wild type (Maurhofer et al. 1994).

Many microbial siderophores including pyoverdines are also involved in directly improving the iron nutrition of various Strategy I and Strategy II plant species (Crowley et al. 1988, 1992; Hordt et al. 2000). An increase in iron content and uptake has been shown in case of strategy I plants such as peanut, cotton, cucumber, and *Arabidopsis* supplemented with Fe-pyoverdine (Bar-Ness et al. 1991). Vansuyt et al. (2007) showed that iron chelated to pyoverdine was taken up by *A. thaliana* plants in a manner independent of reductase activity, leading to an increased plant growth.

### 6.6.2 Role of Rhizobial Siderophores in Plant Growth Promotion

It is not known clearly what advantages are conferred by a free-living *Rhizobium* strain possessing machinery for siderophore production and uptake, but evidences are available (Joshi et al. 2008) to show that rhizobial siderophores do play a role in competition in rhizosphere, perhaps in a manner similar to that of siderophores in *Pseudomonas* strains. In the competitive soil environment, plant growth promotion and nodulation by rhizobia is promoted by certain rhizobacteria (Bai et al. 2002; Dahsti et al. 1998; Rao and Pal 2003), which indirectly becomes beneficial for plant growth. However, other plant deleterious bacteria also exist in soil, and if they are capable of rhizospheric colonization, they negatively affect plant growth. In addition to the nitrogen fixation activity performed by rhizobia, they are also reported as effective biocontrol agents for the inhibition of these soil-borne plant pathogens

(Chakraborty and Purkayastha 1984), plant growth enhancement through IAA production, uptake of phosphorous and other minerals, etc. Many species of rhizobia promote plant growth and also inhibit the growth of certain pathogenic fungi. *Rhizobium meliloti* and *B. japonicum* bacterized seeds are known to have reduced *Macrophomina* infection; the mechanism involved being siderophore production which inhibits the growth of *Macrophomina phaseolina* by starving it for iron (Arora et al. 2001; Deshwal et al. 2003).

## 6.7 Conclusions

Role of PS and microbial siderophores in plant iron nutrition and protection from phytopathogens are well documented and covered in many reviews. The plant rhizosphere is a dynamic environment where severe competition amongst microorganisms occurs for various limiting nutrients, one of which is iron. The emphasis in the present chapter apart from the general aspects has been on specific aspects of involvement of siderophores in saprophytic competition between soil microorganisms based on the affinities of the siderophores and cross-utilization abilities of the strains. Siderophores vary in their affinity for iron and many have such a high affinity that they are expected to extract iron from most chemical and organic complexes simply by equilibrium displacements. Differential affinities of siderophore for iron not only enables them to quench iron from soil but also allows mobilization of iron from weaker associated ferric-siderophore complexes. The biocontrol activity of siderophore is mainly based on this principle. Our studies provided a direct proof for this and showed that rhizospheric isolates differed in their ability to cross-utilize siderophores produced by other rhizospheric organisms and the growth of isolates producing siderophores with weak affinity for iron gets inhibited in the presence of isolates producing siderophores having higher affinity for iron. It can therefore be concluded that an organism producing siderophore having high affinity for iron can have an ecological advantage for growth and survival in the rhizosphere. Studies pertaining to siderophore production and siderophore-mediated iron nutrition through utilization of homologous and heterologous siderophore complexes in rhizobia are of significance, since many iron-rich proteins play significant role in the process of nitrogen fixation. Hydroxamate type siderophores are main amongst the types of siderophores found in soil, most of it being ferrichrome type. Since nodule bacteria are reported to be poor producer as well as utilizers of hydroxamate type siderophores, it was conceptualized that engineering a strain with gene for OM receptor for hydroxamate siderophores can have profound effect on its survival in this type of environment. Our studies done with multiple systems and cloning different siderophore receptor genes confirmed this hypothesis and showed that plants inoculated with such constructs did show significant growth stimulation. The growth stimulation was correlated with the affinity the siderophore had for iron, enterobactin with highest affinity showing highest growth stimulation. However, whether siderophore directly contributes to plant growth or indirectly through competitive interactions based on differential siderophore affinities remains to be explored. Detailed structural information on siderophore receptor protein and its interaction with the cognate siderophore would open newer avenues for research in this field and would help answering some of the unresolved question in this field.

**Acknowledgement** The authors thank students from their laboratory whose work has been cited in this chapter. The financial assistance provided by DBT, India and UGC, India to support the research is acknowledged.

## References

Akers HA (1983) Isolation of the siderophore schizokinen from soil of rice fields. Appl Environ Microbiol 45:1704–1706

Amerelle V, O'Brien MR, Fabiano E (2008) ShmR is essential for utilization of heme as a nutritional iron source in *Sinorhizobium meliloti*. Appl Environ Microbial 74:6463–6475

Andrew SC, Robinson AK, Rodriguez-Quinons F (2003) Bacterial iron homeostasis. FEMS Microbiol Rev 27:215–237

Ankenbaur RG, Quan HN (1994) *Fpt* A, the Fe(III)-pyochelin receptor of *Pseudomonas aeruginosa*: a phenolate siderophore receptor homologous to hydroxamate siderophore receptors. J Bacteriol 176:307–319

Archana G (2010) Engineering nodulation competitiveness of rhizobial bioinoculants in soil. In: Knan MS et al (eds) Microbes for legume improvement. Springer, Wien. doi:10.1007/978-3-211-99753-6_8

Archibald F (1983) *Lactobacillus plantarum*, an organism not requiring iron. FEMS Microbiol Lett 19:29–32

Arora NK, Kang SC, Maheshwari DK (2001) Isolation of siderophore producing strains of *Rhizobium meliloti* and their biocontrol potential against *Macrophomina phaseolina* that causes charcoal rot of groundnut. Curr Sci 81:673–677

Bai YM, D'Aoust F, Smith DL, Driscott BT (2002) Isolation of plant growth promoting *Bacillus* strains from soybean nodules. Can J Microbiol 48:230–238

Bar-Ness E, Chen Y, Hadar Y, Marschner H, Römheld V (1991) Siderophores of *Pseudomonas putida* as an iron source for dicot and monocot plants. Plant Soil 130:231–241

Becker JO, Cook RJ (1988) Role of siderophores in suppression of *Pythium* species and production of increased growth response of wheat by fluorescent pseudomonads. Phytopathology 78: 778–782

Benson HP, Boncompagni E, Guerinot ML (2005) An iron uptake operon required for proper nodule development in the *Bradyrhizobium japonicerm* – Soybean symbiosis. Mol Plant Microbe Interact 18:950–959

Boukhalfa H, Crumbliss AL (2002) Chemical aspects of siderophore mediated iron transport. Biometals 15:325–339

Braun V (2005) Bacterial iron transport related to virulence Russel W, Herwald H (eds) Concepts in Bacterial virulence. Contrib Microbiol, Basl, Kaeger pp 210–233

Braun V, Hantke K, Koster W (1998) Bacterial iron transport: mechanisms, genetics and regulation. Met Ions Biol Syst 35:67–145

Brickman TJ, Armstrong SK (1999) Essential role of the iron-regulated outer membrane receptor fauA in alcaligin siderophore mediated iron uptake in *Bordetella* species. J Bacteriol 181: 5958–5966

Buchanan SK, Smith BS, Venkatramani L, Xia D, Esser L, Palnitkar M, Chakraborty R, Van der Helm D, Deisenhofer J (1999) Crystal structure of the outer membrane active transporter Feg A from *E. coli*. Nat struct Biol 6:56–63
Budzikiewicz H (1993) Secondary metabolites from fluorescent pseudomonads. FEMS Microbiol Rev 104:209–228
Buyer JS, Leong J (1986) Iron transport – mediated antagonism between plant growth promoting and plant deleterious Pseudomonas strains. J Biol Chem 261:791–794
Buyer JS, Kratzke MG, Sikora LJ (1993) A method for detection of pseudobactin, the siderophore produced by a plant-growth-promoting *Pseudomonas* strain, in the barley rhizosphere. Appl Environ Microbiol 59:677–681
Byers BR, Araceneaux JE (1998) Microbial iron transport: iron acquisition by pathogenic microorganisms. Met Ions Biol Syst 35:37–366
Byers BR, and Arceneaux JEL (1977) Microbial transport and utilization of iron. In: Weinberg ED, (ed.), Microorganisms and minerals. Marcel Dekker, Inc., New York p. 215–249
Chakraborty U, Purkayastha RP (1984) Role of rhizobiotoxine in protecting soybean roots from *Macrophomina phaseolina* infection. Can J Microbiol 30:285–289
Chen LM, Dick W, Streeter JG, Hoitink HAJ (1998) Fe chelates from compost microorganisms improve Fe nutrition of soybean and oat. Plant soil 200:139–147
Cobessi D, Celia H, Folschweiller N, Schalk IJ, Abdallah MA, Pattus F (2005) The crystal structure of the pyoverdine outer membrane receptor Fpv A from *Pseudomonas aeruginosa* at 3.6 angstroms resolution. J Mol Biol 347(1):121–134
Cornelis P, Matthijs S (2002) Diversity of siderophore mediated iron uptake systems in fluorescent pseudomonads: not only pyoverdins. Environ Microbiol 4:787–798
Cornelissen CN (2003) Transferrin-iron uptake by gram-negative bacteria. Front Biosci 8:836–847
Cornish AS, Page WJ (1998) The catecholate siderophores of *Azotobacter vinelandi:* their affinity for iron and role in oxygen stress management. Microbiology 144:1747–1754
Crosa JH (1989) Genetics and molecular biology of siderophore mediated iron transport in *E. coli*. Microbiol Rev 53:517–530
Crowley DE, Reid CPP, Szaniszlo PJ (1987) Microbial siderophores as iron sources for plants. In: Winklemann G, van der Helm D, Neilands JB (eds) Iron transport in microbes, plants and animals. VCH Publishers, New York, pp 375–386
Crowley DE, Reid CPP, Szaniszlo PJ (1988) Utilization of microbial siderophores in iron acquisition by oat. Plant Physiol 87:685–688
Crowley DE, Ro¨mheld V, Marschner H, Szaniszlo PJ (1992) Root-microbial effects on plant iron uptake from siderophores and phytosiderophores. Plant Soil 142:1–7
Dahsti N, Zhang F, Hynes R, Smith DL (1998) Plant growth promoting rhizobacteria accelerate nodulation and increase nitrogen fixation activity by field grown soybean under short season conditions. Plant Soil 200:205–213
Darrah PR (1993) The rhizosphere and plant nutrition-a quantitative approach. Plant soil 156:1–20
Dean CR, Poole K (1993) Cloning and characterization of the ferric enterobactin receptor gene (pfe A) of *Pseudomonas aeruginosa*. J Bacteriol 175:317–324
Deshwal VK, Dubey RC, Maheshwari DK (2003) Isolation of plant growth promoting strains of *Bradyrhizobium (Arachis)* sp. with biocontrol potential against *Macrophomina phaseolina* causing charcoal rot of peanut. Curr Sci 84:443–448
Dhungana S, Crumbliss AL (2005) Coordination chemistry and redox processes in siderophore-mediated iron transport. Geomicrobiol J 22:87–98
Duijff BJ, Bakker PAHM, Schippers B (1994) Suppression of Fusarium wilt of carnation by *Pseudomonas putida* WCS358 at different levels of disease incidence and iron availability. Biocontrol Sci Technol 4:279–288
Earhart CF, McIntosh MA (1977) Coordinate regulation by iron of the synthesis of phenolate compounds and three outer membrane proteins in *E. coli*. J Bacteriol 131:331–339
Ehteshamul-Haque S, Hashmi RY, Ghaffar A (1992) Biological control of root rot disease of lentil Lens. Newsletter 19:43–45

Elad Y, Baker R (1985) The role of competition for iron and carbon in suppression of chlamydospore germination of *Fusarium* spp. by Pseudomonas spp. Phytopathology 75:1053–1059

Endriss F, Braun V (2004) Loop deletions indicate regions important for Fhu A transport and receptor functions in *E. coli*. J Bacteriol 186:4818–4823

Faraldo-Gomez JD, and Sansom MSP (2003) Acquisition of siderophores in gram-negative bacteria. Nat Rev Mol Cell Bio 4:105–116

Ferguson A, Hofmann E, Coulton JW, Diederichs K, Welte W (1998) Siderophore mediated Iron transport: crystal structure of Flu A with bound dipopolysaccharide. Science 282:2215–2220

Ferguson AD, Chakraborty R, Smith BS, Esser L, Van der Helm D, and Deisenhofer, J (2002) Structural basis of gating by the outer membrane transporter FecA. Science. 295:1715–1719

Guerinot ML (1994) Microbial iron transport. Annu Rev Microbiol 48:743–772

Guerinot ML, Meidl EJ, Plessner O (1990) Citrate as a siderophore in *Bradyrhizobiam japoniam*. J Bacteriol 172:3298–3303

Guiseppi JD, Fridovich I (1982) Oxygen toxicity in *Streptococcus sanguis*. J Biol Chem 257: 4046–4051

Halda-Alija L (2003) Identification of indole-3 acetic acid producing fresh water wetland rhizosphere bacteria associated with *Juncus effuses* L. Can J Microbiol 49:781–787

Hantke K (1981) Regulation of ferric uptake iron transport in *E. coli* K12- isolation of a constitutive mutant. Mol Gen Genet 182:288–293

Hell R, Stephen UW (2003) Iron uptake, trafficking and homeostasis in plants. Planta 216:541–551

Hemantaranjan A, Garg OK (1986) Introduction of nitrogen fixing nodules through iron and zinc fertilization in the non nodule-forming French bean (*Phaseolus vulgaris* L). J Plant Nutr 9: 281–288

Hibbing ME, Fuqua C, Parsek MR, Peterson SB (2010) Bacterial competition: surviving and thriving in the microbial jungle. Nat Rev Microbiol 8:15–25

Higuchi K, Suzuki K, Nakanishi H, Yamaguchi H, Nishizawa NK, Mori S (1999) Cloning of nicotinamide synthase genes, novel genes involved in the biosynthesis of phytosiderophores. Plant Physiol 119:471–479

Higuchi K, Watanabe S, Takahashi M, Kawasaki S, Nakanishi H, Nishizawa NK, Mori S (2001) Nicotinamide synthase gene expression differs in barley and rice under Fe- deificient conditions. Plant J 25:159–167

Hinsinger P, Plassard C, Tang C, Jaillard B (2003) Origins of root-mediated pH changes in the rhizosphere and their responses to environmental constraints: A review. Plant Soil 248:293–303

Hordt W, Ro¨mheld V, Winkelmann G (2000) Fusarinines and dimerum acid, mono- and dihydroxamate siderophores from *Penicillium chrysogenum*, improve iron utilization by strategy I and Strategy II plants. BioMetals 13:37–46

Inoue H, Higuchi K, Takahashi M, Nakanishi H, Mori S, Nishizawa NK (2003) Three rice nicotinamide synthetase genes OsNAS1, OsNAS2 and OsNAS3 are expressed in cells involved in long distance transport of iron and differentaitially regulated by iron. Plant J 36:366–381

Inoue H, Suzuki M, Takahashi M , Nakashini H, Mori S, Nishizawa NK (2004) Rice nicotinamide aminotransferase gene (NAATI) is expressed in cells involved in long distance transport of iron. In: Abstracts of the XII International symposium on iron nutrition and interactions in plants Tokyo, Japan pp 204

Jadhav RS, Desai AJ (1994) Role of siderophore in iron uptake in cowpea Rhizobium GN1 (Peanut isolate): possible involvement of iron repressible outer membrane proteins. FEMS Microbiol Lett 115:185–190

Jadhav R, Thaker NV, Desai A (1994) Involvement of the siderophores of cowpea Rhizobium in the iron nutrition of the peanut. World J Microbiol Biotechnol 10:360–361

Janakiraman A, Slauch JM (2000) The putative iron transport system SitABCD encoded on SP II is required for full virulence of *Salmonella typhimurium*. Mol Microbiol 35:1146–1155

Jin CW, He YF, Tang CX, Wu P, Zhng SJ (2006) Mechanisms of microbial enhanced iron uptake in red clover. Plant Cell Environ 29:888–897

Jin CW, You GY, Zheng SJ (2008) The iron deficiency-induced phenolics secretion plays multiple important roles in plant iron acquisition underground. Plant Signal Behavior 3:60–61
Joshi F, Archana G, Desai A (2006a) Siderophore cross-utilization amongst rhizospheric bacteria and the role of their differential affinities for $Fe^{3+}$ on growth stimulation under iron-limited conditions. Curr Microbiol 53:141–147
Joshi FR, Kholiya SP, Archana G, Desai AJ (2006b) Siderophore cross-utilization amongst nodule isolates of the cowpea miscellany group and its effect on plant growth in the presence of antagonistic organisms. Microbiol Res. doi:10.1016
Joshi F, Chaudhary A, Joglekar P, Archana G, Desai AJ (2008) Effect of expression of Brdyrhizobium japonicum 61A152 fegA gene in Mesorhizobium sp., on its competitive survival and nodule occupancy on Arachis hypogeal. Appl Soil Ecol 40:338–347
Joyner DC, Lindow SE (2000) Heterogeneity of iron bioavailability on plants assessed with a whole-cell GFP-based bacterial biosensor. Microbiology 146:2435–2445
Jurkevitch E, Hadar Y, Chan Y (1992) Differential siderophore utilization and iron uptake by soil and rhizosphere bacteria. Appl Environ Microbial 58:119–124
Kammler M, Schon C, Hantke K (1993) Characterisation of the ferrous iron uptake system of *E. coli*. J Bacteriol 175:6212–6219
Khan A (2010) Utilization of heterologous iron-siderophore complex by Peanut and Pigeon pea Rhizobia: Cloning and expression of cognate receptor gene PhD Thesis The MS University of Baroda, Vadodara INDIA
Khan A, Geetha R, Akolkar A, Pandya A, Archana G, Desai AJ (2006) Differential cross-utilization of heterologous siderophores by nodule bacteria of Cajanus cajan and its possible role in growth under iron-limited conditions. Appl Soil Ecol 34:19–26
Kloepper JW, Leong J, Tientze M, Schroth MN (1980) Pseudomonas siderophores: a mechanism explaining disease-suppressive soils. Curr Microbiol 4:317–320
Kobayashi T, Suzuki M, Inoue H, Nakashini R, Michiko I (2005) Expression of iron-acquisition-related genes in iron-deficient rice is co-ordinately induced by partially conserved iron-deficiency-responsive elements. J Expt Bot 56:1305–1316
Kraemer SM, Crowley DE, Kretzschmar R (2006) Geochemical aspects of phytosiderophore promoted iron acquisition by plants. Adv Agron 91:1–46
Lemanceau P, Expert D, Gaymard D, Bakker PAHM (2009) Role of iron in plant-microbe interactions. Adv Bot Res 51:491–549
Levier K, Guerinot ML (1996) The *Bradyrhizobium japonicum* feg A gene encodes an iron regulated outer membrane protein with similarity to hydroxamate-type siderophore receptors. J Bacteriol 178:7265–7275
Loper JE (1988) Role of fluorescent siderophore production in biological control of Pythium ultimum by a Pseudomonas fluorescens strain. Phytopathology 78:166–172
Loper JE, Buyer JS (1991) Siderophores in microbial interactions on plant surfaces. Mol Plant Microbe Interact 4:5–13
Loper JE, Henkels MD (1997) Availability of iron to *Pseudomonas fluorescens* in rhizosphere and bulk soil evaluated with an ice nucleation reporter gene. Appl Environ Microbiol 63(1):99–105
Loper JE, Lindow SE (1994) A biological sensor for iron available to bacteria in their habitats on plant surfaces. Appl Environ Microbiol 60:1934–1941
Marschner P, Crowley DE (1997) Iron Stress and Pyoverdin Production by a Fluorescent Pseudomonad in the Rhizosphere of WhiteLupine (*Lupinus albus* L.) and Barley (*Hordeum vulgare* L.). Appl Environ Microbiol 63(1):277–281
Masalha J, Kosegarten H, Elmaci O, Mengel K (2000) The central role of microbial activity for iron acquisition in maize and sunflower. Biol Fertil Soil 30:433–439
Maurhofer M, Hase C, Meuwly P, Métraux J-P, Défago G (1994) Induction of systemic resistance of tobacco to tobacco necrosis virus by the root-colonizing Pseudomonas fluorescens strain CHA0: influence of the gacA gene and of pyoverdine production. Phytopathology 84:139–146
Meyer JN (2000) Pyoverdines: pigments siderophores and potential taxonomic markers of fluorescent *Pseudomonas* species. Arch Microbiol 174:135–142

Meyer JM, Abdallah MA (1978) The fluorescent pigment of *Pseudomonas fluorescens*: biosynthesis, purification and physicochemical properties. J Gen Microbiol 107:319–328

Miller RH, May S (1991) Legume inoculation: successes and failures. In: Keister DL, Cregan PB (eds) The Rhizospere and plant growth. Kluwer, New York, pp 123–134

Morrissey J, Guerinot L (2009) Iron uptake and transport in plants: the good, the bad and the ionome. Chem Rev 109(10):4553–4567

Morton D, Whitby P, Jin H, Ren Z, Stull T (1999) Effect of multiple mutations in the hemoglobin- and hemoglobin-hepatoglobin- binding proteins HgpA, HgpB, and HgpC of Haemoplilus influenzae type b. Infect Immun 67:2729–2739

Nakanishi H, Yamaguchi H, Sasakuma T, Nishizawa N, Mori S (2000) Two dioxygenase genes Ids3 and Ids2, from *Hordeum vulgare* are involved in the biosynthesis of mugineic acid family phytosiderophores. Plant Mol Biol 44:199–207

Negishi T, Nakanishi H, Yazaki J, Kishimoto N, Fujii F, Shimbo K, Yamamoto K, Sakata K, Saski T, Kikuchi S, Mori S, Nishizawa NK (2002) cDNA microarray analysis of gene expression during Fe-deficiency stress in barley suggests that polar transport of vesicles is implicated in phytosiderophore secretion in Fe- deficient barley roots. Plant J 30:83–94

Neilands JB (1977) Siderophores: Biochemical ecology and mechanism of iron transport in enterobacteria: In Bio-organic Chemistry II Raymond, K.N. (ed.) American Chemical Society, Washington D.C. pp.3–22

Neilands JB (1952) A crystalline organo-iron pigment from a rust fungus *Ustilagi sphaerogena*. J Am Chem Soc 74:486–487

Neilands JB (1981) Microbial iron compounds. Annu Rev Biochem 50:715–731

Neilands J.B. (1982) Microbial envelope proteins related to iron. Annu Rev Microbiol 36: 285–309

Neilands JB (1995) Siderophores: structure and function of microbial iron transport compounds. J Biol Chem 270:26723–26726

Nelson M, Cooper CR, Crowley DE, Reid CPP, Szaniszlo PJ (1988) An *Escherichia coli* bioassay of individual siderophores in soil. J Plant Nutr 11:915–924

Nienaber A, Hennecke H, Fischer HM (2001) Discovery of a heme uptake system in the soil bacterium *Bradyrhizobium japonicum*. Mol Microbiol 41:787–800

Nozoye T, Itai RN, Nagasaka S, Takahashi M, Nakanishi H, Mori S, Nishizawa NK (2004) Diuranal changes in the expression of genes that participate in phytosiderophore synthesis in rice. Soil Sci Plant Nutr 50:1125–1131

Nye PH (1981) Changes of pH across the rhizosphere induced by roots. Plant Soil 61:7–26

O'Brien IG, Gibson F (1970) The structure of enterochelin and related 2, 3-dihydroxy-N- benzoylserine conjugates from *E. coli*. Biochim Biophys Acta 215:293–402

O'Cuiv P, Clarke P, Lynch D, O'Connell M (2004) Identification of rht X and fpt X, novel genes encoding proteins that show homology and function in the utilization of the *siderophores rhizobactin* 1021 by *Sinorhizobium meliloti* and pyochelin by *Pseudomonas aeruginosa* respectively. J Bacteriol 186:2996–3005

O'Hara GW, Dilworth MJ, Boonkerd N, Parpial P (1988) Iron deficiency specifically limits nodules development in peanut inoculated with *Bradyrhizobium* sp. Newphytology 108:51–57

Patel HN, Chakraborty RN, Desai SB (1994) Effect of iron on siderophore production and on outer membrane proteins of *Rhizobium leguminosarum* IARI 102. Curr Microbiol 28:119–121

Perry RD, San Clemente CL (1979) Siderophore synthesis in *Klebsiella pneumoniac* and *Shigella sonnei* during iron deficiency. J Bacteriol 140:1129–1132

Persmark M, Pittman P, Buyer JS, Schwyn B, Gill R, Neilands JB (1993) Siderophore utilization by Bradyrhizobium japonicum. Appl Environ Microbiol 59:1688–1690

Plessner O, Klapatch T, Guerinot ML (1993) Siderophore utilization by *Bradyrhizobiam Japonicum*. Appl Environ Microbiol 59:1688–1690

Posey JE, Gherardini FC (2000) Lack of a role for iron in the Lyme disease pathogen. Science 288:1651–1653

Powell PE, Cline GR, Reid CPP, Szaniszlo PJ (1980) Occurrence of hydroxamate siderophore iron chelator in soils. Nature 287:833–834

Powell PE, Szaniszlo PJ, Reid CPP (1983) Confirmation of occurrence of hydroxamate siderophores in soil by a novel *Escherichia coli* bioassay. Appl Environ Microbiol 46: 1080–1083

Qian Y, Lee JH, Holmes RK (2002) Identification of a DtxR-regulated operon that is essential for siderophore-dependent iron uptake in *Corynebacterium diptheriae*. J Bacteriol 184: 4846–4856

Raaijmakers JM, Leeman M, Van Oorschot MPM, Van der Sluis I, Schippers B, Bakker PAHM (1995) Dose-response relationships in biological control of fusarium wilt of radish by Pseudomonas spp. Phytopathology 85:1075–1081

Rajendran G, Mistry S, Desai AJ, Archana G (2007) Functional expression of *E. coli fhu*A gene in *Rhizobium* spp. of *Cajanus cajan* provides growth advantage in presence of $Fe^{3+}$: ferrichrome as iron source. Arch Microbiol 187:257–264

Rao DLN, Pal KK (2003) Biofertilizers in oilseeds production: Status and future strategies. National Seminar on Stress Management in Oilseeds for attaining self reliance in vegetable oils. Directorate of Oilseeds Research. Indian Council of Agricultural research, Hyderabad, India. pp 195–220

Raymond KN, Dertz EA (2004) Biochemical and physical properties of siderophores In: Iron transport in bacteria. In: Cross JH, Mey AR, Pyne SM (eds) pp 3–17

Raymond KR, Dertz E, Kim SS (2003) Enterobactin: an archetype for microbial iron transport. Proc Natl Acad Sci USA 100:3584–3588

Reid RT, Live DH, Faulkner DJ, Butler A (1993) A siderophore from a marine bacterium with an exceptional high affinity constant. Nature 366:455–458

Reigh G, O'Conell M (1993) Siderophore mediated iron transport correlates with the presence of specific iron-regulated proteins in the outer membrane of *Rhizobium melildi*. J Bacteriol 175:94–102

Robin A, Mougel C, Siblot S, Vansuyt G, Mazurier S, Lemaneceae P (2006) Effect of ferritin overexpression in tobacco on the structure of bacterial and Pseudomonad communities associated with roots. FEMS Microbiol Ecol 58:492–502

Robin A, Mougel C, Siblot S, Vansuyt G, Mazurier S, Lemaneceae P (2007) Diversity of root associated fluorescent pseudomonads as affected by ferritin overexpression in tobacco. Environ Microbiol 9:1724–1737

Romheld V, Marschner H (1986) Evidence for a specific uptake system for iron phytosiderophores in roots of grasses. Plant Physiol 80:175–180

Rroco E, Kosegarten H, Harizaj F, Imani J, Mengel K (2003) The importance of soil microbial activity for the supply of iro to sorghum and rape. Eur J Agron 19:487–493

Runyan-Janecky LJ, Reeves SA, Gonzales EG, Payne SM (2003) Contribution of the Shigella flexeneri Sit, Iuc, and Feo iron acquisition systems to iron acquisition in vitro and in cultured cells. Infect Immun 71:1919–1928

Schippers B, Bakker AW, Bakker PAHM (1987) Interactions of deleterious and beneficial rhizosphere microorganisms and the effect of cropping practices. Annu Rev Phytopathol 25:339–358

Smith MJ, Schoolery JN, Schwyn B, Holden I, Neilands JB (1985) Rhizobactin a structurally novel siderophore from *Rhizobium meliloti*. J Am Chem Soc 107:1739–1743

Stevens JB, Carter RA, Hussain H, Carson KC, Dilworth MJ, Johnston AWB (1999) The *fhu* genes of *Rhizobium leguminosarum*, specifying siderophore uptake proteins: *fhu* DCB are adjacent to a pseudogene version of *fhuA*. Microbiology 145:593–601

Strecter JG (1994) Failure of inoculant rhizobia to overcome the dominance of indigenous strains for nodule formation. Can J Microbiol 40:513–522

Takagi, S. 1976. Naturally occurringiron-chelating compounds in oat- and rice-rootwashings. I. Activity measurement and preliminary characterization. Soil Sci. Plant Nutr. 22, 423–433.

Takahashi M, Yamaguchi H, Nakanishi H, Shiori T, Nishizawa N, Mori S (1999) Cloning for two genes for nicotinamide aminotransferase, critical enzyme in iron acquisition (strategy II) in graminaceous plant. Plant Physiol 121:947–956

Todd JD, Wexler M, Sawers G, Yeoman KH, Poole PS, Johnston AW (2002) RirA, an iron-responsible regulator in the symbiotic bacterium *Rhizobium leguminosarum*. Microbiology 148:4059–4071

Touti D (2000) Sensing and protecting against superoxide stress in *E.coli*- how many ways are there to trigger SOX RS response? Redox Rep 5(5):287–293

Van der Helm D (1998) The physical chemistry of bacterial outer-membrane siderophore receptor proteins. Met Ions Ions Biol Syst 35:355–401

Van Tiel-Mankvald GJ, Mentjox-Veruurt JN, Oudega B, Graff FK (1982) Siderophore production by *Enterobacter cloacae* and a common receptor protein for the uptake of aerobactin and cloaein DF 13. J Bacteriol 150:490–497

Vansuyt G, Robin A, Briat JF, Curie C, and Lemanceau P (2007) Iron acquisition from Fe-pyoverdine by Arabidopsis thaliana. Mol. Plant-Microbe Interact. 4:441–447.

Verma DPS, Long S (1983) The molecular biology of Rhizobium-legume symbiosis. Int Rev Cytol Snppl 14:211–245

Wandersman C, Delepelaire P (2004) Bacterial iron sources: From siderophores to homophores. Annu Rev Microbiol 58:611–647

Wang Y, Brown HN, Crowley DE, Szaniszlo PJ (2006) Evidence for direct utilization of a siderophore, ferrioxamine B, in axenically grown cucumber. Plant Cell Environ 16:579–585

Wexler M, Todd JD, O'Kolade D, Bellini AM, Hemmings GS, Johnston AWB (2003) Fur is not the global regulator of iron uptake genes in *Rhizobium leguminosarum*. Microbiology 149: 1357–1365

Yang CH, Crowley DE (2000) Rhizosphere microbial community structure in relation to root location and plant iron nutritional status. Appl Environ Microbiol 66:345–351

Zhang F, Romheld V, Marschner H (1991) Role of root apoplasm for iron acquisition by wheat plants. Plant Physiol 97:1302–1305

Zhang H, San Yan Xie X, Kim MS, Dowd SE, Pare PW (2009) A soil bacterium regulates plant acquisition of iron via deficiency-inducible mechanisms. Plant J 58:568–577

# Chapter 7
# Basic and Technological Aspects of Phytohormone Production by Microorganisms: *Azospirillum* sp. as a Model of Plant Growth Promoting Rhizobacteria

**Fabricio Cassán, Diego Perrig, Verónica Sgroy, and Virginia Luna**

## 7.1 Rhizosphere and Rhizobacteria

The soil is the natural support of plants wherein usually a large number of microorganisms, including bacteria and fungi, can proliferate. The term rhizosphere is used to describe the portion of soil in which growth of microorganisms is induced by the presence of the root system (Garate and Bonilla 2000). The rhizosphere bacteria, so-called rhizobacteria, are capable of colonizing the interior or exterior of the roots of many species and can be divided between those which form a symbiotic relationship with the plant and those that do not. The latter, called free-living, are closely associated with the root surface or reside within the roots as endophytic bacteria (Kloepper et al. 1989). When the presence of rhizobacteria benefits plant growth they are named Plant Growth Promoting Rhizobacteria (PGPR). The PGPR group has been divided according to the bacterial promotion mechanism used during interaction. The classification could include the (1) PGPR group, proposed by Kloepper and Schroth (1978), (2) biocontrol-Plant Growth Promoting Bacteria (biocontrol-PGPB) group, proposed by Bashan and Holguín (1998), and Plant Stress-Homeoregulating Rhizobacteria (PSHR) group, proposed by Cassán et al. (2009a). All of them either directly or indirectly facilitate or promote plant growth under optimal (PGPR) and biotic (biocontrol-PGPB) or abiotic (PSHR) stress conditions (Fig. 7.1).

Indirect plant growth promotion induced by biocontrol-PGPB in biotic stress conditions includes a variety of mechanisms by which bacteria prevent the deleterious effects of phytopathogens on plant growth, such as rhizospheric competition, induced systemic resistance (ISR), biosynthesis of stress-related phytohormones

---

F. Cassán (✉) • D. Perrig • V. Sgroy • V. Luna
Laboratorio de Fisiología Vegetal y de la Interacción planta-microorganismo, Universidad Nacional de Río Cuarto, Campus Universitario, CP 5800, Río Cuarto, Córdoba, República Argentina
e-mail: fcassan@exa.unrc.edu.ar

D.K. Maheshwari (ed.), *Bacteria in Agrobiology: Plant Nutrient Management*,
DOI 10.1007/978-3-642-21061-7_7, 

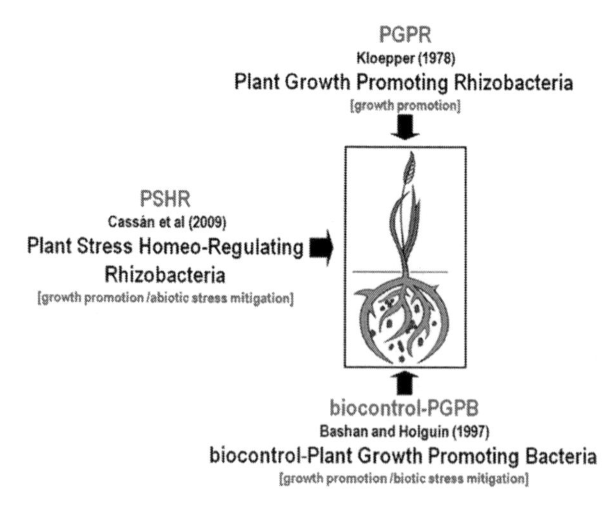


**Fig. 7.1** Plant growth promoting rhizobacteria classification, according to their capacity to interact with plants in either optimal or biotic and abiotic stress conditions

like jasmonic acid (Forchetti et al. 2007), or ethylene and biosynthesis of antimicrobial molecules (Glick and Bashan 1997). Direct growth promoting mechanisms induced by PGPR include: nitrogen fixation phytohormone production such as indole-3-acetic acid (IAA), gibberellic acid ($GA_3$), and cytokinins, e.g., Zeatin (Z) (Boiero et al. 2007; Perrig et al. 2007), iron sequestration by bacterial siderophores (Glick et al. 1999), and phosphate solubilization (de-Bashan and Bashan 2004). Indirect plant growth promotion induced by PSHR in abiotic stress conditions include production of stress-related phytohormones such as abscisic acid (ABA) (Perrig et al. 2007; Cohen et al. 2008), or other plant growth regulators such as cadaverine (Cassán et al. 2009a) and ethylene catabolism-related enzyme, e.g., 1-aminocyclopropane-1-carboxylate (ACC) deaminase, which reduces the level of ethylene production under unfavorable conditions, thus conferring resistance to stress (Glick et al. 1998), among other effects.

## 7.2 *Azospirillum* sp. as Model of Plant Growth Promoting Rhizobacteria

The available literature revealed *Azospirillum* sp. as one of the PGPR most studied at present, due to its ability to colonize more than 100 plant species in the world and significantly improve their growth, development, and in some cases productivity under field conditions (Bashan et al. 2004). One of the principal mechanisms currently proposed for *Azospirillum* sp. to explain plant growth promotion of inoculated plants has been related to its ability to produce and metabolize phytohormones and other plant growth regulator-like molecules (Okon and Labandera-González 1994). From a historic perspective, many studies detailing the beneficial effects of inoculation with PGPR, especially with *Azospirillum* sp., have been undertaken and describe morphological and physiological changes that occur in inoculated plants. However, in many cases the compounds responsible for generating such responses have not been identified and those responses are usually considered within a "black box" model which goes beyond the resulting growth promotion due to the presence of only this organism or the active metabolites in the culture medium or plant tissue. As a starting point for this chapter, we describe phytohormone production and other growth regulating compounds studied in the past three decades in different microorganisms. A significant portion of this flow of knowledge has been focused on the genus *Azospirillum* because (a) a considerable number of regulatory compounds have been identified that could potentially be responsible for modifying plant growth and architecture; (b) genes responsible for the synthesis of these compounds and their regulation under certain environmental conditions have been identified; (c) growth response of inoculated plants has been correlated with the levels of certain phytohormones produced by this microorganism in the culture medium, at rhizosphere level or in colonized plant tissues. It has been shown that the plant response to exogenous application of these compounds mimics inoculation, and finally (d) there is evidence that mutant strains with higher or lower phytohormone production have, respectively, more or less pronounced effects on plant hormone balance and growth promotion than isogenic strains, and this under a variety of experimental conditions. Tien et al. (1979) were the first to suggest that rhizosphere bacteria belonging to the genus *Azospirillum* could enhance plant growth by phytohormone production such as auxins, particularly IAA and cytokinins (CK), but subsequent work determined the capacity of this organism to produce also gibberellins (GAs), ethylene (ET), and ABA, among other molecules.

## 7.3 *Azospirillum* sp. and Their Mechanisms to Promote Plant Growth

The first mechanism proposed for the bacterial plant growth promotion by *Azospirillum* sp. has been associated almost exclusively with the nitrogen status in plants (Okon et al. 1983), through biological fixation or rhizospheric and endophytic nitrate reductase activity. However, these mechanisms have been of less agronomic significance than was initially expected (Bashan et al. 2004). Subsequently, other mechanisms have been studied and proposed for this microbial genus as being responsible for the inoculation response in plants, such as siderophore production, phosphate solubilization (Puente et al. 2004), biocontrol of phytopathogens (Bashan and de-Bashan, 2010), and protection of plants against stress like soil salinity or toxic compounds (Creus et al. 1997). Despite this, one of the most important mechanisms currently proclaiming to explain plant growth promotion would be related to the ability of *Azospirillum* sp. to produce or metabolize phytohormones and other plant growth regulators (Tien et al. 1979). Despite this,

*Azospirillum* sp. modes of action could be better explained by the "additive hypothesis" that allows to explain the plant growth promoting effect due to inoculation. This hypothesis was suggested 20 years ago (Bashan and Levanony 1990) and considers multiple mechanisms rather than one mechanism participating in the association of *Azospirillum* with plants. These mechanisms operate simultaneously or in succession with the contribution of an individual mechanism being less significant when evaluated separately.

## 7.4 Phytohormone Production by *Azospirillum* sp. and Other PGPR

One of the main mechanisms proposed to explain the "additive hypothesis" is related to the ability of *Azospirillum* sp. to produce or metabolize compounds such as phytohormones (Okon and Labandera-González 1994). It is known that about 80% of bacteria isolated from plant rhizosphere are capable of producing such compounds IAA (Cheryl and Glick 1996). However, Tien et al. (1979) were the first to suggest that *Azospirillum* sp. could enhance plant growth by phytohormones excretion. Today, we know that these rhizobacteria has been correlated with production in chemically defined compounds such as auxins (Prinsen et al. 1993), cytokinins (Tien et al. 1979), gibberellins (Bottini et al. 1989), ethylene (Strzelczyk et al. 1994), and other plant growth regulators, such as ABA (Perrig et al. 2007), polyamines like spermidine, spermine and the diamine cadaverine (Cassán et al. 2009a) and nitric oxide (Creus et al. 2005).

### *7.4.1 Auxins*

Auxin is the generic name that represents a group of chemical compounds characterized by its ability to induce cell elongation in the subapical region of the stem and to reproduce the physiological effect of IAA. These compounds have been associated in plants with processes such as (a) gravitropism and phototropism (growth of stems and roots in response to gravity and light, respectively); (b) vascular tissue differentiation; (c) apical dominance; (d) lateral and adventitious root initiation; (e) stimulation of cell division; and (f) stem and root elongation (Ross et al. 2000). The bacterial production and metabolism of auxins captured the attention of researchers at metabolic level due to bacterial ability to produce auxins and regulate their production in different environmental conditions (Costacurta and Vanderleyden 1995), and at morpho-physiological level due to the effect caused by bacterial auxin production on inoculated plants (Falik et al. 1989). Thus, members of the genus *Azospirillum* have provided an excellent experimental model to investigate and understand the physiological and molecular role of this phytohormone on plant and microbial growth, and also in plant–rhizobacteria interaction.

Several naturally occurring auxin-like molecules have been described as a product of bacterial metabolism in *Azospirillum* sp. pure cultures. In addition to the main form IAA, many other indole compounds have been identified in such conditions, such as indole-3-butyric acid (IBA) (Costacurta et al. 1994), indole-3-lactic acid (ILA) (Crozier et al. 1988), indole-3-acetamide (IAM) (Hartmann et al. 1983), indole-3-acetaldehyde (Costacurta et al. 1994), indole-3-ethanol and indole-3-methanol (Crozier et al. 1988), tryptamine, anthranilate and other yet uncharacterized indolic compounds (Hartmann et al. 1983). The physiological function of these compounds remains unknown. ILA is inactive as a phytohormone but it could compete with IAA for auxin-binding sites (Sprunck et al. 1995) (Fig. 7.2).

#### 7.4.1.1 Auxins and Bacterial Biosynthesis

At least six metabolic routes for the IAA biosynthesis have been proposed in bacteria and most of them use tryptophan (Trp) as principal precursor (Fig. 7.3). The pathways have been named indole-3-pyruvate (IPyA), IAM, tryptamine (TAM), tryptophan side-chain oxidase (TSO), indole-3-acetonitrile (IAN), and a tryptophan-independent pathway.

*Indole-3-pyruvate (IPyA) pathway* is the major route for IAA biosynthesis in plants. The IPyA pathway has been clearly described in a broad range of bacterial genera, such as *Bradyrhizobium, Azospirillum, Rhizobium,* and *Enterobacter,* as well as some cyanobacteria. The first step in the pathway is the conversion of Trp to IPyA by an aminotransferase enzyme. Then, IPyA is decarboxylated to indole-3-acetaldehyde (IAAld) by indole-3-pyruvate decarboxylase (IPDC). In the last step, IAAld is oxidized to IAA. The gene encoding the key enzyme IPDC (*ipdC*) has been isolated and characterized from *A. brasilense*, *E. cloacae*, *Pseudomonas putida,*

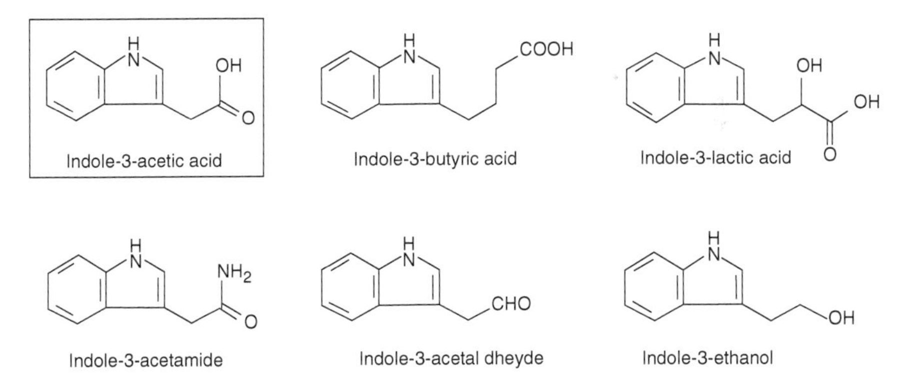


**Fig. 7.2** Various auxin-like molecules produced by *Azospirillum sp.* in pure cultures

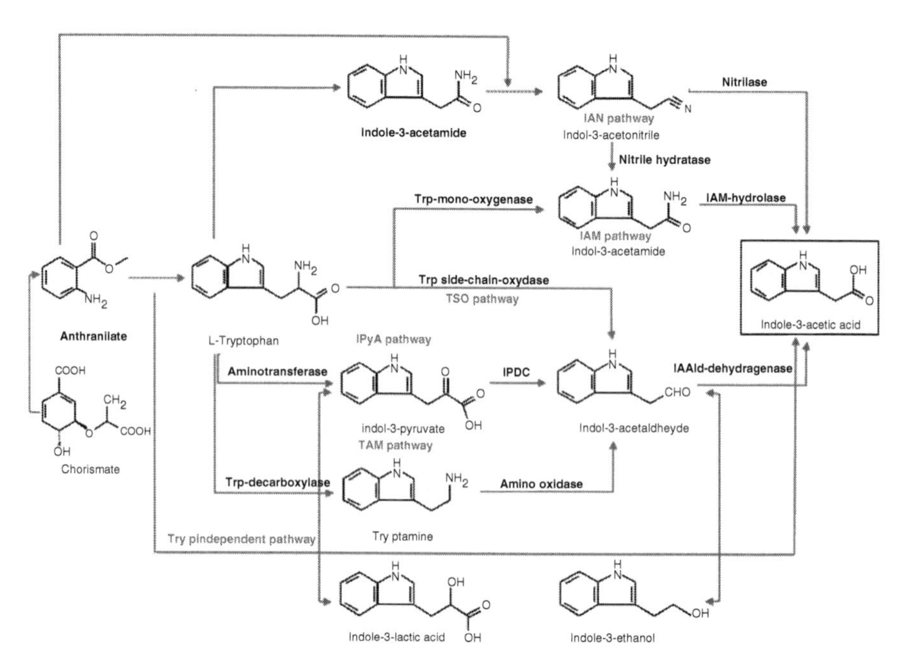


**Fig. 7.3** Hypothetical and actual pathways to synthesize indole-3-acetic acid (IAA) in rhizobacteria. Modified from Spaepen et al. (2008)

and *P. agglomerans* (Costacurta et al. 1994; Brandl and Lindow 1996; Patten et al. 2002).

*Indole-3-acetamide (IAM) pathway* is the best characterized pathway in bacteria. It is a two-step pathway in which tryptophan is first converted to IAM by trp-2-monooxygenase (IaaM), encoded by the *iaaM* gene. Secondly, IAM is converted to IAA by IAM hydrolase (IaaH) encoded by the *IaaH* gene. Both *IaaM* and *IaaH* have been cloned and characterized in various bacteria such as *Agrobacterium tumefaciens, Pseudomonas syringae, Pantoea agglomerans, Rhizobium,* and *Bradyrhizobium* sp. (Sekine et al. 1989; Clark et al. 1993; Morris 1995; Theunis et al. 2004). IAM has been found by GC-MS as an endogenous metabolite of *Arabidopsis thaliana* (Pollmann et al. 2003) which could point to the existence of this pathway in plants.

*Tryptamine (TAM) pathway* has been identified in *Bacillus cereus* by measuring the Trp-decarboxylase activity, catalyzing the decarboxylation of Trp to TAM (Perley and Stowe 1966), and in *Azospirillum brasilense* by conversion of exogenous TAM to IAA via IAAld (Hartmann et al. 1983). Endogenous TAM has also been identified in many plant species and the genes involved have been cloned and characterized.

*Tryptophan side-chain oxidase (TSO) pathway* has only been demonstrated in *Pseudomonas fluorescens* CHA0. In this pathway, Trp is directly converted to indole-3-acetaldehyde (IAAld), bypassing IPyA, and could be oxidized to IAA (Oberhansli et al. 1991). There are no indications that this pathway exist in plants.

*Indole-3-acetonitrile (IAN) pathway* has been extensively studied in plants. In bacteria such as *Alcaligenes faecalis* (Nagasawa et al. 1990; Kobayashi et al. 1993) enzymes that catalyze the conversion of IAN to IAA, named IAA-nitrilases, have been discovered. In both *A. tumefaciens* and *Rhizobium* sp. nitrile hydratase and amidase activity were identified, indicating the conversion of IAN to IAA via IAM (Kobayashi et al. 1995).

*Tryptophan-independent pathway* has been proposed in plants, and particularly in *A. thaliana* (Last et al. 1991). However, no enzyme has been characterized thus far. The Trp-independent pathway could only be demonstrated in *A. brasilense* by feedings experiments with labeled precursors (Prinsen et al. 1993). Because no specific enzymes have been identified yet, their existence remains elusive.

#### 7.4.1.2 IAA Conjugation

In plants, IAA is mostly found conjugated with sugars or amino acids because it is the best way to transport, store and protect it from catabolic enzymes. Conjugates could also control IAA level in the cell through a homeostatic mechanism (Cohen and Bandurski 1982). The only isolated and characterized bacterial gene involved in IAA conjugation is IAA-lysine synthetase (*iaaL*) from *Pseudomonas savastanoi* pv. *savastanoi*. In that strain, the *iaaL* gene codes for an enzyme that converts IAA to IAA-lysine (Glass and Kosuge 1986).

#### 7.4.1.3 Auxins Biosynthesis

Despite this diversity of pathways to produce and metabolize the active phytohormone, prokaryotic IAA-biosynthesis predominantly seems to follow two major routes: the indole 3-acetamide (IAM) and indole 3-pyruvic acid (IPyA) pathways. In accordance with this finding, Lambrecht et al. (2000) have proposed two major representative IAA biosynthesis patterns for plant–microbe interactions, depending on the specific eco-physiological role of the rhizobacteria: phytopathogenic or plant growth promoting. In the case of phytopathogenic *A. tumefaciens*, *A. rhizogenes, P. savastanoi,* and *Erwinia herbicola*, IAA synthesis occurs both in an inducible manner through IAM pathway and constitutively via the indole-3-pyruvic acid (IPyA) route. The gene expression in favor of the indole-3-pyruvic acid pathway is predominant when rhizobacteria live saprophytically on the plant surface, while the gene expression in favor of the IAM pathway becomes predominant when bacteria invade the plant apoplast. In contrast, most of the growth-promoting rhizobacteria such as *Azospirillum* sp. and some species of nonpathogenic *Pseudomonas* or *Bacillus* sp. synthesize IAA predominantly through the IPyA pathway (Patten and Glick 1996). However, in the case of *Azospirillum* sp., Abdel-Salam and Klingmüller (1987) isolated 11 *Azospirillum lipoferum* mutants producing 45–90% less IAA compared to their isogenic wild strains, and the residual synthesis capacity of these mutants suggested two possible explanations: (1) there are several routes for IAA biosynthesis (see below) or (2) there are multiple copies of genes coding for aromatic aminoacid transferases lowering substrate specificity.

#### 7.4.1.4 *Azospirillum* sp. and Auxins Biosynthesis

Until now, at least four different pathways have been described for the *Azospirillum* genus, three Trp-dependent ones named indole-3-pyruvate (IPyA), indole 3-acetamide (IAM), and tryptamine (TAM), and one Trp-independent pathway (Prinsen et al. 1993). The expression magnitude for each one depends mainly on bacterial growth conditions and it is generally considered that synthesis depends mostly on the Trp availability in the substrate. In this sense, evidence confirms that in the presence of the precursor the predominant pathway is the indole-3-pyruvic acid (IPyA) pathway and the IAM pathway is of secondary importance.

*Indole-3-pyruvic acid* route in *Azospirillum* sp. seems to have high resemblance to that described in higher plants (Nonhebel et al. 1993) and it begins with the conversion of Trp to IPyA by an aromatic amino acid transferase enzyme, followed by decarboxylation of indole-3-acetaldehyde by an indole-3-pyruvate decarboxylase (IPDC) and concludes with oxidation to IAA by indole-3-acetaldehyde dehydrogenase (Costacurta et al. 1994). The IPyA pathway was confirmed initially in *A. brasilense* Sp245 when the *ipd*C gene that encodes an IPyA decarboxylase was cloned (Costacurta et al. 1994). The molecular characterization described by Vande Broek et al. (1999) determined that the expression of this gene was regulated upstream by the end product IAA, which represented the first description of a bacterial gene specifically regulated by auxin. The *ipd*C promoter contains an auxin response element (*AuxRE*), which is similar to *AuxRE* found in gene promoters induced by auxin in higher plants (Lambrecht 1999). The key enzyme of this pathway is the indole-3-pyruvate decarboxylase (IPDC) encoded by the *ipdC* gene, since an *ipdC* knock-out mutant is strongly reduced in IAA biosynthesis.

*Indole-3-acetamide* (IAM) pathway has been most studied in phytopathogenic bacteria (Yamada et al. 1985; Klee et al. 1984). The existence of this pathway in *A. brasilense* was suggested by Prinsen et al. (1993) and Bar and Okon (1993) who determined the existence of IAM in cell free supernatants.

*Tryptamine pathway* involves the initial conversion of Trp to tryptamine, catalyzed by Trp-decarboxylase pyridoxal phosphate dependent enzymes, followed by conversion to indole-3-acetaldehyde by amino-oxidases. Although this pathway is present in plants (Conney and Nonhebel 1991) and fungi (Frankenberger and Arshad 1995), very little attention has been paid to bacteria. This pathway was only suggested for two particular species, namely *B. cereus* (Perley and Stowe 1996) and *A. brasilense* (Hartmann et al. 1983) because of their ability to produce IAA from tryptamine in chemically defined culture medium. Subsequently, Ruckdaschel and Klingmüller (1992) detected both intermediates of the pathway in supernatants of *A. lipoferum* confirming this route in other species of the genus. In *A. lipoferum*, the *ipdC* gene is located on the chromosome (Blaha et al. 2005).

*Trp-independent pathway* has been described through labeled precursor experiments by Prinsen et al. (1993), who suggested that the conversion of IAA (in the absence of Trp) in *A. brasilense* has a distribution of 0.1, 10.0, and 90.0% for the IAM, IPyA, and Trp-independent pathway, respectively. This latter route has been discarded recently due to impossibility to isolate the key enzyme or gene.

#### 7.4.1.5 Factors that Modify the IAA Biosynthesis in *Azospirillum* sp.

The factors that modify the IAA biosynthesis in *Azospirillum* sp. are diverse and extensive. Therefore, we will only mention those related to environmental stress and plant signaling (Spaepen et al. 2007). The first group of factors includes acidification, osmotic and matrix stress, and carbon source limitation and the second group of factors is integrated with chemical signals produced by plants in stressed or normal conditions. In *A. brasilense*, IAA production increases under carbon limitation, during reduction of growth rate and acidic pH (Ona et al. 2003, 2005; Vande Broek et al. 2005). Interestingly, carbon limitation and growth rate reduction are related to the physiological state of bacteria that arrive at the stationary growth phase. IAA is produced during all stages of culture growth but increases significantly after stationary phase (Malhotra and Srivastava 2009). This is in agreement with the observation that overproduction of stress-related stationary-phase sigma factor RpoS enhances IAA production in *Enterobacter cloacae* and *Pseudomonas putida* (Patten et al. 2002). The acidic pH regulates the *ipdC* gene expression in *A. brasilense*, with IAA production decreasing in acidic conditions (Vande Broek et al. 2005). In the case of *P. agglomerans*, the gene activity is not regulated by pH. However, osmotic and matrix stress could increase the *ipdC* gene expression more than tenfold (Brandl and Lindow 1996). The effects of these factors were not observed for the *ipdC* gene expression in *A. brasilense*. Concerning plant-derived factors, using an *ipdC*-gusA translation fusion (pFAJ64), Vande Broek et al. (2005) demonstrated that expression of the *ipdC* gene occurs mainly in the stationary growth phase, coinciding with IAA accumulation in the culture medium, and is further enhanced in the presence of IAA through a positive feedback regulation. Recently, Cassán et al. (2010a) showed that in *A. brasilense* Az39 containing plasmid pFAJ64 with *ipdC* and reporter gene, IAA production increased in the presence of osmotically active $PEG_{6000}$, ABA, or *Fusarium oxysporum* filtered supernatant. In contrast, oxidative $H_2O_2$, NaCl and $Na_2SO_4$ salinity, methyl jasmonate (MeJA), hydrolyzed amino acid medium, and *P. syringae* pv. savastanoi supernatant decreased the hormone accumulation under similar experimental conditions. Furthermore, *ipdC* gene expression increased in treatments with $PEG_{6000}$, ABA, MeJA, and *P. syringae* pv. savastanoi supernatant, and decreased in those treated with $H_2O_2$. Part of these results suggest the bacterial capacity to read (understand) the physiological signals produced by plants in stress conditions and then modify bacterial expression to coordinate their response with that of the plant. This phenomenon could be termed "integrated response to stress" or IRS and would be separated from the classical model "plant response to stress" in which plants are observed in experimental conditions without any influence of microorganisms.

#### 7.4.1.6 Physiological Role of Auxins Produced by *Azospirillum* sp.

The primary source of exogenous auxins in higher plants comes from the rhizosphere microbial community, where almost 80% of the established groups of bacteria are able to produce IAA-like compounds in vitro (Cheryl and Glick 1996). Plant response to exogenous IAA can vary from beneficial to deleterious, depending on the concentration incorporated into plant tissues. In the case of beneficial response, some authors believe that the increased hormone content of the soil due to microbial activity could supplement temporarily suboptimal levels in plants and partially modify the host cell metabolism with the consequent growth promotion. This is the case for the genera *Azospirillum, Azotobacter, Bacillus, Rhizobium,* and *Bradyrhizobium* for which auxin production has been described to be at least in part responsible for the growth-promoting capacity, which acts mainly on the development of the root system and on nodule formation in legumes. However, an excessive increase in auxin content will trigger a plant homeostatic mechanism to reduce the concentration of the hormone in plant tissues. This mechanism involves conjugate biosynthesis, xylem translocation from root to shoot (Martens and Frankenberger 1992), and a rapid IAA catabolism mediated by auxin oxidase activity (Scott 1972). IAA homeostasis could be considered a plastic but limited mechanism, which in several conditions is not enough to evade a physiological response generated by auxin excess in tissues. This is the case for infections caused by *A. tumefaciens*, *A. rhizogenes, Erwinia herbicola,* and some strains of *P. syringae* that fall beyond the scope of this review.

The beneficial interaction begins in the rhizosphere, where most of the substrates required for microbial growth and IAA synthesis are produced and released by plants, i.e., amino acids, organic acids, sugars, vitamins, nucleotides, and other biologically active metabolites, including some auxins (Rovira 1970). The presence of IAA and related compounds in plant exudates is enough to increase *Azospirillum* *ipd*C gene expression with consequent increased synthesis of bacterial IAA, provided that the precursor quantities (i.e., tryptophan) are sufficient (Vande Broek et al. 1999). The result will be an increased endogenous phytohormone content that could initiate the cellular response with its receptors on the cell membrane, responsible for initiate a signal transduction casacade that will have the cell wall and nucleus as main targets (Dharmasiri et al. 2005; Kepinski and Leyser 2005).

From a physiological point of view, *Azospirillum*'s capacity to synthesize auxins and release it into the rhizosphere or plant tissues could initiate a response depending on the type of plant inoculated. In legumes, bacterial IAA would initiate changes in the biological nitrogen fixation process at the level of nodule formation and functionality. Most members of the *Rhizobiales* order induce nodule formation on legume roots and these structures provide the plant with fixed atmospheric nitrogen. For over 70 years since Thiman (1936) proposed that auxins play an important role in formation and development of the nodule, many studies have indicated that changes in the balance of this phytohormone are a prerequisite for nodule organogenesis (Mathesius et al. 1997). In this regard, Prinsen et al. (1991) suggest that in *Rhizobium* sp. the synthesis of *nod* factors and IAA is triggered by *nod*-derived flavonoids produced by the plant. Although *Azospirillum* sp. are incapable of inducing nodule formation and fix nitrogen in a symbiotic way, it has been shown that exogenous application of synthetic auxins (i.e., 2,4-D) in higher concentration than physiological concentrations combined with *Azospirillum* sp. inoculation on grass roots could induce formation of tumorous structures called paranodules, which are effectively colonized by *Azospirillum* and in which bacteria could fix nitrogen more efficiently than in normal roots (Christiansen-Weniger 1998). Other studies have shown the beneficial response on biological nitrogen fixation of co-inoculation with *Rhizobium* and *Azospirillum* in legumes (Yahalom et al. 1990), not only by an increase in the number of nodules, but also by greater nitrogenase activity in symbiosomes. On the other hand, Schmidt et al. (1988) demonstrated that co-inoculation with *Rhizobium meliloti* (inefficient IAA producer) and *A. brasilense* (efficient IAA producer) on alfalfa (*Medicago sativa* L.) seeds significantly increased the number of root nodules in the primary root. This increase was directly related to the number of Azospirilla present in the culture medium and this response was mimicked by the addition of exogenous IAA.

The root growth is perhaps the most remarkable parameter changed during grass–PGPR interaction. The rapid seedling establishment in the substrate due to the root growth promotion could be considered a clear advantage for the plant because this increases its ability to anchor into the soil and get water and nutrients in one of the most critical stages of its development. To understand and study this interaction, researchers designed at least three types of strategies: (1) inoculate wildtype strains and mutants altered in their capacity to produce IAA; (2) inoculate wildtype strains or mutants and compare with pure exogenous IAA application; and (3) inoculate under several conditions (i.e., cell number, physiological growth state) and compare with exogenous IAA treatments. In all cases, the aim of these experiments has been to correlate the root growth promotion with the IAA levels in inoculated and noninoculated plants, and to compare the effect with the results obtained in exogenous IAA application (Baca et al. 1994). In this sense, Kolb and Martin (1985) found that inoculation of *Beta vulgaris* sp. with *A. brasilense* increased the number of lateral roots compared to control plants, and this effect was correlated with the high levels of bacterial IAA in pure liquid culture, and mimicked the exogenous application of similar IAA concentrations. In another report, Falik et al. (1989) inoculated seedlings of maize (*Zea mays* L.), and evaluated the levels of IAA and IBA (in both free and conjugate forms) by gas chromatography-mass spectrometry (GC-MS). They found that the levels of free IAA and IBA were higher in inoculated than in noninoculated roots. Simultaneously, Tien et al. (1979) and Hubbell et al. (1979) proved that the exogenous application of IAA, $GA_3$, and kinetin in pearl millet and sorghum produced similar changes in root morphology to those found in seedlings inoculated with *A. brasilense*. Kucey (1988) found that inoculation of wheat with *A. brasilense* simulated the effect of exogenous IAA and $GA_3$ treatment regarding the growth pattern of stems and roots. Also in wheat plants, Zimmer et al. (1988) proved that the exogenous addition of IAA and nitrate were substituted completely or partially, respectively, by the inoculation with *A. brasilense*. In another interesting report, Barbieri et al. (1988) showed that inoculation with a wild strain of *A. brasilense* (IAA-producer) increased the number and length of lateral roots of wheat (*Triticum aestivum* L.). In contrast, the inoculation with a mutant with lower IAA production did not modify the root development like the wildtype. Subsequently, Barbieri et al. (1991) showed that wheat seedlings inoculated with *A. brasilense* M7918 (with low IAA production) did not modify the root growth parameters and this resulted in a decreased ability of the seedlings to take up water and nutrients from the solution. Bothe et al. (1992) demonstrated that inoculation of wheat plants with *A. brasilense* significantly increased the formation of lateral roots and slightly increased dry weight of root and root hair formation, whereas exogenous application of IAA significantly increased root dry weight, but had no effect on the formation of lateral roots. Dubrovsky et al. (1994) demonstrated that inoculation of *A. thaliana* with *A. brasilense* cd significantly increased the length of root hairs up to twofold, compared with control. Dobbelaere et al. (1999) presented significant evidence about the role of IAA in the phytostimulatory effect upon *Azospirillum* sp. inoculation. They suggest inoculation or exogenous IAA treatments in wheat (with extension to other gramineous plants) result in a decrease in root length and root elongation zone and an increase in root hair length and density (Spaepen et al. 2008) (Fig. 7.4).

### 7.4.2 Gibberellins

Gibberellins (GAs) are a large group of tetracyclic diterpene acids that regulate diverse processes in plant growth and development, such as germination, stem elongation, flowering, and fruiting (Davies 1995). The use of unequivocal methodology such as GC-MS or liquid chromatography-mass spectrometry (LC-MS/MS) to identify and quantify phytohormones revealed that gibberellins are a large group of natural products (Mander 1991). There are more than 130 kinds of gibberellin molecules produced by plants, fungi, and bacteria (Hedden and Phillips 2000). From a structural point of view, free gibberellins are divided into two groups: those that possess the full complement of carbon atoms ($C_{20}$-GAs) and those in which the $C_{20}$ is lost ($C_{19}$-GAs) (Fig. 7.5).

All gibberellins are carboxylated at the $C_7$, with the exception of $GA_{12}$-aldehyde, and possess one ($G_4$), two ($GA_1$), three ($GA_8$), or four ($GA_{32}$) hydroxyl functions. The position of the hydroxylation (OH) is very important because this determines their biological activity. Hydroxylation of the $C_3$ and $C_{13}$ in their $\beta$ and $\alpha$ positions, respectively, leads to the activation of the molecule, whereas the hydroxylation in position $\beta$ of $C_2$ has a strong negative effect on its activity (Pearce et al. 1994). In addition to the free forms, conjugated forms have been identified.

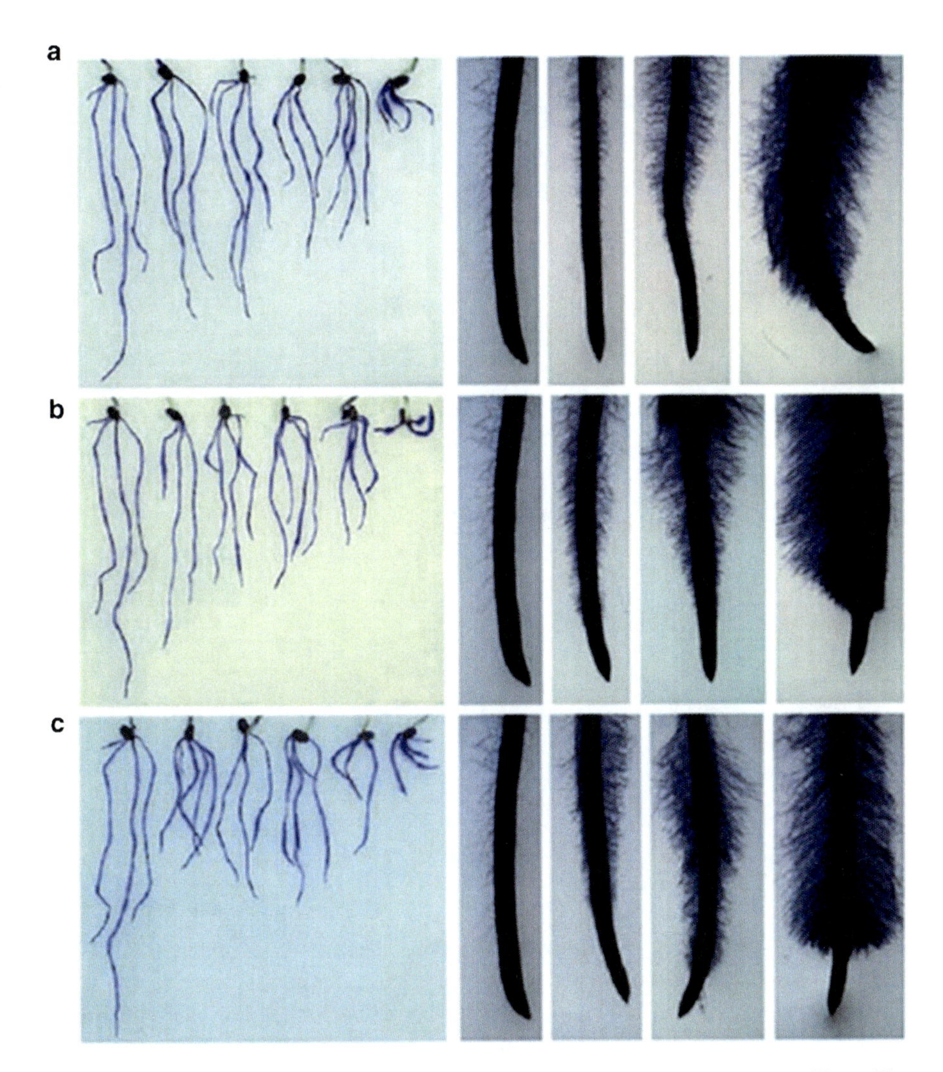


**Fig. 7.4** Effect of inoculation on wheat root morphology of 1-week-old wheat seedlings. The seeds were inoculated with (**a**) *A. brasilense* Sp245, (**b**) Sp245 (pFAJ5002), and (**c**) Sp245 (pFAJ5005). The left pictures represent the effect of inoculation on root length. From left to right: noninoculated plant, seedlings inoculated with $10^5$, $10^6$, $10^7$, $10^8$, and $10^9$ CFU plant$^{-1}$. The right pictures represent the effect of inoculation on root hair formation. From left to right: noninoculated plant, seedlings inoculated with $10^5$, $10^7$, and $10^9$ CFU plant$^{-1}$. Original image and reference, see Spaepen et al. (2008)

These include glycosidic ethers (GA-G), where a glucose molecule is attached to the structure of the GA by a hydroxyl group and glycosidic esters (GA-GE), where glucose binds to the hormone through a carboxyl group on $C_7$ (Sembder et al. 1968). The majority of the conjugated molecules are conjugated with glucose (Schliemann and Schneider 1994), but gibberellins with amino acids attached to

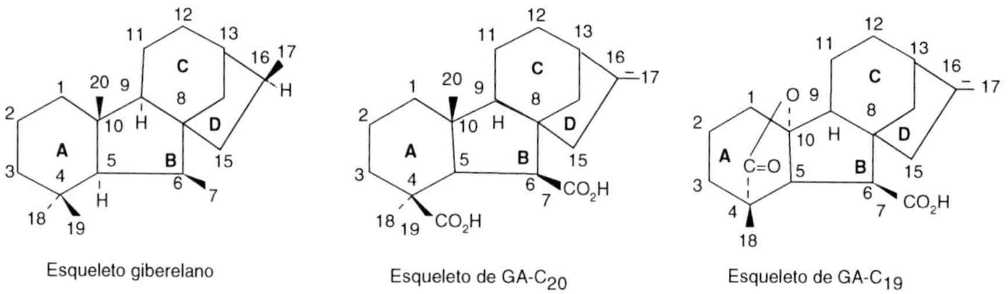

Fig. 7.5 Chemical structures of $C_{19}$-GAs and $C_{20}$-GAs present in nature, compared with a gibberellic skeleton

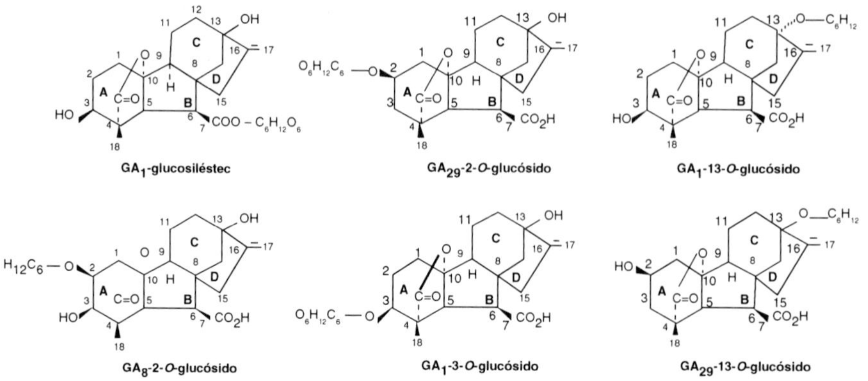

Fig. 7.6 Chemical structures of glucosyl-conjugated GAs. $GA_8$-2-O-glucoside, the first conjugated GA identified in plants, according to Schreiber et al. (1967)

the $C_7$ carboxyl group by a peptide bond have been identified (Sembder et al. 1980). The biochemical and physiological aspects of the GA conjugates have been extensively discussed by Rood and Pharis (1987), who indicated that their main feature is the lack of biological activity and the potential reversibility to the active forms by enzyme activity such as hydrolases (Fig. 7.6).

#### 7.4.2.1 Biosynthesis and Metabolism

The GA synthesis in higher plants begins with the cyclization of a common 20 carbon molecule precursor, the geranylgeranyl pyrophosphate (GGPP). This intermediate is synthesized in plastids starting from mevalonic acid (MA), leading to isopentenyl diphosphate, and combined with glyceraldehyde-3-phosphate or pyruvate coming from the deoxyxylulose 5-phosphate pathway (Litchtenthaler 1999). The first phase of synthesis involves the cyclization of GGPP in protoplasts and results in the formation of ent-kaurene in two successive steps that require the activity of two key enzymes: copalil diphosphate synthase (CPS), which produces the intermediary copalil diphosphate, and ent-kaurene synthase, which synthesizes the final product ent-kaurene. In a second synthesis phase, the ent-kaurene is converted to biologically active GAs by a series of oxidation reactions catalyzed by two types of enzymes. These reactions take place in the extraplastidic membranes with a contraction of ring B from $C_6$ to $C_5$ and formation of the typical gibberellic structure. These reactions are catalyzed by cytochrome P-450 monooxygenases and culminate in the formation of $GA_{12}$, which is considered the first GA in the pathway. In a third stage of synthesis, the intermediate $GA_{12}$ or $GA_{53}$ (product of the 13α-hydroxylation of $GA_{12}$) is metabolized by soluble dioxygenases, which are *2-oxoglutarate* dependent enzymes, and use the 2-oxoglutarate as co-substrate. Two types of dioxygenases are required to convert $GA_{12}$ or $GA_{53}$ in parallel into the active $GA_3$, $GA_1$, $GA_4$, and $GA_7$, namely the $C_{20}$-oxidases and 3β-hydroxylases. A third group of dioxygenases, 2β-hydroxylases, modify the 2β position of the molecule and cause loss of biological activity as is the case in the $GA_1$ to $GA_8$ catabolism (MacMillan 1997) (Fig. 7.7).

#### 7.4.2.2 Biosynthesis and Metabolism of GAs by *Azospirillum* sp. In Vitro

Bottini et al. (1989) confirmed the ability of *A. lipoferum* Op33 to produce $GA_1$, $GA_3$, and iso-$GA_3$-like compounds in chemically defined culture medium. The identification of these molecules was done by GC-MS, whereas the quantification was done by a biological assay in dwarf rice mutants according to Murakami et al.

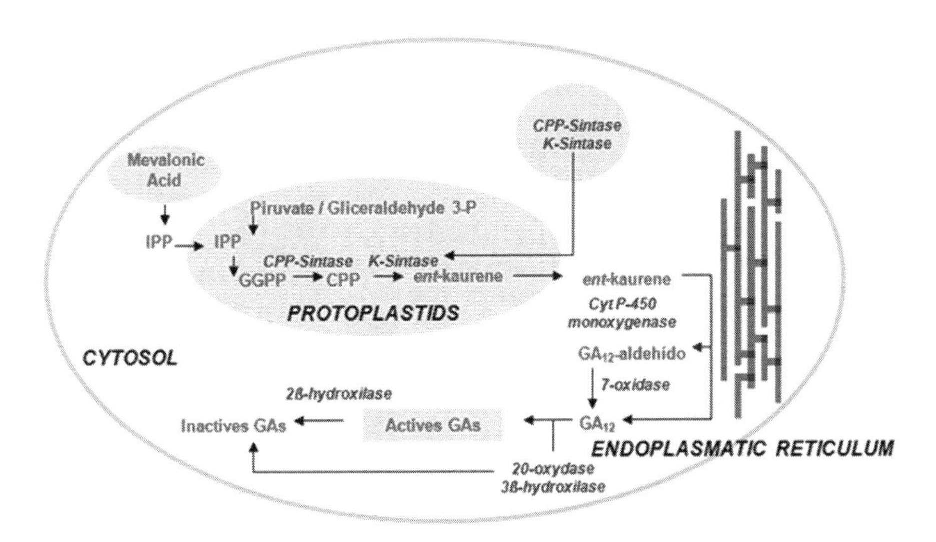


**Fig. 7.7** The gibberellin biosynthetic pathway in the plant cell, according to Kobayashi et al. (1989)

(1972). They found equivalent concentrations of 20 and 40 pg $ml^{-1}$ of $GA_1$ and $GA_3$, respectively. Janzen et al. (1992) analyzed the production of GAs in *A. brasilense* Cd pure culture and in co-culture with *Trichoderma harzianum*, a biocontrol fungus. Identification and quantification of $GA_1$, $GA_3$, and iso-$GA_3$ was performed by GC-MS and found that the amount of gibberellins was $GA_3 >$ iso-$GA_3 > GA_1$. In co-cultures only $GA_1$ was identified. Piccoli and Bottini (1996) highlighted the ability of *A. lipoferum* to produce the immediate precursors $GA_{19}$ and $GA_9$ that allowed speculation on the existence of at least two pathways for the GA synthesis in the bacteria. The first one is related to the early 13α-hydroxylation reactions and the metabolism of $GA_{19}$ to an immediate precursor $GA_{20}$ and the second involved to the 3β and late 13α-hydroxylation and the conversion of $GA_9$ to active molecules. Piccoli and Bottini (1996) and Piccoli et al. (1998) confirmed the ability of *A. lipoferum* Op33 to produce $GA_9$, $GA_{20}$, and $GA_5$ in chemically defined culture medium, which brings new elements to the complex bacterial synthesis scheme. In other reports, Piccoli and Bottini (1994b) demonstrated the ability of *A. lipoferum* Op33 to metabolize $GA_{20}$ to $GA_1$ through an early 13α-hydroxylation pathway in chemically defined medium. Additionally, Piccoli and Bottini (1996) confirmed the ability of the same strain to metabolize $GA_9$ to $GA_3$, another biologically active molecule. These results allow inferring that growth promotion of inoculated plants should be at least in part due to the bacterial ability to produce or metabolize GAs to biologically active forms. In other experiments, Piccoli et al. (1998) successfully established the ability of *A. lipoferum* to hydrolyze the glucose conjugates $GA_{20}$-glycosyl ester and $GA_{20}$-13-*O*-glucoside in chemically defined medium. This report confirmed the capacity of the microorganism to increase the plant active endogenous pool of gibberellins by hydrolysis of inactive glucose conjugates. The complete GAs production by *Azospirillum* was summarized by Bottini et al. (2004).

#### 7.4.2.3 Biosynthesis and Metabolism of GAs by Other PGPR In Vitro

Gutiérrez-Mañero et al. (2001) showed that *B. licheniformis* isolated from the rhizosphere of alder (*Alnus glutinosa* [L.] Gaertn.) produced physiologically active gibberellins $GA_1$ (0.13 μg $ml^{-1}$), $GA_3$ (0.05 μg $ml^{-1}$), and $GA_4$ as well as the precursor $GA_{20}$ and the isomers 3-epi-$GA_1$ and iso-$GA_3$. Additionally, Bastian et al. (1998) confirmed by GC-MS the production of IAA and $GA_1$ and $GA_3$ by *Acetobacter diazotrophicus* and *Herbaspirillum seropedicae*, two free-living nitrogen fixing rhizobacteria commonly used to inoculate several plant species. Recently, Sgroy et al. (2009) confirmed the $GA_3$ production in *Lysinibacillus fusiformis*, *Achromobacter xylosoxidans*, *Bacillus halotolerans*, *B. licheniformis*, *B. pumilus*, and *B. subtilis* strains isolated from roots of the halophytic legume *Prosopis strombulifera*, with a production of 36.5, 50.0, 80.5, 75.5, 21.3, and 10.0 μg $ml^{-1}$, respectively.

#### 7.4.2.4 Biosynthesis and Metabolism of GAs by *Azospirillum* sp. In Vivo

The experiments detailed below were developed in hydroponic culture models, using dwarf mutants as described by Murakami (1968) and Kobayashi et al. (1989). In such conditions, rice (*Oryza sativa* L.) and maize (*Z. mays* L.) dwarf GA-deficient mutants were used for *Azospirillum* sp. inoculation as experimental model. In the particular case of maize, the four mutants *dwarf-1* (d1), *dwarf-2* (D2), *dwarf-3* (d3), and *dwarf-5* (d5), proposed by Phinney and Spray (1988), were used. In the case of rice, the mutants were *dwarf-y* (*dy*) of cultivar Waito C and *dwarf-x* (*dx*) of cultivar Tan-ginbozu (Murakami 1972). Additional experiments included the use of plant growth retardants, synthetic compounds that reduce the plant length by inhibiting the active gibberellin biosynthesis. In this regard, Rademacher (2000) considers four groups of plant growth retardants with capacity to inhibit different steps of the GA biosynthesis pathway: (1) "onion-like" molecules; (2) N-heterocyclic compounds; (3) 2-oxoglutarate-like compounds; and (4) the gibberellins analogous as the 16, 17-dihydro-GAs (Fig. 7.8).

Lucangeli and Bottini (1997) presented direct evidence of dwarfism reversion in a *d1* maize mutant and in *dx* rice mutants by *A. lipoferum* USA 5b and *A. brasilense* Cd inoculation, respectively, similar to that by exogenous application of $GA_3$. The *d1* mutant was root-inoculated with *A. lipoferum* or treated with exogenous solutions of 0.1, 1.0, or 10.0 μg $GA_3$ per plant, while *dx* mutants were pregerminated in a solution of uniconazole S-3370 and inoculated with *A. brasilense* or treated with exogenous solution of 10–3,000 fmol $plant^{-1}$ $GA_3$. Both *A. lipoferum* and *A. brasilense* reversed genetic dwarfism in both mutants, with a similar phenotype to that observed by addition of free $GA_3$, and these findings correlated with the bacterial ability to produce active forms of GAs in roots, stems, and leaves of inoculated plants. Later, Lucangeli and Bottini (1997) found that maize (*Z. mays* L.) seedlings treated with the retardant Uniconazole-P showed a strong physiological dwarfism that could be reversed by a single or combined inoculation with *A. lipoferum* USA 5b and *A. brasilense* Cd or by exogenous treatment with $GA_3$. In a combined inoculation treatment, dwarf reversion was similar to that obtained by exogenous treatment with 0.1 μg $GA_3$ $planta^{-1}$. Similar results were generated by Gutiérrez-Mañero et al. (2001) with the gibberellin-producers *Bacillus pumilus* and *B. licheniformis*, which reversed the dwarf phenotype induced in alder (*Alnus glutinosa* [L.] Gaertn.) seedlings pretreated with the growth retardant paclobutrazole. This response was similar to the one obtained by exogenous application of pure $GA_3$. Cassán et al. (2001b) found that inoculation of *A. lipoferum* USA 5b and *A. brasilense* Cd on rice *dy* mutant seedlings that were preincubated with 1 μg of conjugate [17,17-$^2H_2$]-$GA_{20}$ resulted in the reversion of the genetic dwarfism and this phenotype correlated with [17,17-$^2H_2$]-$GA_1$identification by GC-MS. These results allow us to speculate on the in vivo capacity of *Azospirillum* sp. to produce biologically active $GA_1$ from the immediate precursor $GA_{20}$ through the 3β-hydroxylation path. Furthermore, addition of the plant growth retardant Prohexadine-Ca, which inhibits 3β-hydroxylation, resulted in the disruption of

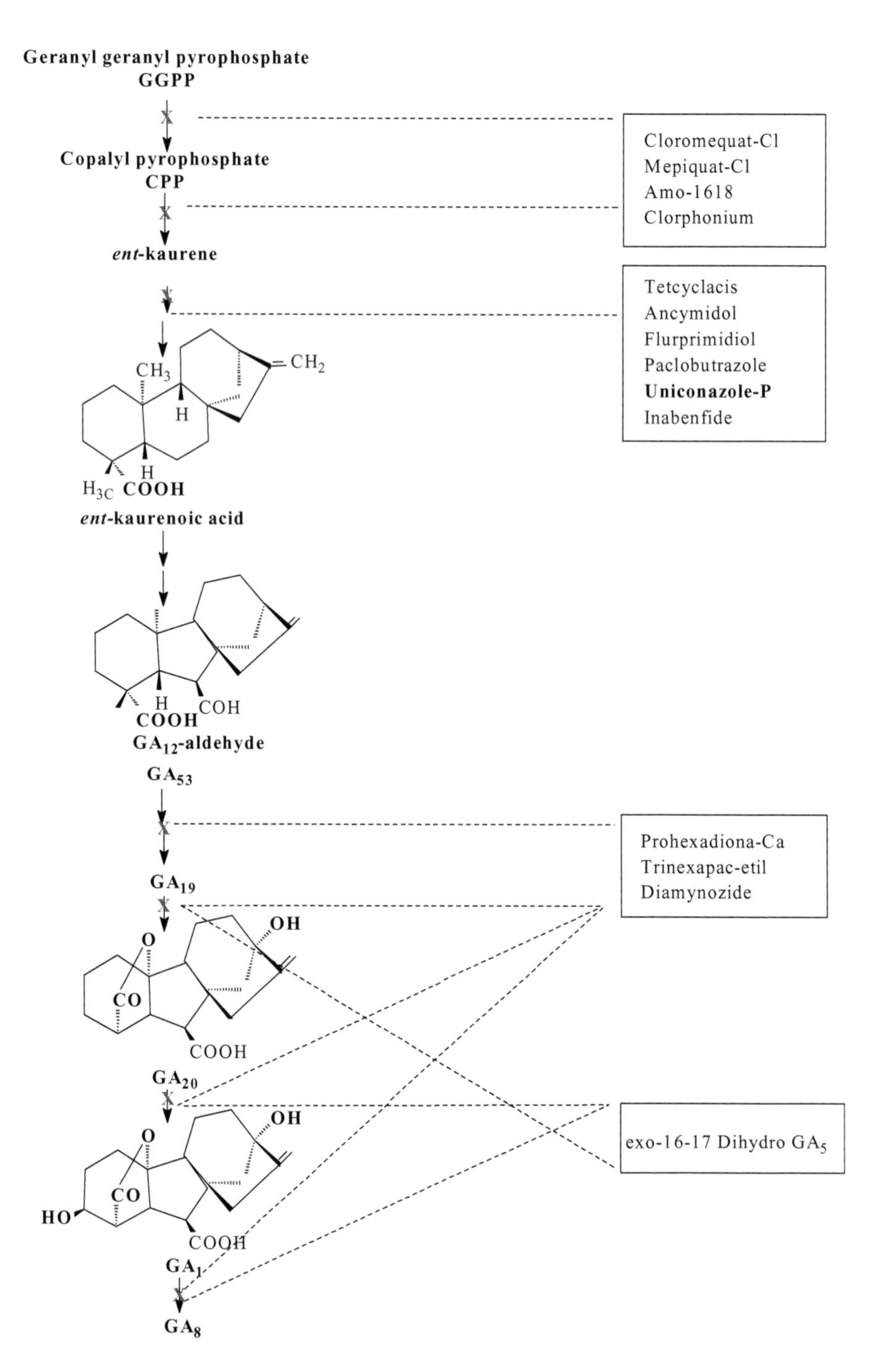


**Fig. 7.8** Simplified scheme of GA biosynthesis pathways affected by the four groups of Plant Growth Retardants (PGR), according to Rademacher (2000)

phenotypic complementation and the failure to identify [17,17-$^2H_2$]-$GA_1$ by GC-MS, which strongly suggested that the enzyme catalyzing the gibberellin activation pathway in *Azospirillum* sp. could be a 2-oxoglutarate-dependent dioxygenase (3β-hydroxylase), similar to those identified in higher plants. Later, Cassán et al. (2001c), demonstrated the ability of *Azospirillum* sp. to 3β-hydroxylate another intermediate precursor [17,17-$^2H_2$] $GA_9$ to the active molecule [17,17-$^2H_2$] $GA_3$. In another study, Cassán (2003) found that inoculation with *Azospirillum* sp. in rice (*O. sativa* L.) dwarf mutants exogenously treated with the precursor [17,17-$^2H_2$] $GA_{12}$ reversed the genetic dwarfism. This result could be explained through bacterial ability to metabolize $^2H_2$-$GA_{12}$ to biologically active gibberellin $^2H_2$-$GA_1$ or $^2H_2$-$GA_3$. However, with addition of the plant growth retardant Uniconazole-P into culture medium, the bacteria lose their ability to reverse the plant dwarfism and to produce active gibberellins. This would imply that $GA_{12}$ metabolism in *Azospirillum* sp. could be regulated by cytochrome $P_{450}$-dependent monooxygenases (Chapple 1998) sensitive to Uniconazole-P, as previously described in a filamentous fungus (Rademacher 2000). Such experiments were conducted with addition of another growth retardant, Prohexadione-Ca, showing similar results. This also would imply that $GA_{12}$ metabolism could be regulated by a 2-oxoglutarate-dependent diooxygenase that is sensitive to Prohexadione-Ca, as in higher plants. An alternative hypothesis would involve the combined existence of both enzyme types in *Azospirillum* sp. that are able to convert a late precursor $GA_{12}$ in a biologically active molecule (Fig. 7.9).

In regard to the conjugate metabolism, Cassán et al. (2001a) found that *Azospirillum* sp. reversed genetic dwarfism in rice (*O. sativa* L.) dwarf mutants treated with [17,17-$^2H_2$] $GA_{20}$-glucosyl ester or [17,17-$^2H_2$] $GA_{20}$-glucosyl ether. In these seedlings, phenotypic complementation was observed and the phenotype correlated with the bacterial ability to hydrolyze GA conjugates into active gibberellins such as [17,17-$^2H_2$] $GA_1$ by 3ß-hydroxylase enzymes. This report confirmed the capacity of *Azospirillum* sp. to increase the plant endogenous pool of active gibberellins by hydrolysis of inactive glucosyl conjugates.

#### 7.4.2.5 Physiological Role of GAs Produced by *Azospirillum* sp.

Of the approximately 130 currently known GAs (Crozier et al. 2000), less than 15 are specific to fungi, around 100 are exclusive to plants, and less than 15 are ubiquitous. Despite this wide distribution and the large number of known forms, only a few GAs are biologically active per se (Hedden and Phillips 2000). Evidence suggests that in the case of microorganisms, GA and auxin production increase rapidly at the beginning of the stationary growth phase suggesting that reduced C sources (Omay et al. 1993) or N sources (Piccoli and Bottini 1994a) in the culture medium may trigger bacterial phytohormone production. It is complex to hypothesize a physiological role for the GAs produced by rhizobacteria, but it seems logical to think that promoting the overall growth of colonized plants could be beneficial to bacteria because of the increased nutrient availability in the rhizosphere

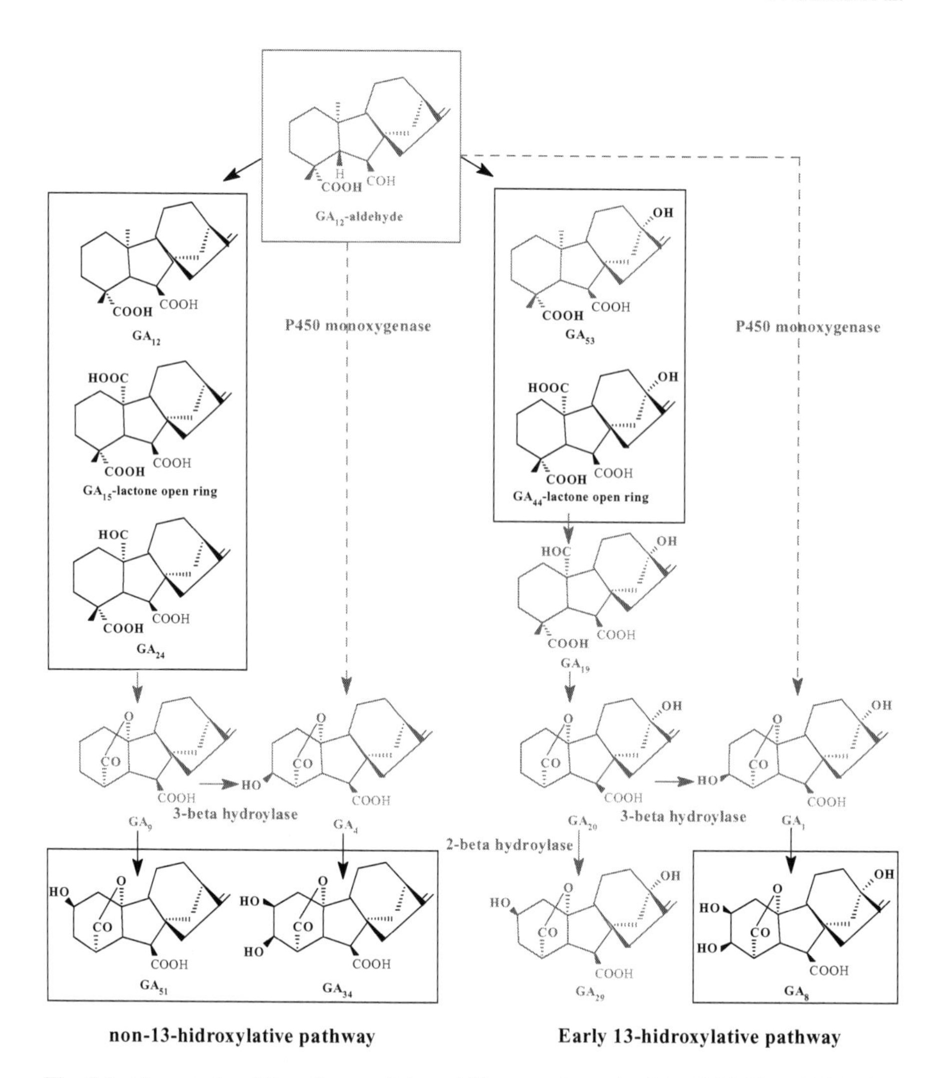


**Fig. 7.9** The putative GA pathway of *Azospirillum* sp. determined by GC-MS and biological assays. The black molecules and arrows correspond to pathways described in higher plants

(Rademacher 1994). However, attributing this phenomenon only to the production of a single substance by the microorganism is somewhat simplistic, especially considering the variable plant responses and the positive responses of many plant species to inoculation. Most of the research conducted in field conditions has focused on increase in the yield due to by inoculation, with an emphasis on nitrogen incorporation and dry matter production. In this regard, Fulchieri and Frioni (1994) found that corn plants (*Z. mays* L.) inoculated with a mixture of three *Azospirillum* strains in Hapludol soil in Argentina showed a significant increase in dry weight of roots, shoots, and grains, compared to uninoculated plants as also reported in

Nepalese wheat varieties inoculated with indigenous *Azospirillum* strains (Bhattarai and Hess 1993). Considering these results, the growth promotion response could perfectly be attributed to at least three bacterial promotion mechanisms: (1) atmospheric nitrogen fixation; (2) auxin and gibberellin-like phytohormone production; and (3) indirectly, through the interaction of the *Azospirillum* cells with the rhizosphere community. Similar results were observed in inoculated wheat and sorghum plants by Pozzo Ardizzi (1982) and in several other commercially interesting plant species (Paredes-Cardona et al. 1988; Sarig et al. 1990). Twenty years of evaluation of field inoculation trials showed that 60–70% of the experiments were successful, with a significant increase from 5 to 30% on agronomical interesting crop yield (Bashan and Holguin 1997). The authors attributed the use of exponential growth phase bacterial cultures for inoculation, as a crucial factor for the effectiveness (Vandenhobe et al. 1993). While these results were interesting for agricultural production, from the ecophysiological point of view but insufficient in establishing the proper mechanism of increased plant growth. In fact, effects of GA production by *Azospirillum* sp. have not been studied in-depth. To evaluate the physiological importance of bacterially produced plant hormones in general and GAs in particular on plant growth promotion, it would be necessary to obtain *Azospirillum* sp. mutants completely deficient in the synthesis of these compounds. However, such mutants are not yet available, although their availability would be the key to further understanding the bacterium's mechanisms for plant growth promotion.

### 7.4.3 Cytokinins

This is a group of natural compounds that regulate cell division and differentiation processes in meristematic tissues of higher plants. Chemically, they are purines, mostly derived from adenine and modified by substitutions on the $N^6$, which also includes their respective ribotides, ribosides, and glycosides. These plant hormones have been associated with many physiological and cellular processes including: senescence delay by chlorophyll accumulation, organ formation in a wide range of tissues, root development, root hair formation, root elongation, stem initiation, and leaf expansion. By definition, these compounds (combined with an optimal auxin concentration) induce cell division in plant tissues. The first cytokinin-like regulator was discovered by Miller et al. (1955) and was named kinetin (K), which was considered an unnatural form of the phytohormone. In 1963, Letham (1963) identified a naturally occurring compound called zeatin (Z) and since then more than 50 molecules and their metabolites have been classified. The bioactivity is not uniform in all cytokinin-like compounds and normally depends on: aminopurine ring in the molecule, substitution of $N^6$ with a simple ribosyl chain isopurine-derived unit, replacement of positions 2 and 9 of the ring for the groups H, $CH_3$-S, or presence of unsaturated side chain (optimally five carbons). The natural and synthetic forms with higher biological activity on tissues or crops are Zeatin (Z),

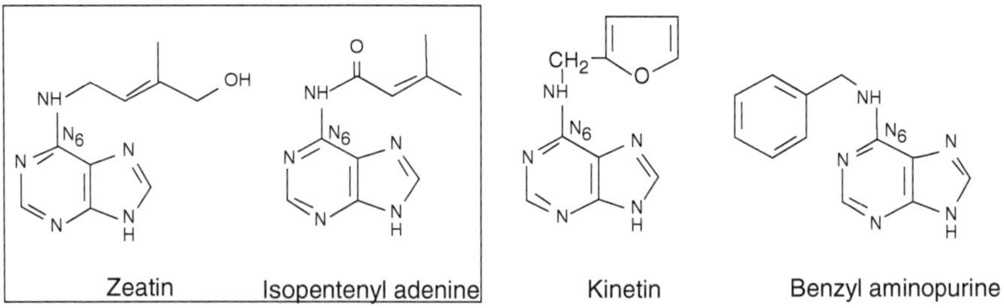

Fig. 7.10 Natural cytokinins described in plants and *Azospirillum* sp. (*rectangle*) and synthetic molecules with cytokinin-like activity

isopentenyl adenine (iP), kinetin (K), and benzylaminopurine (BAP). All of them have a double alkyl bridge at position $N^6$ (Fig. 7.10).

#### 7.4.3.1 Cytokinins and Microbial Production

Many rhizospheric microorganisms, including bacteria and fungi, are capable of synthesizing cytokinins in chemically defined culture medium. Barea et al. (1976) found that at least 90% of bacteria isolated from the rhizosphere of agriculturally interesting crops were able to produce cytokinin-like compounds. As a result of the intimate relationship between the microorganisms and the root surface, exogenous production of this hormone has a profound effect on plant growth. Similar to auxins, the microbial production of cytokinin could supplement plant endogenous content and in some cases promote plant growth or even showed phytotoxicity. It is known that plants respond to exogenous addition of cytokinins, which is an interesting fact because ecological significance of microbial synthesis in plant tissues is not described yet. Although the microbial cytokinin production in higher plants began with phytopathogenic models, nowadays researchers have turned to study this process in groups of PGPB or PGPR. The most studied model in this regard has been the *rhizobia*-legume symbiosis, where the bacterial cytokinin synthesis in nodule production has been investigated. In this respect, Nandwall et al. (1981) studied the effect of exogenous addition of K, and noticed that it promoted nodule initiation and increased the content of leghemoglobin in common beans. In other experiments, Yahalom et al. (1990) proved that both, exogenous addition of BAP or co-inoculation of *Rhizobium* and *Azospirillum* sp. increased the number of nodules formed in *Medicago polymorpha*. In other experiments, exogenous application of cytokinins increased nitrogenase activity in *Pisum sativum* root nodules (Jaiswal et al. 1982) and showed that in alfalfa roots and other legumes, cytokinins were responsible for the accumulation of *ENOD2* and *ENOD40* gene transcripts (Hirsch et al. 1997). Other reports highlighted the role of these molecules in *Phaseolus vulgaris* nodule formation (Puppo and Rigaud 1978) and in *Vicia faba* (Henson and Wheeler 1977). Recently, Giraud et al. (2007) proved that in certain legumes, nod factors are not necessary to initiate nodulation because certain *Bradyrhizobium* strains could use an alternative activation pathway, in which a purine derivative (cytokinin) is responsible for triggering nodule formation. In particular, cytokinin production by *Azospirillum* sp. has been widely studied in *A. brasilense* in chemically defined culture medium. In this respect, Tien et al. (1979) used different types of chromatography (HPLC and TLC) and an inoculation bioassay in pearl millet (*Pennisetum americanun* L.) to demonstrate the ability of *A. brasilense* for the production of cytokinin-like molecules (Cacciari et al. 1989) but partially purified compounds were not characterized. Horemans et al. (1986) modified the chromatography extraction (Sephadex LH-20) and included the technique of radio immunoassay (RIA) to prove that *A. brasilense* produced iP, (9R) iP, and Z in chemically defined culture medium. The lack of information for cytokinin synthesis in *Azospirillum* cultures is due to the complexity of analysis of these hormones. The last reference to cytokinin production by *Azospirillum* (Strzelczyk et al. 1994), using culture medium supplemented with different C sources. Tien et al. (1979) showed that inoculation with *A. brasilense* caused significant changes in pearl millet seedling root morphology by increasing the number of lateral roots and the root hair density. The exogenous application of auxins, cytokinins, and gibberellins produced changes in root morphology comparable to those obtained by inoculation. In other trials, Cacciari et al. (1989) found that the mixed culture of *A. brasilense* and *Arthrobacter giacomelloi* showed high gibberellin and cytokinin contents, compared to their individual cultures, which could be of great physiological importance because the presence of different microbial species in the rhizosphere induces production of cytokinins and other phytohormones in PGPR.

The ability of several PGPR to synthesize cytokinins has been described by several workers (Akiyoshi et al. 1987; Nieto and Frankenberger 1989; Salamone et al. 2001; Taller and Wong 1989; Timmusk et al. 1999). Arkhipova et al. (2005) evaluated cytokinin production by inoculating lettuce plants with *B. subtilis* that grew in chemically defined culture media, making use of specific antibodies. Zeatin riboside (ZR) was shown to be the main cytokinin (1.2 $\mu g\ ml^{-1}$ of ZR) present in bacterial cultural medium. Also, inoculation of lettuce plants increased the endogenous cytokinin content, resulting in enhanced plant shoot and root weight compared to control (without inoculum). Finally, Sgroy et al. (2009) confirmed the Z production in *B. subtilis*, *Brevibacterium halotolerans*, *B. pumilus*, and *P. putida* isolated from roots of the halophylic shrub *Prosopis strombulifera*.

### 7.4.4 Ethylene

Ethylene is an important hormone in plant growth and development (Burg 1962). Due to its gaseous state under physiological conditions, for a long time it was not accepted as a phytohormone, but various studies showed that its synthesis and action was critical for certain physiological processes. Although there are many publications related to the synthesis of this hormone in higher plants (Mattoo and Suttle 1991), few studies have been published on the microbial biosynthesis

(Arshad and Frankenberger 1993) and even less in *Azospirillum* sp. The ethylene molecule is very simple and symmetrical, composed of two carbon atoms (joined by a double bond) and four H atoms. Ethylene is soluble in water at about 140 ppm, 25°C, and 760 mm Hg (15 times more than oxygen) and is quite active and can exert its physiological effects at very low concentrations in plant tissues (0.1 ppm). In higher plants, all tissues have the capacity to synthesize this hormone, but in general the concentration is associated with the growth state and developmental phase of the plant, with a higher concentration in those tissues involved in active cell division, which are under stressful conditions or in a senescence stage (Burg and Burg 1968). Since Gane (1934) first reported the ability of plants to synthesize ethylene, a variety of compounds, including methionine, linoleic acid, propanol, β-alanine, ethionine, ethanol, glycerol, organic acids, glucose, and sucrose (Yang 1974) have been proposed as possible precursors for this hormone. However, in a nonenzymatic in vitro study, Abeles and Rubinstein (1964) showed that methionine was the natural precursor for ethylene. In the case of microbial synthesis, proposed precursors varied widely (Fukuda and Ogawa 1991) but L-methionine has most often been described as substrate in bacterial cultures. The regulation of ethylene production in higher plants largely depends on "key" enzymes for ethylene biosynthesis: S-adenosyl-L-methionine synthase (SAM), 1-aminocyclopropane-1-carboxylic acid (ACC) synthase, and ACC oxidase. Genes coding for these enzymes and their regulatory elements have been characterized, modified, and re-introduced into agriculturally interesting plants (Fluhrer and Mattoo 1996). The SAM synthase catalyzes the first reaction from methionine, increasing the content of SAM for various metabolic pathways, among them the ethylene (Fluhrer and Mattoo 1996) and polyamines (Even-Chen et al. 1982) are the main biosynthesis pathways. The second reaction is catalyzed by ACC synthase resulting in the hydrolysis of SAM to form ACC and 5′-methylthioadenosine (MTA) (Kende 1989). Finally, the ACC oxidase catalyzes the ACC conversion to ethylene, $CO_2$, and cyanide. Among the factors that induce ethylene production, the developmental stage of the organ influences this hormone's synthesis rate, with the rate being higher in stages where cells are dividing (Burg and Burg 1968). In general, there is a direct association between high respiration rates, senescent or damaged tissues, and high ethylene content. High doses of auxin application can stimulate the ethylene synthesis. Moreover, ethylene formation is directly related to a stress condition in tissues (Beyer et al. 1984). Low temperatures, excessive heat, flood, and drought stimulate ethylene production and form an interesting response network with other regulators such as ABA, jasmonic acid (JA), and auxin (IAA). Several studies have presented evidence that this hormone could have a decisive role in establishing symbiotic relationships, such as nodule formation and mycorrhizae. In the interaction with *Rhizobium*, the exogenous application of ethylene has shown a negative effect on nodule formation and function. Grobbelaar et al. (1971) showed that nodulation was reduced by 90% in alfalfa explants treated exogenously with 0.4 ppm of ethylene.

#### 7.4.4.1 *Azospirillum* sp. and Ethylene Production

There is little information related to ethylene production by *Azospirillum* sp. and its effect on plant growth. Primrose and Dilworth (1976) determined the ability of the free-living bacteria *Azotobacter* sp. and *Bacillus* sp. to produce ethylene in chemically defined culture medium. Strzelczyk et al. (1994) tested *Azospirillum*'s ability to produce this compound in culture medium modified with different carbon sources. Their results showed that the bacteria could synthesize ethylene and that the production depends on the presence of L-methionine in the culture medium. Recently, Krumpholz et al. (2006) evaluated tomato seedlings inoculated with *A. brasilense* FT326 (an IAA overproducer), and found a correlation between the significant increase in root number and length and the ethylene production, which was up to ten times higher than the controls accompanied by increased ACC synthase activity. According to Peck and Kende (1995), the rate limiting step for ethylene biosynthesis is the conversion of S-adenosylmethionine (SAM) to 1-aminocyclopropane-1-carboxylic acid (ACC), catalyzed by ACC synthase. The expression and activity of this enzyme, as well as ethylene production, is enhanced by the addition of exogenous IAA. This fact indicates that the ethylene increase is at least partially due to cross-talk between the bacterially produced IAA and the ethylene synthesis as suggested by Rahman et al. (2002). In contrast, the ability of some bacteria to promote plant growth could be indirectly related to bacterial expression of 1-aminocyclopropane-1-carboxylate deaminase (ACC deaminase), key enzyme in ethylene metabolism hydrolyzes 1-aminocyclopropane-1-carboxylate (ACC) which is the immediate hormone precursor (Glick et al. 1995). Bacteria that possess this enzyme can cleave ACC to ammonia and alpha-ketobutyrate, preventing the ethylene accumulation and its toxic effects (Glick et al. 1998). Even though *Azospirillum* promotes plant growth, the members of this genus do not produce ACC deaminase and therefore, cannot regulate the ethylene levels in plant tissue. In addition, Holguin and Glick (2001) found that the transfer of the ACC deaminase gene from *E. cloacae* to *A. brasilense* resulted in a significant improvement in plant growth of inoculated plants.

### *7.4.5 Abscisic Acid*

The ABA is involved in different physiological growth and developmental processes, such as bud formation and seed dormancy, fruit ripening and homeostatic regulation under abiotic stress. It is formed in both leaf and root plastids and begins its biosynthesis from MA via farnesyl farnesyl pyrophosphate (FFPP) through terpene biosynthesis. The metabolic pathway continues with carotenoid synthesis and concludes with their cleavage to xantoxine and ABA. This pathway is active in higher plants, especially under abiotic stress conditions such as water and saline stress. ABA confers the ability to higher plants to adapt to stress through a variety of physiological and molecular processes that include osmotic adjustment, stomatal closure, biosynthesis of stress-related proteins, and regulation of gene expression. From a physiological point of view, ABA supports water economy in plants due to its regulatory effect on stomata (Davies 1995). One could say that ABA is considered the true root signal in water stress conditions.

#### 7.4.5.1 Abscisic Acid Production by *Azospirillum* sp. and Other PGPR

There is very little information on ABA identification in chemically defined *Azospirillum* sp. culture medium and its correlation with plant growth and development. Most experiments were conducted in crop systems by inoculating fungi or plant pathogenic bacteria, or in symbiotic associations with root nodule-inducing bacteria. Kolb and Martin (1985) first reported the ABA production by *A. brasilense* Ft326 on defined culture medium. However, identification was done by radioimmunoassay (RIA), an insensitive technique compared to mass spectrometry, which is more commonly used today. Recently, Cohen et al. (2008) reported the characterization of the stress-related plant hormone ABA by GC-EIMS in cultures of *A. brasilense* Sp245 in chemically defined media supplemented with NaCl to generate a moderate stress condition (100 mM NaCl). *A. brasilense* produced higher amounts of ABA when NaCl was incorporated in the culture medium. Inoculation of *A. thaliana* with *A. brasilense* Sp245 enhanced twofold increases in the plant's ABA contents. PGPR synthesizing ABA has been described by Forchetti et al. (2007) who showed that *B. pumilus* strain SF3 and SF4 and *Achromobacter xyloxosidans* strain SF2, isolated from sunflower (*Helianthus annuus* L.) roots, produced significant amount of ABA in chemically defined medium. Later, Sgroy et al. (2009) confirmed the ABA production in chemically defined media for *Lysinibacillus fusiformis, B. subtilis, Brevibacterium halotolerans, Bacillus licheniformis, B. pumilus, Achromobacter xylosoxidans,* and *P. putida*. However, production was significantly higher in *B. subtilis* and *P. putida*, which produced 1.8 and 4.2 $\mu g\ ml^{-1}$, respectively as, compared to *L. fusiformis*, which produced 0.3 $\mu g\ ml^{-1}$, others strains produced less than 0.2 $\mu g\ ml^{-1}$.

### *7.4.6 Polyamines*

The newest compound involved in promoting plant growth by *Azospirillum* sp. is the polyamine cadaverine synthesized from the precursor L-lysine. Polyamines are low molecular weight organic compounds having two or more primary amino groups. Polyamines serve as growth regulating compounds (Kuznetsov et al. 2006). One example is cadaverine, which has been correlated with root growth promotion in pine and soybean (Niemi et al. 2001), response to osmotic stress in turnip (Aziz et al. 1997), and controlling stomatal activity in *Vicia faba* beans

(Liu et al. 2000). *A. brasilense* strain Az39, which is widely used as a wheat and maize inoculant in Argentina, is known to produce polyamines such as spermidine, spermine (Perrig et al. 2007), and putrescine (Thuler et al. 2003) in culture, and also produces cadaverine in chemically defined medium supplemented with the precursor L-lysine and on inoculated rice plants. In rice, exogenous application of cadaverine or inoculation with *A. brasilense* mitigated its water stress, improving water status and decreasing ABA production by seedlings (Cassán et al. 2009a, b).

## *7.4.7 Nitric Oxide*

Nitric oxide (NO) is a volatile, lipophilic free radical that participates in metabolic, signaling, defense, and developmental pathways in plants (Cohen et al. 2010; Lamattina and Polacco 2007). NO plays a major role in the IAA signaling pathway and this participation leads to lateral and adventitious root formation wherein NO acts as an intermediary in IAA-induced root development (Correa-Aragunde et al. 2006).

### 7.4.7.1 Nitric Oxide and *Azospirillum* sp. Production

*A. brasilense* Sp245 can produce NO in vitro, under anoxic and oxic (or aerobic) conditions (Creus et al. 2005). The latter can be achieved by multiple possible pathways, such as aerobic denitrification and heterotrophic nitrification. NO is produced during the middle and late logarithmic phases of growth (Molina-Favero et al. 2007, 2008). The NO production in *A. brasilense* Sp245 induces morphological changes in tomato roots regardless of the full bacterial capacity to synthesize IAA. An IAA-attenuated mutant of this strain, producing up to 10% of the IAA level compared with the wild-type strain (Dobbelaere et al. 1999), had the same physiological characteristics with slightly less effect on root development. These results provided further evidence of an NO-dependent promoting activity of tomato root branching, regardless of the bacterium's capacity to synthesize IAA (Molina-Favero et al. 2008), a phenomenon that occurs in other inoculation systems lacking IAA activity. The relationship between NO and *A. brasilense* showed that, in addition to the well-established connection between NO production and defense responses to pathogenic microorganisms (Modolo et al. 2005), it seems that NO metabolism plays a role in the beneficial close association of *Azospirillum* sp. with roots.

## *7.4.8 Jasmonates*

Jasmonates (JAs) belong to the family of oxygenated fatty acid derivatives, collectively called oxylipins, which are produced via the oxidative metabolism of polyunsaturated fatty acids. The synthesis and signal transduction pathway of these compounds in plants have been well described by Wasternack and Kombrink (2010). The initial substrates are α-linolenic acid (α-LeA; C18:3) or hexadecatrienoic acid (C16:3) released from plastidial galactolipids by phospholipases. The subsequent step is the oxidation of α-LeA by lipoxygenase (LOX) to 13(S)-hydroperoxyoctadecatrienoic acid (13(S)-HPOT) or (9S)-hydroperoxyoctadecatrienoic acid (9(S)-HPOT). The corresponding products with α-LeA as the substrate are (13S)-hydroperoxyoctadecadienoic acid (13-HPOD) and (9S)-hydroperoxyoctadecadienoic acid (9-HPOD). The first committed step of JA biosynthesis is the conversion of the LOX product to the allene oxide 12,13(S)-epoxyoctadecatrienoic acid (12,13(S)-EOT) by allene oxide synthase (AOS). This unstable compound can be enzymatically cyclized by allene oxide cyclase (AOC) to *cis*-(+)-12-oxophytodienoic acid (9S,13S)-OPDA). This compound is the end product of the plastid-localized part of the JA biosynthesis pathway and carries the same stereochemical configuration similar to that of naturally occurring (+)-7-*iso*-JA. Translocation of OPDA into peroxisomes, where the second half of JA biosynthesis occurs, is mediated by the ABC transporter COMATOSE and/or an ion trapping mechanism (Theodoulou et al. 2005). The final step is the reduction of the cyclopentenone ring, catalyzed by a peroxisomal OPDA reductase (OPR), to yield JA.

#### 7.4.8.1 Jasmonates and PGPR Production

JAs have not been described as products of *Azospirillum* sp. biosynthesis. However, many recent reports correlate JA production by PGPR to the plant response of inoculated plants. Accordingly, Forchetti et al. (2007) isolated endophytic rhizobacteria from sunflower (*Helianthus annuus* L.) grown under irrigation and water stress conditions and evaluated the bacterial ability to produce JAs in chemically defined medium. They discovered that *B. pumilus* strains SF2 and SF3 and *Achromobacter xylosoxidans* SF4 grown in controlled medium produced jasmonic acid (JA) and 12-oxo-phytodienoic acid (OPDA) and the former increased under drought conditions. In relation to inoculated plants, during bacteria–legume interaction, Rosas et al. (1998) reported for the first time that exogenous jasmonic acid (JA) caused induction of *nod* genes in *Rhizobium leguminosarum.* Later, Mabood and Smith (2005) showed that JA and methyl-jasmonate (MeJA) strongly induced the expression of nodulation genes of *B. japonicum* and that preincubation with JAs enhanced nodulation, N fixation, and plant growth in soybean under controlled environmental conditions. Likewise, Mabood et al. (2006) reported that inoculation of soybean plants with bradyrhizobial cells incubated with MeJA alone or in combination with genistein (GE) increased nodule number, nodule dry matter (DM) per plant, and seasonal $N_2$ fixation, as compared to *B. japonicum* cells that were not incubated with those compounds. Moreover, it was shown that JAs effectively induced the production and secretion of the NOD factors lipo-chitooligosaccharides, LCOs, from *B. japonicum.*

## 7.5 *Azospirillum* sp., Phytohormones and Plant Growth

The biochemical, molecular, physiological, and functional analysis of phytohormone interactions in higher plant has re-emerged in the last 10 years and because of this revival, the bacterial phytohormones are not exempt from the same analysis. The amount of new "cross-talk" interactions described in literature for thousands of plant species may give new insights into the simple phytohormonal growth promotion-dependent model described for *Azospirillum* sp. In this regard, there is circumstantial evidence of the interaction between the phytohormones produced by *Azospirillum* sp. and the hormonal background of inoculated plants, but a detailed analysis of this interaction may reveal specific interactions that could result in their PGP effect. In this regard, Fulchieri et al. (1993) found that maize seedlings (*Z. mays* L.) inoculated with three *A. lipoferum* strains showed significantly improved root and shoot growth. In these trials, $GA_3$ was identified in the free acid fraction of plant extract and these results could be explained by bacterial ability to increase the whole pool of biologically active gibberellins in the roots of inoculated plants. Ross and O'Neill (2001) suggested that auxin could promote, at least in part, stem elongation by increasing endogenous levels of 3β-hydroxylated gibberellins, which could be directly related to the results obtained by Fulchieri et al. (1993), and that neither the endogenous IAA content or 3β-hydroxylases genes expression was quantified in bacterial culture medium. Another factor to be considered is that at least two of the strains used for inoculation, namely *A. lipoferum* Op33 and *A. lipoferum* iaa320, were IAA producers, which led us to speculate that at least in part, the endogenous $GA_3$ increased in the root may be due to a "cross-talk" between IAA and the GA pool in the root. At least part of the growth response observed in aerial and underground plant tissues could be a result of the GAs produced by different strains of *A. lipoferum* or by the GAs produced by the seedlings induced by bacterial IAA (Yaxley et al. 2001; Ford et al. 2002; Inada and Shimmen 2000). Krumpholz et al. (2006) evaluated the growth response of tomato seedlings inoculated with *A. brasilense* FT326 (an IAA-overproducer), correlating plant phenotypes and ethylene production. The increase in ethylene production was significantly higher accompanied by increased activity of ACC synthase in plant tissues. According to Peck and Kende (1995), the rate limiting step for ethylene biosynthesis is S-adenosylmethionine (SAM) conversion to 1-aminocyclopropane-1-carboxylic acid (ACC) catalyzed by ACC synthase, and the expression of this enzyme, as well as ethylene production, was enhanced by the addition of exogenous IAA. These results indicate that the ethylene increase is at least in part due to a "cross-talk" between the bacterially produced IAA and the ethylene synthesis by the plant (Rahman et al. 2002).

## 7.6 *Azospirillum* sp. Based Inoculants and Plant Growth Promotion

Since 1981–1996, the Instituto de Microbiología y Zoología Agrícola (IMYZA), INTA-Castelar, from República Argentina, developed an intensive program. The main objectives of the project were to select and identify strains of *Azospirillum* sp. and to evaluate their ability to promote plant growth on different crop species. The information generated showed a more pronounced effect of *A. brasilense* on most evaluated plant species than the effect obtained by *A. lipoferum* strains, and allowed bioprospecting of strain Az 39 of *A. brasilense* based on their ability to increase growth and yield of evaluated crops by 13.0–33.0%. Considering the information generated in this program, the SENASA (Servicio Nacional de Sanidad Agropecuaria) proclaimed a nationwide recommendation of the native strain Az39 of *A. brasilense* for inoculant production for maize, wheat, and other nonlegume species.

From a physiological point of view, the plant growth promotion capacity of *A. brasilense* Az39 has been confirmed because of the effectiveness in increasing the productivity of inoculated crops in thousands of assays in field conditions, during last 30 years. However, there was no detailed description of the plant growth promoting mechanisms operating in this strain. Part of the work of Laboratorio de Fisiología Vegetal y de la Interacción planta-microorganismo from the Universidad Nacional de Río Cuarto, Argentina was to elucidate all mechanisms responsible for the growth promotion effect in crops after inoculation with *A. brasilense* Az39. As part of these results, Perrig et al. (2007) proved that *A. brasilense* Az39, the most commonly used strain for inoculant formulation in Argentina, together with the strain Cd, one of the most commonly used for basic research in the world, has the ability to produce and release plant growth regulators such as IAA, Z, $GA_3$, ABA, and ethylene. In this regard, IAA production in *A. brasilense* Az39 was comparable to that previously reported by other authors for *A. lipoferum* (4.1 $\mu g\ ml^{-1}$) and other strains of *A. brasilense* (4.5 $\mu g\ ml^{-1}$) in chemically defined medium (Crozier et al. 1988). The Cd strain showed a major production of IAA (10.8 $\mu g\ ml^{-1}$) compared to strain Az39. In other trials, Dobbelaere et al. (1999) found that inoculation of wheat cv. Soisson with $1 \times 10^8$ cfu $ml^{-1}$ of *A. brasilense* Sp245 was comparable to the exogenous application of IAA. Zeatin (Z) production for strains Cd and Az39 was determined to be 2.37 and 0.75 $\mu g\ ml^{-1}$, respectively. Horemans et al. (1986) detected similar amounts of Z and other cytokinins in *A. brasilense* in chemically defined medium by the radioimmunoassay (RIA) technique. Furthermore, Tien et al. (1979) used liquid chromatography (HPLC) to demonstrate that *A. brasilense* produced cytokinin-like compounds in chemically defined medium with biological activity equivalent to that of K (kinetin). Pan et al. (1999) reported that exogenous application of 0.2 $\mu g\ ml^{-1}$ of kinetin in corn seeds, together with the inoculation of *Serratia liquefaciens*, increased size and weight of the roots. $GA_3$ production was significantly higher in strain Cd (0.66 $\mu g\ ml^{-1}$) than in Az39 (0.30 $\mu g\ ml^{-1}$). In addition, Bottini et al. (1989) evaluated the ability of *A. lipoferum* to produce $GA_1$

and $GA_3$ in the NFB chemically defined medium and the concentration in the supernatant of the culture media was estimated by biological tests to be 20 and 40 pg $ml^{-1}$ of $GA_3$ equivalents, respectively, for a bacterial titer of $10^9$ cfu $ml^{-1}$ Additionally, Janzen et al. (1992) evaluated the $GA_1$ and $GA_3$ production in *A. brasilense* Cd in chemically defined medium and $GA_3$ production reached 5.0 μg $ml^{-1}$ for $10^7$ cfu $ml^{-1}$. ABA production was significantly higher in Az39 (77.50 ng $ml^{-1}$) than for Cd (0.65 ng $ml^{-1}$). There are very few reports on ABA identification in chemically defined medium or in inoculated plants with *Azospirillum* sp. Kolb and Martin (1985) reported the ABA production by *A. brasilense* Ft326, but the detection method used was less sensitive (RIA). The bacterial contribution to the role of ABA in plant–microbe interactions is uncertain and there is no direct evidence that this phytohormone promotes or regulates plant growth, but in restrictive soils (e.g., saline soils), ABA could contribute, together with other bacterial molecules such as cadaverine (Aziz et al. 1997), in regulation of plant homeostasis and the response to stress. This is an emerging research line in our group that is particularly focused on free-living rhizobacteria that could be re-classified as a third group of beneficial bacteria to plants, called "Plant Stress Homeostasis-regulating Rhizobacteria" (PSHR) as proposed by Cassán et al. (2009a). Ethylene biosynthesis in culture medium modified by the addition of L-methionine was found for both strains in the presence or absence of the precursor, but major production was determined in the medium modified by the addition of precursor in strain Cd (3.94 ng ml $h^{-1}$). In contrast to our results, Strzelczyk et al. (1994) found that ethylene production in chemically defined medium was completely dependent on the presence of L-methionine. From the physiological point of view, the growth promoting effect of *Azospirillum* sp. at root level may depend on ethylene production by the plant (bacterial-mediated auxin biosynthesis) or on bacterial ethylene production (Fig. 7.11).

Cassán et al. (2009a) showed that *A. brasilense* Az39 and *B. japonicum* E109, inoculated singly or in combination, have the capacity to promote seed germination and early seedling growth in soybean and corn, and this capacity was correlated with the bacterial phytohormone biosynthesis in culture medium. In this respect, Az39 and E109 were able to excrete gibberellic acid ($GA_3$), zeatin (Z), and also IAA in culture medium at a sufficient concentration to produce morphological and physiological changes in seeds and young seed tissues. Recently, Cassán et al. (2009b) showed that both *B. japonicum* E109 and *A. brasilense* Az39 possess the capacity to promote germination and early seedling growth in maize, wheat, and soybean, and that such capacity is not only dependent on the bacterial cell number, but also on the concentration of gibberellic acid ($GA_3$), zeatin (Z), and IAA released into culture medium during exponential (EP) or stationary (SP) growth phase (Cassán et al. 2010b). The concentration of Z and IAA increased considerably in SP (0.78 and 14.2 μg $ml^{-1}$, respectively) compared to EP (0.36 and 9.1 μg $ml^{-1}$, respectively), while the $GA_3$ maintained similar concentrations in both phases (1.70 and 1.2 μg $ml^{-1}$, respectively). In relation to the capacity to promote plant growth, all parameters evaluated were significantly increased when seeds were treated with SP culture medium of *A. brasilense* Az39 and *B. japonicum* E109. Our results

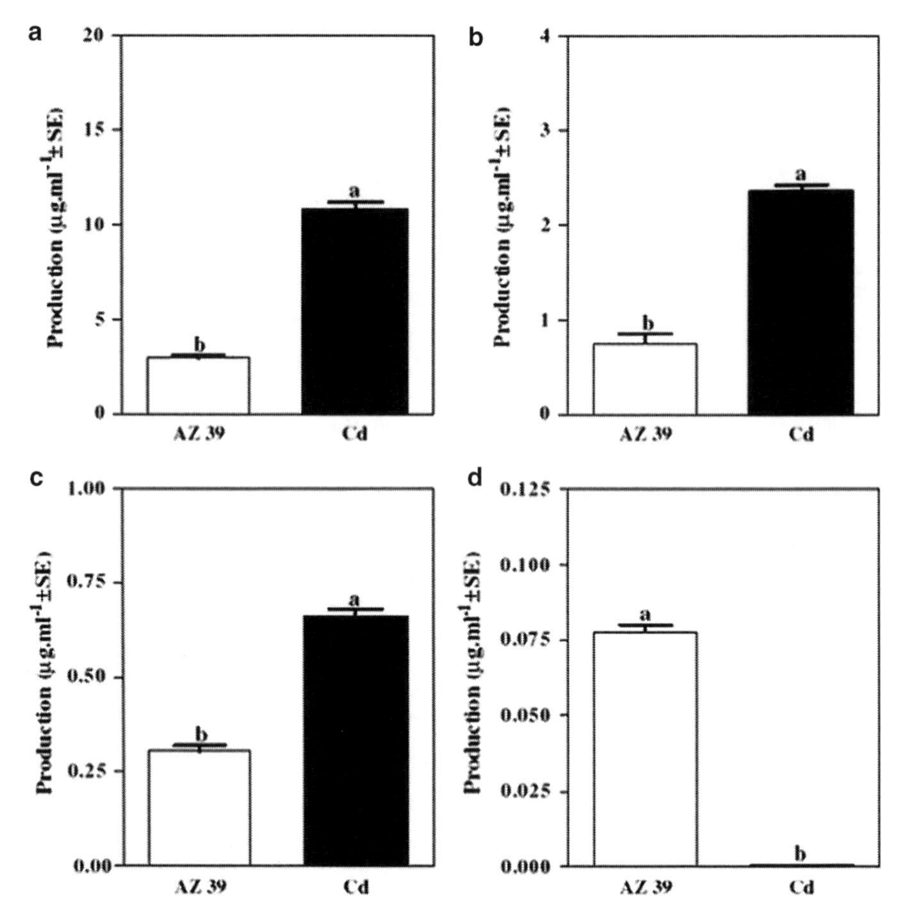


**Fig. 7.11** Identification and quantification of phytohormones by HPLC-UV-GC-MS SIM, obtained from chemically defined medium of *A. brasilense* Az39 and Cd. (**a**) IAA, (**b**) Z, (**c**) $GA_3$, and (**d**) ABA. Published by Perrig et al. (2007)

indicate that *A. brasilense* Az39 has the potential capacity to promote plant growth through phytohormone production, as an additional mechanism to the well-known biological nitrogen fixation. Az39 strain showed intrinsic ability to produce and release several phytohormones in chemically defined medium. This capacity was correlated with the bacterial capacity to promote germination and early growth in maize and soybean seeds and seedlings in single inoculation or in co-inoculation with *B. japonicum*. Also, we established that stationary cultures of Az39 produce more phytohormones than exponential ones, and accumulation of these compounds in culture medium also modify the inoculant capacity to promote the early growth parameters in seeds or seedlings of soybean, maize, and wheat. We defined this capacity as the "phytohormonal effect of inoculation" and the mentioned parameters could have great value on the classification of strains and as an evaluation tool in strain selection for agricultural use. There are, however, more than 30 years of research and hundreds of publications related to the use of *A. brasilense* Az39 in field and greenhouse conditions (Diaz-Zorita and Fernández-Canigia 2009) which confirm largely the results presented in this chapter.

## 7.7 Conclusion and Future Prospects

We know relatively little about the phytohormones in plant growth-promoting rhizobacteria, compared with plants. Accurately known its biosynthetic routes, enzymes and genes responsible for synthesis, catabolism and conjugation, must be considered the initial step to understand its capacity to regulate the plant growth and development, as well as productivity in agricultural conditions. Only in the case of auxins, and particularly for IAA, has been described its physiological and molecular functionality in both, chemically defined medium and plant–microbe interaction. In the case of gibberellins and particularly for gibberellic acid and $GA_1$, the model has been described since a physiological but not molecular point of view. Finally, for cytokinins, ABA, ethylene, JAs, polyamines, and nitric oxide, we only know about the ability of several microorganisms to produce it in chemically defined medium or in less case in plant–microbe interaction. Integration of both, microbial and plant models since a phytohormonal and physiological point of view, could be the beginning of a new understanding about the natural process includes into a new kind of physiology named "plant-microbe physiology" or simply "integrative physiology." This proposal not only provides a semantic denomination, but on the contrary, seeks to establish the concept that sterile plants only grown in our laboratories, except in those where microbiologists works fortunately! As we have seen throughout this chapter, several rhizobacteria, but principally *Azospirillum* sp. have the potential capacity to modify growth, development, and behavior of several plants even in stress conditions. Knowing that, only we must generate an elementary question: Our study models could be considered inside of an integrative physiology model? If your answer is not clear at this point, you should read this chapter one more time.

**Acknowledgements** We would like to thank the Consejo Nacional de Investigaciones Científico-Tecnológicas (CONICET), the Agencia Nacional de Promoción Científica y Tecnológica (ANPCyT) and the Ministerio de Ciencia y Tecnología from República Argentina. Universidad Nacional de Río Cuarto. Special thanks to Yoav Bashan (CBNOR), Jos Vanderleyden and Stijn Spaepen (KUL), Cecilia Creus (INTA-UNMdP) and Guillermina Abdala (UNRC) to facilitate some information included here to complete the phytohormonal model of *Azospirillum* sp. and to Stijn Bossuyt (KUL) for helping us with language editing of the chapter.

## References

Abdel-Salam M, Klingmüller W (1987) Transposon *Tn5* mutagenesis in *Azospirillum lipoferum*: isolation of indole acetic acid mutants. Mol Gen Genet 210:165–170

Abeles F, Rubinstein B (1964) Cell-free ethylene evolution from etiolated pea seedlings. Biochem Biophys Acta 93:675–677

Akiyoshi D, Regier A, Gordon M (1987) Cytokinin production by *Agrobacterium* and *Pseudomonas* spp. J Bacteriol 169:135–140

Arkhipova T, Veselov S, Melentiev A, Martynenko E, Kudoyarova G (2005) Ability of bacterium *Bacillus subtilis* to produce cytokinins and to influence the growth and endogenous hormone content of lettuce plants. Plant Soil 272:201–209

Arshad M, Frankenberger Jr (1993) Microbial production of plant growth regulators. In: Meetin B (ed) Soil microbial ecology. Marcel Dekker, New York, pp 307–347

Aziz A, Martin-Tanguy J, Larher F (1997) Plasticity of polyamine metabolism associated with high osmotic stress in rape leaf discs and with ethylene treatment. Plant Growth Regul 21:153–163

Baca B, Soto-Urzua L, Xochiua-Corona Y, Cuervo-García A (1994) Characterization of two aromatic aminoacid aminotransferases and production of indoleacetic acid in *Azospirillum* strains. Soil Biol Biochem 26:57–63

Bar T, Okon Y (1993) Tryptophan conversion to indole-3-acetic acid via indole-3-acetamide in *Azospirillum brasilense* Sp7. Can J Microbiol 39:81–86

Barbieri P, Bernardi A, Galli E, Zanetti G (1988) Effects of inoculation with different strains of A. *brasilense* on wheat roots development. In: Klingmüller W (ed) *Azospirillum* IV. Genetics physiology ecology. Springer, Heidelberg, pp 181–188

Barbieri P, Baggio C, Bazzicalupo M, Galli E, Zanetti G, Nuti M (1991) *Azospirillum*-gramineae interaction: effect on indole-3-acetic acid. Dev Plant Soil Sci 48:161–168

Barea J, Navarro M, Montoya E (1976) Production of plant growth regulators by rhizosphere phosphate-solubilizing bacteria. J Appl Bacteriol 40:129–134

Bashan Y, de-Bashan L (2010) How the plant growth-promoting bacterium *Azospirillum* promotes plant growth. A critical assessment. Adv Agron 108:77–136

Bashan Y, Holguin G (1997) *Azospirillum*-plant relationships: environmental and physiological advances. Can J Microbiol 43:103–121

Bashan Y, Holguín G (1998) Proposal for the division of plant growth-promoting rhizobacteria into two classifications: biocontrol-PGPB (plant growth promoting bacteria) and PGPB. Soil Biol Biochem 30:1225–1228

Bashan Y, Levanony H (1990) Current status of *Azospirillum* inoculation technology: *Azospirillum* as a challenge for agriculture. Can J Microbiol 36:591–608

Bashan Y, Holguin G, de-Bashan L (2004) *Azospirillum*-plant relationships: physiological, molecular, agricultural, and environmental advances (1997–2003). Can J Microbiol 50:521–577

Bastian F, Cohen P, Piccoli V, Luna R, Baraldi S, Bottini R (1998) Production of indole 3-acetic acid and gibberellins A1 y A3 by *Acetobacter diazotrophicus* and *Herbaspirillum seropedicae* in chemically-defined culture media. Plant Growth Regul 24:7–11

Beyer E, Morgan P, Yang S (1984) Ethylene. In: Wilkins M (ed) Advance plant physiology. Pitman Publishing, London, pp 111–126

Bhattarai T, Hess (1993) Yield responses of Nepalese spring wheat (*Triticum aestivum* L.) cultivars to inoculation with *Azospirillum spp.* of Nepalese origin. Plant Soil 151:67–76

Blaha D, Sanguin H, Robe P, Nalin R, Bally R, Moenne-Loccoz Y (2005) Physical organization of phytobeneficial genes nifH and ipdC in the plant growth-promoting rhizobacterium *Azospirillum lipoferum* 4 V(I). FEMS Microbiol Lett 244:157–163

Boiero L, Perrig D, Masciarelli O, Penna C, Cassán F, Luna V (2007) Phytohormone production by three strains of *Bradyrhizobium japonicum*, and possible physiological and technological implications. Appl Microbiol Biotechnol 74(4):874–880

Bothe H, Körsgen H, Lehmaher T, Hundeshagen B (1992) Differential effects of *Azospirillum*, auxin and combined nitrogen on the growth of the roots of wheat. Symbiosis 13:167–179

Bottini R, Fulchieri M, Pearce D, Pharis R (1989) Identification of gibberellins $A_1$, $A_3$, and Iso-$A_3$ in cultures of *A. lipoferum*. Plant Physiol 90:45–47

Bottini R, Cassán F, Piccoli P (2004) Gibberelin production by bacteria and its involvement in plant growth promotion and yield increase. Appl Microbiol Biotechnol 65(5):497–503

Brandl M, Lindow S (1996) Cloning and characterization of a locus encoding an indolepyruvate decarboxylase involved in indole-3-acetic acid synthesis in *Erwinia herbicola*. Appl Environ Microbiol 62:4121–4128

Burg S (1962) The physiology of ethylene formation. Annu Rev Plant Physiol 13:265–302

Burg S, Burg P (1968) Ethylene formation in pea seedlings: its relation to the inhibition of bud growth caused by indole-3-acetic acid. Plant Physiol 43:1069–1073

Cacciari I, Lippi D, Pietrosanti T (1989) Phytohormone-like substances produced by single and mixed diazotrophic cultures of *Azospirillum spp*. and *Arthrobacter*. Plant Soil 115:151–153

Cassán F (2003) Activación de giberelinas in vivo por bacterias endofíticas a través de la deconjugación de glucosil conjugados y la 3ß-hidroxilación. Tesis Doctoral. Hemeroteca de la UNRC pp 140

Cassán F, Bottini R, Piccoli P (2001a) In vivo gibberellin $A_9$ metabolism by *Azospirillum* sp in *dy* dwarf rice mutants seedlings. Proc Plant Growht Regul 28:124–129

Cassán F, Bottini R, Schneider G, Piccoli P (2001b) *Azospirillum brasilense* and *Azospirirllum lipoferum* hidrolize conjugates of $GA_{20}$ and metabolize the resultant aglycones to $GA_1$ in seedlings of rice dwarf mutants. Plant Physiol 125:2053–2058

Cassán F, Lucangeli C, Bottini R, Piccoli P (2001c) *Azospirillum spp*. metabolize [17,17-$^2H_2$] gibberellin $A_{20}$ to [17,17-$^2H_2$] gibberellin $A_1$ in vivo in *dy* rice mutant seedlings. Plant Cell Physiol 42:763–767

Cassán F, Maiale S, Masciarelli O, Vidal A, Luna V, Ruiz O (2009a) Cadaverine production by *Azospirillum brasilense* and its possible role in plant growth promotion and osmotic stress mitigation. Eur J Soil Biol 45:12–19

Cassán F, Perrig D, Sgroy V, Masciarelli O, Penna C, Luna V (2009b) *Azospirillum brasilense* Az39 and *Bradyrhizobium japonicum* E 109 promote seed germination and early seedling growth, independently or co-inoculated in maize (*Zea mays* L.) and soybean (*Glycine max* L.). Eur J Soil Biol 45:28–35

Cassán F, Spaepen S, and Vanderleyden J (2010a) Indole-3-acetic acid biosynthesis by *Azospirillum brasilense* Az39 and its regulation under biotic and abiotic stress conditions. Abstract of 20th International Conference on Plant Growth Substances p 85

Cassán F, Torres D, Ribas A, Penna C and Luna V (2010b) Exponential and stationary cultures of *Azospirillum brasilense* and *Bradyrhizobium japonicum* differentially promote early growth of soybean and maize seedlings. Abstract of 20th International Conference on Plant Growth Substances p 85

Chapple C (1998) Molecular-genetic analysis of plant cytochrome P450 monooxygenases. Annu Rev Plant Physiol Plant Mol Biol 49:311–343

Cheryl P, Glick B (1996) Bacterial biosynthesis of indole-3-acetic acid. Can J Microbiol 42:207–220

Christiansen-Weniger C (1998) Endophytic establishment of diazotrophic bacteria in auxin-induced tumors of cereal crops. Crit Rev Plant Sci 17:55–76

Clark E, Manulis S, Ophir Y, Barash I, Gafni Y (1993) Cloning and characterization of iaaM and iaaH from *Erwinia herbicola* pathovar gypsophilae. Phytopathology 83:234–240

Cohen J, Bandurski R (1982) Chemistry and physiology of the bound auxins. Annu Rev Plant Physiol 33:403–430

Cohen A, Bottini R, Piccoli P (2008) *Azospirillum brasilense* Sp 245 produces ABA in chemically-defined culture medium and increases ABA content in arabidopsis plants. Plant Growth Regul 54:97–103

Cohen M, Lamattina L, Yamasaki H (2010) Nitric oxide signaling by plant-associated bacteria. In: Hayat S, Mori M, Pichtel J, Ahmad A (eds) Nitric oxide in plant physiology. Wiley-VCH, Weinheim, Germany, pp 161–172

Conney T, Nonhebel H (1991) Biosynthesis of indole-3-acetic acid in tomato shoots: measurement, mass-spectral identification and incorporation of $^2H$ from $^2H_2O$ into indole-3-acetic acid, D- and L-tryptophan, indole-3-pyruvate and tryptamine. Planta 184:368–376

Correa-Aragunde N, Graziano M, Chevalier C, Lamattina L (2006) Nitric oxide modulates the expression of cell cycle regulatory genes during lateral root formation in tomato. J Exp Bot 57:581–588

Costacurta A, Vanderleyden J (1995) Synthesis of phytohormones by plant-associated bacteria. Crit Rev Microbiol 21:1–18

Costacurta A, Keijers V, Vanderleyden J (1994) Molecular cloning and sequence analysis of an *Azospirillum brasilense* indole-3-pyruvate decarboxylase. Mol Gen Genet 243:463–472

Creus C, Sueldo R, Barassi C (1997) Shoot growth and water status in *Azospirillum*-inoculated wheat seedlings grown under osmotic and salt stresses. Plant Physiol Biochem 35:939–944

Creus C, Graziano M, Casanovas E, Pereyra A, Simontacchi M, Puntarulo S, Barassi C, Lamattina L (2005) Nitric oxide is involved in the *Azospirillum brasilense*-induced lateral root formation in tomato. Planta 221:297–303

Crozier A, Arruda P, Jasmim JM, Monteiro AM, Sandberg G (1988) Analysis of indole-3-acetic acid and related indoles in culture medium from *Azospirillum lipoferum* and *Azospirillum brasilense*. Appl Environ Microbiol 54:2833–2837

Crozier A, Malcom J, Graebe J (2000) Biochemistry and molecular biology of plants. Academic, Cambridge, p 17, 850–864

Davies P (1995) Plant hormones. Physiology, biochemistry and molecular biology. Kluwer Acad Press, Dordrecht, p 833

De-Bashan L, Bashan Y (2004) Recent advances in removing phosphorus from wastewater and its future use as fertilizer. Water Res 38:4222–4246

Dharmasiri N, Dharmasiri S, Estelle M (2005) The F-box protein TIR1 is an auxin receptor. Nature 435:441–446

Diaz-Zorita M, Fernández-Canigia M (2009) Field performance of a liquid formulation of *Azospirillum brasilense* on dryland wheat productivity. Eur J Soil Biol 45:3–11

Dobbelaere S, Croonenborghs A, Thys A, Vande Broek A, Vanderleyden J (1999) Phytostimulatory effect of *Azospirillum brasilense* wild type and mutant strains altered in IAA production on wheat. Plant Soil 212:155–164

Dubrovsky J, Puente M, Bashan Y (1994) *Arabidopsis thaliana* as a model system for the study of the effect of inoculation by *Azospirillum brasilense* Sp-245 on root hairs growth. Soil Biol Biochem 26:1657–1664

Even-Chen Z, Matoo A, Goren R (1982) Inhibition of ethylene biosynthesis by aminoethoxyvinylglycine and by polyamines shunts label from 3,4-($^{14}C$)methionine into spermidine in aged orange peels disc. Plant Physiol 69:385–388

Falik E, Okon Y, Epstein E, Goldman A, Fischer M (1989) Identification and quantification of IAA and IBA in *Azospirillum brasilense*-inoculated maize roots. Soil Biol Biochem 21:147–153

Fluhrer R, Mattoo A (1996) Ethylene-biosynthesis and perception. Crit Rev Plant Sci 15:479–523

Forchetti G, Masciarelli O, Alemano S, Alvarez D, Abdala G (2007) Endophytic bacteria in sunflower (*Helianthus annuus* L.) isolation, characterization, and production of jasmonates and abscisic acid in culture medium. Appl Microbiol Biotechnol 76:1145–1152

Ford Y, Taylor J, Blake P, Marks P (2002) Gibberellin $A_3$ stimulates adventitious rooting of cuttings from cherry (*Prunus avium*). Plant Growth Regul 37:127–133

Frankenberger W, Arshad M (1995) Phytohormones in soil. Marcel Dekker, New York, pp 1–135

Fukuda H, Ogawa T (1991) Microbial ethylene production. In: Matoo K, Suttle J (eds) The plant hormone ethylene. CRC Press, Boca Raton, FL, pp 279–292

Fulchieri M, Frioni L (1994) *Azospirillum* inoculation on maize (*Zea mays* L.): effects on yield in a field experiment in central Argentina. Soil Biol Biochem 26:921–923

Fulchieri M, Lucangeli C, Bottini R (1993) Inoculation with A. *lipoferum* affects growth and gibberellin status of corn seedlings roots. Plant Cell Physiol 34:1305–1309
Gane R (1934) Production of ethylene by some ripening fruit. Nature 134:1008
Garate A, Bonilla I (2000) Nutrición mineral y producción vegetal. In: Fundamentos de fisiología vegetal. Azcón-Bieto and Talón E (eds) McGraw-Hill Interam, Madrid pp 113–130
Giraud E, Moulin L, Vallenet D, Barbe D, Cytryn E (2007) Legumes symbioses: absence of Genes in *Nod* in photosynthetic Bradyrhizobia. Science 316:1307
Glass N, Kosuge T (1986) Cloning of the gene for indoleacetic acid-lysine synthetase from *Pseudomonas syringae* subsp savastanoi. J Bacteriol 166:598–603
Glick B, Bashan Y (1997) Genetic manipulation of plant growth promoting bacteria to enhance biocontrol of fungal phytopathogens. Biotechnol Adv 15:353–378
Glick B, Karaturovic D, Newell P (1995) A novel procedure for rapid isolation of plant growth promoting pseudomonads. Can J Microbiol 41:533–536
Glick B, Shah S, Li J, Penrose D, Moffatt B (1998) Isolation and characterization of ACC deaminase genes from two different plant growth-promoting rhizobacteria. Can J Microbiol 44:833–843
Glick B, Patten C, Holguin G, Penrose D (1999) Biochemical and genetic mechanisms used by plant growth promoting bacteria. Imperial College Press, London, UK
Grobbelaar N, Clarke B, Hough M (1971) The modulation and nitrogen fixation of isolated roots of *Phaseolus vulgaris* L. Plant Soil (Special vol.) 215–223
Gutiérrez-Mañero F, Ramos-Solano B, Probanza A, Mehouachi J, Tadeo F, Talon M (2001) The plant-growth-promoting rhizobacteria *Bacillus pumilus* and *Bacillus licheniformis* produce high amounts of physiologically active gibberellins. Physiol Plantarum 111:206–211
Hartmann A, Singh M, Klingmüller W (1983) Isolation and characterization of *Azospirillum* mutants excreting high amounts of indoleacetic acid. Can J Microbiol 29:916–923
Hedden P, Phillips A (2000) Gibberellin metabolism: new insigths revealed genes. Trends Plant Sci 5:523–530
Henson J, Wheeler H (1977) Hormones in plant bearing nitrogen-fixing root nodules: distribution and seasonal changes in levels of cytokinins in *Alnus glutinosa* (L.) Gaernt. J Exp Bot 28:205–214
Hirsch A, Fang Y, Asad S, Kapultnik Y (1997) The role of phytohormones in plant-microbe symbioses. Plant Soil 194:171–184
Holguin G, Glick B (2001) Expression of the ACC deaminase gene from *Enterobacter cloacae* UW4 in *Azospirillum brasilense*. Microb Ecol 41:281–288
Horemans S, Koninck K, Neuray J, Hermans R, Vlassak K (1986) Production of plant growth substances by *Azospirillum* sp. and other rhizophere bacteria. Symbiosis 2:341–346
Hubbell D, Tien T, Gaskins M, Lee J (1979) Physiological interaction in the *Azospirillum*-grass root association. In: Vose P, Ruschel A (eds) Associative $N_2$-fixation. CRC Press, Boca Raton, FL, pp 1–6
Inada S, Shimmen T (2000) Regulation of elongation growth by gibberellin in root segments of *Lemna minor*. Plant Cell Physiol 41:932–939
Jaiswal V, Rizvi H, Mukerji D, Mathur S (1982) Nitrogenase activity in root nodules of *Vigna mungo*: The role of nodular cytokinins. Angew Bot 56:143–148
Janzen R, Rood S, Dormar J, McGill W (1992) *Azospirillum brasilense* produces gibberellins in pure culture and chemically-medium and in co-culture on straw. Soil Biol Biochem 24:1061–1064
Kende H (1989) Enzymes of ethylene biosynthesis. Plant Physiol 91:1–4
Kepinski S, Leyser O (2005) The *Arabidopsis* F-box protein TIR1 is an auxin receptor. Nature 435:446–451
Klee H, Montoya A, Horodyski F, Lichenstein C, Garfinkel D, Fuller S, Flores C, Peschon J, Nester E, Gordon M (1984) Nucleotide sequence of the tms genes of the pTiANC octopine C plasmid: two genes products involved in plants tumorogenesis. Proc Natl Acad Sci USA 81:1728–1732

Kloepper J, Schroth M (1978) Plant growth-promoting rhizobacteria in radish. In: Proceedings of the 4th International Conference on Plant Pathogenic Bacteria. vol 2 INRA, Angers, France pp 879–882
Kloepper J, Lifshitz R, Schroth M (1989) Pseudomonas inoculants to benefit plant production. ISI Atlas Sci Anim Plant Sc 8:60–64
Kobayashi M, Sakurai A, Saka A, Takahashi N (1989) Quantitative analysis of endogenous gibberellins in normal and dwarf cultivars of rice. Plant Cell Physiol 30:963–969
Kobayashi M, Izui H, Nagasawa T, Yamada H (1993) Nitrilase in biosynthesis of the plant hormone indole-3-acetic acid from indole-3-acetonitrile – cloning of the *Alcaligene* gene and site directed mutagenesis of cysteine residues. Proc Natl Acad Sci USA 90:247–251
Kobayashi M, Suzuki T, Fujita T, Masuda M, Shimizu S (1995) Occurrence of enzymes involved in biosynthesis of indole-3-acetic acid from indole-3-acetonitrile in plant-associated bacteria, *Agrobacterium* and *Rhizobium*. Proc Natl Acad Sci USA 92:714–718
Kolb W, Martin P (1985) Response of plant roots to inoculation with *Azospirillum brasilense* and to application of indoleacetic acid. In: Klingmüller W (ed) *Azospirillum* III: genetics, physiology, ecology. Springer, Berlin, pp 215–221
Krumpholz E, Ribaudo C, Cassán F, Bottini R, Cantore M, Curá A (2006) *Azospirillum* sp. promotes root hair development in tomato plants through a mechanism that involves ethylene. J Plant Growth Regul 25:175–185
Kucey R (1988) Alteration of size of wheat root systems and nitrogen fixation by associative nitrogen-fixing bacteria measured under field conditions. Can J Microbiol 34:735–739
Kuznetsov V, Radyukina N, Shevyakova N (2006) Polyamines and stress: biological role, metabolism, and regulation. Russ J Plant Physiol 53:583–604
Lamattina L, Polacco J (2007) Nitric oxide in plant growth development and stress physiology. Springer, Berlin, p 283
Lambrecht M (1999) The ipdC promoter auxin-responsive element of *Azospirillum brasilense*, a prokaryotic ancestral form of the plant *AusxRE*. Mol Microbiol 32:889–890
Lambrecht M, Okon Y, Vande Broek A, Vanderleyden J (2000) Indole-3-acetic acid: a reciprocal signalling molecule in bacteria-plant interactions. Trends Microbiol 8:298–300
Last R, Bissinger P, Mahoney D, Radwanski E, Fink G (1991) Tryptophan mutants in *Arabidopsis* – the consequences of duplicated tryptophan synthase beta genes. Plant Cell 3:345–358
Letham D (1963) Zeatin, a factor inducing cell division from *Zea mays*. Life Sci 8:569–573
Litchtenthaler H (1999) The 1-deoxy-d-xylulose-5-phosphate pathway of isoprenoid biosynthesis in plants. Annu Rev Plant Physiol Plant Mol Biol 50:47–65
Liu K, Fu H, Bei Q, Luan S (2000) Inward potassium channel in guard cells as a target for polyamine regulation of stomatal movements. Plant Physiol 124:1315–1325
Lucangeli C, Bottini R (1997) Effects of *Azospirillum* spp. on endogenous gibberellin content and growth of maize (*Zea mays* L.) treated with Uniconazole. Symbiosis 23:63–72
Mabood F, Smith L (2005) Pre-incubation of *Bradyrhizobium japonicum* with jasmonates accelerates nodulation and nitrogen fixation in soybean (*Glycine max* L.) at optimal and suboptimal root zone temperatures. Plant Physiol 125:311–323
Mabood F, Souleimanov A, Khan W, Smith D (2006) Jasmonates induce Nod factor production by *Bradyrhizobium japonicum*. Plant Physiol Biochem 44:759–765
MacMillan J (1997) Biosynthesis of the gibberellin plant hormones. Nat Prod Rep 14:221–243
Malhotra M, Srivastava S (2009) Stress-responsive indole-3-acetic acid biosynthesis by *Azospirillum brasilense* SM and its ability to modulate plant growth. Eur J Soil Biol 45:73–80
Mander L (1991) Recent progress in the chemistry and biology of gibberellins. Sci Progress 75:33–50, Oxford
Martens D, Frankenberger J (1992) Assimilation of 3′-$^{14}$C-indole-3-acetic acid and tryptophan by wheat varieties from nutrient media. In: Proceedings 19$^{th}$ Annual Meeting Plant Growth Regulator Society of America, San Francisco, CA, pp 99–100
Mathesius U, Shalaman H, Meijer D, Lugtenberg B, Spaink H, Weinman J, Rodam L, Sautter C, Rolfe B, Djordjevic M (1997) New tools for investigating nodule initiation and ontogeny: spot

inoculation and microtargeting of transgenic with clover roots shows auxin involvement and suggest a role for flavonoids. In: Stacey G, Mullin B, Gresshoff P (eds) Advances in molecular genetics of plant-microbe interactions. Kluwer Academic Publishers, Dordrecht

Mattoo A, Suttle J (1991) The Plant hormone ethylene. CRC Press, Boca Raton, FL, 337

Miller C, Skoog F, Von Saltza M, Strong F (1955) Kinetin, a cell division factor from deoxyribonucleic acid. J Am Chem Soc 77:1392

Modolo L, Augusto O, Almeida I, Magalhaes J, Salgado I (2005) Nitrite as the major source of nitric oxide production by *Arabidopsis thaliana* in response to *Pseudomonas syringae*. FEBS Lett 579:3814–3820

Molina-Favero C, Creus C, Lanteri M, Correa-Aragunde N, Lombardo M, Barassi C, Lamattina L (2007) Nitric oxide and plant growth promoting rhizobacteria: common features influencing root growth and development. Adv Bot Res 46:1–33

Molina-Favero C, Creus C, Simontacchi M, Puntarulo S, Lamattina L (2008) Aerobic nitric oxide production by *Azospirillum brasilense* Sp245 and its influence on root architecture in tomato. Mol Plant Microbe Interact 21:1001–1009

Morris R (1995) Genes specifying auxin and cytokinin biosynthesis in prokaryotes. Plant Hormones (Davies PJ, eds), pp. 318–339. Kluwer Academic Publishers, Dordrecht gical role, metabolism, and regulation. Russian J Plant Physiol 53: 583–604

Murakami Y (1968) A new rice seedling bioassay for gibberellins, microdrop method and its use for testing extracts of rice and morning glory. Bot Mag 81:3–43

Murakami Y (1972) Dwarfing genes in rice and their relation to gibberellin biosynthesis. In: Carr D (ed) Plant growth substances 1970. Springer, Berlín, pp 164–174

Nagasawa T, Mauger J, Yamada H (1990) A novel nitrilase, arylacetonitrilase, of *Alcaligenes faecalis* JM3 – purification and characterization. Eur J Biochem 194:765–772

Nandwall A, Bharti S, Garg O, Ram P (1981) Effects of indole acetic acid and kinetin on nodulation and nitrogen fixation in pea (*Pisum sativum* L.). Ind J Plant Physiol 24:47–52

Niemi K, Haggman H, Sarjala T (2001) Effects of exogenous diamines on the interaction between ectomycorrhizal fungi and adventitious root formation in Scots pine *in vitro*. Tree Physiol 22:373–38

Nieto K, Frankenberger W (1989) Biosynthesis of cytokinins produced by *Azotobacter chroococcum*. Soil Biol Biochem 21:967–972

Nonhebel H, Cooney T, Simpson R (1993) The route, control and compartmentation of auxin synthesis. Aust J Plant Physiol 20:527–539

Oberhansli T, Defago G, Haas D (1991) Indole-3-acetic-acid (IAA) synthesis in the biocontrol strain CHA0 of *Pseudomonas fluorescens* – role of tryptophan side-chain oxidase. J Gen Microbiol 137:2273–2279

Okon Y, Labandera-González C (1994) Agronomic applications of *Azospirillum*: an evaluation of 20 years worldwide field inoculation. Soil Biol Biochem 26:1591–1601

Okon Y, Heytler P, Hardy W (1983) $N_2$ Fixation by *Azospirillum brasilense* and its incorporation into host *Setaria italica*. Appl Environ Microbiol 46:694–697

Omay S, Schmidt W, Martin P, Bangerth F (1993) Indoleacetic acid production by the rhizosphere bacterium *Azospirillum brasilense* Cd under in vitro conditions. Can J Microbiol 39:187–192

Ona O, Smets I, Gysegom P, Bernaerts K, Impe J, Prinsen E, Vanderleyden J (2003) The effect of pH on indole-3-acetic acid (IAA) biosynthesis of *Azospirillum brasilense* sp7. Symbiosis 35:199–208

Ona O, van Impe J, Prinsen E, Vanderleyden J (2005) Growth and indole-3-acetic acid biosynthesis of *Azospirillum brasilense* Sp245 is environmentally controlled. FEMS Microbiol Lett 246:125–132

Pan B, Bai Y, Leibovitch S, Smith D (1999) Plant-growth-promoting rhizobacteria and kinetin as ways to promote corn growth and yield in a short-growing-season area. Eur J Agron 11:179–186

Paredes-Cardona E, Carcaño-Montiel MG, Mascarúa-Esparza MA, Caballero-Mellado J (1988) Respuesta del maíz a la inoculación con *Azospirillum brasilense*. Rev Latinoamericana de Microbiol 30:351–355

Patten C, Glick B (1996) Bacterial biosynthesis of indole 3-acetic acid. Can J Microbiol 42:207–220

Patten C, Land B, Glick B (2002) Role of *Pseudomonas putida* indole acetic acid in development of the host plant root system. Appl Environ Microbiol 68:3745–3801

Pearce D, Koshioka M, Pharis R (1994) Chromatography of gibberellins. J Chromatogr A 658:91–122

Peck S, Kende H (1995) Sequential induction of the ethylene biosynthetic enzymes by indole-3-acetic acid in etiolated peas. Plant Mol Biol 28:298–301

Perley J, Stowe B (1966) On the ability of *Taphrina deformans* to produceindoleacetic acid from tryptophan by way of tryptamine. Plant Physiol 41:234–237

Perley J, Stowe B (1996) The production of tryptamine from tryptophan by *Bacillus cereus*. Biochem J 100:169–174

Perrig D, Boiero L, Masciarelli O, Penna C, Cassán F, Luna V (2007) Plant growth promoting compounds produced by two agronomically important strains of *Azospirillum brasilense*, and their implications for inoculant formulation. Appl Microbiol Biotechnol 75:1143–1150

Phinney B, Spray C (1988) Dwarf mutants of maize-research tools for the analysis of growth. In: Pharis R, Rood S (eds) Plant growth substances 1988. Springer, Berlín, pp 65–73

Piccoli P, Bottini R (1994a) Effect of C/N ratio, N content, pH and incubation time on growth and gibberellin production by *Azospirillum lipoferum* cultures. Symbiosis 21:263–264

Piccoli P, Bottini R (1994b) Metabolism of 17,17-[$^{2}H_2$]-gibberellin $A_{20}$ to 17,17-[$^{2}H_2$]-gibberellin $A_1$ by *A. lipoferum* cultures. AgriScientiae 11:13–15

Piccoli P, Bottini R (1996) Gibberellins production in A. *lipoferum* cultures y enhanced by ligth. Biocell 20:185–190

Piccoli P, Lucangeli C, Schneider G, Bottini R (1998) Hydrolisis of 17,17-[$^{2}H_2$]-Gibberellin $A_{20}$-Glucoside and 17,17-[$^{2}H_2$]-Gibberellin $A_{20}$-Glucosyl Esther by *Azospirillum lipoferum* cultured in nitrogen-free biotin-based chemycally-definded medium. Plant Growth Regul 23:179–182

Pollmann S, Neu D, Weiler E (2003) Molecular cloning and characterization of an amidase from Arabidopsis thaliana capable of converting indole-3-acetamide into the plant growth hormone, indole-3-acetic acid. Phytochemistry 62:293–300

Pozzo Ardizzi M (1982) Non-symbiotic nitrogen fixing bacteria from Patagonia. In: Technology of Tropical Agriculture (Graham, P. and Harris, S. Eds). NiFTAL. Columbia. USA pp 599–601

Primrose S, Dilworth M (1976) Ethylene production by bacteria. J Gen Microbiol 93:177–181

Prinsen E, Chauvaux N, Schmidt J, John M, Wieneke U, Degreef J, Schell J, Vanonckelen H (1991) Stimulation of indole-3-acetic acid production in *Rhizobium* by flavonoids. FEBS Lett 282:53–55

Prinsen E, Costacurta A, Michiels K, Vanderleyden J, Van Onckelen H (1993) *Azospirillum brasilense* indole-3-acetic acid biosynthesis: evidence for a non-tryptophan dependent pathway. Mol Plant Microb Interact 6:609–615

Puente M, Li C, Bashan Y (2004) Microbial populations and activities in the rhizoplane of rock-weathering desert plants II. Growth promotion of cactus seedlings. Plant Biol 6:643–650

Puppo A, Rigaud J (1978) Cytokinins and morphological aspects of French-bean roots in presence of *Rhizobium*. Physiol Plant 42:202–206

Rademacher W (1994) Gibberellin formation in microorganisms. Plant Growth Regul 15:303–314

Rademacher W (2000) Growth retardants: effects on gibberellin biosynthesis and other metabolic pathways. Annu Rev Plant Physiol Plant Mol Biol 51:501–531

Rahman A, Hosokawa S, Oono Y, Amakawa T, Goto N, Tsurumi S (2002) Auxin and ethylene response interactions during *Arabidopsis* root hair development diss. Plant Physiol 130:1908–1917

Rood S, Pharis R (1987) Evidence for reversible conjugation of gibberellins in higher plants. In: Schreiber H, Schutte H, Semder G (eds) Conjugates plants hormons. structure, metabolism and function. VEB Deustcher Verlag der Wissenschaften, Berlin, pp 183–190

Rosas S, Soria R, Correa N, Abdala G (1998) Jasmonic acid stimulates the expression of nod genes in *Rhizobium*. Plant Mol Biol 38:1161–1168

Ross J, O'Neill D (2001) New interactions between classical plant hormones. Trends Plant Sci 6:2–4

Ross J, O'Neill D, Smith J, Kerckhoffs Elliot R (2000) Evidence that auxin promotes the gibberellin $A_1$ biosynthesis in pea. Plant J 21:547–552

Rovira A (1970) Plant root exudates and their influence upon soil microorganisms. In: Baker K, Synder W (eds) Ecology of soil-borne plant pathogens. University of California Press, Berkeley, pp 170–186

Ruckdaschel E, Klingmüller W (1992) Analysis of IAA biosynthesis in *Azospirillum lipoferum* and Tn5 induced mutants. Symbiosis 13:123–131

Salamone G, Hynes R, Nelson L (2001) Cytokinin production by plant growth promoting rhizobacteria and selected mutants. Can J Microbiol 47:404–411

Sarig S, Okon Y, Blum A (1990) Promotion of leaf area development and yield in *Sorghum bicolor* inoculated with *Azospirillum brasilense*. Symbiosis 9:235–245

Schliemann W, Schneider G (1994) Gibberellins conjugates: an overview. Plant Growth Regul 15:247–260

Schmidt W, Martin P, Omay H, Bangerth F (1988) Influence of *Azospirillum* brasilense on nodulation of legumes. In: Klingmuller W (ed) *Azospirillim* IV. Genetics, physiology, ecology. Springer, Heidelberg, pp 92–100

Schreiber K, Weiland J, Sembdner G (1967) Isolierung und Struktur eines Gibberel-linglucosisds. Tetrahedron Lett:4285–4288

Scott T (1972) Auxins and roots. Annu Rev Plant Physiol 23:235–258

Sekine M, Watanabe K, Syono K (1989) Molecular cloning of a gene for indole-3-acetamide hydrolase from *Bradyrhizobium japonicum*. J Bacteriol 171:1718–1724

Sembder G, Weiland J, Aurich O, Schreiber K (1968) Isolation structure and metabolism of a gibberellin glucoside. In: Mac Millan J (ed) Plant growth regulators. SCI Monographs, London, pp 70–86

Sembder G, Gross D, Liebisch H, Schneider G (1980) Biosynthesis and metabolism of plant hormones. In: Mac Millan J (ed) Encyclopedia of plant physiology, new series. Springer, Berlin, pp 281–444

Sgroy V, Cassán F, Masciarelli O, Del Papa M, Lagares A, Luna V (2009) Isolation and characterization of endophytic plant growth-promoting (PGPB) or stress homeostasis regulating (PSHB) bacteria associated to the halophyte *Prosopis strombulifera*. Appl Microbiol Biotechnol 85:371–381

Spaepen S, Vanderleyden J, Remans R (2007) Indole-3-acetic acid in microbial and microorganism-plant Signaling. FEMS Microbiol Rev 31:425–448

Spaepen S, Dobbelaere S, Croonenborghs A, Vanderleyden J (2008) Effects of *Azospirillum brasilense* indole-3-acetic acid production on inoculated wheat plants. Plant Soil 312:15–23

Sprunck S, Jacobsen H, Reinard T (1995) Indole-3-lactic acid is a weak auxin analogue but not an anti-auxin. J Plant Growth Regul 14:191–197

Strzelczyk E, Kamper M, Li C (1994) Cytocinin-like-substances and ethylene production by *Azospirillum* in media with different carbon sources. Microbiol Res 149:55–60

Taller B, Wong T (1989) Cytokinins in *Azotobacter vinelandii* culture medium. Appl Environ Microbiol 55:266–267

Theodoulou F, Job K, Slocombe S, Footitt S, Holdsworth M, Baker A, Larson T, Graham I (2005) Jasmonic acid levels are reduced in COMATOSE ATP-binding cassette transporter mutants. Implications for transport of jasmonate precursors into peroxisomes. Plant Physiol 137:835–840

Theunis M, Kobayashi H, Broughton W, Prinsen E (2004) Flavonoids, NodD1, NodD2, and nod-box NB15 modulate expression of the y4wEFG locus that is required for indole-3-acetic acid synthesis in Rhizobium sp. strain NGR234. Mol Plant Microbe Interact 17:1153–1161

Thiman K (1936) On the physiology of the formation of nodule in legumes roots. Proc Natl Acad Sci 22:511–514

Thuler D, Floh E, Handro W, Barbosa H (2003) Plant growth regulators and amino acids released by *Azospirillum* sp in chemically defined media. Lett Appl Microbiol 37:174–178

Tien T, Gaskins M, Hubbell D (1979) Plant growth substances produced by *Azsopirillum brasilense* and their effect on the growth of pearl millet (*Pennisetum americanum* L.). Appl Environ Microbiol 37:1016–1024

Timmusk S, Nicander B, Granhall U, Tillberg E (1999) Cytokinin production by *Paenibacillus polymyxa*. Soil Biol Biochem 31:1847–1852

Vande Broek A, Lambrecht M, Kristel E, Vanderleyden J (1999) Auxins upregulate expression of the indole-3-pyruvate decarboxylase gene in *Azospirillum brasilense*. J Bacteriol 181:1338–1342

Vande Broek A, Gysegom P, Ona O, Hendrickx N, Prinsen E, Van Impe J, Vanderleyden J (2005) Transcriptional analysis of the *Azospirillum brasilense* indole-3-pyruvate decarboxylase gene and identification of a cis-acting sequence involved in auxin responsive expression. Mol Plant Microbe Interact 18(4):311–323

Vandenhobe H, Merckx R, van Steenberg VK (1993) Microcalorimetric characterization, physiological satages and survival ability of *Azospirillum brasilense*. Soil Biol Biochem 25:513–519

Wasternack C, Kombrink E (2010) Jasmonates- Structural requirements for lipid-derived signals active in plant stress responses and development. ACS Chem Biol 5(1):63–77

Yahalom E, Okon Y, Dovrat A (1990) Possible mode of action of *Azospirillum brasilense* strain Cd on the roots morphology and nodule formation in burr medic (*Medicago polymorpha*). Can J Microbiol 36:10–14

Yamada T, Palm C, Brooks B, Kosuge T (1985) Nucleotide sequences of the *Pseudomonas savastanoi* indoleacetic acid genes show homology with *Agrobacterium tumefaciens* T-DNA. Proc Natl Acad Sci USA 82:6522–6526

Yang S (1974) The biochemistry of ethylene: biogenesis and metabolism. In: Sondheimer C, Walton E (eds) The chemistry and biochemistry of plant hormones. Recent Advances in Phytochemistry. Academic, New York, pp 131–178

Yaxley J, Ross J, Sherriff L, Reid J (2001) Gibberellin biosynthesis mutations and root development in pea. Plant Physiol 125:627–633

Zimmer W, Roeben K, Bothe H (1988) An alternative explanation for plant growth promotion by bacteria of the genus *Azospirillum*. Planta 176:333–342

# Chapter 8
# ACC Deaminase Containing PGPR for Potential Exploitation in Agriculture

**Venkadasamy Govindasamy, Murugesan Senthilkumar, Pranita Bose, Lakkineni Vithal Kumar, D. Ramadoss, and Kannepalli Annapurna**

## 8.1 Introduction

The growth of plants in the field is determined by the numerous and diverse interactions among its physical, chemical, and biological components of soil as modulated by the prevalent environmental conditions. In particular, the varied genetic and functional activities of the extensive microbial populations have a critical impact on soil functions and plant growth, based on the fact that microorganisms are a driving force for fundamental metabolic processes involving specific enzyme activities (Nannipieri et al. 2003). The crop production in general and productivity in particular is inhibited by a large number of both biotic and abiotic stresses. These stresses include extremes of temperature, high light, flooding, drought, the presence of toxic metals and environmental organic contaminants, radiation, wounding, insect predation, high salt, and various pathogens including viruses, bacteria, fungi, and nematodes (Abeles et al. 1992). As agricultural production intensified over the past few decades, management of these biotic and abiotic stresses on crop plants with increased use of chemical inputs caused several harmful effects on the environmental quality.

A major strategy to counteract the rapid decline in environmental quality is to promote sustainable agriculture with the gradual reduction in the use of fertilizers and pesticides and greater use of the biological and genetic potential of plant and microbial species, which may help to sustain high production in agriculture. In addition to the plants ability to modify its physiology and metabolism, certain soil microorganisms present in the rhizosphere can help the plants to either avoid or partially overcome these environmental stresses. Many microbial–plant interactions are regulated by specific molecules/signals (Pace 1997) and are responsible for key environmental processes such as the biogeochemical cycling of nutrients and matter

V. Govindasamy • M. Senthilkumar • P. Bose • L.V. Kumar • D. Ramadoss • K. Annapurna (✉)
Division of Microbiology, Indian Agricultural Research Institute, New Delhi 110012, India
e-mail: annapurna93@yahoo.co.in

D.K. Maheshwari (ed.), *Bacteria in Agrobiology: Plant Nutrient Management*,
DOI 10.1007/978-3-642-21061-7_8, 

and maintenance of plant health and soil quality (Bowen and Rovira 1999; Barea et al. 2005; Govindasamy et al. 2008b). During the last couple of years, the application of plant growth promoting rhizobacteria (PGPR) as biofertilizers and biocontrol agents is being considered as an alternative or supplemental way of reducing the use of chemicals in crop production (Kloepper et al. 1989; Vessey 2003; Maheshwari 2010).

The beneficial free-living soil bacteria present in the plant rhizosphere are usually referred to as PGPR (Kloepper et al. 1989) and can affect plant growth in a variety of ways (Glick 1995; Glick et al. 1999). They increase the seed germination and viability, root respiration, and formation; stimulate root growth and whole plant growth by accelerating cell division; and increase plant membrane permeability for most efficient nutrient uptake. Interestingly, certain PGPR strains possess the enzyme 1-aminocyclopropane-1-carboxylic acid (ACC) deaminase (Jacobson et al. 1994; Glick et al. 1997; Shah et al. 1998), and this enzyme can cleave the plant ethylene precursor ACC, and thereby lower the level of ethylene in a developing seedling or stressed plant (Sheehy et al. 1991; Mayak et al. 2004a). For many plants, a burst of ethylene is required to break seed dormancy but following germination, a sustained high level of ethylene would inhibit root elongation (Jacobson et al. 1994). The stress hormone ethylene is synthesized in plant tissues from the precursor ACC during biotic and abiotic stress conditions, which in turn retarded root growth and caused senescence in crop plants (Abeles et al. 1992; Sheehy et al. 1991; Ma et al. 2002). By facilitating the formation of longer roots through the action of ACC deaminase, these PGPR may enhance the survival of seedlings under various abiotic and biotic stresses (Grichko and Glick 2001a; Wang et al. 2000). The present chapter is an attempt to document the existing literature on ACC deaminase containing PGPR and its potential exploitation in crop production.

## 8.2 Rhizosphere and Rhizobacteria

Plant roots offer a niche for the proliferation of soil microbes. Many studies have demonstrated that soil-borne microbes interact with plant roots and soil constituents at the root–soil interface, which leads to the development of a dynamic environment known as the rhizosphere, i.e., the zone of soil as influenced by roots through the release of substrates in the form of root exudates that affect the microbial activity (Lynch 1990). The differing physical, chemical, and biological properties of the root-associated soil compared with those of the root free bulk soil, are responsible for changes in microbial diversity and for increased numbers and activity of microorganisms in the rhizosphere microenvironment (Kennedy and Smith 1995). The release of root exudates and decaying plant material provides sources of carbon and energy for the root-associated microbiota. Whereas microbial activity in the rhizosphere affects rooting patterns and the supply of available nutrients to plants, thereby modifying the quality and quantity of root exudates (Bowen and Rovira 1999).

A large number of different microorganisms are commonly found in the rhizosphere soil including bacteria, fungi, actinomycetes, protozoa, and algae (Paul and Clark 1989). Of these, bacteria are by far the most common type of soil microorganisms possibly because they can grow rapidly and have the ability to utilize a wide range of substances as either carbon or nitrogen sources, whereas many of the bacteria found in rhizosphere are bound to the surface of soil particles and are found in soil aggregates. In fact, the concentration of bacteria (per gram of soil) that is found around the roots of plants is generally much greater than the bacterial density or concentration that is found in the rest of the soil (Lynch 1990). The high population densities of bacteria in the rhizosphere stimulate nutrient delivery and uptake by plant roots.

The interaction between bacteria and the roots of plants may be beneficial, harmful, or neutral for the plant, and sometimes the effects of a particular bacterium may vary as a consequence of soil conditions (Lynch 1990). Kloepper and Schroth (1981) first highlighted the importance of bacteria responsible for the inhibition of plant growth and they proposed the term "deleterious rhizobacteria or DRB." Subsequently the term "deleterious rhizosphere microorganisms" or DRMO has been used to include pathogenic fungi (Cherrington and Elliott 1987). Some of the rhizosphere bacteria that colonize the plant root tissues and interact positively by promoting the plant growth are called as PGPR.

### 8.2.1 Plant Growth-Promoting Rhizobacteria

Now, it is established that bacteria that provide some benefit to plants are of two general types, those that form a symbiotic relationship with the plant and those that are free-living in the soil, but are often found near, on, or even within the roots of plants. The PGPR are defined by three intrinsic characteristics:

(a) They must be able to colonize the root
(b) They must survive and multiply in microhabitats associated with the root surface, in competition with other microbiota, at least for the time needed to express the plant promotion and protection activities
(c) They must promote the plant growth and development (Kloepper et al. 1989)

A number of different bacteria may be considered to be PGPR; these bacterial genera include *Acinetobacter*, *Agrobacterium*, *Arthrobacter*, *Azotobacter*, *Azospirillum*, *Bacillus*, *Burkholderia*, *Bradyrhizobium*, *Frankia*, *Pseudomonas*, *Rhizobium*, *Serratia*, *Thiobacillus*, and others (Vessey 2003). These PGPR genera may employ various mechanisms to stimulate plant growth and development.

### 8.2.2 Mechanisms of Plant Growth Promotion by PGPR

Several mechanisms have been postulated to explain how PGPR may affect the growth and development of inoculated plants. These can be broadly categorized as either direct or indirect mechanisms of growth stimulation/promotion. Direct growth promotion occurs when a rhizobacterium produces metabolites, that is, phytohormones or improves nutrient availability that directly promotes plant growth (Kloepper et al. 1989, 1991). In contrast, indirect promotion of plant growth occurs when rhizobacteria decrease or prevent some of the deleterious effects of phytopathogenic organisms by acting as biological disease control agents. However, Kloepper (1993) explained that there is often no clear separation between direct growth promotion and biological disease control promoting indirect growth, rather, growth promotion and biological control of disease should be viewed as the same.

There are several ways in which PGPR may directly facilitate the proliferation of plant growth. The PGPR-mediated processes involved in nutrient availability include those related to nonsymbiotic nitrogen fixation. Several studies have shown that diazotrophic free-living bacteria as well as rhizobial strains can promote the growth of several cereal and leguminous plants by contributing to N-economy through their ability to fix atmospheric nitrogen. Malik et al. (1997) investigated the nitrogen-fixing ability of *Azospirillum* strain N-4 in rice and found a significant contribution of nitrogen fixed by the PGPR. Similarly, Biswas et al. (2000) reported that biological nitrogen fixation by diazotrophic PGPR strains may be the contributing factor of rice growth promotion in addition to other mechanisms. Solubilization of mineral nutrients such as phosphorus and iron by PGPR via releasing organic acids and siderophores, respectively makes them more readily available for plant uptake. (Kloepper et al. 1987; Richardson 2001). Several studies reported a similar role for other PGPR in Zn, Cu, Mn, and other mineral nutrient solubilization, enhancing nutrient uptake and subsequently, the promotion of plant growth (Belimov et al. 1995; Noel et al. 1996).

Many PGPR can affect plant growth and development by the production of plant hormones namely auxins, gibberellins, cytokinins, ethylene, and abscisic acid (Frankenberger and Arshad 1995). Okon (1994) showed the production of auxin-type of phytohormones by *Azospirillum* that affect the root morphology and plant growth. *Azospirillum* species are considered to be PGPR due to this phytohormone production, which is more important, than their nitrogen-fixing activity (Bashan 1999; Zahir et al. 2004). Under gnotobiotic conditions, Noel et al. (1996) showed the direct involvement of IAA and cytokinin production by PGPR in the growth of canola and lettuce. Similarly Young et al. (1997) screened the PGPR strains of *Pseudomonas* and *Serratia* for phytohormones and observed a good correlation between induction of rice root elongation and phytohormone production. PGPR can modify plant growth by producing and releasing substances other than IAA such as ethylene and abscisic acid (Barazani and Friedman 1999). Glick and his co-workers (1995) have suggested the involvement of an enzyme ACC deaminase produced by *Pseudomonas putida* GR12-2 in modifying the root growth of different plants. They found that this bacterium hydrolyses ACC, the immediate precursor of ethylene in higher plants thereby lowering the endogenous levels of ACC and ethylene, which subsequently results in plant growth promotion.

### 8.2.3 ACC Deaminase Containing PGPR

The cyclopropanoid amino acid ACC was originally identified as a natural product in the juices of several fruits (Burroughs 1957) and is now regarded as a key intermediate in the biosynthesis of the plant hormone ethylene (Adams and Yang 1979). Honma and Shimomura (1978) reported for the first time that soil bacteria *Pseudomonas* sp. strain ACP and the yeast *Hansenula saturnus* were capable of utilizing the ACC as a nitrogen source owing to induction of the enzyme ACC deaminase in these organisms. Later, it is considered as the one of the mechanisms that a number of PGPR use to facilitate plant growth and development by lowering of plant ethylene concentration (Glick 1995). Lifshitz et al. (1986) isolated cold tolerant- and nitrogen fixing-species of *Pseudomonas* from the rhizosphere and reported that one of these strains *P. putida* GR12-2 directly promoted the canola growth and yield in the absence of plant pathogens and deleterious microorganisms. It was found to be a highly competitive root colonizer of canola grown in field soils. It was shown that *P. putida* GR12-2 could stimulate root elongation of canola under gnotobiotic conditions (Hong et al. 1991). Subsequently, Glick et al. (1994) reported that the ability of *P. putida* GR12-2 to stimulate root elongation is at least in part due to ACC deaminase activity, which hydrolyses the ethylene precursor ACC thereby lowering the level of ethylene in a developing plant. The early development of seedlings was reported when treating canola seeds with *P. putida* GR12-2 as compared to its mutants lacking ACC deaminase activity (Glick et al. 1997; Li et al. 2000).

Microorganisms capable of degrading ACC can be readily isolated from soil (Glick et al. 1995). Klee et al. (1991) screened 600 microorganisms for ACC-degrading ability and characterized two pseudomonads, that is, *Pseudomonas* sp. 3F2 and 6G5 with ACC deaminase activity. ACC deaminase containing PGPR are relatively common in soil and most of them are mesophilic in nature. Glick and co-workers (1995) isolated some *Pseudomonas* species and *Enterobacter cloacae* strains CAL2 and UW4 from the soil samples based on their ability to utilize ACC as a sole source of nitrogen and one strain *P. putida* UW4 was isolated from the rhizosphere of Reed plant found to be cold tolerant. Campbell and Thomson (1996) screened soil bacteria for ACC-degrading ability and reported two *Pseudomonas* strains 15 and 17, which displayed high enzyme activity under in vitro conditions. Burd et al. (1998) reported a newly isolated ACC-utilizing bacterium *Kluyvera ascorbata* SUD 165. This bacterium was found to improve the growth of canola, tomato, and Indian mustard seedlings treated with a toxic concentration of nickel, lead, and zinc (Burd et al. 2000). It was shown that ACC deaminase activity is much more widely distributed in soil bacteria belonging to genera *Alcaligens*, *Variovorax*, *Rhodococcus,* and *Bacillus* and to different species of *Pseudomonas*, which were isolated from the rhizosphere of pea and Indian mustard grown in different soils and a long-standing sewage sludge contaminated with heavy metals (Belimov et al. 2001, 2005). Phyllosphere methylobacteria of rice were found positive for ACC deaminase activity (Chinnadurai et al. 2009). One of the isolate identified as *Methylobacterium radiotolerans* showed 98% *accD* gene sequence homology to *Rhizobium leguminosarum*. Foliar spray of this bacterium enhanced root and shoot length of rice and tomato under gnotobiotic conditions and lower ethylene level to 60–80% in the rice.

ACC deaminase containing *Rhizobium* species are prevalent and promote the nodulation of legume plants. Ma et al. (2003a) documented the presence of ACC deaminase in 5 of the 13 different rhizobial strains that were examined. One of the positive strain *R. leguminosarum* bv. *viciae* 128C53K containing ACC deaminase activity enhances the nodulation of pea plants (Ma et al. 2003b). Similarly, ACC deaminase was studied in other rhizobial strains of *Lotus japanicus* and *Macroptilium atropurpureum*, which lowered the ethylene levels and increased nodulation and subsequence biomass formation by 25–40% (Nukui et al. 2000). Studies in our lab have shown the presence of *acdS* gene in rhizobia of different legumes. The enzyme activity varied in these rhizobial isolates (Vikas 2010). Ghosh et al. (2003) reported three Gram-positive bacterial strains belonging to bacilli such as *Bacillus circulans* DUCl, *Bacillus firmus* DUC2, and *Bacillus globisporus* DUC3 promoting plant growth with the ability to catabolize ACC. PGPR having ACC deaminase activities were isolated from soil samples taken from the Arava region of Southern Israel. One of these strains was *Archromobacter piechaudii* ARV 8 that promotes growth of tomatoes and peppers under water stress (Mayak et al. 2004a) and to salt stress (Mayak et al. 2004b). An endophyte-harboring ACC deaminase activity was isolated from root nodules of *Mimosa pudica* (Pandey et al. 2005). Similarly, Dell'Amico et al. (2008) isolated and characterized four bacterial isolates from heavy metal-polluted rhizosphere. The strains *Pseudomonas tolaasii* ACC 23, *Pseudomonas fluorescens* ACC 9 *Alcaligens* sp. ZN4 and *Mycobacterium* sp. ACC 14 showed ACC deaminase activity and improved Canola (*Brassica napus*) growth under cadmium stress (Martínez-Viveros et al. 2010).

## 8.3 Role of Ethylene on Plant Growth and Development

Ethylene the simplest unsaturated hydrocarbon having biological activity was discovered by Russian plant physiologist Dimitri Neljubov (1789–1926) in etiolated pea seedlings (Abeles et al. 1992). Most plant tissues produce the ethylene and it is formed from methionione via S-adenosyl-L-methionine (SAM) and the cyclic non-protein amino acid ACC. Adams and Yang (1979) proposed that the ethylene biosynthetic pathway begins with the enzyme ACC synthase that converts SAM to ACC and 5′ methyl thioadenisine (MTA), which is recycled to L-methionine through Yang cycle (Fig. 8.1). The next step is the conversion of ACC to ethylene by enzyme ACC oxidase, which is present in most tissues at very low levels with several isoforms that are active under different physiological conditions (Arshad and Frankenberger 2002).

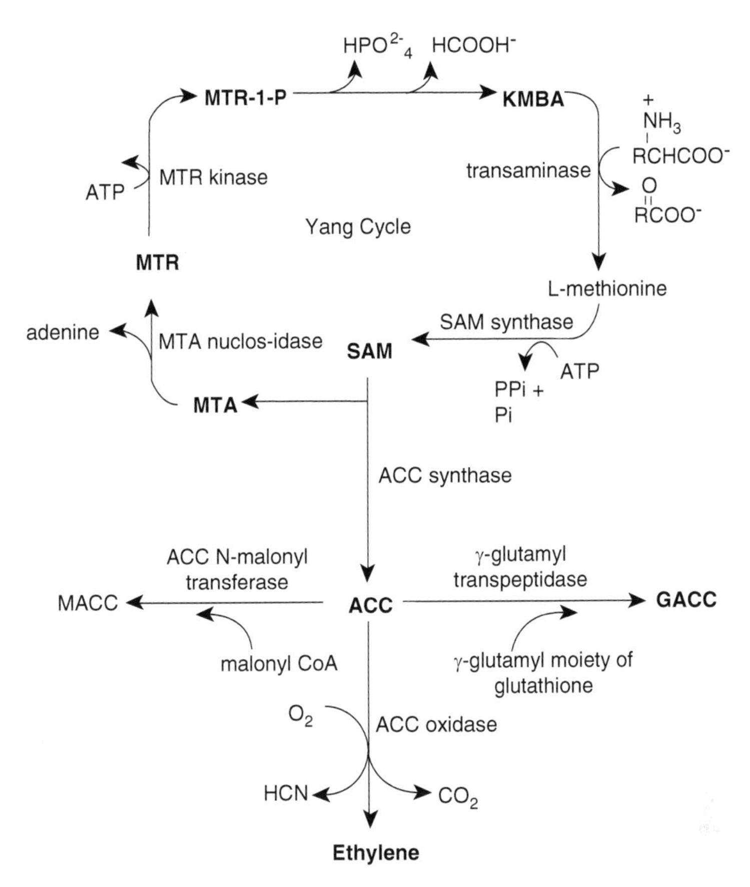


**Fig. 8.1** Yang cycle and ethylene biosynthesis pathway (Arshad and Frankenberger 2002). *ACC* 1-aminocyclopropane-1-carboxylic acid, *GACC* 1-(glutamyl)-ACC, *KMBA* 2-keto-4-methylthiobytyric acid, *MACC* 1-(malonyl)-ACC, *MTA* 5Vmethylthioadenosine, *MTR* 5V-methylthioribose, *MTR-1-P* 5V-methylthioribose-1-phosphate, *SAM* S-adenosylmethionine

Ethylene is essential for proper plant development, growth, and survival and responsible for signaling changes in seed germination; growth of vegetative organs such as roots, stems, petioles, flower, and fruit development; plant–microbial interactions including rhizobial nodulation of legumes (Ma et al. 2002); and a number of interactions with other plant hormones (Mattoo and Shuttle 1991) and the onset of most of the plant defense mechanisms including plant response to biotic and abiotic stresses. Ethylene has the ability to trigger exaggerated disease symptoms and exacerbate an environmental pressure. High levels of ethylene are produced at the beginning of the ripening stage in climacteric fruits. Except for fruit ripening and lateral root initiation, high levels of ethylene are usually deleterious to plant growth and health.

The term “stress ethylene” originally coined by Abeles, describes the acceleration of ethylene biosynthesis associated with biological and environmental stresses including pathogen attack (Hyodo 1991; Robison et al. 2001a). Pierik et al. (2006) proposed the model that explains the contradictory effects of stress ethylene when the plant is subjected to stresses and ethylene is synthesized in two peaks. The first peak of ethylene is generally small and occurs within a few hours of onset of the stresses and initiate protective response by the plant such as transcription of PR genes and acquired resistance (Van Loon and Glick 2004). The second peak of ethylene is large, usually occurs in 1–3 days, and initiates processes such as senescence, chlorosis, abscission, and inhibits the plant's survival. The second peak of ethylene production occurs as a consequence of increased transcription of ACC synthase genes triggered by environmental and developmental cues (Yang and Hoffman 1984). A selective decrease in the second but not the first ethylene peak would appear to be advantageous for plant growth and may be achieved by using PGPR having ACC deaminase enzyme or to engineer the plant for modulation of ethylene production.

## 8.4 Mechanisms of Lowering Plant Ethylene by ACC-Containing PGPR

A model was previously proposed by which PGPR can lower plant ethylene levels and in turn facilitate plant growth (Glick et al. 1998). In this model, (Fig. 8.2) the PGPR bind to the surface of a plant (usually seeds or roots, although ACC deaminase-producing bacteria may also be found on leaves and flowers). In response to tryptophan and other small molecules in the plant exudates, the bacteria synthesize and secrete indole-3-acetic acid (IAA), some of which is taken up by the plant. This IAA together with endogenous plant IAA can stimulate plant cell proliferation, plant cell elongation, or induce the transcription of ACC synthase, which is the enzyme that catalyzes the formation of ACC. Some of the ACC is exuded from seeds, roots, or leaves (Grichko and Glick 2001a) along with other small molecules normally present in these exudates and may be taken up by the bacteria and subsequently cleaved by the enzyme, ACC deaminase, to ammonia and $\alpha$-ketobutyrate.

In this model, the bacterium acts as a sink for plant ACC and as a result lowering either the endogenous or the IAA-stimulated ACC level and hence, the amount of ethylene in the plant is also reduced. Penrose et al. (2001) reported that in the presence of the ethylene inhibitor (AVG) or treating the plants with ACC deaminase-containing bacteria reduces the amount of ACC – the immediate precursor of ethylene in the root tissues. The ACC-related compounds $\alpha$- and $\gamma$-aminobutyric acid, both known to stimulate ethylene production were also found reduced in both the root exudates and root tissues of the *E. cloacae* CAL3-treated canola seed (Penrose and Glick 2001). As a direct consequence of lowering plant ethylene levels, plant growth-promoting bacteria that possess the enzyme ACC deaminase can reduce the extent of ethylene inhibition of plant growth following a wide range of stresses. Thus, plants grown in association with these bacteria should have longer roots and

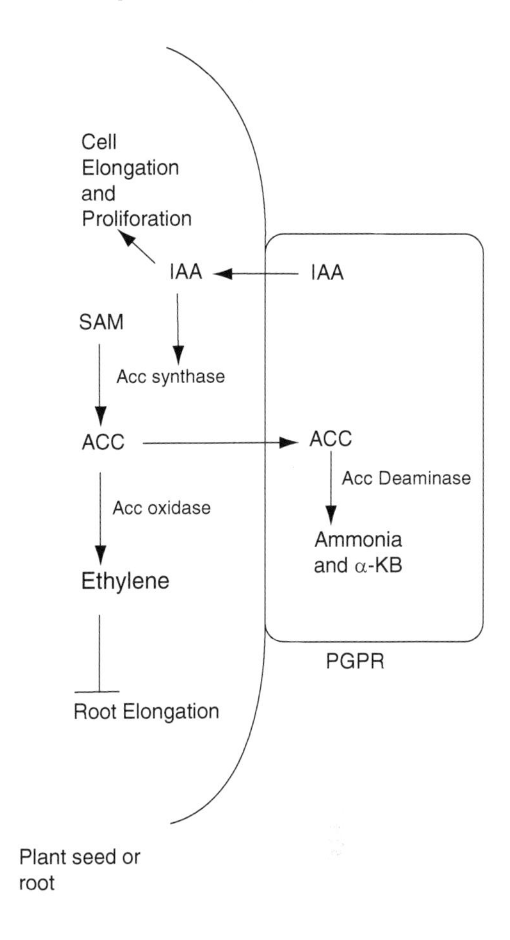


**Fig. 8.2** Schematic model to explain how ACC deaminase containing PGPR lowers the ethylene concentration thereby preventing ethylene inhibition of root elongation (Glick et al. 1998). *IAA* indole aceticacid, *ACC* 1-aminocyclo-propane-1-carboxylic acid, *SAM* S-adenosyl methionine, *α-KB* α-ketobutyrate

shoots and be more resistant to growth inhibition by a variety of ethylene-inducing stresses. The question arises, as to how bacterial ACC deaminase can selectively lower deleterious ethylene levels but not affect the small peak of ethylene that is thought to activate some plant defense responses (Fig. 8.3a).

ACC deaminase is generally present in bacteria at a low level until it is induced, and the induction of enzyme activity is a relatively slow and complex process. Immediately following an environmental stress, the pool of ACC in the plant is low as is the level of ACC deaminase in the associated bacterium. Following the relatively rapid induction of a low level of ACC oxidase in the plant, it is likely that there is increased flux through this enzyme resulting in the first small peak of ethylene, which is of sufficient magnitude to induce a protective/defensive response in the plant. With time, bacterial ACC deaminase is induced (by the increasing amounts of ACC that ensue from the induction of ACC synthase in the plant) so that the magnitude of the second, deleterious, ethylene peak is decreased significantly (Fig. 8.3b). The second ethylene peak may be reduced dramatically, but it is never completely abolished since ACC oxidase has a much higher affinity for ACC than

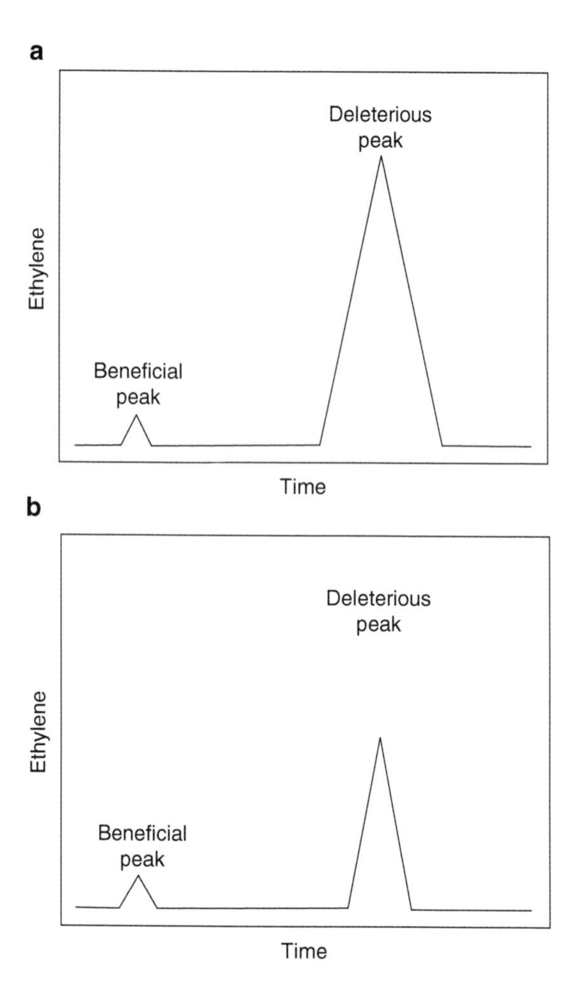


**Fig. 8.3** Plant ethylene production as a function of time following an environmental stress. (**a**) In the absence of any exogenous bacteria. (**b**) In the presence of an ACC deaminase producing plant growth-promoting bacterium (Van Loon and Glick 2004)

does ACC deaminase (Glick et al. 1998). Thus, when ACC deaminase-producing bacteria are present, ethylene levels are ultimately dependent upon the ratio of ACC oxidase to ACC deaminase (Glick et al. 1998).

Two physiological effects of ACC containing *Rhizobium* were proposed in the establishment of legume–*Rhizobium* symbiosis. One of these effects is enhancement of early infection events on host roots and the other effect is a positive effect on establishment and/or maintenance of mature nodules by accumulation of ACC deaminase protein and lowering the ethylene level around bacteroids (Okazaki et al. 2004). Hontzeas et al. (2004a) demonstrated that at the genetic level, the ACC deaminase containing plant growth-promoting bacteria *E. cloacae* UW4 changes the gene expression in treated canola roots. Some of the genes are upregulated, which included a cell division cycle protein 48-homolog and a eucaryotic translation initiation factor 3-subunit 7-gene homolog. The downregulated genes include one encoding a glycine-rich RNA binding gene during ethylene-induced stress, which is expressed only in roots, and another gene thought to be involved in a defense-signaling pathway.

## 8.5 Biochemistry of ACC Deaminase

In 1978, an enzyme capable of degrading ACC was isolated from *Pseudomonas* sp. strain ACP (Honma and Shimomura 1978) and from the fungus *Penicillium citrinum* (Honma 1993). Since then, ACC deaminase (EC 4.1.99.4) has been detected in the yeast *H. saturnus* (Minami et al. 1998) and in a number of other bacterial strains (Klee and Kishore 1992; Jacobson et al. 1994; Glick et al. 1995; Campbell and Thomson 1996) all of which originated in the soil. Enzymatic activity of ACC deaminase is assayed by monitoring the production of either ammonia or α-ketobutyrate, the products of ACC hydrolysis (Honma and Shimomura 1978). ACC deaminase has been found only in microorganisms, and there are no microorganisms that synthesize ethylene via ACC (Fukuda et al. 1993). However, there is strong evidence that the fungus, *P. citrinum*, produces ACC from SAM via ACC synthase, one of the enzymes of plant ethylene biosynthesis, and degrades the ACC by ACC deaminase. It appears that the ACC, which accumulates in the intracellular spaces, can induce ACC deaminase (Jia et al. 2000).

Very few ACC deaminase proteins have been purified and biochemically characterized; and their crystal structures have been produced for only two organisms (Tables 8.1 and 8.2). ACC deaminase has been purified to homogeneity from *Pseudomonas* sp. strain ACP (Honma and Shimomura 1978), *H. saturnus* (Minami et al. 1998), and *P. citrinum* (Jia et al. 1999) and partially purified from *Pseudomonas* sp. strain 6G5 (Klee et al. 1991) and *P. putida* GR12-2 (Jacobson et al. 1994). Enzyme activity is localized exclusively in the cytoplasm of the bacterium (Jacobson et al. 1994). The molecular mass and form is similar for the bacterial ACC deaminases. The enzyme is a trimer, the size of the holoenzyme is approximately 104–105 kDa and the subunit mass is approximately 36.5 kDa (Honma and Shimomura 1978; Jacobson et al. 1994). Similar subunit sizes were predicted from nucleotide sequences of cloned ACC deaminase genes from *Pseudomonas* sp. strains ACP (Sheehy et al. 1991) and 6G5 (Klee et al. 1991), and *E. cloacae* UW4 (Shah et al. 1997). The molecular mass of the holoenzymes and subunits from *H. saturnus* (69 and 40 kDa, respectively) and *P. citrinum* (68 and 41 kDa, respectively) suggests that these ACC deaminases are dimers (Minami et al. 1998; Jia et al. 1999). Thus it would appear that in nature there are two types of ACC deaminase.

Pyridoxal phosphate is a tightly bound cofactor of ACC deaminase in the amount of approximately 3 mol of enzyme-bound pyridoxal phosphate per mole of enzyme, or 1 mol per subunit. $K_m$ values for the binding of ACC by ACC deaminase have been estimated for enzyme extracts from 12 microorganisms at pH 8.5. These values ranged from 1.5 to 17.4 mM (Honma and Shimomura 1978; Klee and Kishore 1992; Honma 1993) indicating that the enzyme does not have

**Table 8.1** List of ACC deaminase containing microorganisms of different genera and species reported so far

| Sl. No. | Microorganisms (bacteria/fungi) | Source of isolation | References |
|---|---|---|---|
| 1. | *Pseudomonas* sp. ACP | Soil | Honma and Shimomura (1978) |
| 2. | *Hansenula saturnus* (yeast) | Soil | Minami et al. (1998) |
| 3. | *Penicillium citrinum* (fungus) | | Honma (1993) |
| 4. | *Pseudomonas putida* GR12-2 | Rhizospheric soil of cold regions | Lifshitz et al. (1986), Hong et al. (1991), Glick et al. (1994) |
| 5. | *Enterobacter cloacae* CAL2 and UW4 | Soils of cold region | Glick et al. (1995) |
| 6. | *Kluyvera ascorbata* SUD 165 | Heavy metal contaminated soils | Burd et al. (2000) |
| 7. | *Alcaligens* sp., *Variovorax* sp., *Rhodococcus* sp. | Pea and Indian mustard rhizophere (sewage irrigated soils) | Belimov et al. (2001, 2005) |
| 8. | *Rhizobium leguminosarum* bv. *viciae* 128C53K | Root nodules of pea | Ma et al. (2003b) |
| 9. | *Bacillus circulans* DUCl, *Bacillus firmus* DUC2, *Bacillus globisporus* DUC3 | Indian mustard rhizosphere | Ghosh et al. (2003) |
| 10. | *Klebsiella oxytoca* 10MKR 7 | Soil | Babalola et al. (2003) |
| 11. | *Archromobacter piechaudii* ARV 8 | Soils of Arava region of Southern Israel | Mayak et al. (2004a, b) |
| 12. | *Azospirillum braziliensis* | Rhizosphere soil | Blaha et al. (2006) |
| 13. | *Methylobacterium oryzae* sp. *nov* strain CBMB20T | Stem tissues of rice | Madhaiyan et al. (2007) |
| 14. | *Pseudomonas tolaasii* ACC23, *Pseudomonas fluorescens* ACC9 *Alcaligens* sp. ZN4 and *Mycobacterium* sp. ACC 14 | Heavy metals polluted rhizosphere | Dell'Amico et al. (2008) |
| 15. | *Achromobacter* spp. *Achromobacter* sp. GKA-1 and *Pseudomonas stutzeri* GKA-13 | Soils, wheat rhizosphere | Hontzeas et al. (2005), Govindasamy et al. (2008a) |
| 16. | *Pyrococcus horikoshii* OT3 | Hyperthermophillic sea | Fujino et al. (2004) |

a particularly high affinity for ACC (Glick et al. 1999). ACC deaminase activity has been induced in both *Pseudomonas* sp. strain ACP and *P. putida* GR 12-2 by ACC, at the levels as low as 100 nM, (Honma and Shimomura 1978; Jacobson et al. 1994); both bacterial strains were grown on a rich medium and then transferred to a minimal medium containing ACC as its sole nitrogen source. The rate of induction, similar for the enzyme from the two bacterial sources, was relatively slow; complete induction required 8–10 h. Enzyme activity increased only approximately tenfold over the basal level of activity even when the concentration of ACC increased up to 10,000-fold (Honma 1985). In the fungus *P. citrinum*, it was also described that the ACC accumulated in the intracellular spaces induces the expression of ACC deaminase (Jia et al. 1999). In addition to ACC, D-serine as well as

**Table 8.2** Biochemical characteristics of some of the purified ACC deaminases (EC 4.1.99.4)

| Characteristics | *Pseudomonas putida* UW4 | *Pseudomonas putida* GR12-2 | *Pseudomonas* sp. ACP | *Hansenula saturnus* | *Penicillium citrinum* |
|---|---|---|---|---|---|
| Subunit molecular mass (Daltons) | 36,874 | 35,000 | 36,500 | 40,000 | 41,000 |
| Estimated subunits | 3 | 3 | 3 | 2 | 2 |
| $P^H$ optimum | 8.0 | 8.5 | 8.5 | 8.5 | 8.5 |
| $\Delta T_m$ | 60.2°C | nd | nd | nd | nd |
| $\Delta K_m$ | 3.4 mM | nd | 1.5 mM | 2.6 mM | 4.8 mM |
| $\Delta G^{\#}$ at 298 K | 69.6 KJ/mol | nd | nd | nd | nd |
| $\Delta H^{\#}$ | 46.8 KJ/mol | nd | nd | nd | nd |
| $\Delta S^{\#}$ | −78 J/mol K | nd | nd | nd | nd |
| Crystal structure | No | No | Yes | Yes | No |
| References | Hontzeas et al. (2004b) | Jacobson et al. (1994) | Karthikeyan et al. (2004) | Minami et al. (1998) | Jia et al. (2000) |

*nd* not determined
# denotes free energy (G,S,H denote Law of Thermodynamics)

other ACC-related substrates such as di-coronamic acid and dimethyl-ACC acts as an active substrate for this enzyme.

In the early 1980s, Walsh et al. (1981) proposed two main reaction mechanisms for all the pyridoxal 5′-phosphate (PLP)-dependent enzymes. The first mechanism illustrates the possibility of initial hydrogen abstraction opening the cyclopropane ring by $Lys^{51}$, followed by a series of hydrolytic reactions (Ose et al. 2003). The second mechanism involves a nucleophillic attack by an amino acid residence on the pro-S-β carbon of ACC to initiate the cyclopropane ring opening, followed by β-proton abstraction at the pro-R carbon by a basic residue most likely $Lys^{51}$ (Zhao et al. 2003). In both proposed mechanisms, the final products are α-ketobutyrate and ammonium. However, experiments performed with the modified ACC deaminase substrate resulted in the specific inactivation of the enzyme and a specific turnover product, suggesting that ACC deaminase is unique amongst PLP dependent enzymes showing nucleophillic addition as the major mechanism utilized in the reaction pathway (Hontzeas et al. 2006).

## 8.6 Molecular Biology of ACC Deaminase Enzyme

The enzyme ACC deaminase (EC 4.1.99.4) is a common component of many soil microorganisms, both bacteria and fungi. It has also been suggested, based largely on sequence similarities, that some plants may contain ACC deaminase genes. However, it has not yet been unequivocably demonstrated that these putative ACC deaminase genes encode an enzyme with ACC deaminase activity. ACC deaminase is a multimeric enzyme (homodimeric or homotrimeric) with a subunit molecular mass of approximately 35–42 kDa. It is a sulfhydryl enzyme in which one molecule of the essential co-factor pyridoxal phosphate (PLP) is tightly bound to each subunit.

### *8.6.1 Structural Organization of ACC Deaminase Gene*

Sheehy et al. (1991) studied the complete amino acid sequence of purified ACC deaminase from the soil bacterium *Pseudomonas* sp. strain ACP. The sequence information was used to design the oligonucleotide primers and amplified the portion of ACC deaminase gene from a 6-Kb *EcoR*I fragment of *Pseudomonas* sp. strain ACP DNA in *Escherichia coli* through PCR. They reported that the DNA sequence analysis of *EcoR*I–*Pst*I subclone contained an open reading frame (ORF) encoding a polypeptide with a deduced amino acid sequence identical to the protein sequence determined chemically and a predicted molecular mass of 36,674 Da. The ORF also contained an additional 72 bp upstream sequence not predicted by the amino acid sequence. Klee et al. (1991) constructed the genomic DNA library of *Pseudomonas* sp. strain 6G5 and introduced it into *E. coli*. The subsequent subcloning and selection of clones for growth on minimal ACC-containing medium permitted them to identify 2.4-kb DNA fragment containing the ACC-degrading gene. This fragment was subjected to DNA analysis, which revealed a single ORF of 1,017 nucleotides encoding a protein with a molecular weight of 36,800 Da similar to that of *Pseudomonas* sp. strain ACP.

Campbell and Thomson (1996) designed the primers to the most conserved regions of the ACC deaminase nucleotide sequences from *Pseudomonas* sp. strain 6G5 and ACP. PCR amplification from DNA extracts showed the expected 817-bp product and that only *Pseudomonas* sp. strain F17 had an ACC deaminase gene similar to those of strains 65G and ACP. Southern blot analysis of *Pvu*II-digested genomic DNA hybridized to an 817-bp ACC deaminase gene probe confirmed that only strain F17 had an ACC deaminase gene homologous to that of strains 6G5 and ACP. Thanananta et al. (1997) cloned the ACC deaminase gene in *E. coli* strain XL-1-blue from the chromosomal DNA of *P. fluorescens* strain J2. Restriction mapping showed no *BamH*I, *EcoR*1, *Hind*III, *Kpn*I, and *Pst*I sites except for the sites of *EcoR*V and *Xho*I in the plasmid inserts. Hybridization results indicated that all the plasmid clones contained an overlapping region of ACC deaminase gene from the *P. fluorescens* strain J2 genome.

Shah et al. (1998) isolated the genes for ACC deaminase from two PGPR strains *E. cloacae* CAL 2 and UW4. Both of these strains have an ORF of 1,014 nucleotides, which encodes a protein that is 338 amino acids long. They are highly homologous to each other and to the ACC deaminase genes of *Pseudomonas* sp. strains 6G5, 3F2, and F17. At the nucleotide level UW4, CAL2, 6G5, 3F2, and F17 were 85–95% identical to each other and most of the dissimilarities were in the Wobble position. On the other hand, the DNA sequences of the two genes showed only 74–75% homology with the sequence of the ACC deaminase gene from *Pseudomonas* sp. strain ACP. At the amino acid level, the sequence of strains

UW4, CAL2, 6G5, 3F2, and F17 were approximately 96–99% identical to each other and 81–82% identical to ACP. Southern hybridization experiments suggest that there is a single copy of the ACC deaminase gene in *E. cloacae* strain UW4 and CAL 2 and that there may be several different types of ACC deaminase gene in different microbes.

Minami et al. (1998) isolated and sequenced the cDNA encoding the ACC deaminase from the yeast, *H. saturnus* cells incubated in the α-amino isobutyrate medium and the sequence results showed that the ORF codes for the enzyme of 341 amino acid residues of which 60–63% are identical to those reported for *Pseudomonas* enzymes. Jia et al. (2000) isolated and analyzed cDNA of *Penicillium citrium* encoding ACC deaminase enzyme. The sequencing of 1,233 bp cDNA showed to consist of an ORF of 1,080 bp encoding ACC deaminase with 360 amino acids, which is longer than other reported sources. The deduced amino acids from the cDNA are identical by 52 and 45% to those of enzymes of *Pseudomonas* sp. ACP and *H. saturnus*. Li et al. (2001) isolated a unique ACC deaminase like gene from one of the clone *E. coli* DH5α/pWA2 containing *E. cloaceae* UW4 genome. The clone was with a plasmid containing an insert size of approximately 0.8 kb that conferred ACC deaminase activity. Sequence analysis revealed that this DNA fragment contains an ORF of 696 nucleotides predicted to encode a protein of 232 amino acids. It was found that the sequence of pWA2 ORF and at least one of three genes was 42.4% and shorter than ORFs of other genes. The deduced amino acid sequence was one of only 17.5% identity with normal ACC deaminase sequence and no conserved regions were found.

Babalola et al. (2003) isolated the ACC deaminase gene from three rhizobacteria through PCR amplification using degenerate gene-specific primers. It was observed that not all the isolates of PGPR were having ACC deaminase gene and the expected amplification product of 0.73 kb was observed only for *Pseudomonas* sp. 4MKS 8 and *Klebsiella oxytoca* 10MKR 7. Govindasamy et al. (2008a) isolated the complete *acdS* gene of ~1,017 bp from ten different rhizobacterial isolates and partial *acdS* genes of ~800 bp size isolated from two more isolates *Achromobacter* sp. GKA-1 and *Pseudomonas stutzeri* GKA-13. In addition to this, 261 rhizobacteria from various crop rhizospheres were screened initially for ACC deaminase enzyme activity. However, the presence of *acdS* gene was detected in 11 more rhizobacterial isolates using nucleic acid dot blot and southern hybridization techniques (Govindasamy et al. 2009). In addition to these ACC deaminases, several ORFs in fully decoded organisms are annotated as ACC deaminase homologues with significant sequence identity. One of them, the protein PH0054 isolated from the hyperthermophillic archaebacterium *Pyrococcus horikoshii* OT3 was predicted to be ACC deaminase. The function of these protein ORFs in archaebacteria remains to be elucidated (Fujino et al. 2004).

Hontzeas et al. (2005) isolated ACC deaminase gene from a range of Gram positive and Gram-negative bacterial species using PCR with degenerate primers. The bacterial colonies can be quickly screened for the presence of the ACC deaminase genes by amplification of a fragment of approximately 750 bp. In the phylogenetic tree, *Pseudomonas* sp. and *Achromobacter* sp. the ACC deaminase genes are distributed throughout the tree and proposed that some ACC deaminase genes have evolved through horizontal multiple gene transfer pattern, which have not evolved in the same manner as 16S ribosomal RNA genes. Blaha et al. (2006) studied the distribution and phylogeny of the ACC deaminase encoding structural gene *acdS* and reported that *acdS* is present in several pathogenic proteobacteria including opportunistic human pathogens and therefore the gene is not restricted to plant beneficial taxa. The *acdS* gene was also reported in the PGPR genus *Azospirillum* that was earlier considered as negative for ACC deaminase activity. Phylogenetic analysis evidenced that there were three main *acdS* phylogenetic groups; the group I and II gathered strains belonging to the Beta and Gamma proteobacteria subdivision, whereas group III was composed of α-proteobacteria.

### 8.6.2 Regulation of ACC Deaminase Gene

Analysis of the DNA sequence upstream of the *E. cloacae* UW4 ACC deaminase gene was found to contain several features (Fig. 8.4) that may be involved in the transcriptional regulation of this gene including half of a cAMP receptor protein (CRP)- binding site, a fumarate-nitrate reduction regulatory protein (FNR) binding site, a leucine responsive regulatory protein (LRP) binding site, and an LRP-like protein coding region. The experimental results showed that these potential regulatory regions are involved either partially or completely in the regulation of the *E. cloacae* UW4 ACC deaminase gene (Grichko and Glick 2000). It was speculated that ACC binds to the LRP-like protein and this complex activates one of the weak promoters just upstream of *acdS*. One of these promoters overlaps with a putative FNR-box and might be active during anaerobic conditions, while the other

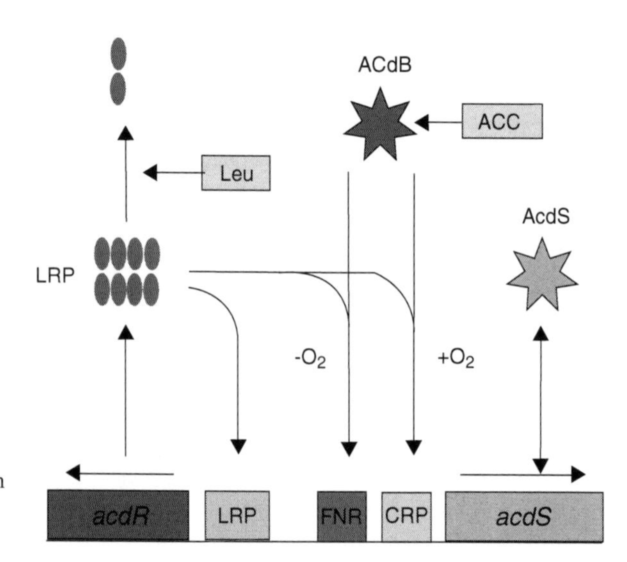


**Fig. 8.4** Model of the transcriptional regulation of ACC deaminase expression in *Pseudomonas putida* UW4 (Li and Glick 2001)

promoter overlaps with a putative CRP-box and might function under higher oxygen levels. When the bacterium has a surfeit of branched chained amino acids, leucine may interact with the LRP-like protein and this complex may bind to the LRP-like box preventing additional LRP-like protein from being synthesized (Li and Glick 2001).

In *Bradyrhizobium japonicum* and *R. leguminosarum* bv. *viciae* 128C53K, the *acdS* gene is also probably regulated by an LRP-like protein and a sigma-70 promoter (Kaneko et al. 2002; Ma et al. 2003b). However, in *Mesorhizobium loti* the *acdS* gene and other symbiotic genes are positively regulated by the symbiotic nitrogen-fixing regulator gene *nifA2* and the mode of gene expression suggests that *M. loti acdS* participates in the establishment and/or maintenance of native nodules by interfering with the production of ethylene, which induces negative regulation of nodulation (Nukui et al. 2006).

## 8.7 Exploitation of ACC Deaminase Containing PGPR

Initially, the enzyme ACC-deaminase isolated from soil bacteria was utilized to perturb ACC levels and ethylene biosynthesis, thus leading to understand the role of plant hormone ethylene (Sheehy et al. 1991). Later, it was proposed that the PGPR containing ACC deaminase is an important trait of plant growth stimulation and to overcome a variety of environmental stresses on growing plants (Glick 1995).

### *8.7.1 Improvement of Crop Yield and Quality*

This PGPR trait is exploited in two ways to increase crop yield and productivity, one by strain improvement of existing PGPR biofertilizers and another by direct application of ACC deaminase containing PGPR in the crop field as biofertilizers. The microbial inoculation with ACC deaminase containing PGPR known to promote root growth by lowering of plant ethylene was well documented in different crops such as tomato, lettuce, wheat, and canola (Lifshitz et al. 1987; Hall et al. 1996). It was recently observed that when canola seeds treated with *E. coli* cells expressing a cloned *E. cloaceae* ACC deaminase gene; these were able to promote root elongation (Shah et al. 1998).

Holguin and Glick (2001) introduced the *acdS* gene from *E. cloaceae* UW4 into *Azospirillum brasilense* strain cd and sp245. Indeed, the introduced *acdS* gene was readily expressed in the transformants similar to that of *E. cloacae* UW4. The expression of ACC deaminase improved/enhanced the existing growth-promoting ability and other phytobeneficial effects of *Azospirillum* in tomato, canola, and wheat seedlings (Holguin and Glick 2003). Similarly, the ACC deaminase structural gene (*acdS*) and its upstream regulatory gene from *R. leguminosarum* bv. *viciae* 128C53K were introduced into *Sinorhizobium meliloti*, which does not produce this enzyme. The resulting ACC deaminase producing *S. meliloti* strains showed more competitive ability and 35–40% greater efficiency in nodulating alfa-alfa (*Medicago sativa*) than the wild-type strain (Ma et al. 2004).

Dey et al. (2004) reported that enhancement of root length by ACC-deaminase activity of the inoculated PGPR strains had a significant effect on the majority of the growth, yield and nutrient uptake parameters in peanut cultivars GG2 under both pot and field conditions. Shaharoona et al. (2006a) reported that inoculation of PGPR containing ACC deaminase promotes the root growth and yield of maize under axenic conditions. The results also showed that coinoculation of PGPR containing ACC-deaminase along with competitive *Rhizobia* increased the nodulation and yield of mungbean. In another study, Shaharoona et al. (2006b) assessed the performance of PGPR containing ACC deaminase for improving growth and yield of maize grown in nitrogenous fertilizer amended soil, and results showed that even in the presence of optimum levels of nitrogenous fertilizers inoculation with rhizobacteria containing ACC deaminase could be effective in improving the growth and yield of inoculated crop plants.

The ACC deaminase containing PGPR trait is useful for examining the role of ethylene in many developmental processes in plants as well as for extending the shelf-life of fruit and vegetables whose ripening is mediated by ethylene. Klee et al. (1991) developed the transgenic tomato plants expressing bacterial ACC deaminase enzyme and results showed that the reduction in ethylene synthesis in these transgenic plants did not cause any apparent changes in vegetative phenotypic observations. However, fruits from these plants exhibited significant delays in ripening and the mature fruits remained firm for at least 6 weeks longer than the nontransgenic control fruits. Nayani et al. (1998) reported that incubating the carnation petals with ACC-deaminase containing PGPR bacterial suspension, the life time of the petals increased by approximately 4–5 days compared to the petals of untreated flowers. Thus the results imply that applications of ACC-deaminase PGPR to cut flowers have great potential to replace the chemical ethylene inhibitors used at present.

### *8.7.2 Improvement of Plant Fitness Against Environmental Stresses*

The PGPR containing ACC deaminase have been exploited to overcome the detrimental effects of various environmental stresses induced by biotic and abiotic factors either partially or completely and making the crop plants more resistance against these stresses.

Phytopathogens not only directly inhibit plant growth; they also cause the plant to synthesize stress ethylene. It is well known that exogenous ethylene often increases severity of a phytopathogen infection, whereas some ethylene synthesis inhibitors significantly decrease the disease severity of phytopathogens (Van Loon et al. 2006). Hence, treating plant seeds or roots with biocontrol bacteria containing

ACC deaminase is more effective in preventing phytopathogen infection. Wang et al. (2000) transformed the IAA-producing strains *P. fluorescens* CHAO with ACC deaminase gene from *E. cloaceae* UW4. The results showed that the transformation increased the ability of *P. fluorescens* CHAO to protect potato tubers from *Erwinia carotovora* mediated soft rot and cucumber from *Pythium* damping-off. In addition, the transgenic plants expressing a bacterial ACC deaminase gene were developed, which significantly protected the damage caused by *Verticillium* wilt in tomatoes (Robison et al. 2001b).

Belimov et al. (2007) reported the phytopathogenic bacterium *P. brassicacearum* strain AM3 possessing ACC deaminase activity. This ACC utilizing strain increased in vitro root elongation and root biomass of tomatoes at low bacterial concentrations than its mutant strain T8, but had negative effects on in vitro root elongation at higher bacterial concentrations. The results suggested that *P. brassicacearum* AM3 could have growth promoting, neutral or phytopathogenic effects on growth of a single plant cultivar according to the dose and environmental conditions. The outcome of this *P. brassicacearum*–tomato interaction is influenced by bacterial ACC deaminase activity, as decreasing this activity modified the root growth response of tomatoes to inoculation.

Treatment of tomato plants with ACC-deaminase containing PGPR (Grichko and Glick 2001a) or genetically engineered plants to express this microbial enzyme significantly decreased the damage caused in these plants as a result of flooding (Grichko and Glick 2001b). Mayak et al. (2004a) reported that treating peppers and tomato plants with ACC-deaminase containing PGPR significantly reduced the growth inhibition caused by drought stress. In another study, they showed that treating tomato plants with this bacterium prevented the growth inhibition by high concentrations of salt (Mayak et al. 2004b). Saravanakumar and Samiyappan (2007) reported that inoculation of *P. fluorescens* TDK1 possessing ACC deaminase activity enhanced the salinity resistance in groundnut by lowering stress induced ethylene level and showed the significant improvement in root elongation, plant growth parameters and yield of groundnut under both pot and field conditions. Similarly, Tank and Saraf (2010) have reported that increase in the salinity is directly proportional to the ACC deaminase activity which increases survival rate in saline soils. As the uptake and hydrolysis of ACC by the PGPR decreases the ACC levels in plants, the biosyhthesis of the stress ethylene is impeded. Zahir et al. (2009) reported a significant increase in plant height, root length, grain yield, 100-grain weight and straw yield upto 52, 60, 76, 19 and 67%, respectively over uninoculated control in wheat grown under saline conditions when inoculated with ACC deaminase producing *P. putida*. They hypothesized that the enzyme activity of *P. putida* may have caused reduction in the synthesis of stress induced inhibitory levels of ethylene.

The presence of high levels of heavy metals and environmental contaminants in soil, induce most plants to synthesize growth inhibiting stress ethylene and treating these plants with ACC-deaminase and siderophore producing PGPR can help the plants to overcome many of the effects caused by these environmental stresses (Burd et al. 1998, 2000; Ganesan 2008). Similarly transgenic plants expressing bacterial ACC-deaminase gene under root specific promoter showed more resistance to the toxic effects of metals than the non-transformed plants (Grichko et al. 2000; Nie et al. 2002; Stearns et al. 2005). The ACC deaminase containing PGPR trait was exploited for the phytoremediation of organic contaminants such as oil spills, polycyclic aromatic hydrocarbons (PAHs) and polycyclic biphenyls (Reed and Glick 2005). In an interesting study Hao et al. (2010) reported the presence of an active ACC deaminase in *Agrobacterium tumefaciens* increased the transformation efficiency of three commercial canola cultivars. A similar observation was also made by Nonaka et al. (2008).

## 8.8 Conclusion

Our understanding of some of the mechanisms used by PGPR has come a long way in the past one to two decades, and with this increased understanding there has been a concomitant increase in the commercial application of these organisms in agriculture. When comparing developed countries, in many of the less developed countries of the world where agricultural productivity is not as high, relatively cheap labor and high chemical costs provide a situation where the use of PGPR provides an attractive commercial possibility. In particular, ACC deaminase containing PGPR can help plants to tolerate a range of biotic and abiotic stresses through lowering stress ethylene synthesis in plants. A number of labs have reported decreasing plant ethylene production by suppressing the expression of some of the enzymes of the ethylene biosynthetic pathway. However, the activity of many of these enzymes also affects processes other than ethylene synthesis in plants. There are some advantages in utilizing rhizobacterial enzyme ACC deaminase for the above purpose. The exploitation of this trait showed promising potential in crop production, developing *Agrobacterium* mediated transgenics and phyto-remediation of contaminated soils. The use of ACC deaminase containing PGPR in agriculture will be a "green strategy" to address some of the environmental issues raised by agrochemicals usage.

## References

Abeles FB, Morgan PW, Saltveit ME Jr (eds) (1992) Regulation of ethylene production by internal, environmental and stress factors. In: Ethylene in plant biology, 2nd edn. Academic, San Diego, pp 56–119

Adams DO, Yang SF (1979) Ethylene biosynthesis: identification of 1-aminocyclopropane-1-carboxylic acid as an intermediate in the conversion of methionine to ethylene. Proc Natl Acad Sci USA 76:170–174

Arshad M, Frankenberger WT Jr (2002) Ethylene: agricultural sources and applications. Kluwer/Plenum, Dordrecht

Babalola OO, Osir EO, Sanni AI, Odhiambo GD, Bulimo WD (2003) Amplification of 1-amino-cyclopropane-1-carboxylic (ACC) deaminase from plant growth promoting rhizobacteria in striga-infested soil. Afr J Biotechnol 2:157–160

Barazani O, Friedman J (1999) Is IAA the major root growth factor secreting from plant growth mediating bacteria? J Chem Ecol 25:2397–2406

Barea JM, Pozo MJ, Azcon R, Azcon-Aguilar C (2005) Microbial cooperation in the rhizosphere. J Exp Bot 56:1761–1778

Bashan Y (1999) Interactions of *Azospirillum* spp. in soils: a review. Biol Fert Soils 29:246–256

Belimov AA, Kojemiakov AP, Chuvarliyeva CV (1995) Interaction between barley and mixed cultures of nitrogen fixing and phosphate-solubilizing bacteria. Plant Soil 173:29–37

Belimov AA, Safronova VI, Sergeyeva TA, Egorova TN, Matveyeva VA, Tsyganov VE, Borisov AY, Tikhonovich IA, Kluge C, Preisfeld A, Dietz KJ, Stepanok VV (2001) Characterization of plant growth promoting rhizobacteria isolated from polluted soils and containing 1-aminocyclopropane-1-carboxylate deaminase. Can J Microbiol 47:242–252

Belimov AA, Hontzeas N, Safronova VI, Demchinskaya SV, Piluzza G, Bullitta S, Glick BR (2005) Cadmium-tolerant plant growth-promoting rhizobacteria associated with the roots of Indian mustard (*Brassica juncea* L. Czern). Soil Biol Biochem 37:241–250

Belimov AA, Dodd IC, Safronova VI, Hontzeas N, Davies WJ (2007) *Pseudomonas brassicacearum* strain Am3 containing 1-aminocyclopropane-1-carboxylate deaminase can show both pathogenic and growth-promoting properties in its interaction with tomato. J Exp Bot 58:1485–1495

Biswas JC, Ladha JK, Dazzo FB (2000) *Rhizobia* inoculation improves nutrient uptake and growth of lowland rice. Soil Sci Soc Am J 64:1644–1650

Blaha D, Prigent-Combaret C, Mirza MS, Moenne-Loccoz Y (2006) Phylogeny of the 1-aminocyclopropane-1-carboxylic acid deaminase-encoding gene *acdS* in phytobeneficial and pathogenic proteobacteria and relation with strain biogeography. FEMS Microbiol Ecol 56:455–470

Bowen GD, Rovira AD (1999) The rhizosphere and its management to improve plant growth. Adv Agron 66:1–102

Burd GI, Dixon DG, Glick BR (1998) A plant growth promoting bacterium that decreases nickel toxicity in seedlings. Appl Environ Microbiol 64:3663–3668

Burd GI, Dixon DG, Glick BR (2000) Plant growth-promoting bacteria that decrease heavy metal toxicity in plants. Can J Microbiol 46:237–245

Burroughs LF (1957) 1-Aminocyclopropane-l-carboxylic acid: a new aminoacid in Perry pears and Cider apples. Nature 179:360–361

Campbell BG, Thomson JA (1996) 1-Aminocyclopropane-1-carboxylate deaminase genes from *Pseudomonas* strains. FEMS Microbiol Lett 138:207–210

Cherrington CA, Elliott LF (1987) Induction of inhibitory pseudomonads in the pacific North-West. Plant Soil 101:159–165

Chinnadurai C, Balachandar D, Sundaram SP (2009) Characterization of 1-aminocyclopropane-1-carboxylate deaminase producing methylobacteria from phyllosphere of rice and their role in ethylene regulation. World J Microbiol Biotechnol 25:1403–1411

Dell'Amico E, Cavalca L, Andreoni V (2008) Improvement of *Brassica napus* growth under cadmium stress by cadmium-resistant rhizobacteria. Soil Biol Biochem 40:74–84

Dey R, Pal KK, Bhatt DM, Chauhan SM (2004) Growth promotion and yield enhancement of peanut (*Aracis hypoggaea* L.) by application of plant growth promoting rhizobacteria. Microbiol Res 159:371–394

Frankenberger WT Jr, Arshad M (1995) Phytohormones in soil: microbial production and function. Dekker, New York, p 503

Fujino A, Ose T, Yao M, Tokiwano T, Honma M, Watanabe N, Tanaka I (2004) Structural and enzymatic properties of 1-aminocyclopropane-1-carboxylate deaminase homologue from *Pyrococcus horikoshii*. J Mol Biol 341:999–1013

Fukuda H, Ogawa T, Tanase S (1993) Ethylene production by microorganisms. Adv Microb Physiol 35:275–306

Ganesan V (2008) Rhizoremediation of cadmium soil using a cadmium-resistant plant growth-promoting rhizopseudomonad. Curr Microbiol 56:403–407

Ghosh S, Penterman JN, Little RD, Chavez R, Glick BR (2003) Three newly isolated plant growth-promoting bacilli facilitate the seedling growth of canola, *Brassica campestris*. Plant Physiol Biochem 41:277–281

Glick BR (1995) The enhancement of plant growth by free-living bacteria. Can J Microbiol 41:109–117

Glick BR, Jacobson CB, Schwarze MMK, Pasternak JJ (1994) 1-Aminocyclopropane-1-carboxylic acid deaminase mutants of the plant growth promoting rhizobacterium *Pseudomonas putida* GR12-2 do not stimulate canola root elongation. Can J Microbiol 40:911–915

Glick BR, Karaturovc DM, Newell PC (1995) A novel procedure for rapid isolation of plant growth-promoting pseudomonads. Can J Microbiol 41:533–536

Glick BR, Liu C, Ghosh S, Dumbroff EB (1997) The effect of the plant growth promoting rhizobacterium *Pseudomonas putida* GR12-2 on the development of canola seedlings subjected to various stresses. Soil Biol Biochem 29:1233–1239

Glick BR, Penrose DM, Li J (1998) A model for the lowering of plant ethylene concentrations by plant growth promoting bacteria. J Theor Biol 190:63–68

Glick BR, Patten CL, Holguin G, Penrose DM (1999) Biochemical and genetic mechanisms used by plant growth promoting bacteria. Imperial College Press, London

Govindasamy V, Senthilkumar M, Gaikwad K, Annapurna K (2008a) Isolation and characterization of ACC deaminase gene from two plant growth promoting rhizobacteria. Curr Microbiol 57:312–317

Govindasamy V, Senthilkumar M, Upendra Kumar, Annapurna K (2008b) PGPR-biotechnology for management of abiotic and biotic stresses in crop plants. In: Maheshwari DK, Dubey RC (eds) Potential microorganisms for sustainable agriculture, IK International, New Delhi, pp 26–48

Govindasamy V, Senthilkumar M, Mageshwaran V, Annapurna K (2009) Detection and characterization of ACC deaminase in plant growth promoting rhizobacteria. J Plant Biochem Biotechnol 18:71–76

Grichko VP, Glick BR (2000) Identification of DNA sequences that regulate the expression of the *Enterobacter cloacae* UW4 1-aminocyclopropane-1-carboxylate deaminase gene. Can J Microbiol 46:1159–1165

Grichko VP, Glick BR (2001a) Amelioration of flooding stress by ACC deaminase-containing plant growth-promoting bacteria. Plant Physiol Biochem 39:11–17

Grichko VP, Glick BR (2001b) Flooding tolerance of transgenic tomato plants expressing the bacterial enzyme ACC deaminase controlled by the 35S, rolD or PRB-1b promoter. Plant Physiol Biochem 39:19–25

Grichko VP, Filby B, Glick BR (2000) Increased ability of transgenic plants expressing the bacterial enzyme ACC deaminase to accumulate Cd, Co, Cu, Ni, Pb and Zn. J Biotechnol 81:45–53

Hall JA, Peirson D, Ghosh S, Glick BR (1996) Root elongation in various agronomic crops by the plant growth promoting rhizobacterium *Pseudomonas putida* GR12-2. Isr J Plant Sci 44:37–42

Hao Y, Charles TC, Glick BR (2010) ACC deaminase increases the *Agrobacterium tumifaciens*-mediated transformation frequency of commercial canola cultivars. FEMS Microbiol Lett 307:185–190

Holguin G, Glick BR (2001) Expression of the ACC deaminase gene from *Enterobacter cloacae* UW4 in *Azospirillum brasilense*. Microb Ecol 41:281–288

Holguin G, Glick BR (2003) Transformation of *Azospirillum brasilense* Cd with an ACC deaminase gene from *Enterobacter cloacae* UW4 fused to the Tetr gene promoter improves its fitness and plant growth promoting ability. Microb Ecol 46:122–133

Hong Y, Glick BR, Pasternak JJ (1991) Plant microbial interaction under gnotobiotic conditions: a scanning electron microscope study. Curr Microbiol 23:111–114

Honma M (1985) Chemically reactive sulfhydryl groups of 1-aminocyclopropane-1-carboxylate deaminase. Agric Biol Chem 49:567–571

Honma M (1993) Stereoscopic reaction of 1-aminocyclopropane-1-carboxylate deaminase. In: Pach JC, Latche A, Balague C (eds) Cellular and molecular aspects of the plant hormone ethylene. Kluwar Academic, Dordrecht, The Netherlands, pp 111–116

Honma M, Shimomura T (1978) Metabolism of l-aminocyclopropane-1-carboxylic acid. Agric Biol Chem 42:1825–1831

Hontzeas N, Saleh S, Glick BR (2004a) Changes in gene expression in canola roots induced by ACC-deaminase-containing plant-growth-promoting bacteria. Mol Plant Microbe Interact 17:951–959

Hontzeas N, Zoidakis J, Glick BR, Abu-Omar MM (2004b) Expression and characterization of 1-aminocyclopropane-1-carboxylate deaminase from the rhizobacterium *Pseudomonas putida* UW4: a key enzyme in bacterial plant growth promotion. Biochim Biophys Acta 1703:11–19

Hontzeas N, Richardson AO, Belimov AA, Safranova VI, Abu-Omar MM, Glick BR (2005) Evidence for horizontal gene transfer (HGT) of ACC deaminase genes. Appl Environ Microbiol 71:7556–7558

Hontzeas N, Hontzeas CE, Glick BR (2006) Reaction mechanisms of the bacterial enzyme ACC deaminase. Biotechnol Adv 24:420–426

Hyodo H (1991) Stress/wound ethylene. In: Mattoo AK, Suttle JC (eds) The plant hormone ethylene. CRC, Boca Raton, pp 65–80

Jacobson CB, Pasternak JJ, Glick BR (1994) Partial purification and characterization of 1-aminocyclopropane-1-carboxylate deaminase from the plant growth promoting rhizobacterium *Pseudomonas putida* GR12-2. Can J Microbiol 40:1019–1025

Jia YJ, Kakuta Y, Sugawara M, Igarashi T, Oki N, Kisaki M, Shoji T, Kanetuna Y, Horita T, Matsui H, Honma M (1999) Synthesis and degradation of 1-aminocyclopropane-1-carboxylic acid by *Penicillium citrinum*. Biosci Biotechnol Biochem 63:542–549

Jia YJ, Ito H, Matsui H, Honma M (2000) 1-Aminocyclopropane-1-carboxylate (ACC) deaminase induced by ACC synthesized and accumulated in *Penicillium citrinum* intracellular spaces. Biosci Biotechnol Biochem 64:299–305

Kaneko T, Nakamura Y, Sato S, Minamisawa K, Uchiumi T, Sasamoto S, Watanabe A, Idesawa K, Iriguchi M, Kawashima K, Kohara M, Matsumoto M, Shimpo S, Tsuruoka H, Wada T, Yamada M, Tabata S (2002) Complete genomic sequence of nitrogen-fixing symbiotic bacterium *Bradyrhizobium japonicum* USDA110. DNA Res 9:189–197

Karthikeyan S, Zhou Q, Zhao Z, Kao CL, Tao Z, Robinson H, Liu HW, Zhang H (2004) Structural analysis of *Pseudomonas* 1-aminocyclopropane-1-carboxylate deaminase complexes: insight into the mechanism of a unique pyridoxal-5-phosphate dependent cyclopropane ring-opening reaction. Biochemistry 43:13328–13339

Kennedy AC, Smith KL (1995) Soil microbial diversity and the sustainability of agricultural soils. Plant Soil 170:75–86

Klee HJ, Kishore GM (1992) Control of fruit ripening and senescence in plants. US Patent 5,702,933

Klee HJ, Hayford MB, Kretzmer KA, Barry GF, Kishore GM (1991) Control of ethylene synthesis by expression of a bacterial enzyme in transgenic tomato plants. Plant Cell 3:1187–1193

Kloepper JW (1993) Plant growth promoting rhizobacteria as biological control agents. In: Metting FB Jr (ed) Soil microbial ecology. Dekker, New York, pp 255–274

Kloepper JW, Schroth MN (1981) Plant growth-promoting rhizobacteria and plant growth under gnotobiotic conditions. Phytopathology 71:642–644

Kloepper JW, Hume DJ, Schez FM, Singleton C, Tipping B, Lalibert EM, Fraulay K, Kutchaw T, Simonson C, Lifshitz R, Zaleska L, Lee L (1987) Plant growth promoting rhizobacteria on canola (rapeseed). Phytopathology 71:42–46

Kloepper JW, Lifshitz R, Zablotowicz RM (1989) Free living bacterial inocula for enhancing crop productivity. Trends Biotechnol 7:39–43

Kloepper JW, Zablotowick RM, Tipping EM, Lifshitz R (1991) Plant growth promotion mediated by bacterial rhizosphere colonizers. In: Keister DL, Cregan PB (eds) The rhizosphere and plant growth. Kluwer, Dordrecht, pp 315–326

Li J, Glick BR (2001) Transcriptional regulation of the *Enterobacter cloacae* UW4 1-aminocyclopropane-1-carboxylate (ACC) deaminase gene (*acdS*). Can J Microbiol 47:259–267

Li J, Ovakim DH, Charles TC, Glick BR (2000) An ACC deaminase minus mutant of *Enterobacter cloacae* UW4 no longer promotes root elongation. Curr Microbiol 41:101–105

Li J, Shah S, Moffatt BA, Glick BR (2001) Isolation and characterization of an unusual 1-aminocyclopropane-1-carboxylic acid (ACC) deaminase gene from *Enterobacter cloacae* UW4. Antonie Van Leeuwenhoek 80:255–261

Lifshitz R, Kloepper JW, Scher FM, Tipping EM, Laliberte M (1986) Nitrogen-fixing Pseudomonads isolated from roots of plants grown in the Canadian High Arctic. Appl Environ Microbiol 51:251–255

Lifshitz R, Kloepper JW, Kozlowski M, Simonson C, Carlson J, Tipping EM, Zaleska I (1987) Growth promotion of canola (rapeseed) seedlings by a strain of *Pseudomonas putida* under gnotobiotic conditions. Can J Microbiol 33:390–395

Lynch JM (1990) The rhizosphere. Wiley, New York

Ma W, Penrose DM, Glick BR (2002) Strategies used by rhizobia to lower plant ethylene levels and increase nodulation. Can J Microbiol 48:947–954

Ma W, Guinel FC, Glick BR (2003a) *Rhizobium leguminosarum* biovar *viciae* 1-aminocyclopropane-1-carboxylate deaminase promotes nodulation of pea plants. Appl Environ Microbiol 69:4396–4402

Ma W, Sebestianova SB, Sebestian J, Burd GI, Guinel FC, Glick BR (2003b) Prevalence of 1-aminocyclopropane-1-carboxylate deaminase in *Rhizobium* spp. Antonie Van Leeuwenhoek 83:285–291

Ma W, Charles TC, Glick BR (2004) Expression of an exogenous 1-aminocyclopropane-1-carboxylate deaminase gene in *Sinorhizobium meliloti* increases its ability to nodulate alfalfa. Appl Environ Microbiol 70:5891–5897

Madhaiyan M, Kim BY, Poonguzhali S, Kwon SW, Song MH, Ryu JH, Go SJ, Koo BS, Sa TM (2007) *Methylobacterium oryzae* sp. *nov*., an aerobic, pink-pigmented, facultatively methylotrophic, 1-aminocyclopropane-1-carboxylate deaminase-producing bacterium isolated from rice. Int J Syst Evol Microbiol 57:326–331

Maheshwari DK (2010) Plant growth and health promoting bacteria. In: Maheshwari DK (ed) Microbiology monographs. Springer, Heidelberg, pp 99–116

Malik KA, Bilol R, Mehnaz S, Rasul G, Mirza MS, Ali S (1997) Association of nitrogen fixing, plant growth promoting rhizobacteria (PGPR) with Kallar grass and rice. Plant Soil 194:37–44

Martínez-Viveros O, Jorquera MA, Crowley DE, Mora Gajardo G (2010) Mechanisms and practical considerations involved in plant growth promotion by rhizobacteria. J Soil Sci Plant Nutr 10:293–319

Mattoo AK, Shuttle JC (1991) The plant hormone ethylene. CRC, Boca Raton

Mayak S, Tirosh T, Glick BR (2004a) Plant growth promoting bacteria that confer resistance to water stress in tomato and pepper. Plant Sci 166:525–530

Mayak S, Tirosh T, Glick BR (2004b) Plant growth promoting bacteria that confer resistance in tomato to salt stress. Plant Physiol Biochem 42:565–572

Minami R, Uchiyama K, Murakami T, Kawai J, Mikami K, Yamada T, Yokoi D, Ito H, Matsui H, Honma M (1998) Properties, sequence, and synthesis in *Escherichia coli* of 1-aminocyclopropane-1-carboxylate deaminase from *Hansenula saturnus*. J Biochem 123:1112–1118

Nannipieri P, Ascher J, Ceccherini MT, Landi L, Pietramellara G, Renella G (2003) Microbial diversity and soil functions. Eur J Soil Sci 54:655–670

Nayani S, Mayak S, Glick BR (1998) The effect of plant growth promoting rhizobacteria on the senescence of flower petals. Ind J Exp Biol 36:836–839

Nie L, Shah S, Burd GI, Dixon DG, Glick BR (2002) Phytoremediation of arsenate contaminated soil by transgenic canola and the plant growth-promoting bacterium *Enterobacter cloacae* CAL2. Plant Physiol Biochem 40:355–361

Noel TC, Sheng C, Yost CK, Pharis RP, Hynes MF (1996) *Rhizobium leguminosarum* as a plant growth promoting rhizobacterium: direct growth promotion of canola and lettuce. Can J Microbiol 42:279–283

Nonaka S, Sugawara M, Minimasawa K, Yuhashi K, Ezura H (2008) 1-Aminocyclopropane-1-carboxylate deaminase enhances *Agrobacterium tumefaciens*-mediated gene transfer into plant cells. Appl Environ Microbiol 74:2526–2528

Nukui N, Ezura H, Yuhashi KI, Yasuta T, Minamisawa K (2000) Effect of ethylene precursor and inhibitors for ethylene biosynthesis and perception on nodulation in *Lotus japonicus* and *Macroptilium atropurpureum*. Plant Cell Physiol 41:893–897

Nukui N, Minamisawa K, Ayabe SI, Aoki T (2006) Expression of the 1-aminocyclopropane-1-carboxylic acid deaminase gene requires symbiotic nitrogen-fixing regulator gene *nifA2* in *Mesorhizobium loti* MAFF303099. Appl Environ Microbiol 72:4964–4969

Okazaki S, Nukui N, Sugawara M, Minamisawa K (2004) Rhizobial strategies to enhance symbiotic interactions: rhizobitoxine and 1-aminocyclopropane-1-carboxylate deaminase. Microbes Environ 19:99–111

Okon Y (1994) Azospirillum/plant associations. CRC, Boca Raton

Ose T, Fujino A, Yao M, Watanabe N, Honma M, Tanaka I (2003) Reaction intermediate structures of 1-aminocyclopropane-1-carboxylate deaminase. J Biol Chem 278:41069–41076

Pace NR (1997) A molecular view of microbial diversity in the biosphere. Science 276:734–740

Pandey P, Kang SC, Maheshwari DK (2005) Isolation of endophytic plant growth promoting *Burkholderia* sp. MSSP from root nodules of *Mimosa pudica*. Curr Sci 89:117–180

Paul EA, Clark FE (1989) Soil microbiology and biochemistry. Academic, San Diego, CA

Penrose DM, Glick BR (2001) Levels of 1-aminocyclopropane-1-carboxylic acid (ACC) in exudates and extracts of canola seeds treated with plant growth-promoting bacteria. Can J Microbiol 47:368–372

Penrose DM, Moffatt BA, Glick BR (2001) Determination of 1-aminocyclopropane-1-carboxylic acid (ACC) to assess the effects of ACC deaminase-containing bacteria on roots of canola seedlings. Can J Microbiol 47:77–80

Pierik R, Tholen D, Poorter H, Visser EJW, Voesenek LACJ (2006) The janus face of ethylene: growth inhibition and stimulation. Trends Plant Sci 11:176–183

Reed MLE, Glick BR (2005) Growth of canola (*Brassica napus*) in the presence of plant growth-promoting bacteria and either copper or polycyclic aromatic hydrocarbons. Can J Microbiol 51:1061–1069

Richardson AE (2001) Prospects for using soil microorganisms to improve the acquisition of phosphorus by plants. Aust J Plant Physiol 28:897–906

Robison MM, Griffith M, Pauls KP, Glick BR (2001a) Dual role of ethylene in susceptibility of tomato to *Verticillium* wilt. J Phytopathol 149:385–388

Robison MM, Shah S, Tamot B, Pauls KP, Moffatt BA, Glick BR (2001b) Reduced symptoms of *Verticillium* wilt in transgenic tomato expressing a bacterial ACC deaminase. Mol Plant Pathol 2:135–145

Saravanakumar D, Samiyappan R (2007) ACC deaminase from *Pseudomonas fluorescens* mediated saline resistance in groundnut (*Arachis hypogea*) plants. J Appl Microbiol 102:1283–1292

Shah S, Li J, Moffatt BA, Glick BR (1997) ACC deaminase genes from plant growth promoting rhizobacteria. In: Ogoshi A, Kobayashi K, Hemma Y, Kodema F, Kondo N, Akino S (eds) Plant growth-promoting rhizobacteria. Present status and future prospects. Organization for Economic Cooperation and Development, Paris, pp 320–324

Shah S, Li J, Moffatt BA, Glick BR (1998) Isolation and characterization of ACC deaminase genes from two different plant growth-promoting rhizobacteria. Can J Microbiol 44:833–843

Shaharoona B, Arshad M, Zahir ZA (2006a) Effect of plant growth promoting rhizobacteria containing ACC-deaminase on maize (*Zea mays* L.) growth under axenic conditions and on nodulation in mung bean (*Vigna radiata* L.). Lett Appl Microbiol 42:155–159

Shaharoona B, Arshad M, Zahir ZA, Khalid A (2006b) Performance of *Pseudomonas* spp. containing ACC-deaminase for improving growth and yield of maize (*Zea mays* L.) in the presence of nitrogenous fertilizer. Soil Biol Biochem 38:2971–2975

Sheehy RE, Honma M, Yamada M, Sasaki T, Martineau B, Hiatt WR (1991) Isolation, sequence, and expression in *Escherichia coli* of the *Pseudomonas* sp. strain ACP gene encoding 1-aminocyclopropane-1-carboxylate deaminase. J Bacteriol 173:5260–5265

Stearns JC, Shah S, Dixon DG, Greenberg BM, Glick BR (2005) Tolerance of transgenic canola expressing 1-aminocyclopropane-carboxylic acid deaminase to growth inhibition by nickel. Plant Physiol Biochem 43:701–708

Tank N, Saraf M (2010) Salinity resistant PGPR ameliorates NaCl stress on tomato plants. J Plant Interact 5:51–58

Thanananta T, Engkagul A, Peyachoknagul S, Pongtongkam P, Apisitwanich S (1997) Cloning of 1-aminocyclopropane-1-carboxylate deaminase gene from soil microorganism. Thammasat Int J Sci Technol 2:56–60

Van Loon LC, Glick BR (2004) Increased plant fitness by rhizobacteria. In: Sandermann H (ed) Molecular ecotoxicology of plants. Springer, Berlin, pp 177–205

Van Loon LC, Geraats BPJ, Linthorst HJM (2006) Ethylene as a modulator of disease resistance in plants. Trends Plant Sci 11:184–191

Vessey JK (2003) Plant growth promoting rhizobacteria as biofertilizers. Plant Soil 255:571–586

Vikas S (2010) To study the prevalence of *acdS* gene in rhizobia of tropical legumes. M.Sc. thesis, PG School I.A.R.I, New Delhi-12, India

Walsh C, Pascal RA Jr, Johnston M, Raines R, Dikshit D, Krantz A, Honma M (1981) Mechanistic studies on the pyridoxal phosphate enzyme 1-aminocyclopropane-1-carboxylate deaminase from *Pseudomonas* sp. Biochemistry 20:7509–7519

Wang C, Knill E, Glick BR, Défago G (2000) Effect of transferring 1-aminocyclopropane-1-carboxylic acid (ACC) deaminase genes into *Pseudomonas fluorescens* strain CHA0 and its *gacA* derivative CHA96 on their growth-promoting and disease-suppressive capacities. Can J Microbiol 46:898–907

Yang SF, Hoffman NE (1984) Ethylene biosynthesis and its regulation in higher plants. Annu Rev Plant Physiol 35:155–189

Young S, Pharis RP, Reid D, Reddy MS, Lifshitz R, Brown G (1997) PGPR: is there a relationship between plant growth regulators and the stimulation of plant growth or biological control activity? In: Keel C, Koller B, Defago G (eds) Second international workshop on plant growth promoting rhizobacteria. Bull Srop Interlaken, Switzerland, pp 102–103

Zahir ZA, Arshad M, Frankenberger WT (2004) Plant growth promoting rhizobacteria: applications and perspectives in agriculture. Adv Agron 81:97–168

Zahir AZ, Usman G, Muhammad N, Sajid MN, Hafiz NA (2009) Comparative effectiveness of *Pseudomonas* and *Serratia* sp. containing ACC deaminase for improving growth and yield of wheat (*Triticum aestivum* L.) under salt-stressed conditions. Arch Microbiol 191:415–424

Zhao Z, Chen H, Li K, Du W, He S, Liu HW (2003) Reaction of 1-amino-2-methylenecyclopropane-1-carboxylate with 1-aminocyclopropane-1-carboxylate deaminase: analysis and mechanistic implications. Biochemistry 42:2089–2103

# Chapter 9
# Quorum-Sensing Signals as Mediators of PGPRs' Beneficial Traits

**Leonid S. Chernin**

## 9.1 Introduction

Expression of many phenotypic characteristics in bacteria is a cell-density-dependent phenomenon mediated by cell-to-cell communication in a process known as quorum sensing (QS). In both Gram-negative and Gram-positive bacteria, including beneficial and pathogenic plant-associated species and strains (von Bodman et al. 2003; Pierson and Pierson 2007; Boyer and Wisniewski-Dye 2009; Faure et al. 2009), QS involves the production and detection of extracellular signaling molecules called autoinducers. Several chemically distinct families of QS signal molecules employed by bacteria for intercellular communication including *N*-acyl-homoserine lactones (AHLs) among proteobacteria (Fuqua and Greenberg 2002; Waters and Bassler 2005) and oligopeptides among Gram-positive microbes (Lyon and Novick 2004). In Gram-negative bacteria, the main QS system, known as LuxIR, operates to control the response, mainly via production of AHL signal molecules, which differ in the structure of their fatty *N*-acyl side chains. These signals are synthesized by members of the LuxI family of proteins, identified first in a marine bacterium *Vibrio fischeri*, where AHL signals are responsible for QS control of bioluminescence (Fuqua and Greenberg 2002; Waters and Bassler 2005). AHL signaling system has been extensively studied over the last 30 years and has been identified in more than 70 Gram-negative genera belonging to the Proteobacteria (Ng and Bassler 2009).

LuxIR systems consist of the genes involved in the production and sensing of the AHL signal, *luxI* and *luxR* respectively, and the genes that are controlled by the AHL system that are responsible for the observed QS-controlled phenotype

L.S. Chernin (✉)
Department of Plant Pathology and Microbiology and The Otto Warburg Center for Biotechnology in Agriculture, The Robert H. Smith Faculty of Agriculture, Food and Environment, The Hebrew University of Jerusalem, Rehovot 76100, Israel
e-mail: chernin@agri.huji.ac.il

D.K. Maheshwari (ed.), *Bacteria in Agrobiology: Plant Nutrient Management*,
DOI 10.1007/978-3-642-21061-7_9, 

(e.g., production of antibiotics or and virulence factors). At low cell density or under conditions of limited diffusion, the *luxI* gene is expressed at a low rate, which leads to a basal level of AHL synthesis by LuxI. As the cell density increases, AHLs accumulate in the growth medium. On reaching a critical threshold concentration, the AHL molecule binds to its cognate receptor, a LuxR protein, which in turn activates or represses the expression of target genes. In addition to the genes responsible for luminescence and virulence factor production, the LuxR–AHL complex also binds to the *luxI* promoter and induces the expression of the AHL synthase so that the concentration of the AHL is amplified via an autoinduction loop. Thus, the enhanced signal production can lead to a massive increase in the local concentration of the AHL, which leads to induction of gene expression within the bacterial population. This induction is reflected in a coordinated expression of the phenotypic genes. LuxI catalyzes the synthesis of AHL by combining the cellular precursors, a fatty acyl-acyl carrier protein (acyl-ACP) and *S*-adenosyl-methionine (SAM), via an amide bond to form highly conserved AHL signals, in which SAM donates the homoserine lactone moiety and the acyl-ACP provides the fatty acyl side chain. Differences in AHLs depend on which fatty acyl side chains are available and recognized as appropriate substrates by corresponding AHL synthases [reviewed by Whitehead et al. (2001), Withers et al. (2001), Fuqua and Greenberg (2002), Waters and Bassler (2005), and Ng and Bassler (2009)].

The basic structure of the AHL signal molecule family consists of a homoserine lactone moiety adjoined with a *N*-acyl chain, ranging in length from 4 to 14 carbons and which may be saturated or unsaturated and may or may not contain a hydroxy- or oxo-group at the 3-carbon position (Table 9.1). The length and nature of the substitution at C3 of the acyl side chain confer signal specificity. Many Gram-negative bacteria employing QS systems produce multiple AHL molecules belonging to the same and/or different chemical classes. These additional signals may be due to the presence of multiple QS systems or they may be the products of a single AHL synthase (Holden et al. 1999) or different LuxI homologous proteins might catalyze the synthesis of a range of specific AHL signals. Therefore, some bacterial strains have more than one LuxI synthase and others produce more than one AHL signal molecules by the same LuxI. For example, QS regulates many virulence and colonization traits in *Pseudomonas aeruginosa* an opportunistic human pathogen known also as plant-associated bacterium, which is able to induce plant resistance to various pathogens (van Loon et al. 1998), and its regulation is highly complex and hierarchical. Production of AHL signaling molecules in *P. aeruginosa* strain PAO1 requires two AHL synthases, LasI and RhlI, both of LuxI-type, and each of which directs, respectively, two autoinducers, PAI-1 (*N*-(3-oxododecanoyl)-L-homoserine lactone; 3-oxo-C12-AHL, OdDHL) and PAI-2 (*N*-butyryl-L-homoserine lactone; C4, BHL) (Table 9.1), interacting with their cognate transcriptional activator proteins LasR and RhlR, respectively, and forming a hierarchical cascade for the regulation of multiple structural and regulatory genes (Withers et al. 2001).

Although AHL-mediated QS has been the most intensively investigated bacterial intercellular signaling mechanism, not all bacteria communicate using AHLs and Gram-negative bacteria also utilize QS signal molecules unrelated to the AHLs.

**Table 9.1** Examples of AHLs and related QS signal molecules

| Type of molecule/ abbreviations | Structure |
|---|---|
| *N*-butanoyl (butyryl)-AHL/C4, BHL | 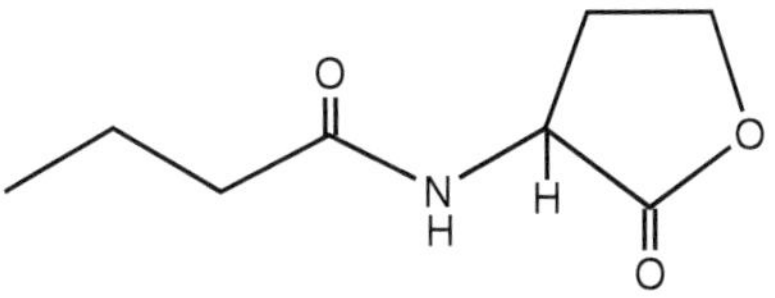 |
| *N*-hexanoyl-AHL/C6, HHL | 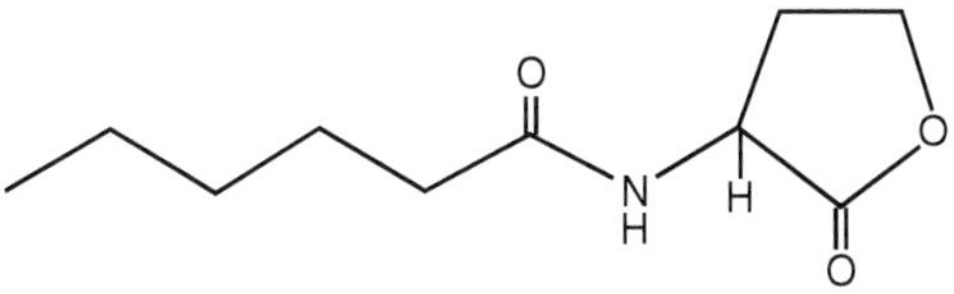 |
| *N*-octanoyl-AHL/C8, OHL | 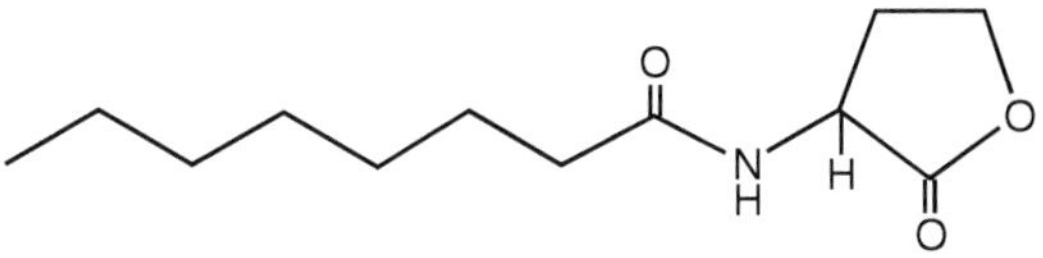 |
| *N*-decanoyl-AHL/C10, DHL | 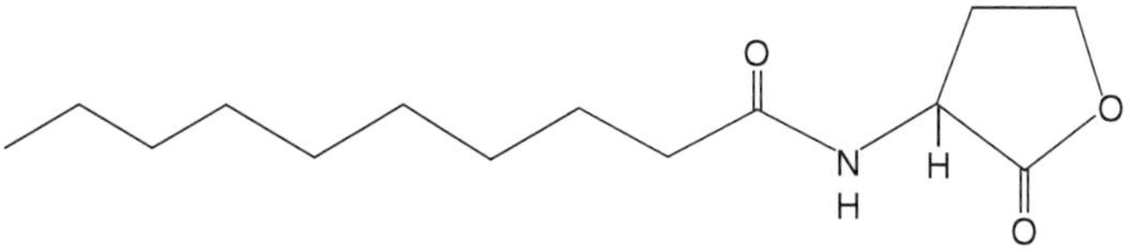 |
| *N*-dodecanoyl-AHL/C12, dDHL | 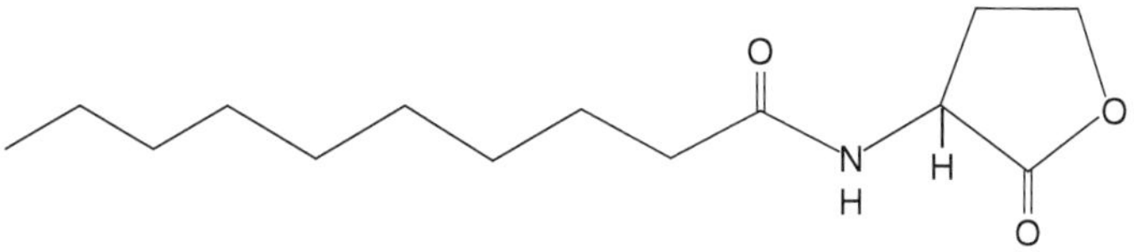 |
| *N*-tetradecanoyl-HSL/C14, TDHL | 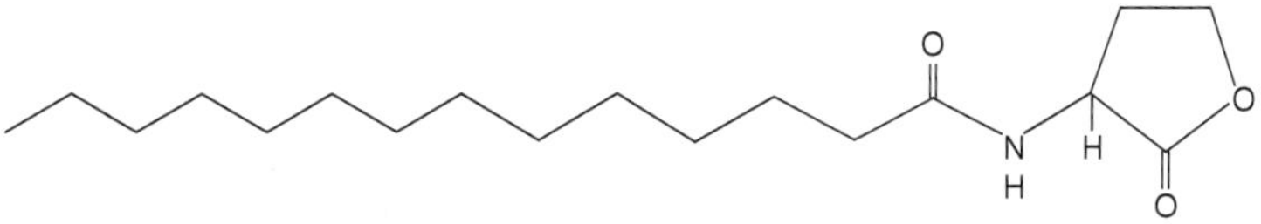 |
| *N*-(3-oxo-hexanoyl)-AHL/OC6, OHHL | 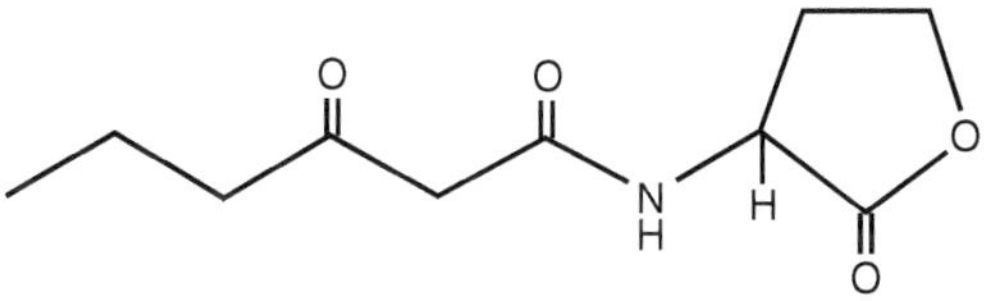 |
| *N*-(3-oxo-octanoyl)-AHL/OC8, OOHL | 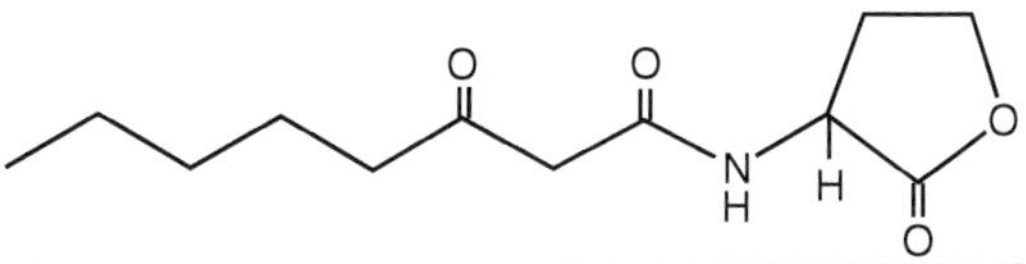 |

(continued)

**Table 9.1** (continued)

| Type of molecule/ abbreviations | Structure |
|---|---|
| *N*-(3-hydroxy-hexanoyl)-AHL/ HOC6, HHHL | 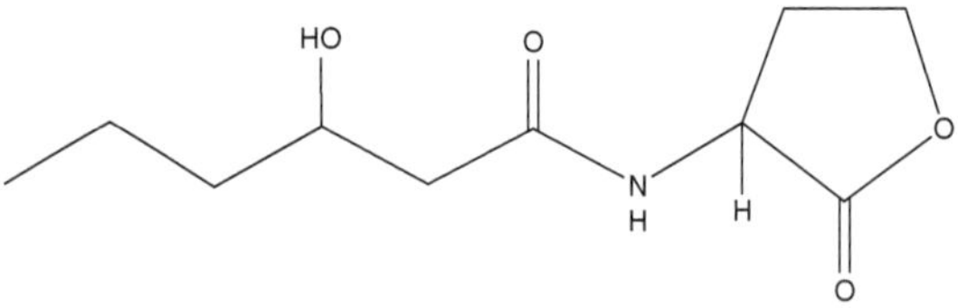 |
| *N*-(3-hydroxy-octanoyl)-AHL/ HOC8, HOHL | 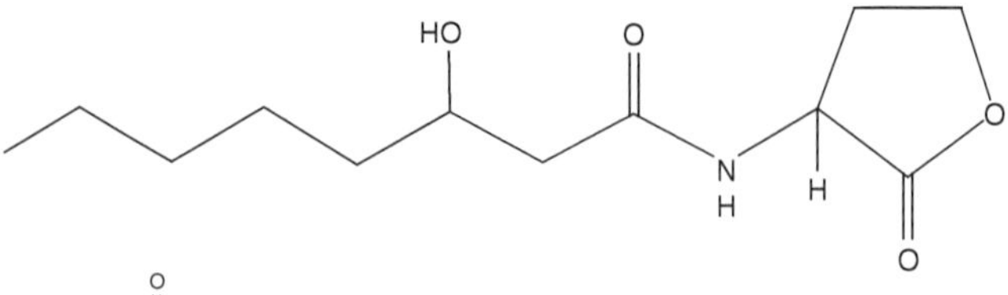 |
| 2-heptyl-3-hydroxy-4-quinoline (PQS) | 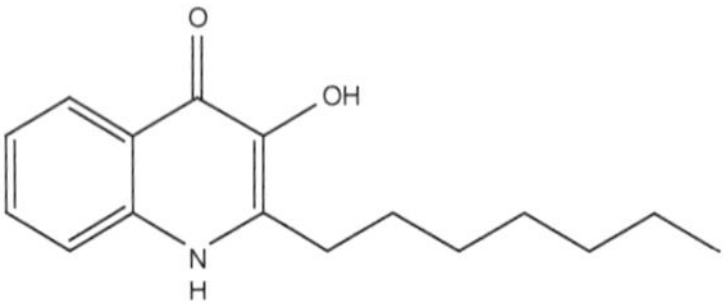 |
| Cyclo(Δala-L-Val) | 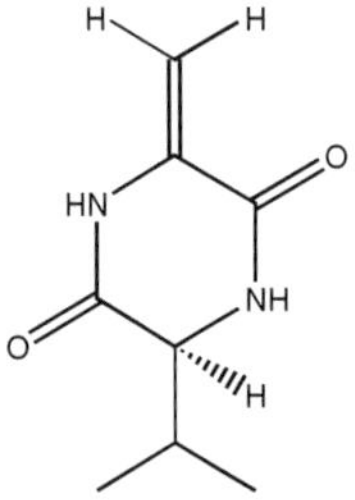 |

For example, a new signal molecule produced by *P. aeruginosa* PAO1, termed the *Pseudomonas* quinolone signal (PQS), whose chemical structure is 2-heptyl-3-hydroxy-4-quinolone (Table 9.1), was identified as a further component of the QS cascade hierarchy. This nonacyl-HSL cell-to-cell signal controlled the expression of *lasB*, which encodes for the major virulence factor, LasB elastase. The synthesis and bioactivity of PQS are mediated by the *P. aeruginosa lasR* and *rhlR* QS systems, respectively. However, it is not clear whether PQS signaling is unique to *P. aeruginosa* or ubiquitous throughout the bacterial world (Pesci et al. 1999). In addition to AHLs and PQS, *P. aeruginosa* has been shown to produce another family of putative QS signal molecules, the diketopiperazines (DKPs) cyclo (Δ Ala-L-Val), and cyclo(L-Pro-L-Tyr) (Table 9.1). These cyclic dipeptides were identified as a consequence of their ability to activate biosensors previously considered specific for acyl-HSLs. Production of DKPs is not limited to *P. aeruginosa*, as other Gram-negative bacteria produce the same or related molecules. Thus, a third DKP, which was chemically characterized as cyclo (L-Phe-L-Pro), was isolated from strains of *P. fluorescens* and *P. alcaligenes* (Holden et al. 1999). Although the DKPs are structurally quite distinct, at high concentrations they are able to crossactivate acyl-HSL-dependent reporter constructs based on several different LuxR homologues. The physiological role of these DKPs has yet to be established; however, these molecules ability to mimick the action of acyl-HSLs, presumably by interacting with LuxR, has been suggested (Holden et al. 2000). Both PQS and DKPs were discovered because of their ability to activate traditional AHL assays. Boron-containing AI-2 (luxS) signal also unrelated to the AHLs widespreads in Gram-negative and Gram-positive bacteria carrying LuxS-type synthase responsible for the production of a AI-2 and first identified in *V. harveyi* (Ng and Bassler 2009 for recent review). AI-2 is a QS signal that represents the only QS system shared by Gram-positive and Gram-negative bacteria and considered a unique, "universal" signal that may be used by a variety of bacteria for communication among and between species (De Keersmaecker et al. 2006).

Functional analyses of differentially expressed proteins can show their direct or indirect involvement in plant-growth promotion. Various molecular biological techniques are now available, including microarrays and proteomics, which have improved our knowledge of the QS-encoding and controlling gene(s) and pathways induced during the host-plant-growth-promoting rhizobacteria (PGPR) interaction. In addition, these approaches have enabled the identification and characterization of various global signal-transduction pathways and posttranslational modifications of proteins involved in the expression of PGPR traits that are beneficial to sustainable agriculture (Nannipieri et al. 2008; Gross and Loper 2009).

The experimental detection of QS signals is important for a better understanding of many genetic and biochemical traits in bacteria and at least some eukaryotic organisms, including plants. In principle, a reporter gene fused to any QS target gene can detect the presence of exogenous signal molecules. However, these molecules are produced at very low concentrations, hence, cannot usually be detected by conventional methods. In Gram-negative bacteria, the detection of autoinducers (mainly AHLs and, more recently, DKPs, PQS and AI-2-type molecules) has been greatly facilitated by the development of sensitive bioassays that allow fast screening of microorganisms for diffusible signal molecules. Most of the reporters are mutants that cannot synthesize their own QS signal, so the wild-type phenotype is only expressed upon addition of an exogenous one. Some microorganisms may produce signals that are not detected by one of the reporters or they may produce molecules at levels below the threshold of sensitivity of the reporter. Thus, the utilization of several bioreporters with different sensitivities and specificities is imperative. Several assays to monitor and quantify autoinducers produced by different Gram-negative bacteria have been developed. The most widely used have generally been dependent upon the use of *lacZ* reporter fusions in an *E. coli* or *Agrobacterium tumefaciens* genetic background providing production of β-galactosidase, on the induction or inhibition of the purple pigment violacein in *Chromobacterium violaceum* or on *lux*-based reporter fusions providing bioluminescence. For detailed protocols, see reviews by Camara et al. (1998), Ravn et al. (2001), Rice et al. (2004), and McLean et al. (2008).

In addition to bioluminescence, violacein production, or β-galactosidase activity determination, other markers have been used to design AHL bioreporter strains. For example, a bioreporter plasmid employing a red pigment prodigiosin-based bioassay, was described for the determination of OHHL (Thomson et al. 2000), whereas the production of phenazine antibiotics was used as a marker to detect HHL in *Pseudomonas aureofaciens* strain 30–84 (Wood and Pierson 1996). AHL-producing microorganisms can be detected by the use of the green fluorescent protein (GFP)-based reporter strains. The AHL detection capability of these strains is due to the presence of plasmids, which carry the *V. fischeri* region encoding *luxR* and the *PluxI* promoter fused to the gene encoding GFP. The GFP-based monitor strain JB357-*gfp* is extremely sensitive to OHHL, whereas tenfold less sensitive to HHL, OHL, and OdDHL, and at least 100-fold less sensitive to BHL (Anderson et al. 2001). Some other methods for screening of QS signals and quorum sensing inhibitors were developed (Rasmussen et al. 2005). A simple β-galactosidase bioassay for detection of a wide range of AHL signals through induction of a lasB′–lacZ transcriptional fusion was described by Ling et al. (2009). Monitoring β-galactosidase activity from this bioassay showed that *P. aeruginosa* strain carrying reporter plasmid pEAL01 could detect the presence of eight AHLs tested at physiological concentrations.

## 9.2 Quorum Sensing in PGPRs

Plant-associated bacteria able to stimulate plant growth and/or to suppress pathogenic organisms are generally designated as PGPR. Mechanisms providing direct plant growth promotion, such as biofertilization, production of phytohormones (e.g., auxins) or enzymes (e.g., ACC-deaminase) that stimulate of root growth, rhizosphere competence, competition for colonization sites, enhancement of nutrient and minerals availability, plant stress control can have beneficial effects on crops productivity (Glick et al. 2007; Lugtenberg and Kamilova 2009; Spaepen et al. 2009). On the other hand, production of antibiotics, siderophores, cell wall lytic enzymes, and induction of systemic resistance are considered to be important prerequisites for the optimal performance of PGPR for biocontrol of plant pathogens (Whipps 2001; Chernin and Chet 2002; van Loon 2007; Weller et al. 2007; Berg 2009; De Vleesschauwer and Hofte 2009). Rhizosphere bacteria multiply to high densities on plant root surfaces where root exudates and root cell lysates provide sufficient nutrients. As a consequence, the number of bacteria surrounding plant roots is 10–100 times higher than in bulk soil (Campbell and Greaves 1990; Spaepen et al. 2009). Such relatively high concentration of rhizosphere bacteria provides a basis for the cells density-dependent control of many beneficial traits. QS might have a general role in root colonization, as they find a higher proportion of AHL-producing bacteria in the rhizosphere than in bulk soil (Cha et al. 1998; Elasri et al. 2001; Veselova et al. 2003; Deangelis et al. 2007). Besides the signals concentration necessary for the activation or repression of QS-controlled genes, the local environment and spatial distribution of cells are also important contributing factors, indicating that environmental factors might overrule density dependence in final QS response (Hense et al. 2007; Duan and Surette 2007).

AHL molecules are also perceived by plants, which in turn often specifically respond to these bacterial signals. Bacterial infection of plants often depends on the exchange of QS signals between nearby bacterial cells. Thus, multiple QS systems in plant-growth promoting rhizobia also show influence of the ability of bacteria to infect root hairs and/or form nodules (Daniels et al. 2002; Zheng et al. 2006; Sanchez-Contreras et al. 2007; Edwards et al. 2009; Gurich and Gonzalez 2009). *Rhizobium leguminosarum bv. viciae*, a soil bacterium that can nodulate several legumes possesses four QS systems, with each synthase being able to make a specific set of AHLs. The *cinR/cinI*/3OH, C14:1-AHL system, at the top of a regulatory cascade, regulates all other QS loci and appears to act as an overall switch potentially influencing many aspects of rhizobial physiology, such as survival, transfer of symbiotic plasmid, and association with specific legumes (Boyer and Wisniewski-Dye 2009).

Plants, in turn, can perceive and often specifically respond to AHL molecules. Several legumes have been shown to secrete compounds that can interfere with bacterial QS. AHL-mimicking compounds and AHL-inhibitory activities were reported in exudates from pea roots (Teplitski et al. 2000). Altering root exudation from pea seedlings was shown to interfere with QS regulation in the bacteria or the expression of plant defense genes in tomato (Schuhegger et al. 2006). *Medicago truncatula* responded differentially with regards to root exudation and protein expression to AHLs produced by its symbiont *Sinorhizobium meliloti* and an opportunistic pathogen *P. aeruginosa*. Treatment of *M. truncatula* roots with AHLs led to significant changes in the accumulation of over 150 proteins, depending on AHL structure, concentration, and time of exposure (Gao et al. 2003; Mathesius et al. 2003).

AHLs among other small diffusible molecules produced by bacteria are widespread components of the mixed microbial communities inhabiting phyllosphere environments. The phyllosphere harbors many diverse types of bacteria that utilize AHL-mediated QS, and AHLs are believed to be common components of the leaf surface environment (Karamanoli and Lindow 2006; Pierson and Pierson 2007). Also transgenic *Nicotiana tabacum* plants constitutively expressing bacterial AHL synthases possessed an abundant amount of short- (3OC6, OHHL) and long-chain (3OC12, OdDHL), which have been recovered from leaf and root surfaces, root exudates and rhizosphere soil (Scott et al. 2006), while exogenous AHLs applied to plants are stable enough to influence plant gene expression (Mathesius et al. 2003; Bauer and Mathesius 2004; Schuhegger et al. 2006).

QS regulation of PGPR's beneficial traits has been most comprehensively studied in plant-associated pseudomonads that are known to utilize one or more

QS systems to regulate multiple traits, some of which affect their persistence and viability on plant surfaces (Faure et al. 2009; Badri et al. 2009). Steidle et al. (2001, 2002) have developed and characterized novel GFP-based monitor strains of *P. putida* IsoF and *Serratia liquefaciens* MG1, capable of colonizing tomato roots that allow in situ visualization of AHL-mediated communication between individual cells in the tomato plant rhizosphere. Additionally, an AHL sensor cassette that responds to the presence of long-chain AHLs with the expression of GFP was integrated into the chromosome of AHL-negative *P. putida* strain F117. This monitor strain was used to demonstrate that the indigenous bacterial community colonizing the roots of tomato plants growing in nonsterile soil produces AHL molecules. The results strongly support the view that AHL signal molecules serve as a universal language for communication between the different bacterial populations of the rhizosphere consortium. AHLs are produced by the bacterial consortium that naturally colonies the roots of plants and probably coordinates the functions of the different populations within the rhizosphere community (Steidle et al. 2002). Other studies on the rice rhizosphere isolate *P. aeruginosa* PUPa3P establishing beneficial associations with plants revealed that this bacterium AHL systems known as RhlI/R and LasI/R (see Sect. 9.1) are involved in the regulation of plant growth-promoting traits being important for the colonization of the rice rhizosphere (Steindler et al. 2009). At the same time, a systematic study of AHL QS on a set of 50 rice rhizosphere *Pseudomonas* isolates implicated a lack of conservation and an unpredictable role played by AHL QS in this group of bacteria (Steindler et al. 2008).

## 9.2.1 Quorum Sensing Regulation of Antibiotic Production

Antibiotics are major determinants of antagonism plant pathogens by various PGPRs (Raaijmakers et al. 2002; Weller et al. 2007; Gross and Loper 2009). Many strains produce one or more potent metabolites with antifungal activity. Antibiotics have been best studied in antagonistic Gram-negative biocontrol bacteria include phenazines (PHZ) (Pierson et al. 1998; Chin-A-Woeng et al. 2003; Mavrodi et al. 2006, 2010); 2,4-diacetylphloroglucinol (DAPG) (Weller et al. 2007), pyoluteorin (Nowak-Thompson et al. 1999), 3-chloro-4-(2′-nitro-3′-chlorophenyl)-pyrrole (pyrrolnitrin, PRN) (Chernin et al. 1996; Kalbe et al. 1996; Hammer et al. 1997; Costa et al. 2009), and the volatile compound HCN (Haas and Defago 2005). Recently it was shown that a number of volatile organic compounds (VOCs) produced by rhizobacteria are involved in the interaction with pathogenic fungi as well as with host plants (Vespermann et al. 2007; Kai et al. 2009; Wenke et al. 2010). However, only a few examples of the role of QS systems in the regulation of these secondary metabolites' production have been described (Table 9.2).

Table 9.2 Examples of PGPRs' traits controlling by QS signal molecules

| Bacteria/autoinducer/type of AHL synthase | Trait |
|---|---|
| *P. aeruginosa* PUPa3P/OC12, OdDHL/LasI/C4, BHL/RhlI | Colonization of the rice rhizosphere (Steindler et al. 2009) |
| *P. chlororaphis* (*aureofaciens*) 30-84/HC6, HHHL/HC8, HOHL/HC10, HDHL/C6, HHL/C8, OHL/HOC7, HHpHL/PhzI;C4, BHL/C6, HHL/CsaI[a] | Production of phenazines (PCA, 2-OH-PCA), biofilm formation (Wood and Pierson 1996; Maddula et al. 2006; Khan et al. 2007) |
| *P. chlororaphis* 449/C4, BHL/C6, HHL/OC6, OHHL/HC6, HHHL/HC8, HOHL/PhzI/CsaI | Production of phenazines (PCA, 2-OH-PCA) (Veselova et al. 2008; Chernin et al. unpublished) |
| *P. chlororaphis* PCL1391/C4, BHL/C6, HHL/C8, OHL/OC6, OHHL/HC6, HHHL/HC8, HOHL/HC10, HDHL/C6, HHL/C8, OHL/HOC7, HHpHL/PhzI[a] | Production of phenazines (PCN) (Chin-A-Woeng et al. 2001; Khan et al. 2007) |
| *P. fluorescens* 2-79/HC6, HHHL/HC8, HOHL/HC10, HDHL/C6, HHL/C8, OHL/HOC7, HHpHL/PhzI | Production of phenazines (PCA, 2-OH-PCA) (Khan et al. 2005, 2007) |
| *P. putida* IsoF/OC6, OHHL/OC8, OOHL/OC10, ODHL/OC12, OdDHL/PpuI | Biofilm formation (Steidle et al. 2001, 2002) |
| *Pseudomonas* sp. M18/*VqsR*/C4, BHL/C8, OHL/RhlI | Production of pyoluteorin/phenazines (PCA) (Huang et al. 2008) |
| *Rhizobium etli* CNPAF512/short-chain HSLs/3-hydroxy long chain-AHLs/RaiI/RaiR/CinI/ | Nitrogen fixation, nodulation, growth inhibition (Daniels et al. 2002) |
| *R. leguminosarum bv. viciae* 8401pRL1JI//HHL, OHL/RhiI | Rhizosphere genes, nodulation efficiency, plasmid transfer (Edwards et al. 2009) |
| *S. liquefaciens* MG1/C4, BHL/C6, HHL/SwrI | Swarming motility, production of proteases and chitinases, biofilm formation, induced systemic resistance (Eberl et al. 1996, 1999; Labbate et al. 2004; Schuhegger et al. 2006) |
| *S. plymuthica* HRO-C48/C4, BHL/C6, HHL/OC6, OHHL/SplI | Production of pyrrolnitrin, chitinases, protease, indolyl-3-acetic acid, VOCs (Liu et al. 2007; Műller et al. 2009) |

[a]See Khan et al. (2007) for updated AHLs pattern of strains *P. chlororaphis* 30-84 and PCL1391

#### 9.2.1.1 Quorum Sensing Regulation of Phenazines Production

In biocontrol bacteria, most known cases involve AHL control of phenazine antibiotics production by rhizospheric pseudomonads. PHZs, of which the major ones are phenazine-1-carboxylic acid (PCA), 2-hydroxy-PCA (2-OH-PCA), and phenazine-1-carboxamide (PCN), are water-soluble, nitrogen-containing aromatic pigments, which have antibiotic activity due to their ability to interfere with respiratory chain function. PHZ are among major determinants of biological control of soilborne plant pathogens by various strains of fluorescent *Pseudomonas* spp. PHZs are responsible for the suppression of several plant pathogenic fungi (Pierson et al. 1998; Weller et al. 2002; Chin-A-Woeng et al. 2003) and contributes to the survival of PHZ-producing strains in soil habitats. The "core" biosynthetic pathway includes seven enzymes for the synthesis of PCA, which is then modified by a variety of other enzymes to yield the diversity of PHZ produced by pseudomonads (Mavrodi et al. 2006, 2010). Production of PHZs in rhizosphere bacteria appears to be controlled by complex sensory networks that may include one or more QS systems and other, regulatory mechanisms (Walsh et al. 2001; Raaijmakers et al. 2002; Chin-A-Woeng et al. 2003; Lapouge et al. 2008).

For example, *P. chlororaphis* strain 30-84 is a root-associated, plant-beneficial bacterium that is able to control take-all disease of wheat caused by the fungal pathogen *Gaeumannomyces graminis* var. *tritici* (Pierson et al. 1998). Take-all, caused by this ascomycetous fungus is an important root and crown rot disease of wheat worldwide. Strain produces mainly two PHZs, PCA and 2-OH-PCA. PHZ production is the primary mechanism of pathogen inhibition and contributes to persistence of strain 30-84 in the rhizosphere (Pierson et al. 1998). Loss of 2-OH-PCA resulted in a significant reduction in the inhibition of the fungus (Maddula et al. 2008). Production of PHZs is regulated by the PhzI/PhzR AHL response system. PhzI is a member of the LuxI family and is responsible for the production of the specific AHL signal molecule *N*-hexanoyl-homoserine lactone (C6, HHL), whereas PhzR is its cognate receptor. Inactivation of PhzI or PhzR resulted in the complete loss of phenazine production, whereas introduction of PhzI in trans into PhzI null mutants restored phenazine production (Wood and Pierson 1996). The second QS system operating in strain 30-84, CsaR/CsaI was supposed to influence cell-surface properties. In combination, these systems regulate the features that contribute to the bacterium's persistence on plants and to their effectiveness as biological control agents (Zhang and Pierson 2001). The same AHLs pattern was detected in *P. chlororaphis* strain 449, isolated from rhizosphere in Russia and able to suppress a number of plant pathogenic fungi due to the production of Phz. Like its "American analogue," strain 449 produces PCA and 2-OH-PCA and synthesizes at least three types of AHL: *N*-butanoyl-L-homoserine lactone (C4, BHL), *N*-hexanoyl-L-homoserine lactone (C6, HHL), and *N*-(3-oxo-hexanoyl)-L-homoserine lactone (OC6, OHHL) (Veselova et al. 2008) demonstrating a high conservation of both antibiotic and QS traits operating in these strains of very different origin (Mavrodi et al. 2010).

Another example of AHLs' control of Phz production is the tomato rhizosphere isolate *P. chlororaphis* PCL1391 that exhibits biocontrol activity against *Fusarium oxysporum* f. sp. *radicis-lycopersici*, the causal agent of tomato foot and root rot. The production of PCN, including the last step, conversion of PCA to PCN, was shown to be essential for the biocontrol activity of strain PCL1391 (Chin-A-Woeng et al. 2003). In vitro production of PCN is observed only at high-population densities, suggesting that production is under the regulation of quorum sensing. The main autoinducer molecule produced by PC L1391 was identified structurally as *N*-hexanoyl-L-homoserine lactone (C6, HHL). The two other autoinducers that were produced comigrate with *N*-butanoyl-L-homoserine lactone (C4, BHL) and *N*-octanoyl-L-homoserine lactone (C8, OHL). Two PC L1391 mutants lacking production of PCN were defective in the genes *phzI* and *phz*R, respectively, the nucleotide sequences of which were determined completely. Production of PCN by the *phzI* mutant could be complemented by the addition of exogenous synthetic HHL, but not by BHL, OHL, or any other AHL tested. Expression analyses of Tn*5luxAB* reporter strains of *phz*I, *phz*R, and the *phz* biosynthetic operon clearly showed that *phzI* expression and PCN production is regulated by HHL in a population density-dependent manner (Chin-A-Woeng et al. 2001).

Strain *P. fluorescens* 2-79 was isolated from wheat roots grown in a take-all decline soil, one that develops natural disease suppressiveness after long-term wheat monoculture. It aggressively colonizes wheat roots and suppresses take-all disease in the field. Production of PCA in the rhizosphere is a primary mechanism of biocontrol by strain 2-79 accounting for up to 90% of disease suppression (Weller et al. 2002). The production of the antibiotic by strain 2-79 is under the regulation of QS networks. The strain *P. fluorescens* 2-79 produces at least six AHL signals, including *N*-3-hydroxy-hexanoyl-L-homoserine lactone (HHHL, HC6); *N*-3-hydroxy-octanoyl-L-homoserine lactone (HOHL, HC8); *N*-3-hydroxy-decanoy)-L-homoserine lactone (HDHL, HC10); hexanoyl-homoserine lactone (HHL, C6); octanoyl homoserine lactone (OHL, C8); and 3-OH-heptanoyl (C7)-HSL (HHpHL, OC7) (Khan et al. 2005) with PhzI directing the synthesis of all the six AHL signals. Of these, *N*-(3-hydroxyhexanoyl)-AHL is the biologically relevant ligand for PhzR (Khan et al. 2007). This is in contrast to the quorum-sensing systems of *P. chlororaphis* strains 30-84 and PCL1391 have been reported to produce and respond to C6 (HHL) signal molecule. These differences were of interest since PhzI and PhzR of strain 2-79 share almost 90% sequence identity with orthologs from strains 30-84 and PCL1391. Recently the major species produced by *P. chlororaphis* 30-84 were identified by mass spectrometry as 3-OH-AHLs with chain lengths of 6, 8, and 10 carbons (Khan et al. 2007). Heterologous bacteria expressing cloned *phzI* from strain 30-84 produced the four 3-oxy-AHLs in amounts similar to those seen for the wild type. Strain 30-84, but not strain 2-79, also produced *N*-(butanoyl)-AHL (C4, BHL). A second AHL synthase of strain 30-84, CsaI, is responsible for the synthesis of this short-chain signal. Strain 30-84 accumulated *N*-(3-OH-hexanoyl)-AHL to the highest levels, more than 100-fold greater than that of *N*-(hexanoyl)-HSL. In titration assays, PhzR in strain 30-84 responded to both *N*-(3-OH-hexanoyl)- and *N*-(hexanoyl)-HSL with equal sensitivities. However, only the 3-OH-hexanoyl signal is produced by strain 30-84 at levels high enough to activate PhzR. The results suggest that strains 2-79, 30-84, and PCL1391 use *N*-(3-OH-hexanoyl)-AHL (OHHL) to activate PhzR (Khan et al. 2007) (Table 9.2).

#### 9.2.1.2 Quorum Sensing Regulation of Pyrrolnitrin and Pyoluteorin Production

PHZ are not the only group of biocontrol-related antibiotics whose production is regulated by QS systems. PRN, a tryptophan-derived secondary metabolite produced by a narrow range of Gram-negative bacteria. The PRN biosynthesis by rhizobacteria presumably has a key role in their life strategies and in the biocontrol of plant diseases. The biosynthetic operon that encodes the pathway that converts tryptophan to PRN is composed of four genes, *prnA* through *prnD* (Hammer et al. 1997). Recent studies suggest that the PRN biosynthetic operon is mobile (Costa et al. 2009), the fact which could explain high similarity between *prn* genes in various strains of PGPRs that belong to genera *Burkholderia*, *Pseudomonas*, and *Serratia*. Production of PRN was shown subjected to AHLs-mediated QS regulation in rhizosphere strains of *Serratia plymuthica* (Liu et al. 2007;Műller et al. 2009) and *Burkholderia* species (Schmidt et al. 2009).

In strains of genus *Serratia*, including *S. marcescens* and *S. plymuthica* several QS-regulated traits, such as production of nuclease, lipase, protease, and chitinase, swarming motility, biofilm formation, the *lipB*-secretion system, production of the red pigment prodigiosin, the β-lactam antibiotic 1-carbapen-2-em-3-carboxylic acid, and VOC 2,3-butanediol fermentation have been identified (Eberl et al. 1999; Thomson et al. 2000; Horng et al. 2002; Christensen et al. 2003). At least four different LuxRI/AHL QS systems have been described in *Serratia* (Wei and Lai 2006; van Houdt et al. 2007). Additionally, LuxS/autoinducer-2 (AI-2) has been reported in *S. marcescens* as a second QS system shown to control some phenotypes, including production of the antibiotic prodigiosin, jointly with the LuxIR systems (Coulthurst et al. 2004), whereas the LuxS system described in *S. plymuthica* strain RVH1 from a food-processing environment was shown to participate in the bacterial growth regulation (van Houdt et al. 2006). However, much less is known about QS systems operating in rhizospheric *Serratia* isolates with the potential to serve as biocontrol agents of plant pathogens.

*S. plymuthica* is a ubiquitous inhabitant of the rhizosphere of different plant species and includes strains antagonistic to soilborne pathogens (Berg 2000, 2009; De Vleesschauwer and Hofte 2007). These isolates produce a range of antifungal compounds, including chitinases and PRN (Chernin et al. 1995, 1996; Kalbe et al. 1996; Frankowski et al. 1998; Kamensky et al. 2003; Ovadis et al. 2004). *S. plymuthica* strain HRO-C48, isolated from the rhizosphere of oilseed rape and described as a chitinolytic bacterium with a wide range of activity against plant-pathogenic fungi (Kalbe et al. 1996; Frankowski et al. 1998, 2001), was shown rhizosphere competent and capable of suppressing pathogenic fungus *Verticillium dahliae* in strawberry, oilseed rape, and olive, and *Rhizoctonia solani* on lettuce (Kurze et al. 2001; Grosch et al. 2005; Műller and Berg 2008). The strain is the first representative of *S. plymuthica* commercially used as a biocontrol agent (Rhizo-Star®, E-nema GmbH Raisdorf, Germany). It provides a cell density-dependent effect on plant growth promotion (Kurze et al. 2001) and reduced biocontrol effects at low cell densities under field conditions, suggesting that QS may play a role in the regulation of biocontrol activity in situ.

Liu et al. (2007) for the first time demonstrated QS regulation of PRN biosynthesis in *S. plymuthica* strain HRO-C48, producing several AHLs, including C4 (BHL), C6 (HHL), and 3-OH-C6 (OHHL). The QS genes *splI* and *splR*, which are analogues of *luxI* and *luxR* genes from other Gram-negative bacteria, were cloned and sequenced. The insertion mutant AHL-4 (*splI*::miniTn5) was simultaneously deficient in the production of AHLs and Prn, as well as in its ability to suppress the growth of several fungal plant pathogens in vitro. However, Prn production could be restored in this mutant by introduction of the *splIR* genes cloned into a plasmid or by addition of the conditioned medium from strain C48 or OHHL standard to the growth medium (Liu et al. 2007). To investigate further the influence of AHL-mediated communication on the biocontrol activity of strain HRO-C48, the enzyme lactonase AiiA (Dong et al. 2001), which hydrolyses AHL molecules, was heterologously expressed in the strain, resulting in abolished AHL production (Műller et al. 2009). This approach is commonly referred to as quorum quenching (QQ) (see Sect. 9.3). The quenched strain as well as the *splI* mutant AHL-4 were phenotypically characterized and compared with the wild-type strain. Expression of the AiiA lactonase in strain HRO-C48 greatly reduced PRN production, which is in full agreement with a previous study that showed that the SplI/SplR QS system is controlling PRN production in this bacterium (Liu et al. 2007). Consequently, inactivation of the AHL signal molecules by AiiA lactonase completely abolished the ability of the strain to protect oilseed rape plants against *V. dahliae* in greenhouse experiments (Műller et al. 2009).

Members of the genus *Burkholderia* are also known for their ability to suppress soil-borne fungal pathogens by the production of various antibiotic compounds. Recently the role of AHL-dependent QS in the expression of antifungal traits in strains of several *Burkholderia* species was investigated (Schmidt et al. 2009). Using a similar QQ approach, that is, by heterologous expression of the *Bacillus* sp. AiiA lactonase, the expression of antifungal activities was shown as AHL dependent in the large majority of the investigated strains belonging to various *Burkholderia* species. In certain strains of *B. ambifaria*, *B. pyrrocinia*, and *B. lata*, one of the QS-regulated antifungal agents is PRN. To investigate the underlying molecular mechanisms of AHL-dependent PRN production in strain *B. lata* 383 in better detail, the genes *cepI* or *cepR* encoding the AHL synthase and the cognate AHL receptor protein were inactivated. Both QS mutants no longer produced PRN and as a consequence were unable to inhibit growth of *R. solani*. Expression of the *prnABCD* operon, which directs the synthesis of PRN, is positively regulated by CepR AHL receptor protein at the level of transcription.

Evidence for AHLs regulation of antibiotic pyoluteorin (Plt) production was recently obtained in studies of fluorescent *Pseudomonas* sp. M18 rhizobacterium strain that can suppress diseases caused by pathogenic fungi in crop plants. This biocontrol capacity largely depends on the production of its two kinds of antibiotics: PCA and Plt. Mutation in a *luxR*-type QS regulatory gene, *vqsR* located immediately downstream of the Plt gene cluster in strain M18 led to a significant decrease in the production of Plt and its biosynthetic gene expression. At the same time, the *vqsR* mutation did not exert any obvious influence on the production of PCA and C4 and C8 AHLs and their biosynthetic gene rhlI expression, suggesting that VqsR specifically regulates Plt production in this bacterium (Huang et al. 2008).

### 9.2.2 Quorum Sensing Regulation of ISR Provided by PGPRs

Colonization roots by selected strains of nonpathogenic bacteria triggered in plants mechanism of defense toward pathogens known as rhizobacteria-induced systemic resistance (ISR), in order to differentiate it from the pathogen-induced and salicylate-mediated systemic acquired resistance (SAR). Bacterial traits operative in triggering ISR have been identified, including flagella, cell envelope components such as lipopolysaccharides, and secreted metabolites like siderophores, cyclic lipopeptides, volatiles, antibiotics, phenolic compounds, and QS molecules. In addition, various hormone-dependent signaling pathways may manage the induced resistance phenotype depending on the rhizobacterium and the plant–pathogen system used (van Loon et al. 1998; van Loon 2007; De Vleesschauwer and Hofte 2009). Enhanced basal resistance of plants depends on the signaling compounds jasmonic acid (JA) and ethylene (ET) in case of ISR and salicylic acid (SA) in case of SAR, and pathogens are differentially sensitive to the resistances activated by each of these signaling pathways. In contrast to the suppression of soilborne pathogens by biocontrol bacteria based on competition for nutrients and production of antimicrobial compounds or lytic enzymes, systemic resistance can be induced when the inducing bacteria and the challenging pathogen remained spatially separated excluding direct interactions (van Loon et al. 1998). Many rhizobacteria of the genus *Pseudomonas* or *Serratia* produce SA under iron-limiting conditions as a precursor of siderophores [reviewed by De Vleesschauwer and Hofte (2009)].

The AHL-dependency was described in a well-characterized system of rhizobacteria-induced resistance between *Arabidopsis thaliana* and *P. fluorescens* strain WCS417r (van Loon et al. 1998). Strain *S. liquefaciens* MG1 was used as another model organism to evaluate the role of AHL signaling in ISR (Schuhegger et al. 2006). This strain produces two AHL molecules BHL and HHL encoded by AHL synthase SwrI (Eberl et al. 1996). QS in this strain regulates its swarming motility, a special form of bacterial surface locomotion, the production of extracellular proteolytic and chitinolytic activity, as well as the ability to form biofilms on abiotic surfaces (Eberl et al. 1999; Riedel et al. 2001; Labbate et al. 2004). Schuhegger et al. (2006) reported that AHL signals produced by the root colonizers *S. liquefaciens* MG1 in the rhizosphere increase systemic resistance of tomato plants against the fungal leaf pathogen *Alternaria alternata* and systemically induce salicylic acid (SA)- and ethylene (ET)-dependent defense genes. SA levels were increased in leaves when AHL-producing bacteria colonized the rhizosphere suggesting that SA, known as an important signal for SAR, might be involved in systemic resistance against *A. alternata* induced by strain MG1. No effects were observed when isogenic AHL-negative mutant derivatives were used in these experiments. The AHL-negative mutant was less effective in reducing symptoms and *A. alternata* growth as compared to the wild type. Furthermore, macroarray and Northern blot analysis revealed that AHL molecules systemically induce SA- and ethylene-dependent defense genes. In addition, expression of plant defense genes was analyzed after treatment with AHL molecules. AHL application to tomato roots led to an induction of SA in roots and, to a lower extent, in leaves suggesting that

AHL-induced SA production is an additional new route to systemic signal transduction in tomato. The results suggest that tomato plants are able to detect the presence of typical rhizobacteria using AHL molecules, and to respond by systemic induction of defense genes. Together, these data support the view that AHL molecules play a role in the biocontrol activity of rhizobacteria through the induction of systemic resistance to pathogens and suggest that AHL molecules play an important role in the biocontrol activity of *S. liquefaciens* and other rhizobacteria in tomato, and act as mediators of communication between prokaryotes and eukaryotes. Even though SA contents were upregulated systemically by inoculation with AHL-producing rhizobacteria, the pattern of AHL-induced defense genes was not confined to salicylate-dependent genes. In addition, the accumulation of specific transcripts depended on the AHL species. AHLs may therefore be considered as potential candidates for a new group of general elicitors for plant defense because they are produced in the tomato rhizosphere in effective amounts, induce SA systemically, lead to enhanced gene expression for typical defense-related proteins, and result in increased resistance against fungal pathogens.

The AHL-dependency of induced resistance between *A. thaliana* and *P. fluorescens* strain WCS417r (van Loon et al. 1998) and the effects of strain *S. liquefaciens* MG1 are quite different (Schuhegger et al. 2006). In contrast to MG1, strain WCS417r does not produce AHLs in detectable amounts. Secondly, ISR in *A. thaliana* is not accompanied by elevated levels of plant signaling molecules, whereas SA contents are systemically induced by *S. liquefaciens* MG1 and AHL molecules. Thirdly, root application of two AHL molecules led to marked systemic inductions of defense gene expression, especially of pathogenesis-related proteins, whereas gene expression of defense-related proteins is not markedly altered in response to ISR-inducing rhizobacteria (van Loon et al. 1998).

Transgenically modified tobacco and potato plants producing AHLs are able to decrease the pathogenicity of *Pectobacterium carotovora* used for plants challenging via interference with AHL-QS system regulating production of toxins and other pathogenic factors when the bacteria rich critical concentration on the plant tissues (Mäe et al. 2001; Toth et al. 2004). The ability to generate bacterial QS signaling molecules in the plant offers novel opportunities for disease control and for manipulating plant/microbe interactions. AHL QS signaling molecules produced by transgenic plants mediate the ability of some PGPR strains to promote plant growth and to induce protection against salt stress, but this effect is dependent on AHL profile of the specific strains as well as the knowledge of which particular beneficial trait was investigated (Barriuso et al. 2008).

### *9.2.3 Quorum Sensing Regulation of Biofilm Formation by PGPRs*

Plant-associated bacteria are common inhabitants of leaves, roots, and soil, and the majority of natural isolates attach to plant surfaces in the form of multicellular communities known as biofilms. The ability to produce biofilms and to colonize the rhizosphere are important factors for plant–microbe interaction. Cell–cell communication between bacteria in rhizosphere or phylosphere environment is supposed to be a prerequirement of biofilm development and the resulting interactions with plants (Morris and Monier 2003; Ramey et al. 2004; Danhorn and Fuqua 2007). Formation of bacterial biofilms and interactions between bacteria in plant environment are QS-dependent processes. Several studies provide example of AHL regulation of biofilm formation by PGPRs. QS-dependent biofilm formation has been shown for various plant-associated strains of the genera *Pseudomonas*, *Burkholderia*, and *Serratia* (Jayaraman and Wood 2008). Thus, the root-associated biocontrol agent *P. fluorescens* strain 2P24 requires AHLs for biofilm formation and therefore controls of take-all disease on wheat as well as on the colonization of the rhizosphere (Wei and Zhang 2006). In biocontrol bacterium *P. chlororaphis* 30-84 (see Sect. 9.2.1.1) QS can indirectly influence biofilm formation via regulation of phenazine antibiotics production. The production of PCA and 2-OH-PCA by strain 30-84 is not only the primary mechanism of pathogen inhibition but also contributes to the persistence of strain 30-84 in the rhizosphere. Both PhzR/PhzI QS mutants of strain 30-84 and a mutant defective only in PHZ biosynthesis was equally impaired in biofilm formation (Maddula et al. 2006). Derivatives of strain 30-84 that produced only PCA or overproduced 2-OH-PCA were found to differ from the wild type in initial attachment, mature biofilm architecture, and dispersal from biofilms (Maddula et al. 2008). Thus, increased 2-OH-PCA production promoted initial attachment and altered the three-dimensional structure of the mature biofilm relative to the wild type. Additionally, deficiency in PCA or 2-OH-PCA both promoted thicker biofilm development and lowered dispersal rates compared to the wild type. These results clearly demonstrated that the effect of AHL signaling on the formation of biofilms is realized via regulation of phenazine antibiotics playing an important role in this process. Inhibition of biofilm formation in pathogenic bacteria by AHL-degrading enzyme AHL acylase probably makes biocontrol easier (Shepherd and Lindow 2009) and might help to clarify the input of biofilms in bacteria maintenance and behavior in plant environment.

### 9.2.4 Quorum Sensing Regulation of Other Plant Growth Promotion and Biocontrol-Related Traits

As was described above (Sect. 9.2.1.2), AHLs signaling influences production of antibiotic PRN in the rhizosphere biocontrol strain *S. plymuthica* HRO-C48. Along with PRN, strain HRO-C48 emits a broad spectrum of VOCs that are involved in antifungal activity and whose relative abundances are influenced by QS control as well (Műller et al. 2009). VOCs emitted from the strain HRO-C48 negatively influence the mycelial growth of the soilborne phytopathogenic fungus *R. solani* (Kai et al. 2007) assuming that QS system may influence biocontrol activity of strain HRO-C48 via stimulation of antifungal volatiles production (Table 9.2). Further comparative analysis of the wild type and AHL-negative mutants led to the identification of more AHL-regulated phenotypes in this bacterium. Thus, indole-3-acetic acid (IAA) is a phytohormone, produced by plant-associated bacteria as well as by the plants themselves, which is responsible for the stimulation of plant growth (Patten and Glick 2008). Műller et al. (2009) for the first time demonstrated the regulatory function of AHLs in the synthesis of IAA in vitro. IAA was found upregulated in the AHL-deficient derivative of *S. plymuthica* HRO-C48 suggesting that at low population density the bacteria excrete higher amounts of IAA and thus stimulate plant growth, which in turn provides the bacterial population with additional nutrients to reach a critical size at which the QS system is triggered. At high population density, IAA production of the bacteria is reduced because high levels of IAA may cause damage and promote disease. At the same time, the increased concentrations of AHL signal molecules can stimulate IAA production in the plant (Hartmann personal communication cited by Műller et al. 2009). Moreover, AHL molecules were shown directly able to promote plant growth (Rothballer et al. 2008).

Chitinases, proteases, and other cell wall lytic enzymes are important antifungal factors (Chernin and Chet 2002). Positive control of chitinase production by the AHL signaling was described in *P. aeruginosa* (Winson et al. 1995), *C. violaceum* (Chernin et al. 1998), *Serratia proteamaculans* (Christensen et al. 2003) and in plant-beneficial strain *P. chlororaphis* PCL1391 (Chin-A-Woeng et al. 2003). Strain *S. plymuthica* HRO-C48 produces three chitinases and an unknown number of proteases (Frankowski et al. 2001). Both chitinolytic and proteolytic activity was found to be AHL regulated in HRO-C48 (Műller et al. 2009). This finding is in agreement with previous results obtained for another strain of *S. plymuthica* (van Houdt et al. 2007). Additionally, swimming motility was found to be negatively controlled by AHLs. In contrast, production of extracellular hydrolytic enzymes is shown to be positively AHL regulated (Műller et al. 2009).

Pang et al. (2009) found that AHL-mediated QS signaling is responsible for some other biocontrol-related traits of strain HRO-C48, such as protection of cucumbers against *Pythium apahnidermatum* damping-off disease, ISR to *Botrytis cinerea* gray mold in bean and tomato plants, and root colonization of bean. The results prove that QS regulation may be generally involved in interactions between plant-associated bacteria, fungal pathogens, and host plants and that QS provides regulation of diverse mechanisms that afford plant-beneficial properties of various Gram-negative inhabitants of the rhizosphere. The observed decrease of root colonizing activity of AHL-mutant of strain HRO-C48 (Pang et al. 2009) may be related with downregulation of AHL-dependent production of antibiotics, including volatile compounds, critical for the competition with other microbes in the plant environment. Taken together, the data indicate that AHL signaling regulates the beneficial interactions between strain HRO-C48, phytopathogens, and host plants. Further studies of this regulation may open new approaches to improve the biocontrol ability of rhizobacteria via manipulating the QS pathways.

## 9.3 Quorum Quenching and PGPRs Performance

The ability to disrupt QS networks called quorum quenching (QQ) is an important mechanism of competition between bacteria that may give one bacterial species an advantage over another one. QQ, which might be useful in controlling virulence of many pathogenic bacteria as well as in beneficial performance of various PGPRs, can be achieved by inhibiting the production, diffusion, inactivation, and perception of QS signals molecules, as well as their interaction with signal receptors (Persson et al. 2005; Dong et al. 2007; Boyer and Wisniewski-Dye 2009; Riaz et al. 2008; Raina et al. 2009; Uroz et al. 2009). Bacteria which commonly occur in soil or plant environment have the capacity to inactivate AHLs via the production of AHL-degrading enzymes. Most of such bacteria are nonpathogenic. Currently known AHL-inactivating enzymes include AHL lactonases and AHL acylases (syn. AHL amidases). AHL laconases have broad AHL substrate specificities and catalyze the opening of the homoserine lactone ring via hydrolysis of the ester bond to produce an inactive open-chain acyl homoserine and hence making the signal molecules inactive to induce cell-to-cell communication (Dong et al. 2000, 2001). AHL acylases that liberate a free homoserine lactone and a fatty acid AHL acylases are *N*-terminal nucleophile (Ntn) hydrolases that inactivate signals by cleaving the acyl chain from the homoserine lactone via nucleophilic attack on the carboxy carbon in the amide linkage. AHL acylases display high substrate specificities based on the lengths of the AHL acyl chains, and like other Ntn hydrolases they are transcribed as large propolypeptides that undergo autoproteolytic cleavage to reach maturation (e.g., Diby et al. 2009; Tait et al. 2009; Uroz et al. 2009). AHL-inactivating oxidoreductase represents a different AHL-modifying activity that is not degradation but conversion of 3-oxo-AHL to 3-hydroxy-AHL. Since the substitution at C3 is crucial for signal specificity, the oxidoreductase leads to a change in or loss of the signaling capability of the QS molecules (Uroz et al. 2005, 2008). The activity of all these enzymes results in silencing the QS-regulated processes, as degradation products cannot act as appropriate signal molecules. AHL-degrading enzymes might modulate QS by recycling AHLs once QS is achieved (Zhang et al. 2002), or they might enable cells to degrade AHLs produced by competing species (Park et al. 2005) and even utilize AHL breakdown products as carbon sources (Flagan et al. 2003; Huang et al. 2006). The presence of multiple AHL-degrading enzymes in a single strain adds complexity to their potential roles.

Since QS controls the expression of many traits important for a bacterium's lifestyle, including the expression of virulence factors, AHL-degrading enzymes such as AHL-lactonase, encoding by *aiiA* gene from various *Bacillus* strains (Dong et al. 2001) and its analogues in many other bacteria, it has become recently an important tool to investigate involvement of AHL in expression of many traits in Gram-negative bacteria (Roche et al. 2004; Dong and Zhang 2005) and to protect plants against pathogenic bacteria. Expression of heterologous AHL lactonase or coinoculation with bacteria producing this or other AHL-degrading enzymes were implemented as a novel biocontrol approach toward plant-pathogenic *Erwinia carotovora* (Dong et al. 2001; Molina et al. 2003), *E. amylovora* (Molina et al. 2005) and *A. tumefaciens* (Molina et al. 2003) whose pathogenicity is regulated by QS signals. Moreover, some of such bacteria demonstrated even plant growth-promoting effect (Cirou et al. 2007).

The potential utility of an engineered AHL-degrading bacterial strain as a biocontrol agent was illustrated when the AHL lactonase AiiA was introduced into *Burkholderia* sp. strain KJ006, a nonpathogenic bacterial endophyte of rice (Cho et al. 2007). The engineered *Burkholderia* sp. strain degraded the QS signal of pathogenic *B. glumae* and reduced the incidence of disease when two strains were coinoculated. Additionally, secretion of AiiA from the biocontrol strain *Bacillus thuringiensis* increased the strain's effectiveness at controlling plant disease by *Pectobacterium* (*Erwinia*) *carotovora* (Zhang et al. 2007). Another example is AHL-degrading bacteria isolated from the leaf surface of *Solanum tuberosum*, which are able to inactivate both short- and long-chain AHLs. Two of these isolates, identified as *Microbacterium testaceum*, showed putative AHL-lactonase activity. These two strains interrupted quorum-sensing dependent bacterial infection by plant pathogen *P. carotovorum* subsp. *carotovorum*. *M. testaceum* strains, which were isolated in this study, might be useful in the biocontrol of plant diseases (Morohoshi et al. 2009).

The list of QQ compounds from other than bacteria natural sources includes halogenated furanone from red algae *Delisea pulchra* able to inhibit formation of bacterial biofilms, penicillic acid, and patulin produced by fungi (Rasmussen and Givskov 2006), as well as QS-disruptive chemicals secreted by many plants, which might block undesired QS signals (Karamanoli and Lindow 2006; Degrassi et al. 2007; Adonizio et al. 2008). Strong QQ activity was revealed by garlic extracts (Rasmussen and Givskov 2006; Bodini et al. 2009), by P-coumaric acid, produced by plants as a part of lignin pathway and in response to wounds and nutritional stress (Bodini et al. 2009) and by essential oils from many ornamentals (Khan et al. 2009; Szabó et al. 2010).

Plants also can be engineered specifically to influence their associated bacteria, as exemplified by QQ strategies that suppress the virulence of pathogens of the genus *Pectobacterium* (Barriuso et al. 2008). On the other hand, since QQ can be considered as a natural barrier operating in plants to prevent them being infected by pathogenic bacteria in which QS regulates production of phytotoxic compounds, the coincident negative impact of QQ engineered plants on nontarget bacterial populations, including PGPRs should not be ignored. One possibility to overcome this problem is to use highly specific QQ enzymes or to combine in one strain an AHL-degrading enzyme and antibiotic activity that is not regulated by QS. QQ can also be used as a promising strategy for biocontrol of various bacteria whose plant pathogenicity is dependent on production of other than AHLs signals (Uroz et al. 2009).

## 9.4 Conclusion

Several lines of data presented in this chapter demonstrate that many beneficial traits in PGPRs, such as biofilm formation, plant colonization, plant-growth promotion, induction of plant resistance to pathogens, and production of antimicrobial antibiotics, volatiles, and cell-wall lytic enzymes are regulated by the QS network mediated by AHLs and other signaling molecules which coordinate the functions of the different populations in the rhizosphere and phyllosphere communities. QS regulation appears to be specific, with various signals pattern and distinct QS systems displayed by various strains of the same species. Examples of interkingdom cross talk between eukaryotes (plants, fungi, algae, etc.) and bacteria suggest that eukaryotes have evolved strategies to interfere with bacterial signaling in order to protect themselves from pathogenic bacteria or to help PGPRs perform optimally in the plant environment. Recent technologies such as proteomics, metabolomics, transcriptomics, and secretomics are being used to further highlight these interactions for the benefit of agriculture. Combining the data analyses obtained from these modern approaches will further our knowledge, providing a more complete picture of the role of QS cross talk in multispecies interactions. The ability to disrupt QS networks termed quorum quenching (QQ) is an important mechanism of competition between bacteria which may give one bacterial species an advantage over another one. The distribution and perception of the AHL signals during plant–microbe interactions are highly dependent on abiotic environmental factors, as well as on members of the bacterial community – such as AHL-degrading bacteria and compounds produced by eukaryotes acting as AHL mimics or inhibitors. Genetic engineering approaches based on current knowledge of QS- and QQ systems acting in plant-associated bacteria may offer a new challenge: to protect plants against bacterial pathogens and to achieve highly valuable input from PGPRs that will help support sustainable agriculture.

## References

Adonizio A, Kong K-F, Mathee K (2008) Inhibition of quorum sensing-controlled virulence factor production in *Pseudomonas aeruginosa* by south Florida plant extracts. Antimicrob Agents Chemother 52:198–203

Anderson JB, Heydorn A, Hentzer M, Eberl L, Geisenberger O, Christensen BB, Molin S, Givskov M (2001) gfp-based *N*-acyl homoserine-lactone sensor systems for detection of bacterial communication. Appl Environ Microbiol 67:575–585

Badri DV, Weir TL, van der Lelie D, Vivanco JM (2009) Rhizosphere chemical dialogues: plant-microbe interactions. Curr Opin Biotechnol 20:642–650

Barriuso J, Solano BR, Fray RG, Cámara M, Hartmann A, Gutiérrez Mañero FJ (2008) Transgenic tomato plants alter quorum sensing in plant growth-promoting rhizobacteria. Plant Biotechnol J 6:442–452

Bauer WD, Mathesius U (2004) Plant responses to bacterial quorum sensing signals. Curr Opin Plant Biol 7:429–433

Berg G (2000) Diversity of antifungal and plant-associated *Serratia plymuthica* strains. J Appl Microbiol 88:952–960
Berg G (2009) Plant–microbe interactions promoting plant growth and health: perspectives for controlled use of microorganisms in agriculture. Appl Microbiol Biotechnol 84:11–18
Bodini SF, Manfredini S, Epp M, Valentini S, Santori F (2009) Quorum sensing inhibition activity of garlic extract and p-coumaric acid. Lett Appl Microbiol 49:551–555
Boyer M, Wisniewski-Dye F (2009) Cell–cell signaling in bacteria: not simply a matter of quorum. FEMS Microbiol Ecol 70:1–19
Camara M, Daykin M, Chhabra SR (1998) Detection, purification, and synthesis of *N*-acylhomoserine lactone quorum sensing signal molecules. Meth Microbiol 27:319–330
Campbell R, Greaves MP (1990) Anatomy and community structure of the rhizosphere. In: Lynch JM (ed) The rhizosphere. Wiley, Chichester, pp 11–34
Cha C, Gao P, Chen Y-C, Shaw PD, Farrand SK (1998) Production of acyl-homoserine lactone quorum-sensing signals by gram-negative plant-associated bacteria. Mol Plant Microbe Interact 11:1119–1129
Chernin L, Chet I (2002) Microbial enzymes in biocontrol of plant pathogens and pests. In: Burns R, Dick R (eds) Enzymes in the environment: activity, ecology, and applications. Dekker, New York, pp 171–225
Chernin L, Ismailov Z, Haran S, Chet I (1995) Chitinolytic *Enterobacter agglomerans* antagonistic to fungal plant pathogens. Appl Environ Microbiol 61:1720–1726
Chernin L, Brandis A, Ismailov Z, Chet I (1996) Pyrrolnitrin production by an *Enterobacter agglomerans* strain with a broad spectrum of antagonistic activity towards fungal and bacterial phytopathogens. Curr Microbiol 32:208–212
Chernin LS, Winson MK, Thompson JM, Haran S, Bycroft BW, Chet I, Williams P, Stewart GSAB (1998) Chitinolytic activity in *Chromobacterium violaceum*: substrate analysis and regulation by quorum sensing. J Bacteriol 180:4435–4441
Chin-A-Woeng TFC, van den Broek D, de Voer G, van der Drift KMGM, Tuinman S, Thomas-Oates JE, Lugtenberg B (2001) Phenazine-1-carboxamide production in the biocontrol strain *Pseudomonas chlororaphis* PCL1391 is regulated by multiple factors secreted into the growth medium. Mol Plant Microbe Interact 14:969–979
Chin-A-Woeng TFC, Bloemberg GV, Lugtenberg BJJ (2003) Phenazines and their role in biocontrol by *Pseudomonas* bacteria. New Phytol 157:503–523
Cho HS, Park SY, Ryu CM, Kim JF, Kim JG, Park SH (2007) Interference of quorum sensing and virulence of the rice pathogen *Burkholderia glumae* by an engineered endophytic bacterium. FEMS Microbiol Ecol 60:14–23
Christensen AB, Riedel K, Eberl L, Flodgaard LR, Molin S, Gram L, Givskov M (2003) Quorum-sensing-directed protein expression in *Serratia proteamaculans* B5a. Microbiology 149: 471–483
Cirou A, Diallo S, Kurt C, Latour X, Faure D (2007) Growth promotion of quorum-quenching bacteria in the rhizosphere of *Solanum tuberosum*. Environ Microbiol 9:1511–1522
Costa R, van Aarle IM, Mendes R, van Elsas JD (2009) Genomics of pyrrolnitrin biosynthetic loci: evidence for conservation and whole-operon mobility within gram-negative bacteria. Environ Microbiol 11:159–175
Coulthurst SJ, Kurz CP, Salmond GPC (2004) LuxS mutants of *Serratia* defective in autoinducer-2-dependent quorum sensing show strain-dependent impacts on virulence and production of carbapenem and prodigiosin. Microbiology 150:1901–1910
Danhorn T, Fuqua C (2007) Biofilm formation by plant-associated bacteria. Annu Rev Microbiol 61:401–422
Daniels R, De Vos DE, Desair J, Raedschelders G, Luyten E, Rosemeyer V, Verreth C, Schoeters E, Vanderleyden J, Michiels J (2002) The *cin* quorum sensing locus of *Rhizobium etli* CNPAF512 affects growth and symbiotic nitrogen fixation. J Biol Chem 277:462–468
De Keersmaecker SCJ, Sonck K, Vanderleyden J (2006) Let LuxS speak up in AI-2 signaling. Trends Microbiol 14:114–119

De Vleesschauwer D, Hofte M (2007) Using *Serratia plymuthica* to control fungal pathogens of plants. CAB reviews: perspectives in agriculture, veterinary science, nutrition and natural resources 2, No. 046, 12 pp.

De Vleesschauwer D, Hofte M (2009) Rhizobacteria-induced systemic resistance. In: van Loon LC (ed) Plant innate immunity book series: advances in botanical research, vol 51. Academic, San Diego, CA, pp 223–281

Deangelis KM, Firestone MK, Lindow SE (2007) Sensitive whole-cell biosensor suitable for detecting a variety of *N*-acyl homoserine lactones in intact rhizosphere microbial communities. Appl Environ Microbiol 73:3724–3727

Degrassi G, Devescovi G, Solis R, Steindler L, Venturi V (2007) *Oryza sativa* rice plants contain molecules that activate different quorum-sensing *N*-acyl homoserine lactone biosensors and are sensitive to the specific AiiA lactonase. FEMS Microbiol Lett 269:213–220

Diby P, Young Sam K, Kweon PK, Hyang J (2009) Application of quorum quenching to inhibit biofilm formation. Environ Eng Sci 26:1319–1324

Dong YH, Zhang LH (2005) Quorum sensing and quorum-quenching enzymes. J Microbiol 43:101–109

Dong YH, Xu JL, Li XZ, Zhang LH (2000) AiiA, an enzyme that inactivates the acyl-homoserine lactone quorum-sensing signal and attenuates the virulence of *Erwinia carotovora*. Proc Natl Acad Sci USA 97:3526–3531

Dong YH, Wang LH, Xu JL, Zhang HB, Zhang XF, Zhang LH (2001) Quenching quorum-sensing-dependent bacterial infection by an *N*-acyl homoserine lactonase. Nature 411:813–817

Dong YH, Wang LY, Zhang LH (2007) Quorum-quenching microbial infections: mechanisms and implications. Philos Trans R Soc Lond B Biol Sci 362:1201–1211

Duan K, Surette MG (2007) Environmental regulation of *Pseudomonas aeruginosa* PAO1 Las and Rhl quorum-sensing systems. J Bacteriol 189:4827–4836

Eberl L, Winson MK, Sternberg C, Stewart GSAB, Christiansen G, Chhabra SR, Bycroft BW, Williams P, Molin S, Givskov M (1996) Involvement of *N*-acyl-L-homoserine lactone autoinducers in controlling the multicellular behaviour of *Serratia liquefaciens*. Mol Microbiol 20:127–136

Eberl L, Molin S, Givskov M (1999) Surface motility of *Serratia liquefaciens* MG1. J Bacteriol 181:1703–1712

Edwards A, Frederix M, Wisniewski-Dyé F, Jones J, Zorreguieta A, Downie JA (2009) The *cin* and *rai* quorum-sensing regulatory systems in *Rhizobium leguminosarum* are coordinated by ExpR and CinS, a small regulatory protein coexpressed with CinI. J Bacteriol 191:3059–3067

Elasri M, Delorme S, Lemancean P, Stewart G, Laue B, Glickmann E, Oger PM, Dessaux Y (2001) Acyl-homoserine lactone production is more common among plant-associated *Pseudomonas* spp. than among soilborne *Pseudomonas* spp. Appl Environ Microbiol 67:1198–1209

Faure D, Vereecke D, Leveau JHJ (2009) Molecular communication in the rhizosphere. Plant Soil 321:279–303

Flagan S, Ching WK, Leadbetter JR (2003) *Arthrobacter* strain VAI-A utilizes acyl-homoserine lactone inactivation products and stimulates quorum signal biodegradation by *Variovorax paradoxus*. Appl Environ Microbiol 69:909–916

Frankowski J, Berg G, Bahl H (1998) Mechanisms involved in the antifungal activity of the rhizobacterium *Serratia plymuthica*. IOBC Bull 21:45–50

Frankowski J, Lorito M, Schmid R, Berg G, Bahl H (2001) Purification and properties of two chitinolytic enzymes of *Serratia plymuthica* HRO-C48. Arch Microbiol 176:421–426

Fuqua C, Greenberg EP (2002) Listening in on bacteria: acyl-homoserine lactone signaling. Nat Rev Mol Cell Biol 3:685–695

Gao M, Teplitski M, Robinson JB, Bauer WD (2003) Production of substances by *Medicago truncatula* that affect bacterial quorum-sensing. Mol Plant Microbe Interact 16:827–834

Glick B, Todorovic B, Czarny J, Cheng Z, Duan J, McConkey B (2007) Promotion of plant growth by bacterial ACC deaminase. Crit Rev Plant Sci 26:227–242

Grosch R, Faltin F, Lottmann J, Kofoet A, Berg G (2005) Effectiveness of three antagonistic bacterial isolates to suppress *Rhizoctonia solani* Kuhn on lettuce and potato. Can J Microbiol 51:345–353

Gross H, Loper JE (2009) Genomics of secondary metabolite production by *Pseudomonas* spp. Nat Prod Rep 26(11):1408–1446

Gurich N, Gonzalez JE (2009) Role of quorum-sensing in *Sinorhizobium meliloti*-alfalfa symbiosis. J Bacteriol 191:4372–4382

Haas D, Defago G (2005) Biological control of soil-borne pathogens by fluorescent pseudomonads. Nat Rev Microbiol 3:307–319

Hammer PE, Hill DS, Lam ST, Van-Pee KH, Ligon JM (1997) Four genes from *Pseudomonas fluorescens* that encode the biosynthesis of pyrrolnitrin. Appl Environ Microbiol 63:2147–2154

Hense BA, Kuttler C, Műller J, Rothballer M, Hartmann A, Kreft JU (2007) Does efficiency sensing unify diffusion and quorum sensing? Nat Rev Microbiol 5:230–239

Holden MT, Chhabra SR, de Nys R, Stead P, Bainton NJ, Hill PJ, Manefield M, Kumar N, Labatte M, England D, Rice S, Givskov M, Salmond GP, Stewart GS, Bycroft BW, Kjelleberg S, Williams P (1999) Quorum-sensing cross talk: isolation and chemical characterization of cyclic dipeptides from *Pseudomonas aeruginosa* and other gram-negative bacteria. Mol Microbiol 33:1254–1266

Holden M, Swift S, Williams P (2000) New signal molecules on the quorum-sensing block. Trends Microbiol 8:101–103

Horng YT, Deng SC, Daykin M, Soo PC, Wei JR, Luh KT, Ho SW, Swift S, Lai HC, Williams P (2002) The LuxR family protein SpnR functions as a negative regulator of *N*-acylhomoserine lactone-dependent quorum sensing in *Serratia marcescens*. Mol Microbiol 45:1655–1671

Huang JJ, Peterson A, Whiteley M, Leadbetter JR (2006) Identification of QuiP, the product of gene PA1032, as the second acyl-homoserine lactone acylase of *Pseudomonas aeruginosa* PA01. Appl Environ Microbiol 72:1190–1197

Huang X, Zhang X, Xu Y (2008) Positive regulation of pyoluteorin biosynthesis in Pseudomonas sp. M18 by quorum-sensing regulator VqsR. J Microbiol Biotechnol 18:828–836

Jayaraman A, Wood TK (2008) Bacterial quorum sensing: signals, circuits, and implications for biofilms and disease. Annu Rev Biomed Eng 10:145–167

Kai M, Effmert U, Berg G, Piechulla B (2007) Volatiles of bacterial antagonists inhibit mycelial growth of the plant pathogen *Rhizoctonia solani*. Arch Microbiol 187:351–360

Kai M, Haustein M, Molina F, Petri A, Scholz B, Piechulla B (2009) Bacterial volatiles and their action potential. Appl Microbiol Biotechnol 81:1001–1012

Kalbe C, Marten P, Berg G (1996) Strains of the genus Serratia as beneficial rhizobacteria of oilseed rape. Microbiol Res 151:4400–4433

Kamensky M, Ovadis M, Chet I, Chernin L (2003) Soil-borne strain IC14 of *Serratia plymuthica* with multiple mechanisms of antifungal activity provides biocontrol of *Botrytis cinerea* and *Sclerotinia sclerotiorum* diseases. Soil Biol Biochem 35:323–331

Karamanoli K, Lindow SE (2006) Disruption of *N*-acyl homoserine lactone-mediated cell signaling and iron acquisition in epiphytic bacteria by leaf surface compounds. Appl Environ Microbiol 72:7678–7686

Khan SR, Mavrodi DV, Jog GJ, Suga H, Thomashow LS, Farrand SK (2005) Activation of the phz operon of *Pseudomonas fluorescens* 2-79 requires the LuxR homolog PhzR, *N*-(3-OH-Hexanoyl)-L-homoserine lactone produced by the LuxI homolog PhzI, and a cis-acting *phz* box. J Bacteriol 187:6517–6652

Khan SR, Herman J, Krank J, Serkova NJ, Churchill MEA, Suga H, Farrand SK (2007) *N*-(3-hydroxyhexanoyl)-L-homoserine lactone is the biologically relevant quormone that regulates the phz operon of *Pseudomonas chlororaphis* strain 30-84. Appl Environ Microbiol 73:7443–7455

Khan MS, Zahin M, Hasan S, Husain FM, Ahmad I (2009) Inhibition of quorum sensing regulated bacterial functions by plant essential oils with special reference to clove oil. Lett Appl Microbiol 49:354–360

Kurze S, Dahl R, Bahl H, Berg G (2001) Biological control of soil-borne pathogens in strawberry by *Serratia plymuthica* HRO-C48. Plant Dis 85:529–534

Labbate M, Queek SY, Koh KS, Rice SA, Givskov M, Kjelleberg S (2004) Quorum sensing-controlled biofilm development in *Serratia liquefaciens* MG1. J Bacteriol 186:692–698
Lapouge K, Schubert M, Allain FH, Haas D (2008) Gac/Rsm signal transduction pathway of gamma-proteobacteria: from RNA recognition to regulation of social behaviour. Mol Microbiol 67:241–253
Ling EA, Ellison ML, Pesci EC (2009) A novel plasmid for detection of *N*-acyl homoserine lactones. Plasmid 62:16–21
Liu X, Bimerew M, Ma Y, Műller H, Ovadis M, Eberl L, Berg G, Chernin L (2007) Quorum-sensing signaling is required for production of the antibiotic pyrrolnitrin in a rhizospheric biocontrol strain of *Serratia plymuthica*. FEMS Microbiol Lett 270:299–305
Lugtenberg B, Kamilova F (2009) Plant-growth-promoting rhizobacteria. Annu Rev Microbiol 63:541–556
Lyon GJ, Novick RP (2004) Peptide signaling in *Staphylococcus aureus* and other gram-positive bacteria. Peptides 25:1389–1403
Maddula VSRK, Zhang Z, Pierson EA, Pierson LS III (2006) Quorum sensing and phenazines are involved in biofilm formation by *Pseudomonas chlororaphis* (*aureofaciens*) strain 30-84. Microb Ecol 52:289–301
Maddula VSRK, Pierson EA, Pierson LS III (2008) Altering the ratio of phenazines in *Pseudomonas chlororaphis* (*aureofaciens*) strain 30-84: effects on biofilm formation and pathogen inhibition. J Bacteriol 190:2759–2766
Mäe A, Montesano M, Koiv V, Palva ET (2001) Transgenic plants producing the bacterial pheromone *N*-acyl-homoserine lactone exhibit enhanced resistance to the bacterial phytopathogen *Erwinia carotovora*. Mol Plant Microbe Interact 14:1035–1042
Mathesius U, Mulders S, Gao M, Teplitski M, Caetano-Anolles G, Rolfe BG, Bauer WD (2003) Extensive and specific responses of a eukaryote to bacterial quorum-sensing signals. Proc Natl Acad Sci USA 100:1444–1449
Mavrodi DV, Blankenfeldt W, Thomashow LS (2006) Phenazine compounds in fluorescent *Pseudomonas* spp. biosynthesis and regulation. Annu Rev Phytopathol 44:417–445
Mavrodi DV, Peever TL, Mavrodi OV, Parejko JA, Raaijmakers JM, Lemanceau P, Mazurier S, Heide L, Blankenfeldt W, Weller DM, Thomashow LS (2010) Diversity and evolution of the phenazine biosynthesis pathway. Appl Environ Microbiol 76:866–879
McLean RJC, Bryant SA, Dhiraj A, Vattem DA, Givskov M, Rasmussen TB, Balaban N (2008) Detection in vitro of quorum-sensing molecules and their inhibitors. In: Balaban N (ed) Control of biofilm infections by signal manipulation. Springer, Berlin, pp 39–50, doi:10.1007/7142_2007_008
Molina L, Constantinescu F, Michel L, Reimmann C, Duffy B, Defago G (2003) Degradation of pathogen quorum-sensing molecules by soil bacteria: a preventive and curative biological control mechanism. FEMS Microbiol Ecol 45:71–81
Molina L, Rezzonico F, Défago G, Duffy B (2005) Autoinduction in *Erwinia amylovora*: evidence of an acyl-homoserine lactone signal in the fire blight pathogen. J Bacteriol 187:3206–3213
Morohoshi T, Someya N, Ikeda T (2009) Novel *N*-acylhomoserine lactone-degrading bacteria isolated from the leaf surface of *Solanum tuberosum* and their quorum-quenching properties. Biosci Biotechnol Biochem 73:2124–2127
Morris CE, Monier JM (2003) The ecological significance of biofilm formation by plant-associated bacteria. Annu Rev Phytopathol 41:429–453
Műller H, Berg G (2008) Impact of formulation procedures on the effect of the biocontrol agent *Serratia plymuthica* HRO-C48 on *Verticillium* wilt in oilseed rape. Biocontrol 53:305–316
Műller H, Westendorf C, Leitner E, Chernin L, Riedel K, Schmidt S, Eberl L, Berg G (2009) Quorum-sensing effects in the antagonistic rhizosphere bacterium *Serratia plymuthica* HRO-C48. FEMS Microbiol Ecol 67:468–478
Nannipieri P, Ascher J, Ceccherini MT, Guerri G, Renella G, Pietramellara G (2008) Recent advances in functional genomics and proteomics of plant associated microbes. In: Nautiyal CS,

Dion P (eds) Molecular mechanisms of plant and microbe coexistence. Springer, Berlin, pp 215–241

Ng WL, Bassler BL (2009) Bacterial quorum-sensing network architectures. Annu Rev Genet 43:197–222

Nowak-Thompson B, Chaney N, Wing JS, Gould SJ, Loper JE (1999) Characterization of the pyoluteorin biosynthetic gene cluster of *Pseudomonas fluorescens* Pf-5. J Bacteriol 181:2166–2174

Ovadis M, Liu X, Gavriel S, Ismailov Z, Chet I, Chernin L (2004) The global regulator genes from biocontrol strain *Serratia plymuthica* IC1270: cloning, sequencing, and functional studies. J Bacteriol 186:4986–4993

Pang Y, Liu X, Ma Y, Chernin L, Berg G, Gao K (2009) Induction of systemic resistance, root colonization and biocontrol activities of rhizospheric strain of *Serratia plymuthica* are dependent on *N*-acyl homoserine. Eur J Plant Pathol 124:261–268

Park SY, Kang HO, Jang HS, Lee JK, Koo BT, Yum DY (2005) Identification of extracellular *N*-acylhomoserine lactone acylase from a *Streptomyces* sp. and its application to quorum quenching. Appl Environ Microbiol 71:2632–2641

Patten CL, Glick BR (2008) Plant growth-promoting bacteria: significance in horticulture. In: Ray RC, Ward OP (eds) Microbial biotechnology in horticulture, vol 2. Science, Enfield, pp 86–111

Persson T, Givskov M, Nielsen J (2005) Quorum sensing inhibition: targeting chemical communication in gram-negative bacteria. Curr Med Chem 12:3103–3115

Pesci EC, Milbank JBJ, Pearson JP, McKnight S, Kende AS, Greenberg EP, Iglewski BH (1999) Quinolone signaling in the cell-to-cell communication system of *Pseudomonas aeruginosa*. Proc Natl Acad Sci USA 96:11229–11234

Pierson LS III, Pierson EA (2007) Roles of diffusible signals in communication among plant-associated bacteria. Phytopathology 97:227–232

Pierson LS III, Wood DW, Pierson EA (1998) Homoserine lactone-mediated regulation in plant-associated bacteria. Annu Rev Phytopathol 36:207–225

Raaijmakers JM, Vlami M, de Souza JT (2002) Antibiotic production by bacterial biocontrol agents. Antonie Van Leeuwenhoek 81:537–547

Raina S, De Vizio D, Odell M, Clements M, Vanhulle S, Keshavarz T (2009) Microbial quorum sensing: a tool or a target for antimicrobial therapy. Biotechnol Appl Biochem 54:65–84

Ramey BE, Koutsoudis M, von Bodman SB, Fuqua C (2004) Biofilm formation in plant-microbe associations. Curr Opin Microbiol 7:602–609

Rasmussen TB, Givskov M (2006) Quorum sensing inhibitors: a bargain of effects. Microbiology 152:895–904

Rasmussen TB, Bjarnsholt T, Skindersoe ME, Hentzer M, Kristoffersen P, Köte M, Nielsen J, Eberl L, Givskov M (2005) Screening for quorum-sensing inhibitors (QSI) by use of novel genetic system, the QSI selector. J Bacteriol 187:1799–1814

Ravn L, Christensen AB, Molin S, Givskov M, Gram L (2001) Methods for detecting acylated homoserine lactones produced by gram-negative bacteria and their application in studies of AHL-production kinetics. J Microbiol Meth 44:239–251

Riaz K, Elmerich C, Moreira D, Raffoux A, Dessaux Y, Faure D (2008) A metagenomic analysis of soil bacteria extends the diversity of quorum-quenching lactonases. Environ Microbiol 10:560–570

Rice SA, Kjelleberg S, Givskov M, De Boer W, Chernin L (2004) Detection of bacterial homoserine lactone quorum sensing signals. In: Kowalchuk GA, de Bruijn FJ, Head IM, Akkermans ADL, van Elsas JD (eds) Molecular microbial ecology manual, 2nd edn. Kluwer, Dordrecht, pp 1629–1649, Chapter 8.04

Riedel K, Ohnesorg T, Krogfelt KA, Hansen TS, Omori K, Givskov M, Eberl L (2001) *N*-acyl-L-homoserine lactone-mediated regulation of the Lip secretion system in *Serratia liquefaciens* MG1. J Bacteriol 183:1805–1809

Roche DM, Byers JT, Smith DS, Glansdorp FG, Spring DR, Welch M (2004) Communications blackout? Do *N*-acylhomoserine-lactone-degrading enzymes have any role in quorum sensing? Microbiology 150:2023–2028

Rothballer M, Eckert B, Schmid M, Klein I, Fekete A, Schloter M, Hartmann A (2008) Endophytic root colonization of gramineous plants by *Herbaspirillum frisingense*. FEMS Microbiol Ecol 66:85–95
Sanchez-Contreras M, Bauer WD, Gao M, Robinson JB, Downie JA (2007) Quorum-sensing regulation in *Rhizobium* and its role in symbiotic interactions with legumes. Phil Trans R Soc B 362:1149–1163
Schmidt S, Blom J, Pernthaler J, Berg G, Baldwin A, Mahenthiralingam E, Eberl L (2009) Production of the antifungal compound pyrrolnitrin is quorum sensing-regulated in members of the *Burkholderia cepacia* complex. Environ Microbiol 11:1422–1437
Schuhegger R, Ihring A, Gantner S, Bahnweg G, Knappe C, Vogg G, Hutzler P, Schmid M, Van Breusegem F, Eberl L, Hartmann A, Langebartels C (2006) Induction of systemic resistance in tomato by *N*-acyl-L-homoserine lactone-producing rhizosphere bacteria. Plant Cell Environ 29:909–918
Scott RA, Weil J, Le PT, Williams P, Fray RG, von Bodman SB, Savka MA (2006) Long- and short-chain plant-produced bacterial *N*-acyl-homoserine lactones become components of phyllosphere, rhizosphere, and soil. Mol Plant Microbe Interact 19:227–239
Shepherd RW, Lindow SE (2009) Two dissimilar n-acyl-homoserine lactone acylases of Pseudomonas syringae influence colony and biofilm morphology. Appl Environ Microbiol 75:45–53
Spaepen S, Vanderleyden J, Okon Y (2009) Plant growth-promoting actions of rhizobacteria. In: van Loon LC (ed) Plant innate immunity book series: advances in botanical research, vol 51. Academic, San Diego, CA, pp 283–320
Steidle A, Sigl K, Schuhegger R, Ihring A, Schmid M, Gantner S, Stoffels M, Riedel K, Givskov M, Hartmann A, Langebartels C, Eberl L (2001) Visualization of *N*-acylhomoserine lactone-mediated cell–cell communication between bacteria colonizing the tomato rhizosphere. Appl Environ Microbiol 67:5761–5770
Steidle A, Allesen-Holm M, Riedel K, Berg G, Givskov M, Molin S, Eberl L (2002) Identification and characterization of an *N*-acylhomoserine lactone-dependent quorum-sensing system in *Pseudomonas putida* strain IsoF. Appl Environ Microbiol 68:6371–6382
Steindler L, Bertani I, De Sordi L, Bigirimana J, Venturi V (2008) The presence, type and role of *N*-acyl homoserine lactone quorum sensing in fluorescent *Pseudomonas* originally isolated from rice rhizospheres are unpredictable. FEMS Microbiol Lett 288:102–111
Steindler L, Bertani I, De Sordi L, Schwager S, Eberl L, Venturi V (2009) LasI/R and RhlI/R quorum sensing in a strain of *Pseudomonas aeruginosa* beneficial to plants. Appl Environ Microbiol 75:5131–5140
Szabó MA, Varga GZ, Hohmann J, Schelz Z, Szegedi E, Amaral L, Molnár J (2010) Inhibition of quorum-sensing signals by essential oils. Phytother Res 24:782–786
Tait K, Williamson H, Atkinson S, Williams P, Camara M, Joint I (2009) Turnover of quorum sensing signal molecules modulates cross-kingdom signaling. Environ Microbiol 11:1792–1802
Teplitski M, Robinson JB, Bauer WD (2000) Plants secrete substances that mimic bacterial *N*-acyl homoserine lactone signal activities and affect population density-dependant behaviors in associated bacteria. Mol Plant Microbe Interact 13:637–648
Thomson NR, Crow MA, McGowan SJ, Cox A, Salmond GPC (2000) Biosynthesis of carbapenem antibiotic and prodigiosin pigment in *Serratia* is under quorum sensing control. Mol Microbiol 36:539–556
Toth IK, Newton JA, Hyman LJ, Lees AK, Daykin M, Ortori C, Williams P, Fray RG (2004) Potato plants genetically modified to produce *N*-acylhomoserine lactones increase susceptibility to soft rot *Erwiniae*. Mol Plant Microbe Interact 17:880–887
Uroz S, Chhabra SR, Cámara M, Williams P, Oger PM, Dessaux Y (2005) *N*-Acylhomoserine lactone quorum-sensing molecules are modified and degraded by *Rhodococcus erythropolis* W2 by both amidolytic and novel oxidoreductase activities. Microbiology 151:3313–3322
Uroz S, Oger PM, Chapelle E, Adeline MT, Faure D, Dessaux Y (2008) A *Rhodococcus qsdA*-encoded enzyme defines a novel class of large-spectrum quorum-quenching lactonases. Appl Environ Microbiol 74:1357–1366

Uroz S, Dessaux Y, Oger P (2009) Quorum sensing and quorum quenching: the Yin and Yang of bacterial communication. Chembiochem 10:205–216
van Houdt R, Moons P, Jansen A, Vanoirbeek K, Michiels CW (2006) Isolation and functional analysis of luxS in *Serratia plymuthica* RVH1. FEMS Microbiol Lett 262:201–209
van Houdt R, Givskov M, Michiels CW (2007) Quorum sensing in *Serratia*. FEMS Microbiol Rev 31:407–424
van Loon LC (2007) Plant responses to plant growth-promoting rhizobacteria. Eur J Plant Pathol 119:243–254
van Loon LC, Bakker PAHM, Pieterse CMJ (1998) Systemic resistance induced by rhizosphere bacteria. Annu Rev Phytopathol 36:453–483
Veselova M, Kholmeckaya M, Klein S, Voronina E, Lipasova V, Metlitskaya A, Mayatskaya A, Lobanok E, Khmel I, Chernin L (2003) Production of *N*-acyl homoserine lactone signal molecules in gram-negative soil-borne and plant-associated bacteria. Folia Microbiol 48:794–798
Veselova MA, Klein S, Bass IA, Lipasova VA, Metlitskaya AZ, Ovadis MI, Chernin LS, Khmel IA (2008) Quorum sensing systems of regulation, synthesis of phenazine antibiotics, and antifungal activity in rhizospheric bacterium *Pseudomonas chlororaphis* 449. Russ J Genet 44:1400–1408
Vespermann A, Kai M, Piechulla B (2007) Rhizobacterial volatiles affect the growth of fungi and *Arabidopsis thaliana*. Appl Environ Microbiol 73:5639–5641
von Bodman SB, Bauer WD, Coplin DL (2003) Quorum sensing in plant-pathogenic bacteria. Annu Rev Phytopathol 41:455–482
Walsh UF, Morrissey JP, O'Gara F (2001) *Pseudomonas* for biocontrol of phytopathogens: from functional genomics to commercial exploitation. Curr Opin Biotechnol 12:289–295
Waters CM, Bassler BK (2005) Quorum sensing: cell-to-cell communication in bacteria. Annu Rev Cell Dev Biol 21:319–346
Wei JR, Lai HC (2006) *N*-acylhomoserine lactone-mediated cell-to-cell communication and social behavior in the genus *Serratia*. Int J Med Microbiol 296:117–124
Wei HL, Zhang LQ (2006) Quorum sensing system influences root colonization and biological control ability in *Pseudomonas fluorescens* 2P24. Antonie Van Leeuwenhoek 89:267–280
Weller DM, Raaijmakers JM, McSpadden Gardener BB, Thomashow LS (2002) Microbial populations responsible for suppressiveness to plant pathogens. Annu Rev Phytopathol 15:1–17
Weller DM, Landa BB, Mavrodi OV, Schroeder KL, De La Fuente L, Blouin-Bankhead S, Allende-Molar R, Bonsal RF, Mavrodi DV, Tomashow LS (2007) Role of 2,4-diacetylphloroglucinol-producing fluorescent Pseudomonas spp. in the defence of plant roots. Plant Biol 9:4–20
Wenke K, Kai M, Piechulla B (2010) Belowground volatiles facilitate interactions between plant roots and soil organisms. Planta 231:499–506
Whipps JM (2001) Microbial interaction and biocontrol in the rhizosphere. J Exp Bot 52:487–511
Whitehead NA, Barnard AML, Slater H, Simpson NJL, Salmond GPC (2001) Quorum-sensing in gram-negative bacteria. FEMS Microbiol Rev 25:365–404
Winson MK, Camara M, Latifi A, Foglino M, Chhabra SR, Daykin M, Bally M, Chapon V, Salmond GPC, Bycroft BW, Lazdunski A, Stewart GSAB, Williams P (1995) Multiple *N*-acyl-L-homoserine lactone signal molecules regulate production of virulence determinants and secondary metabolites in *Pseudomonas aeruginosa*. Proc Natl Acad Sci USA 92:9427–9431
Withers H, Swift S, Williams P (2001) Quorum sensing as an integral component of gene regulatory networks in gram-negative bacteria. Curr Opin Microbiol 4:186–193
Wood DW, Pierson LS III (1996) The *phzI* gene of *Pseudomonas aureofaciens* 30-84 is responsible for the production of a diffusible signal required for phenazine antibiotic production. Gene 168:49–53
Zhang Z, Pierson LS III (2001) A second quorum-sensing system regulates cell surface properties but not phenazine antibiotic production in *Pseudomonas aureofaciens*. Appl Environ Microbiol 67:4305–4315

Zhang HB, Wang LH, Zhang LH (2002) Genetic control of quorum-sensing signal turnover in *Agrobacterium tumefaciens*. Proc Natl Acad Sci USA 99:4638–4643

Zhang L, Ruan L, Hu C, Wu H, Chen S, Yu Z, Sun M (2007) Fusion of the genes for AHL-lactonase and S-layer protein in *Bacillus thuringiensis* increases its ability to inhibit soft rot caused by *Erwinia carotovora*. Appl Microbiol Biotechnol 74:667–675

Zheng H, Zhong Z, Lai X, Chen WX, Li S, Zhu J (2006) A LuxR/LuxI-type quorum-sensing system in plant bacterium, *Mesorhizobium tianshanense*, controls symbiotic nodulations. J Bacteriol 188:1943–1949

# Chapter 10
# Management of Plant Diseases by Microbial Metabolites

**M. Jayaprakashvel and N. Mathivanan**

## 10.1 Introduction

Plant diseases are of serious concern at all times. However, in the present global scenario, management of plant diseases is much needed in order to sustain food security for ever increasing human population. A vast number of plant pathogens/parasites from much simpler forms such as viroids to higher organisms cause diseases in crop plants. The destructive effects of pests and pathogens of crops range from mild symptoms to catastrophes. Catastrophic plant diseases aggravate the current scarcity of food supply, in which at least 800 million people are inadequately fed (Strange and Scott 2005). The twenty-first century is more about the global climate change. All over the world, in all the fields, the effect of global climate change has been under rigorous consideration. Like any other field, climate change is affecting plants in natural and agricultural ecosystems throughout the world and it might have definite influence on the future food security. However, modeling the effects of predicted twenty-first century climate change on plant disease epidemics has been given little attention (Evans et al. 2008). Changing weather parameters such as temperature, rainfall and many others can induce severe disease epidemics in plants such as staple food crops which would really be a threat to future food security (Chakraborty 2005; Evans et al. 2008). Crop protection is one of the basic components of the sustainable agriculture to ensure increased crop

M. Jayaprakashvel
Biocontrol and Microbial Metabolites Lab, Centre for Advanced Studies in Botany, University of Madras, Guindy Campus, Chennai 600025, India

Department of Biotechnology, AMET University, Kanathur, Chennai 603112, India

N. Mathivanan (✉)
Biocontrol and Microbial Metabolites Lab, Centre for Advanced Studies in Botany, University of Madras, Guindy Campus, Chennai 600025, India
e-mail: prabhamathi@yahoo.com

D.K. Maheshwari (ed.), *Bacteria in Agrobiology: Plant Nutrient Management*,
DOI 10.1007/978-3-642-21061-7_10, © Springer-Verlag Berlin Heidelberg 2011

production. Biological control has been actively practiced as a crop protection measure for more than five decades and the history of biocontrol, its successes and failures, have been extensively reviewed (Cook and Baker 1983; Mathre et al. 1999; Cook 2000; Fravel 2005; Mathivanan et al. 2006; Ash 2010). Many saprophytic microorganisms colonizing plant roots have the ability to protect plants from damage caused by parasitic nematodes, bacterial and fungal pathogens, and several strains of them have been used as biocontrol agents (BCAs) (Paulitz and Belanger 2001; Weller et al. 2002; Harman et al. 2004; Haas and Defago 2005). However, fungicides remain vital option for the control of plant diseases, which are estimated to cause yield reductions of almost 20% in the major food and other crops worldwide. Since their introduction in the 1960s, systemic fungicides have gradually replaced the older non-systemic fungicides, establishing higher levels of disease control and developing new fungicide markets (Gullino et al. 2000). However, increased concern for health and environmental hazards associated with the use of these synthetic agrochemicals has resulted in the need for alternative disease control measures to attain sustainability in agriculture. Hence, there is an increasing commercial and environmental interest in the use of microbe-based products as alternatives to, or in combination with, chemicals for controlling the spread and severity of a range of crop diseases (Dowling and O'Gara 1994; Mathivanan et al. 1997). These microbe-based agro-agents are mostly the whole organisms or their metabolites.

Among microbial metabolites, secondary metabolites which include antibiotics, pigments, toxins, alkaloids, etc. proved effectors of ecological competition and symbiosis, pheromones, enzyme inhibitors, immune modulating agents, receptor antagonists and agonists, pesticides, antitumor agents and growth promoters. They play vital role in animal and plant health. Interestingly, some of the microbial metabolites that are used in the plant disease control programs are of great importance. Production of antimicrobial metabolites is commonly observed during microbial interaction of antagonistic microorganisms and pathogens. In many biocontrol systems, one or more antibiotics have been shown to play a role in disease suppression and it has been extensively emphasized by many researchers (Handelsman and Stabb 1996; Yamaguchi 1996; Fravel 1988; Mathivanan et al. 2008). Antibiotics encompass a chemically heterogeneous group of organic, low-molecular weight compounds produced by microorganisms. At low concentrations, antibiotics are lethal to the growth or metabolic activities of other microorganisms (Handelsman and Stabb 1996). The antibiotic secondary metabolites of biological control agents are unique and advantageous over their competitors, the chemical fungicides (Kim and Hwang 2007). Most of the research on these metabolites done in vitro has suggested the existence of various types of metabolites both chemically and functionally. Moreover, the producing microorganisms are also of several types, though bacteria have been studied profoundly. With this background, this chapter approaches the use of microbial secondary metabolites in disease control over the past few decades and its future perception.

## 10.2 Uniqueness and Advantages of Microbial Secondary Metabolites

In order to produce adequate, abundant and quality food, feed and other agro products by the farmers around the world, control of plant diseases is one of the pre-requisites. Different approaches may be used to prevent, mitigate or control plant diseases. Beyond good agronomic practices, farmers often rely heavily on chemical fertilizers and pesticides. Currently, there are approximately 145 pesticides registered for use in India alone, and production has increased to approximately 85,000 metric tons (Gupta 2004). Such inputs to agriculture have contributed significantly to the spectacular improvements in crop productivity and quality in the past. However, the environmental pollution caused by unwarranted use and misuse of agrochemicals, as well as the development of resistance against these chemicals among insect pests and pathogens, have led to considerable changes among researchers and farmers toward the use of pesticides in agriculture. Today, there are strict regulations on chemical pesticide use, and there is pressure to remove the most hazardous chemicals from the market. Consequently, some pest management researchers have focused their efforts on developing alternative inputs to synthetic chemicals for controlling insect pests and diseases. Among these, biological control is considered one of the best viable alternatives (Gullino et al. 2000; Mathivanan et al. 2006).

The term biocontrol was initially used to refer to the biological control of insects as the suppression of insect populations by the actions of their native or introduced enemies. However, the widely quoted and accepted definition of biological control of diseases is "the reduction in the amount of inoculum or disease-producing activity of a pathogen accomplished by or through one or more organisms." The antagonistic or biocontrol activity of BCAs against pathogens can be tested using a simple *in vitro* dual culture technique (Fig. 10.1). Hence in generic terms, biological control can be defined as a population-leveling process, in which the population of one species lowers the number of another species by mechanisms such as competition, parasitism and antibiosis (Cook and Baker 1983; Vasudevan et al. 2002). Bacteria belonging to the genera *Agrobacterium*, *Bacillus*, *Burkholderia*, *Enterobacter*, *Erwinia*, *Lysobacter*, *Pseudomonas* and *Serratia* were successfully used as BCAs against many plant diseases. Some of the fungal biocontrol genera used are *Ampelomyces*, *Aspergillus*, *Coniothyrium*, *Gliocladium*, *Laetisaria*, *Penicillium*, *Phlebiopsis*, *Sporodesmia*, *Talaromyces*, *Tilletiopsis*, *Trichoderma* and *Trichothecium* (Mathivanan and Manibhushanrao 2004; Mathivanan et al. 2006). In addition, several species of actinomycetes belonging to the genera *Streptomyces*, *Actinoplanes*, *Actinomadura*, *Micromonospora*, *Streptosporangium*, *Streptoverticillium* and *Spirillospora* were used as BCAs. Interestingly, they produce biologically active secondary metabolites that are having potential in controlling plant pathogens (Doumbou et al. 2002; El-Tarabily and Sivasithamparam 2006; Prabavathy et al. 2008; Ramesh 2009).

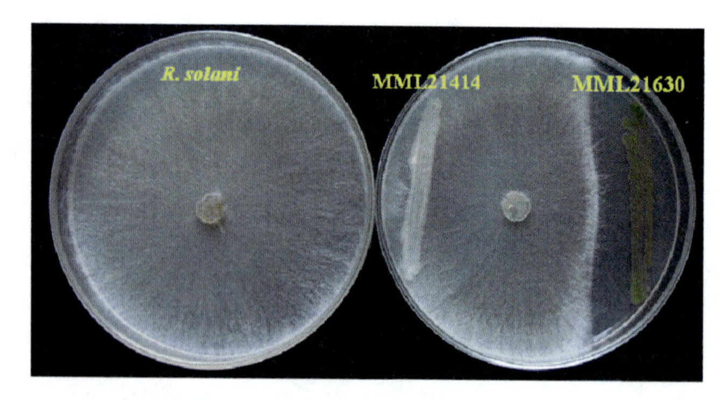


**Fig. 10.1** Antagonism of rhizobacteria against *Rhizoctonia solani*

Biological control of plant pathogens is a complex process involving interactions among the host plant, pathogen and antagonist. Other abiotic factors such as, temperature, humidity, water, soil type, pH, time and place of application of the BCAs have also played a major role in biological control (Cook and Baker 1983; Mathivanan and Manibhushanrao 2004). Though biological control is mediated by different groups of microorganisms, their operational mechanisms fall under some group of mechanisms generally known as antagonism. The antagonism is of three types: competition, antibiosis, and parasitism (Vasudevan et al. 2002; Mathivanan et al. 2008). Competition occurs between BCAs and pathogens for space and nutrition through which the former takes the advantage and suppresses the growth of pathogen by limiting space and nutrition. For example, in Fig. 10.2 *Trichothecium roseum* MML003 has over grown on *R. solani* MML004, a causal organism of sheath blight disease of rice, by utilizing more space in the plate. Similarly, in Fig. 10.3, *T. roseum* MML003 deprived the low iron present in PDA medium by the production of siderophores, which resulted in the enhanced inhibition of *R. solani* MML004 than in iron-amended PDA medium. Parasitism is the most crucial mechanism of action of BCAs against pathogens. The BCAs are capable of producing cell wall lytic enzymes such as chitinases, glucanases and proteases extracellularly and these enzymes help in degrading the pathogen structures thereby limiting the growth of pathogen by direct parasitism. Figure 10.4 depicts the role of parasitism, in which mycelial abnormalities and their degradations by the action of BCAs in dual culture with pathogen are demonstrated. Antibiosis is the most studied and widely found mechanism of BCAs (Gupta et al. 2001). The antibiotics, either volatile or non-volatile compounds produced by BCAs during antagonism, inhibit the growth of pathogens (Figs. 10.5 and 10.6). In addition, during the interaction with the host plant, the BCAs induce the resistance systemically, which ultimately prevent disease development and is referred to as induced systemic resistance (ISR) (van Loon et al. 1998).

Among the biocontrol mechanisms, antibiosis appears to be more promising. It refers to the inhibition of pathogen by the metabolic products produced by the antagonist. These products include volatile compounds, toxic compounds and

**Fig. 10.2** Competition for space: *Trichothecium roseum* MML003 over grows on *Rhizoctonia solani* MML004 on potato dextrose agar

**Fig. 10.3** Competition for nutrients: *T. roseum* MML003 antagonizes *R. solani* MML004 more effectively in iron limited medium due to the production of siderophores

antibiotics, which are deleterious to the growth or metabolic activities of other microorganisms at low concentrations (Fravel 1988). In many biocontrol systems, one or more antibiotics have been shown to play a role in disease suppression as demonstrated by many researchers (Handelsman and Stabb 1996; Fravel 1988; Jayaprakashvel et al. 2010; Shanmugaiah 2007; Shanmugaiah et al. 2010). Among various BCAs, fluorescent pseudomonads (FPs) are found to be the prolific producers of a wide variety of metabolites such as phenazines, pyrrolnitrin, pyoluteorin, oomycin A, viscosinamide and hydrogen cyanide (Dwivedi and Johri 2003). *Bacillus* spp. were found to produce many antibiotics such as zwittermycin A, kanosamine, rhizocticin C, iturins, fungicin and saltavalin and they are also capable of producing thermostable antimicrobial peptides (Emmert and Handelsman 1999; Kavitha et al. 2005). Among fungal BCAs, *Trichoderma* spp. are producing a range of antibiotic metabolites such as trichodermin, peptaibols, pyrones, etc. (Mathivanan et al. 2008). *Streptomyces* is the dominant genera among actinomyectes known to produce numerous types of antibiotics, of which, many of them, for example, validamycin and kasugamycin were commercialized as fungicides (Tanaka and Omura 1993; Mathivanan et al. 2008).

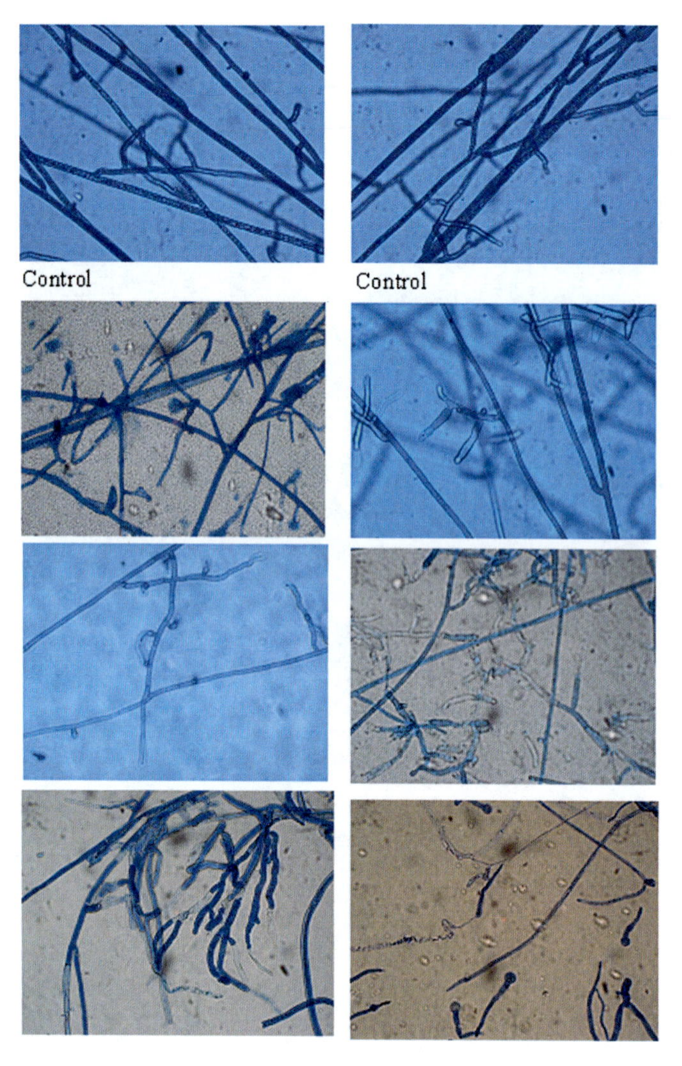


**Fig. 10.4** Effect of parasitism: mycelial abnormalities and degradations of *R. solani* MML004 due to antagonistic activity of *Pseudomonas* sp. MML21926 (40×)

Biocontrol method using whole microbial cells is cumbersome, needs at least little skill and consumes time to be used habitually by the resource-poor farmers in developing countries. In general, the BCAs perform well in laboratory and controlled environment but often fail in the field to suppress the disease. Availability of quality BCAs in viable formulations is still a hurdle in rural areas of developing countries. Moreover, the farmers are fascinated by the immediate cure by the chemical agents instead of the slow acting BCAs. In this scenario, the metabolites produced by BCAs can effectively be used to suppress the disease and can be applied easily similar to that of fungicides in the field. Further, they do not

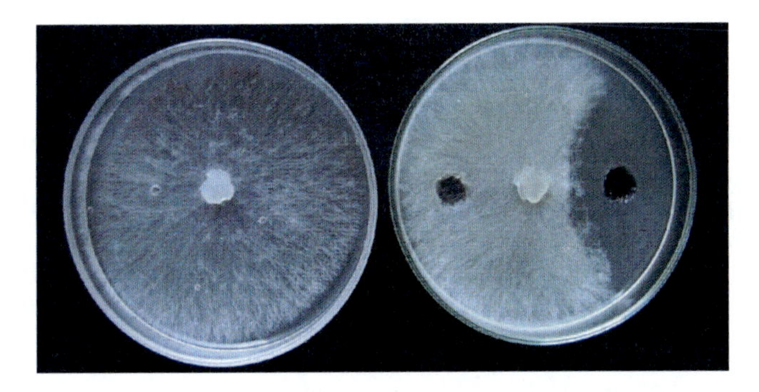

**Fig. 10.5** Antibiosis by non-volatile antibiotics of *T. roseum* MML003 against *R. solani* MML004

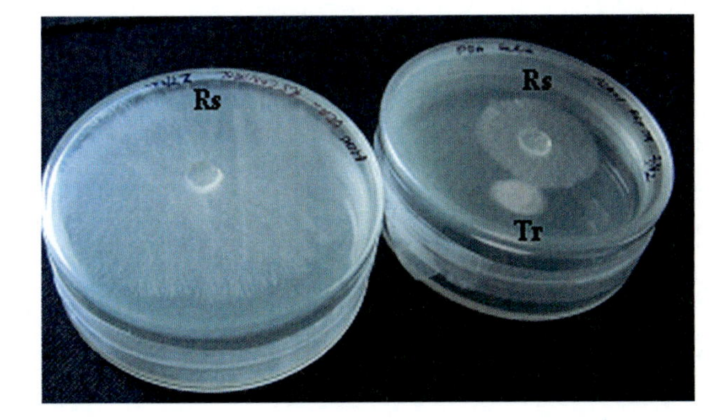


**Fig. 10.6** Antibiosis by volatile antibiotics of *T. roseum* MML003 against *R. solani* MML004 in dual bottom plates experiment

accumulate in the environment and hence, will not pose any environmental problems and health hazard. Therefore, use of microbial metabolites is considered as a wise choice in combating many plant diseases.

Hence, in the recent years, the use of secondary metabolites of microbial origin is gaining momentum in crop protection and such metabolites may be a supplement or an alternative to chemical control (Fravel 1988; Prabavathy et al. 2006; Mathivanan et al. 2008). Since these secondary metabolites are biologically synthesized, they are highly selective for target organism and hence, have little effect on beneficial organisms. Besides, as organic compounds, these metabolites are inherently biodegradable and often do not accumulate in nature and are safe to the environment (Suzni 1992; Yamaguchi 1996; Prabavathy et al. 2008). Thus, worldwide interest in them has been renewed and presently several plant diseases are being managed by the use of microbial metabolites.

Microbial metabolites are having greater advantages to be considered as potent tool in disease management strategies. The major advantages of microbial metabolites are listed below:

1. They are expected to overcome the resistance and pollution that have accompanied the use of synthetic pesticides.
2. Because they are produced by living cells, their production can be maintained in a sustainable manner. As long as the metabolites producing organism is in culture, it could be able to produce the metabolites.
3. The scale up process for the optimal synthesis of secondary metabolites by microorganisms is a very straightforward approach and hence their production can be maximized.
4. The nutritional parameters for enhanced production of metabolites by microorganisms can be optimized. It increases the scope of producing the metabolites at economically viable costs by growing the organisms with cheaper substrates such as agricultural wastes, etc.
5. Since most of the agro-active metabolites are produced as an effect of competitive survival in the natural environment, their degree of action can be superior to the rival, chemically synthesized pesticides.
6. The microbially synthesized agro-active metabolites are mostly organic substances which eventually make them easier for biological degradation at natural environments.

## 10.3 Types of Microbial Secondary Metabolites

Microbial metabolites are associated with BCAs and are used in plant disease control like any other chemical that can be categorized based on various modes. The vast and depth literature survey has suggested us that there is no single mode to characterize the metabolites into different types. There are several parameters to be considered while categorizing the microbial metabolites. But, the following four are the most important and convenient factors for the classification of microbial secondary metabolites and a simplified pattern of classification is emphasized in Fig. 10.7.

### *10.3.1 Chemical Nature of Microbial Metabolites*

The microorganisms are diverse in their nutritional requirements, habitats, morphology, physiology and metabolism; so are the secondary metabolites of microorganisms. Unlike primary metabolites which are meant for growth and reproduction, the secondary metabolites of microorganisms are responsible for interactions within the organisms and the environment and therefore their functional and chemical diversities are enormous. The potential of microorganisms to produce diverse group of secondary metabolites is remarkable. Pyrroles, macrolides, glycosides, lactones, oligopeptides, benzene derivatives, terpenes, terpenoids, alkaloids, pyrrolidones, azole compounds, fatty acids, indole types, blasticidns, polyoxins,

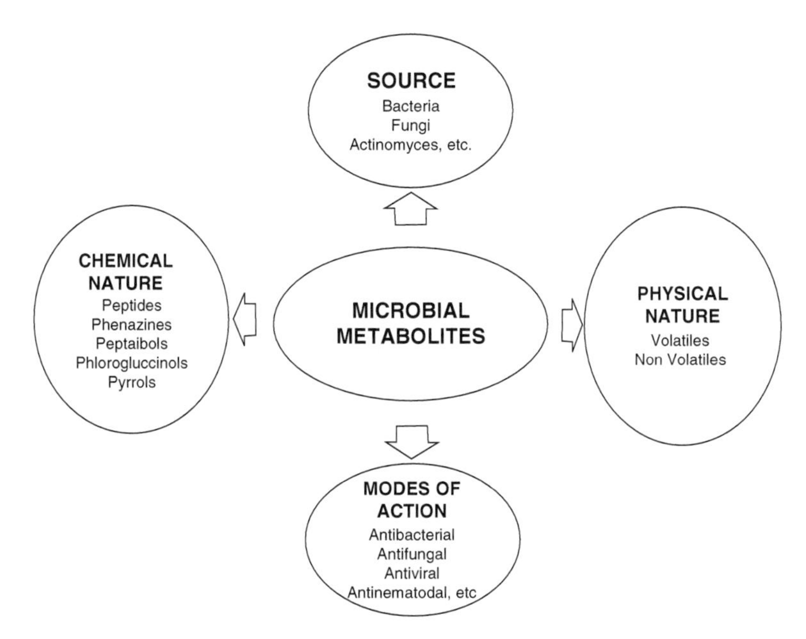


**Fig. 10.7** Classification of microbial metabolites used in the plant disease control programs

phloroglucinols, phenazines, siderophores and strobilurins are some of the chemical groups produced by microorganisms and also used in crop protection programs (Lange and Lopez 1996). Peptides of *Bacillus* spp. were used as antimicrobial metabolites against plant diseases (Kavitha et al. 2005). Cyanides, ammonia and other organic volatile compounds were used against plant pathogen (Fravel 1988; Dwivedi and Johri 2003; Jayaprakashvel and Mathivanan 2009; Jayaprakashvel et al. 2010). Siderophores, the iron-binding molecules produced by many rhizobacteria were responsible for inhibition of many phytopathogens. Examples of siderophores produced by BCAs are pyoverdin and pyochelin, which were also reported to have antimicrobial activity on their own (Arora et al. 2001; Haas and Defago 2005). However, in most of the studies, the siderophores were reported to inhibit the pathogens by iron completion only. In iron deficient medium, the siderophores were produced by BCA and the same has resulted in the increased inhibition of mycelia growth of the pathogen (Jayaprakashvel 2008). Mercado-Blanco et al. (2004) reported the siderophore-mediated suppression of Verticilium wilt by root-associated *Pseudomonas* sp. Idris et al. (2007) have reported that several rhizobacteria showed inhibitory activity toward *Pythium ultimum* by the production of antibiotic metabolites and siderophores. Bano and Musarrat (2004) have also demonstrated the role of siderophores produced by rhizobacteria for the control of *Fusarium* sp. Arora et al. (2001) have also reported siderophore-mediated biocontrol potential of *Rhizobium meliloti* against *Macrophomina phaseolina* that causes charcoal rot of groundnut.

The chemical classes of these microbial metabolites are not kingdom, genus or species specific. Some groups of microbial metabolites have been reported in most of the microorganisms while many of the microorganisms are reported to have produced different groups of antibiotics. Table 10.1 shows that *Pseudomonas* spp. could produce at least two dozen types of metabolite groups with antifungal activity. Likewise, viscosinamide is an antifungal metabolite produced by different genera of bacteria including *Pseudomonas* and *Burkholderia*. In addition, different groups of metabolites have been shown to exhibit similar biological activities. For example, mycosubtilin and zwittermycin A were two different kinds of antibiotics that have similar spectrum of activity against oomycetous pathogens (Table 10.1). Some group of the microbial metabolites such as blasticidin S, kasugamycin, polyoxins, validamycin, pyrrolnitrin and strobilurins have been commercially established as fungicides which also have bactericidal activity (Lange and Lopez 1996). Tables 10.1–10.3 provide a comprehensive list of some of the important classes of microbial metabolites produced by actinomycetes, other bacteria and fungi that are used in plant disease suppression.

### *10.3.2 Physical Nature of Microbial Metabolites*

Several antimicrobial secondary metabolites which are either readily water soluble or soluble in other organic solvents have been reported. Recently, Shanmugaiah et al. (2010) have isolated and purified a water-soluble metabolite, phenazine-1-carboxide from the culture filtrate of *Pseudomonas aeruginosa* MML2212 using solvent extraction methods and purified the same in crystalline form for the first time. The metabolite has been successfully evaluated for its potential against sheath blight disease of rice caused by *R. solani*. Earlier, Prashanth (2007) has isolated, purified and evaluated a water-soluble antifungal antibiotic belonging to the azole group of compounds from *Bacillus licheniformis* MML2501. This azole compound has significantly suppressed the dry root rot disease of groundnut plants caused by *M. phaseolina*.

The volatile metabolites are widely considered in the plant disease control programs. Bacteria, isolated from canola and soybean plants, have produced six effective antifungal organic volatile compounds such as benzothiazole, cyclohexanol, *n*-decanal, dimethyl trisulfide, 2-ethyl-1-hexanol and nonanal. These compounds inhibited sclerotia and ascospore germination, and mycelial growth of *Sclerotinia sclerotiorum*, in vitro and in soils (Fernando et al. 2005). *Bacillus subtilis* PPCB001 has produced a maximum of 21 volatile compounds, of which, 3-hydroxy-2-butanone (acetoin) was the predominant ketone type volatile which has inhibited the post harvest pathogens of citrus such as *Penicillium digitatum*, *Penicillium italicum* and *Penicillium crustosum* (Arrebola et al. 2010). Two antifungal compounds, phenylethyl alcohol and (+)-epi-bicyclesesquiphellandrene were detected in the volatile profile of *Streptomyces platensis* F-1. Consistent fumigation of healthy tissues of rice, oilseed rape and strawberry with volatile

**Table 10.1** Bacterial metabolites used in plant disease management

| Sl. no. | Name of the metabolite | Source | Mode of action | Target disease(s)/pathogen(s) | Reference(s) |
|---|---|---|---|---|---|
| 1 | Soraphen A | *Sorangium cellulosum* (A myxobacterium) | Inhibiting acetyl-CoA-carboxylase function | *Erysiphe* mildew in barley and snow mold in rye. Apple scab and gray mold on grapes | Gerth et al. (1994) |
| 2 | Agrocin 84 | *Agrobacterium radiobacter* | Inhibits protein synthesis | *A. tumefaciens* | Kerr (1980) |
| 3 | Pyrrolnitrin, pseudane | *Burkholderia cepacia* | Impairs mitochondrial function | *R. solani* and *Pyricularia oryzae* | Homma et al. (1989) |
| 4 | Xanthobaccin A | *Lysobacter* sp. strain SB-K88 | Not clear: cell wall synthesis inhibition or membrane disruption? | *Aphanomyces cochlioides* | Islam et al. (2005), Nakayama et al. (1999) |
| 5 | Herbicolin | *Pantoea agglomerans* C9-1 | Plasma membrane de-stability | *Erwinia amylovora* | Sandra et al. (2001) |
| 6 | Bacillomycin D | *Bacillus vallismortis* ZZ185 | Plasma membrane de-stability | *Fusarium graminearum*, *Alternaria alternata*, *Rhizoctonia solani*, *Cryphonectria parasitica* and *Phytophthora capsici* | Zhao et al. (2010) |
| 7 | Bacillomycin, fengycin | *Bacillus amyloliquefaciens* FZB42 | Plasma membrane de-stability | *Fusarium oxysporum* | Koumoutsi et al. (2004) |
| 8 | Iturin A | *Bacillus subtilis* | Plasma membrane de-stability | *Colletotrichum gloeosporioides*, *B. cinerea* and *R. solani* | Paulitz and Belanger (2001), Kim et al. (2010) |
| 9 | Mycosubtilin | *B. subtilis* BBG100 | Forms pores in the plasma membrane | *Pythium aphanidermatum* | Leclere et al. (2005) |
| 10 | Zwittermicin A Kanasomine | *Bacillus cereus* UW85 | May interfere with protein synthesis (?) | *Phytophthora medicaginis* and *P. aphanidermatum* | Smith et al. (1993), Milner et al. (1996) |
| 11 | 2,4-Diacetylphloroglucinol | *Pseudomonas fluorescens* | Impairs mitochondrial function | *Gaeumannomyces gramini* var. *tritici* | Nowak-Thompson et al. (1994), Raaijmakers and Weller (1998) |

(continued)

**Table 10.1** (continued)

| Sl. no. | Name of the metabolite | Source | Mode of action | Target disease(s)/pathogen(s) | Reference(s) |
|---|---|---|---|---|---|
| 12 | Oomycin A | *P. fluorescens* | Unclear | *Pythium ultimum* | Howie and Suslow (1991) |
| 13 | Phenazine-1-carboxylic acid | *P. fluorescens*<br>*Pseudomonas aureofaciens*<br>*Pseudomonas aeruginosa* | Oxidative damage | *Gaeumannomyces gramini* var. *tritici* | Pierson and Pierson (1996), Thomashow et al. (1990) |
| 14 | Phenazine-1-carboxamide | *P. aeruginosa*<br>*Pseudomonas chlororaphis* | Oxidative damage | *Fusarium oxysporum* f. sp. *radicis-lycopersici*<br>*Rhizoctonia solani* | Shanmugaiah et al. (2010) |
| 15 | Pyocyanin | *P. aeruginosa* | Interrupts the cell membrane transport systems | *Septoria tritici* | Baron and Rowe (1981), Hassan and Fridovich (1980), Howell and Stipanovic (1980), Gutterson et al. (1988) |
| 16 | Anthranilate | *P. aeruginosa* | Unclear | *Fusarium oxysporum* f. sp. *ciceris*, *Pythium* | Anjaiah et al. (1998) |
| 17 | Pyrrolnitrin | *Pseudomonas cepacia*<br>*P. fluorescens* | Impairs electron transport chain | *Aphanomyces cochlioides*, *Rhizoctonia solani*, *Pyrenophora tritici-repens* | Arima et al. (1964), Hill et al. (1994), Homma (1989), Burkhead et al. (1994) |
| 18 | Pyoluteorin | *P. fluorescens* | Unclear | *Pythium ultimum* | Howell and Stipanovic (1980) |
| 19 | Hydrogen cyanide | *P. fluorescens*<br>*P. aeruginosa* | Inhibits mitochondrial cytochrome oxidase | *Thielaviopsis basicola*<br>*Rhizoctonia solani*<br>*Meloidogyne javanica* | Voisard et al. (1989), Jayaprakashvel et al. (2010), Siddiqui et al. (2003) |

| | | | | | |
|---|---|---|---|---|---|
| 20 | Ammonia | *P. fluorescens* *Enterobacter* sp. | Unclear | *Thielaviopsis basicola* *Pythium* sp. | Howell et al. (1988), Candole and Rothrock (1997) |
| 21 | Cyclic lipopeptides like viscosinamide tensin, and amphisin | *P. fluorescens* *Burkholderia cepacia* | Inhibits membrane transport | *Rhizoctonia solani* *Pythium ultimum* | Thrane et al. (2000), Nielsen et al. (1999, 2002) |
| 22 | Gluconic acid | *Pseudomonas* strain AN5 | Unclear | *Gaeumannomyces graminis* var. *tritici* | Kaur et al. (2006) |
| 23 | Volatile organic compound | *Pseudomonas* spp., *Serratia* spp., *Stenotrophomonas* spp., etc. | Varies. But most of them inhibit mitochondrial function and cellular respiration | Many soil borne fungal pathogens | Dwivedi and Johri (2003), Kai et al. (2007) |

**Table 10.2** List of actinomycetous metabolites used in plant disease management

| Sl. no. | Name of the metabolite | Source | Mode of action | Target disease(s)/pathogen(s) | Reference(s) |
|---|---|---|---|---|---|
| 1 | Blasticidin S | *Streptomyces griseochromogenes* | Inhibition of peptide-bond formation in the ribosomal machinery | Rice blast | Takeuchi et al. (1958) |
| 2 | Kasugamycin | *Streptomyces kasugaensis* | Inhibitor of protein biosynthesis | Rice blast bacterial diseases of other crops | Umezawa et al. (1965) |
| 3 | Polyoxin B and D | *Streptomyces cacaoi* var. *asoensis* | Chitin synthase inhibitor | Fungal pathogens in fruits, vegetables and ornamentals | Suzuki et al. (1965) |
| 4 | Validamycin A | *Streptomyces hygroscopicus* var. *limoneus* | Strong inhibitor of trehalase | Rice sheath blight | Iwasa et al. (1970, 1971) |
| 5 | Fumaramidmycin | *Streptomyces kurssanovii* | Unknown | Downy mildew of vine | Loubinoux et al. (1991) |
| 6 | Streptomycin | *Streptomyces griseus* | Inhibits protein synthesis | Bacterial diseases | McManus and Stockwell (2001) |
| 7 | Irumamycin | *Streptomyces flavus* sub sp. *irumaensis* | Unclear | Blast of rice | Omura et al. (1984) |
| 8 | Avermectins | *Streptomyces avermitilis* | Dysfunction of nervous system | Insect pests | Lasota and Dybas (1991) |
| 9 | SPM5-1 | *Streptomyces* sp. PM5 | Unknown | Blast and sheath blight of rice | Prabavathy et al. (2006) |

**Table 10.3** List of fungal metabolites used in plant disease management

| Sl. no. | Name of the metabolite | Source | Mode of action | Target disease(s)/pathogen(s) | Reference(s) |
|---|---|---|---|---|---|
| 1 | Peptide antibiotics | *Trichoderma polysporum*, *Trichoderma viride* | Many | *Pythium* | Dennis and Webster (1971) |
| 2 | Gliovirin | *Gliocladium virens* (Synonym: *Trichoderma virens*) | Unclear | *Pythium ultimum* | Howell and Stipanovic (1995), Howell et al. (1993) |
| 3 | Pyrones | *Trichoderma harzianum*, *Trichoderma koningii* | Unclear | *Bipolaris sorokiniana*, *Fusarium oxysporum*, *Gaeumannomyces graminis* var. *tritici*, *Phytophthora cinnamomi*, *Pythium middletonii* and *R. solani* | Claydon et al. (1987), Simon et al. (1988) |
| 4 | Peptaibols | *Trichoderma* spp. | Affects membrane permeability | *Alternaria alternata* | Szekeres et al. (2005), Wiest et al. (2005) |
| 5 | Fusapyrone deoxyfusapyrone | *Fusarium semitectum* | Unclear | *Alternaria alternata*, *Ascochyta rabiei*, *Aspergillus flavus*, *Botrytis cinerea*, *Cladosporium cucumerinum*, *Phoma tracheiphila* and *Penicillium verrucosum*, but they least active against *Fusarium* spp. | Altomare et al. (2000, 2004) |
| 6 | p-Disubstituted | *Fusarioum chalydosporum* | Inhibition of uredospore germination | *P. arachidis* | Mathivanan and Murugesan (1999) |
| 7 | Trichothecin | *Trichothecium roseum* | Inhibits protein synthesis against fungi | Viral infection of bean and tobacco (*Tobacco necrosis virus*, *Tobacco mosaic virus* and *Tomato bushy stunt virus*). Many soil borne fungal pathogens | Bawden and Freeman (1952) |
| 8 | Strobilurin A and B | *Strobilurus tenacellus* | Inhibits mitochondrial respiration | Fungal pathogens in cereal, fruit and vegetables | Anke et al. (1977) |
| 9 | Asperfuran | *Aspergillus oryzae* | Inhibition of chitin synthesis | Fungal pathogens of cereals | Pfeferle et al. (1990) |

substances of *S. platensis* in origin effectively reduced the incidence and/or the severity of leaf blight/seedling blight of rice, leaf blight of oilseed rape and fruit rot of strawberry (Wan et al. 2008). Hydrogen cyanide (HCN) is the most potent volatile compound produced by many soil bacteria. The HCN produced by antagonistic fluorescent pseudomonads (FPs) have been very well proved to have exemplary antifungal activity against phytopathogens. It has been shown in two different studies that volatile fractions of cyanogenic FPs isolated from acidic soils of plantation crops rhizosphere and rhizosphere of coastal sand dune vegetation have inhibited the rice sheath blight pathogen up to 88% (Jayaprakashvel et al. 2006, 2010). Siddiqui et al. (2003) have clearly proved the protective nature of HCN produced by *P. aeruginosa* against *Meloidogyne javanica*, the root-knot nematode in tomato. Some of the cyanogenic strains have also been shown to have weedicidal properties. Kai et al. (2007) have demonstrated that small volatile organic compounds emitted from bacterial antagonists negatively influence the mycelial growth of the soil-borne fungus, *R. solani*. Strong inhibitions of *R. solani* (99–80%) under the test conditions were observed with the bacterial antagonists *Stenotrophomonas maltophilia* R3089, *Serratia plymuthica* HRO-C48, *Stenotrophomonas rhizophila* P69, *Serratia odorifera* 4Rx13, *Pseudomonas trivialis* 3Re2-7, *S. plymuthica* 3Re4-18 and *B. subtilis* B2g. Isolates of *Trichoderma viride* and *Trichoderma harzianum* inhibited the growth of *Fusarium moniliforme* and *Aspergillus flavus* by producing inhibitory volatile compounds (Calistru et al. 1997). The volatile secondary metabolites produced by *Trichoderma pseudokoningii*, *T. viride* and *Trichoderma aureoviride* affected the mycelial growth and protein synthesis in two isolates of *Serpula lacrymans* in varying degrees (Humphris et al. 2002).

### 10.3.3 Stability of Microbial Metabolites

Most of the microbial metabolites are showing potent bioactivities at laboratory conditions and the stability of metabolites to withstand various factors in the natural environments is very important. Among the factors, light, pH and temperature are the most considerable that could either positively or negatively influence the stability of the metabolites. Based on the factors, the secondary metabolites can be categorized as thermostable, photostable and pH stable. Jayaprakashvel et al. (2010) have characterized the crude metabolites of *T. roseum* MML003 for their photo- and thermostability in biological control against sheath blight of rice. Earlier, Kavitha et al. (2005) have isolated and established the thermostability of an antifungal peptide from *B. subtilis* against rice blast pathogen, *Magnaporthe grisea*. The bacillomycin D type metabolites isolated from *Bacillus vallismortis* ZZ185 were relatively thermostable with more than 50% of the antifungal activity even after being held at 121°C for 30 min. Meanwhile, the antifungal activity of the crude metabolites against the growth of *Alternaria alternata* and *Fusarium graminearum* remained almost unchanged (>75%) when the culture was exposed to pH ranging from 1 to 8, but significantly reduced after the filtrate had been exposed to basic conditions (Zhao et al. 2010).

## *10.3.4 Source of Microbial Metabolites*

All kinds of chemical classes of plant protective metabolites are produced by microorganisms. There is no selectivity on the production of metabolites by microorganisms. Some of the familiar metabolites are reported widely in microbial world. However, voluminous works over several decades have convincingly arrived at a conclusion that they can be conveniently categorized as bacterial, fungal and actinomycetous metabolites on the basis of biological origin.

### 10.3.4.1 Secondary Metabolites from Actinomycetes

Actinomycetes are potent producers of a wide variety of secondary metabolites with diverse biological activities, which include therapeutically and agriculturally important compounds (Tanaka and Omura 1993). Over 1,000 secondary metabolites from actinomycetes were discovered during 1988–1992. Most of these compounds are produced by various species of the genus *Streptomyces*. In fact, about 60% of the new insecticides and herbicides reported were originated from *Streptomyces* (Tanaka and Omura 1993). Actinomycetes have produced various antibacterial, antifungal, nematicidal and herbicidal antibiotic compounds and many are used in agriculture. The tetracyclines, chloramphenicol, neomycin, erythromycin, vancomycin, kanamycin, cephalosporin and rifamycin, were few of the antibacterial antibiotics produced by actinomycetes (Prabavathy et al. 2008). Phenazine antibiotics were also extracted from the mycelium of the alkaliphilic *Nocardiopsis* strain OPC-15 (Tsujibo et al. 1988). Actinomyctes have also produced several insecticidal metabolites such as avermectins, tetranactin and milbemycin (Misato 1983; Isono 1990; Tanaka and Omura 1993). Secondary metabolites showing herbicidal activity such as anismycins bialaphos, herbicidines, herbimycins, hydantocidin, cornexistin, phthoxazolin and homoalanosin were reported to be produced by actinomycetes and many of them were used as weedicides (Stephen and Lydon 1987; Zhang et al. 1987; Mio et al. 1991; Nakajima et al. 1991; Shen 1997; Copping 1996). The research progress on the microbial herbicides including the metabolites of actinomycetes has been reviewed recently (Li et al. 2003).

Blasticidin, kasugamycin, polyoxins and validamycins are few of the commercially successful metabolites of actinomycetes used in plant disease control (Takeuchi et al. 1958; Suzuki et al. 1965; Umezawa et al. 1965; Shibata et al. 1970). The origin, modes of action and biologically active spectrum of the major metabolites produced by actinomycetes are summarized in Table 10.2. Two antifungal aliphatic compounds, SPM5C-1 and SPM5C-2 with a lactone and ketone carbonyl unit, respectively, obtained from *Streptomyces* sp. PM5 were evaluated under in vitro and in vivo conditions against major rice pathogens, *Pyricularia oryzae* and *R. solani* (Prabavathy et al. 2006). Several antimicrobial secondary metabolites of *Streptomyces* sp. MML1042 were partially purified and demonstrated for the antifungal activity against many soil-borne fungal phytopathogens (Malarvizhi 2006). Similarly, Ramesh (2009) has isolated and characterized bioactive compounds from *Streptomyces fungicidicus* MML1614 that were having exceptional inhibitory activity against mycelial growth of *R. solani* and *A. alternata*. An antifungal protein from the marine bacterium, *Streptomyces* sp. strain AP77 was found to be inhibitory against *Pythium porphyrae*, a causative agent of red rot disease in *Porphyra* spp. (Woo et al. 2002). In vivo antifungal activity of 5-hydroxyl-5-methyl-2-hexenoic acid from *Actinoplanes* sp. HBDN08 under greenhouse conditions demonstrated that the metabolite could effectively control the diseases caused by *Botrytis cinerea*, *Cladosporium cucumerinum* and *Cladosporium cassiicola* with 71.42, 78.63 and 65.13%, respectively, at 350 mg/L. This strong antifungal activity suggested that 5-hydroxyl-5-methyl-2-hexenoic acid might be a promising candidate for new antifungal agents (Zhang et al. 2010). Though streptomycetes metabolites were extensively studied in relation to plant disease control, no attempt to date has been made to elucidate the production of antifungal antibiotics by non-streptomycetes actinomycetes (NSA). Detailed studies on metabolites of NSA in relation to the growth and pathogenic activities of soil-borne fungal plant pathogens are clearly warranted (El-Tarabily and Sivasithamparam 2006).

#### 10.3.4.2 Secondary Metabolites from Bacteria

*Bacillus* and *Pseudomonas* are the two most important genera among bacteria that are studied quite extensively for antimicrobial metabolites in the plant disease control programs (Table 10.1). Bacillomycins, fengycin, iturin A, mycosubtilin and zwittermicin A are some of the important antibiotic secondary metabolites produced by *Bacillus* spp. and are widely used in plant disease control (Smith et al. 1993; Paulitz and Belanger 2001; Kloepper et al. 2004; Koumoutsi et al. 2004; Leclere et al. 2005; Zhao et al. 2010). Recently, an azole compound produced by *B. licheniformis* MML2501 was completely characterized for its disease control potential against *M. phaseolina*, the causative agent of dry root rot of groundnut. The purified azole compound has exceptional antifungal activity against many soil-borne fungal phytopathogens except *R. solani* (Prashanth 2007). Kavitha et al. (2005) have isolated and purified a thermostable antifungal protein from *Bacillus* sp. which retained antifungal activity against *M. grisea* even after autoclaving. The list of antimicrobial metabolites produced by the genera *Pseudomonas* is quite astonishing. The members of this genus are the most studied bacterial BCAs with their potential to produce diverse range of bioactive secondary metabolites. 2,4-Diacetylphloroglucinol, oomycin A, phenazine-1-carboxylic acid, phenazine-1-carboxamide, pyocyanin, anthranilate, pyrrolnitrin, pyoluteorin, hydrogen cyanide, ammonia, viscosinamide and gluconic acid are the most widely reported antimicrobial metabolites of *Pseudomonas* spp. The extensive production of antimicrobial secondary metabolites by *Pseudomonas* spp. as crop protection agents has been comprehensively reviewed (Dowling and O'Gara 1994; Dwivedi and Johri 2003; Chin-A-Woeng et al. 2003). Other bacteria such as *Agrobacterium radiobacter*, *Burkholderia cepacia*, *Lysobacter* sp. and *Pantoea agglomerans* have been reported to produce various antibiotic compounds such as agrocin 84, pyrrolnitrin, pseudane, xanthobaccin A and herbicolin (Kerr 1980; Homma et al. 1989; Islam et al. 2005; Sandra et al. 2001; Zhao et al. 2010).

#### 10.3.4.3 Secondary Metabolites from Biocontrol Fungi

*Trichoderma*, *Gliocladium*, non-pathogenic *Fusarium* and *Trichothecium* are few fungal genera that are found to produce various antimicrobial metabolites against plant pathogens. Table 10.3 emphasizes the major agro-active metabolites produced by fungal BCAs. Liu and Li (2004) and Mathivanan et al. (2008) have reviewed the major anitimicrobial metabolites produced by these three fungi and others as well. It has been reported that *Trichoderma* spp. produces a wide range of volatile and non-volatile antibiotic secondary metabolites such as trichodermin, viride, gliotoxin and peptaibols (Weindling and Emerson 1936; Sivasithamparam and Ghisalberti 1998; Vyas and Mathur 2002; Wiest et al. 2002). Garret and Robinson (1969) isolated nonanoic acid from *Fusarium oxysporum*, which inhibited the spore germination of *Cunninghamella elegans*. The toxic metabolites produced by *Fusarium chlamydosporum* have effectively inhibited groundnut rust pathogen, *Puccinia arachidis,* and was able to reduce the number of pustules (Mathivanan 1995). Further, an antifungal metabolite of p-disubstituted aromatic nature isolated from the culture filtrate of *F. chlamydosporum* inhibited the uredospore germination at 30 μg/ml concentration (Mathivanan and Murugesan 1999). Further, two pyrones, viz., fusapyrone and deoxyfusapyrone from *Fusarium semitectum* were highly active against *A. alternata*, *Ascochyta rabiei*, *A. flavus*, *B. cinerea*, *C. cucumerinum*, *Phoma tracheiphila* and *Penicillium verrucosum* while they were least active against *Fusarium* spp. (Altomare et al. 2000). Babalola (2010) have reported the improved mycoherbicidal activity of *Fusarium arthrosporioides*. *T. roseum* has been reported to produce many antifungal and antiviral secondary metabolites of agricultural importance. Trichothecin is the major antibiotic produced by *T. roseum* and found to have both antifungal and antiviral activities. Few decades ago, Bawden and Freeman (1952) have reported the effect of two heat stable trichothecene compounds which inhibited viral infection of bean and tobacco. Similarly, Urbasch (1992) reported the toxic effects of water-soluble, heat-resistant metabolites from *T. roseum* against mycelial growth and conidial germination of *Pestalotia funerea*. Trichothecin was used in cottonseeds and crop plants to prevent wilt diseases (Askarova and Ioffe 1962). Recently, Jayaprakashvel et al. (2010) have successfully controlled the sheath blight disease of rice under green house conditions with thermostable, photostable crude metabolites of *T. roseum* MML003.

A number of fungal genera namely, *Aphanocladium album*, *Acremonium obclavatum*, *Myrothecium verrucaria*, *Verticillium chlamydosporium*, *Penicillium brevicompactum*, *Penicillium expansum*, *Penicillium pinophilum* and *Coniothyrium minitans* were reported to produce antimicrobial metabolites against plant pathogens. However, their potential have inadequately been exploited in plant protection (Liu and Li 2004; Mathivanan et al. 2008).

## 10.4 Current Trends of Using Microbial Metabolites for Plant Disease Control

The scope of the usage of microbial metabolites has been widened by the extensive research and developmental activities over the past few decades. One of the major such advancement is yet not so recent but being practiced for years, is the use of microbial metabolites either directly as fungicides or many a times as lead molecules for the synthesis of novel class of fungicides. Blasticidin S, kasugomycin, polyoxin B and D, validamycins, pyrrolnitrin, strobilurins, fumaramidmycin, coniothyriomycin, rhizocticins and gliovirin are the most important examples of such microbial metabolites used either as lead molecules or as fungicides (Lange and Lopez 1996). The microbial metabolites were also used as elicitors for the induction of systemic resistance in plants against pests and diseases (Heil and Bostock 2002). Plant growth promoting rhizobacteria through various mechanisms including their metabolites were used extensively in the induction of plant defense for crop protection against crop pests and pathogens (Ramamoorthy et al. 2001; De Vleesschauwer and Höfte 2009) In recent years, the cellular communication within and among BCAs were considered more in the plant disease control programs (Jayaprakashvel 2008) and Sect. 10.4.1 of this book elaborates the significance of this new trend in plant disease control.

### *10.4.1 Role of Quorum Sensing, Auto Inducers, Quorum Quenching on Plant Disease Control Programs*

In quorum sensing, bacteria monitor the presence of other bacteria in their surroundings by producing and responding to signaling molecules known as autoinducers. There are two general types of bacterial quorum-sensing systems: Gram-negative LuxIR circuits and Gram-positive oligopeptide two-component circuits. Low mol% G + C content Gram-positive bacteria typically use modified oligopeptides as autoinducers. These signals are generically referred to as autoinducing polypeptides (AIPs). The LuxI-type proteins in Gram-negative bacteria catalyze the formation of a specific acyl-homoserine lactone (AHL) autoinducer that freely diffuses into and out of the cell and the concentration increases in proportion to cell population density (Taga and Bassler 2003). In plant-associated bacteria, AHLs are found in pathogenic, symbiotic and biological control strains and they regulate a diverse range of phenotypes including diverse pathogenicity determinants, conjugation, rhizosphere competence and production of antifungal metabolites (Newton and Fray 2004).

Quorum sensing regulates virulence in many human and plant pathogens. Presumably, in an attempt to avoid alerting the host's immune system to their presence, quorum sensing bacteria delay the virulence factor production until a high cell number is reached so that secretion of virulence factors resulted in a productive infection (Miller and Bassler 2001). For example, pathogens such as *P. aeruginosa* and *Erwinia carotovora* (*Pectobacterium caratovorum*) use AHL-mediated quorum sensing to activate their virulence genes in specific animal and plant hosts, respectively (de Kievit and Iglewski 2000). One striking example of this type of host response occurs in the seaweed, *Delisea pulchra*. This organism produces a number of halogenated furanones and enones that interfere with homoserine lactone (HSL) mediated processes such as swarming in *Serratia liquefaciens* (Miller and Bassler 2001). Since plant pathogenic bacteria such as *Erwinia* spp. are regulating their pathogenicity by AHL type autoinducers, the AHL antagonist produced by other microbes can be used to control the diseases caused by Erwinias (Dong et al. 2000). This kind of degrading pathogen quorum sensing signals has been considered as preventive and curative biocontrol mechanism (Molina et al. 2003). Jayaprakashvel (2008) has utilized the principles of quorum sensing and quorum quenching for the development of a synergistically performing consortium of BCAs against sheath blight disease of rice. Though research on this microbial communication with relevance to human pathology has achieved greater pace, the significance of such studies in plant disease control programs is yet to be realised.

## 10.5 Future Perspectives of Microbial Metabolites in Plant Disease Control

High throughput screening methods are revolutionizing the discovery of novel metabolites of microbial origin with novel bioactivities (Zhang et al. 2007). However, the successful use of this advanced technology in identifying novel agroactive metabolites is yet to be followed vigorously. This would save the time required for discovering and developing agroactive molecules into a commercial scale. Microbial metabolic engineering is another field where the generation of well-characterized parts and the formulation of biological design principles in synthetic biology are laying the foundation for more complex and advanced engineering of synthetic microbial metabolism (McArthur and Fong 2010). The research on metabolic system biology to map out the regulation networks that control microbial metabolism is always a challenging one (Heinemann and Sauer 2010). However, in metabolic engineering, people use the microbial genomic information with metabolic pathways, networks and regulation information in databases to create strategies for genomic engineering that are carried out to optimize the microbial production of metabolites.

The earth is going greener in the recent years and in all the aspects, environmental safety is considered as a vital issue. Though synthetic chemicals such as fungicides offer satisfactory control over plant diseases, their environmental toxicity, persistence in nature and resistance development have urged us to dedicate more interests on the microbial metabolites for various purposes; here, for plant disease control. Almost all microbial metabolites in plant disease control have well been accepted to be environment friendly. Their organic nature makes them the more desirable disease control agents. Identification of biologically produced antibiotic secondary metabolites has always been a difficult task. The chemical complexity of these metabolites makes them potential candidates against the resistance developing pathogens. However, the same chemical complexity makes them really hard molecules to solve the structures. As structures are related with function, they need to be concentrated more. Moreover, structural studies are essential in the preliminary screening of identifying the molecules whether novel or not. For such studies, development of a comprehensive database for holding the structures of all purified metabolites in relation to their function has to be made. Though currently many such structural databases are available, the relevance of structure and function is yet to be considered.

## 10.6 Conclusion

Biological control using the microbial metabolites produced by antagonistic microorganisms is an attractive alternative for chemically synthesized pesticides. Their biodegradability and specific action make them the prime choice for environment-friendly plant disease management. Such plant protective metabolites are produced by many microorganisms consisting of diverse chemical groups. Besides, use of these microbial metabolites as lead molecules for the synthesis of plant protective chemicals opens up newer avenues for entrepreneurs and industrialists.

## References

Altomare C, Perrone G, Zonno MC, Evidente A, Pingue R, Fanti F, Polonelli L (2000) Biological characterization of fusapyrone and deoxyfusapyrone, two bioactive secondary metabolites of *Fusarium semitectum*. J Nat Prod 63:1131–1135

Anjaiah V, Koedam N, Nowak-Thompson B, Loper JE, Höfte M, Tambong JT, Cornelis P (1998) Involvement of phenazines and anthranilate in the antagonism of *Pseudomonas aeruginosa* PNA1 and Tn5-derivatives towards *Fusarium* sp. and *Pythium* sp. Mol Plant Microbe Interact 11:847–854

Anke T, Oberwinkler F, Steglich W, Schramm G (1977) The strobilurins-new antifungal antibiotics from the basidiomycete *Strobilurus tenacellus*. J Antibiot (Tokyo) 30:806–810
Arima K, Imanaka I, Kousaka M, Fukuta A, Tamura G (1964) Pyrrolnitrin, a new antibiotic substance, produced by *Pseudomonas*. Agric Biol Chem 28:575–576
Arora NK, Kang SC, Maheshwari DK (2001) Isolation of siderophore-producing strains of *Rhizobium meliloti* and their biocontrol potential against *Macrophomina phaseolina* that causes charcoal rot of groundnut. Curr Sci 81:673–677
Arrebola E, Sivakumar D, Korsten L (2010) Effect of volatile compounds produced by *Bacillus* strains on postharvest decay in citrus. Biol Control 53:122–128
Ash GJ (2010) The science, art and business of successful bioherbicides. Biol Control 52:230–240
Askarova SA, Ioffe RI (1962) On the possibility of the use of the fungicidal preparation trichothecin in the control of *Fusarium decemcellulare* on *Puccinia psidii* in guava (*Psidium guajava*). Antibiotics 7:929–930
Babalola OO (2010) Improved mycoherbicidal activity of *Fusarium arthrosporioides*. Afr J Microbiol Res 4:1659–1662
Bano N, Musarrat J (2004) Characterization of a novel carbofuran degrading *Pseudomonas* sp. with collateral biocontrol and plant growth promoting potential. FEMS Microbiol Lett 231:13–17
Baron SS, Rowe JJ (1981) Antibiotic action of pyocyanin. Antimicrob Agents Chemother 20:814–820
Bawden FC, Freeman GG (1952) The nature and behavior of inhibitors of plant viruses produced by *Trichothecium roseum* Link. J Gen Microbiol 7:154–168
Burkhead KD, Schisler DA, Slininger PJ (1994) Pyrrolnitrin production by biocontrol agent *Pseudomonas cepacia* B37w in culture and in colonized wounds of potatoes. Appl Environ Microbiol 60:2031–2039
Calistru C, McLean M, Berjak P (1997) In vitro studies on the potential for biological control of *Aspergillus flavus* and *Fusarium moniliforme* by *Trichoderma* species. Mycopathologia 139:115–121
Candole BL, Rothrock CS (1997) Characterization of the suppressiveness of hairy vetch-amended soils to *Thielaviopsis basicola*. Phytopathology 87:197–202
Chakraborty S (2005) Potential impact of climate change on plant–pathogen interactions Australian. Plant Pathol 34:443–448
Chin-A-Woeng TFC, Bloomberg GV, Lugtenberg BJJ (2003) Phenazines and their role in Biocontrol by *Pseudomonas* bacteria. New Phytol 157:503–523
Claydon N, Allan M, Hanson JR, Avent AG (1987) Antifungal alkyl pyrones of *Trichoderma harzianum*. Trans Br Mycol Soc 88:505–513
Cook RJ (2000) Advances in plant health management in the 20th century. Annu Rev Phytopathol 38:95–116
Cook RJ, Baker KF (1983) The nature and practice of biological control of plant pathogens. American Phytopathological Society, St. Paul, MN
Copping LG (1996) Crop protection agents from nature: natural products and analogues. The Royal Society of Chemistry, Cambridge, p 501
de Kievit TR, Iglewski BH (2000) Bacterial quorum sensing in pathogenic relationships. Infect Immun 68:4839–4849
De Vleesschauwer D, Höfte M (2009) Rhizobacteria-induced systemic resistance. Adv Bot Res 51:223–281
Dennis C, Webster J (1971) Antagonistic properties of species-groups of Trichoderma. III. Hyphal interaction. Trans Br Mycol Soc 57:363–369
Dong YH, Xu JL, Li XZ, Zhang LH (2000) AiiA, an enzyme that inactivates the acylhomoserine lactone quorum-sensing signal and attenuates the virulence of *Erwinia carotovora*. Proc Natl Acad Sci USA 97:3526–3531
Doumbou CL, Hamby Salove MK, Crawford DL, Beaulieu C (2002) Actinomycetes, promising tools to control plant diseases and to promote plant growth. Phytoprotection 82:85–102

Dowling DN, O'Gara F (1994) Metabolites of *Pseudomonas* involved in the biocontrol of plant disease. Trends Biotechnol 12:133–141

Dwivedi D, Johri BN (2003) Antifungals from fluorescent pseudomonads: biosynthesis and regulation. Curr Sci 12:1693–1703

El-Tarabily KA, Sivasithamparam K (2006) Non-streptomycete actinomycetes as biocontrol agents of soil-borne fungal plant pathogens and as plant growth promoters. Soil Biol Biochem 38:1505–1520

Emmert EAB, Handelsman J (1999) Biocontrol of plant disease: a (gram-) positive perspective. FEMS Microbiol Lett 171:1–9

Evans N, Baierl A, Semenov MA, Gladders P, Fitt BDL (2008) Range and severity of a plant disease increased by global warming. J R Soc Interface 5:525–531

Fernando WG, Ramarathnam R, Krishnamoorthy AS, Savchuk SC (2005) Identification and use of potential bacterial organic antifungal volatiles in biocontrol. Soil Biol Biochem 37:955–964

Fravel D (1988) Role of antibiosis in the biocontrol of plant diseases. Annu Rev Phytopathol 26:75–91

Fravel D (2005) Commercialization and implementation of biocontrol. Annu Rev Phytopathol 43:337–359

Garret MK, Robinson PM (1969) A stable inhibitor of spore germination produced by fungi. Arch Microbiol 67:370–377

Gerth K, Bedorf N, Irschik H, Höfle G, Reichenbach H (1994) The soraphens: a family of novel antifungal compounds from *Sorangium cellulosum* (*Myxobacteria*). I. Soraphen A1 alpha: fermentation, isolation, biological properties. J Antibiot 47:23–31

Gullino ML, Leroux P, Smith C (2000) Uses and challenges of novel compounds for plant disease control. Crop Prot 19:1–11

Gupta PK (2004) Pesticide exposure-Indian scene. Toxicology 198:83–90

Gupta CP, Dubey RC, Kang SC, Maheshwari DK (2001) Antibiosis mediated necrotrophic effect of *Pseudomonas* $GRC_2$ against two fungal pathogens. Curr Sci 81:91–94

Gutterson N, Ziegle JS, Warren GJ, Layton TJ (1988) Genetic determinants for catabolite induction of antibiotic biosynthesis in *Pseudomonas fluorescens* HV 37a. J Bacteriol 170:380–385

Haas D, Defago G (2005) Biological control of soil-borne pathogens by fluorescent pseudomonads. Nat Rev Microbiol 3:307–319

Handelsman J, Stabb EV (1996) Biocontrol of soil-borne plant pathogens. Plant Cell 8:1855–1869

Harman GE, Howell CR, Vitarbo A, Chet I, Lorito M (2004) *Trichoderma* species – opportunistic, avirulent plant symbionts. Nat Rev Microbiol 2:43–56

Hassan HM, Fridovich I (1980) Mechanism of the antibiotic action of pyocyanine. J Bacteriol 141:156–163

Heil M, Bostock R (2002) Induced systemic resistance (ISR) against pathogens in the context of induced plant defences. Ann Bot 89:503–512

Heinemann M, Sauer U (2010) Systems biology of microbial metabolism. Curr Opin Microbiol 13:337–343

Hill DS, Stein JI, Torkewitz NR, Morse AM, Howell CR, Pachlatko JP, Decker JO, Ligon JM (1994) Cloning of genes involved in the synthesis of pyrrolnitrin from *Pseudomonas fluorescens* and role of pyrrolnitrin synthesis in biological control of plant disease. Appl Environ Microbiol 60:78–85

Homma Y, Sato Z, Hirayama F, Konno K, Shirahama H, Suzui T (1989) Production of antibiotics by *Pseudomonas cepacia* as an agent for biological control of soil borne pathogens. Soil Biol Biochem 21:723–728

Howell CR, Stipanovic RD (1980) Suppression of *Pythium ultimum*-induced damping-off of cotton seedlings by *Pseudomonas fluorescens* and its antibiotic, pyoluteorin. Phytopathology 70:712–715

Howell CR, Stipanovic RD (1995) Mechanisms in the biocontrol of *Rhizoctonia solani*-induced cotton seedling disease by *Gliocladium virens*: antibiosis. Phytopathology 85:469–472

Howell CR, Beier RC, Stipanovic RD (1988) Production of ammonia by *Enterobacter cloacae* and its possible role in the biological control of *Pythium* pre-emergence damping-off by the bacterium. Phytopathology 78:1075–1078

Howell CR, Stipanovic RD, Lumsden RD (1993) Antibiotic production by strains of *Gliocladium virens* and its relation to the biocontrol of cotton seedling diseases. Biocontrol Sci Technol 3:435–441

Howie WJ, Suslow TV (1991) Role of antibiotic biosynthesis in the inhibition of *Pythium ultimum* in the cotton spermosphere and rhizosphere by *Pseudomonas fluorescens*. Mol Plant Microbe Interact 4:393–399

Humphris SN, Bruce A, Buultjens E, Wheatley RE (2002) The effects of volatile microbial secondary metabolites on protein synthesis in *Serpula lacrymans*. FEMS Microbiol Lett 210:215–219

Idris AH, Labuschagne N, Korsten L (2007) Screening rhizobacteria for biological control of Fusarium root and crown rot of sorghum in Ethiopia. Biol Control 40:97–106

Islam MdT, Hashidoko Y, Deora A, Ito T, Tahara S (2005) Suppression of damping-off disease in host plants by the rhizoplane bacterium *Lysobacter* sp. strain SB-K88 is linked to plant colonization and antibiosis against soil-borne peronosporomycetes. Appl Environ Microbiol 71:3786–3796

Isono K (1990) Antibiotics as non-pollution agricultural pesticides. Comments Agr Food Chem 2:123–142

Iwasa T, Yamamoto H, Shibata M (1970) Studies on validamycins, new antibiotics. I: *Streptomyces hygroscopicus* var. *limoneus* nov. var. validamycin-producing organism. J Antibiot (Tokyo) 23:595–602

Iwasa T, Higashide E, Yamamoto H, Shibata M (1971) Studies on validamycins, new antibiotics. II: production and biological properties of validamycins A and B. J Antibiot (Tokyo) 24: 107–113

Jayaprakashvel M (2008) Development of a synergistically performing bacterial consortium for sheath blight suppression in rice. Ph.D. thesis, University of Madras, Madras, India

Jayaprakashvel M, Mathivanan N (2009) Biological control and its implications on rice diseases management. In: Ponmurugan P, Deepa MA (eds) Role of biocontrol agents for disease management in sustainable agriculture. Scitech, Chennai, India, pp 440–455

Jayaprakashvel M, Ramesh S, Mathivanan N, Baby UI (2006) Prevalence of fluorescent pseudomonads in the rhizosphere of plantation crops and their antagonistic properties against certain phytopathogens. J Plantation Crops 34:728–732

Jayaprakashvel M, Muthezhilan R, Srinivasan R, Jaffar Hussain A, Gopalakrishnan S, Bhagat J, Kaarthikeyan N, Muthulakshmi R (2010) Hydrogen cyanide mediated biocontrol potential of *Pseudomonas* sp. AMET1055 isolated from the rhizosphere of coastal sand dune vegetation. J Adv Biotechnol 9:39–42

Kai M, Effmert U, Berg G, Piechulla B (2007) Volatiles of bacterial antagonists inhibit mycelial growth of the plant pathogen *Rhizoctonia solani*. Arch Microbiol 187:351–360

Kaur R, Macleod J, Foley W, Nayudu M (2006) Gluconic acid: an antifungal agent produced by *Pseudomonas* species in biological control of take-all. Phytochemistry 67:595–604

Kavitha S, Senthilkumar S, Gnanamanickam SS, Inayathullah M, Jayakumar J (2005) Isolation and partial characterization of antifungal protein from *Bacillus polymyxa* strain VLB16. Process Biochem 40:3236–3243

Kerr A (1980) Biological control of crown gall through production of agrocin 84. Plant Dis 64:25–30

Kim BS, Hwang BK (2007) Microbial fungicides in the control of plant diseases. J Phytopathol 155:641–653

Kim PI, Ryu J, Kim YH, Chi YT (2010) Production of biosurfactant lipopeptides Iturin A, fengycin and surfactin A from *Bacillus subtilis* CMB32 for control of *Colletotrichum gloeosporioides*. J Microbiol Biotechnol 20:138–145

Kloepper JW, Ryu CM, Zhang S (2004) Induced systemic resistance and promotion of plant growth by *Bacillus* spp. Phytopathology 94:1259–1266

Koumoutsi A, Chen XH, Henne A, Liesegang H, Gabriele H, Franke P, Vater J, Borris R (2004) Structural and functional characterization of gene clusters directing nonribosomal synthesis of bioactive lipopeptides in *Bacillus amyloliquefaciens* strain FZB42. J Bacteriol 186:1084–1096

Lange L, Lopez CS (1996) Micro-organisms as a source of biologically active secondary metabolites. In: Copping LG (ed) Crop protection agents from nature: natural products and analogues. The Royal Society of Chemistry, London, pp 1–26

Lasota JA, Dybas RA (1991) Avermectins, a novel class of compounds: implications for use in arthropod pest control. Ann Rev Entomol 36:91–117

Leclere VM, Bechet A, Adam JS, Guez B, Wathelet M, Ongena P, Thonart F, Gancel M, Chollet-Imbert Jacques P (2005) Mycosubtilin overproduction by *Bacillus subtilis* BBG100 enhances the organism's antagonistic and biocontrol activities. Appl Environ Microbiol 71:4577–4584

Li Y, Sun Z, Zhuang X, Xu L, Chen S, Li M (2003) Research progress on microbial herbicides. Crop Prot 22:247–252

Liu X, Li S (2004) Fungal secondary metabolites in biological control of crop pests. In: An Z (ed) Handbook of industrial mycology. CRC, New York, pp 723–747

Loubinoux B, Gérardin P, Kunz W, Herzog J (1991) Activity of fumaramidmycin mimics against oomycetes. Pestic Sci 33:263–269

Malarvizhi K (2006) Biodiversity and antimicrobial potential of soil actinomycetes from south India: isolation, purification and characterization of an antifungal metabolite produced by *Streptomyces* sp. MML1042. Ph.D. thesis, University of Madras, Madras, India

Mathivanan N (1995) Studies on extracellular chitinase and secondary metabolites produced by *Fusarium chlamydosporum*, an antagonist to *Puccinia arachidis*, the rust pathogen of groundnut. Ph.D. thesis, University of Madras, Madras, India

Mathivanan N, Manibhushanrao K (2004) An overview of current strategies on biological control of soil-borne pathogens. In: Prakash HS, Niranjana RS (eds) Vistas in applied botany. Department of Applied Botany and Biotechnology, University of Mysore, Mysore, India, pp 119–148

Mathivanan N, Murugesan K (1999) Isolation and purification of an antifungal metabolite from *Fusarium chlamydosporum*, a mycoparasite to *Puccinia arachidis*, the rust pathogen of groundnut. Indian J Exp Biol 37:98–101

Mathivanan N, Srinivasan K, Chelliah S (1997) Evaluation of *Trichoderma viride* and Carbendazim and their integration for the management of root diseases in cotton. Indian J Microbiol 37:107–108

Mathivanan N, Manibhushanrao K, Murugesan K (2006) Biological control of plant pathogens. In: Anand N (ed) Recent trends in botanical research. University of Madras, Chennai, India, pp 275–323

Mathivanan N, Prabavathy VR, Vijayanandraj VR (2008) The effect of fungal secondary metabolites on bacterial and fungal pathogens. In: Karlovsky P (ed) Secondary metabolites in soil ecology. Springer, Berlin, pp 129–140

Mathre DE, Cook RJ, Callan NW (1999) From discovery to use: traversing the world of commercializing biocontrol agents for plant disease control. Plant Dis 83:972–983

McArthur GH IV, Fong SS (2010) Toward engineering synthetic microbial metabolism. J Biomed Biotechnol. doi:10.1155/2010/459760

McManus PS, Stockwell VO (2001) Antibiotic use for plant disease management in the United States. Plant Health Progr. doi:10.1094/PHP-2001-0327-01-RV

Mercado-Blanco J, Rodrguez Jurado D, Hervas A, Jimenez Daza RM (2004) Suppression of Verticillium wilt in olive planting stocks by root associated fluorescent *Pseudomonas* spp. Biol Control 30:474–486

Miller MB, Bassler BL (2001) Quorum sensing in bacteria. Annu Rev Microbiol 55:165–199

Milner JL, Silo-Suh LA, Lee JC, He H, Clardy J, Handelsman J (1996) Production of kanosamine by *Bacillus cereus* UW85. Appl Environ Microbiol 62:3061–3065

Mio S, Sano H, Shindou M, Honma T, Sugai S (1991) Synthesis and herbicidal activity of deoxy derivatives of (+) hydantocidin. Agric Biol Chem 55:1105–1109

Misato T (1983) Recent status and future aspects of agricultural antibiotics. In: Miyamoto J, Kaerney PC (eds) Pesticide chemistry: human welfare and the environment. Pergamon, Oxford, pp 241–246

Molina L, Constantinescu F, Michel L, Reimmann C, Duffy B, Defago G (2003) Degradation of pathogen quorum-sensing molecules by soil bacteria: a preventive and curative biological control mechanism. FEMS Microbiol Ecol 45:71–81

Nakajima M, Itoi K, Takamatsu Y, Sato S, Furukawa Y (1991) Cornexistin: a new fungal metabolite with herbicidal activity. J Antibiot 44:1065–1072

Nakayama T, Homma Y, Hashidoko Y, Mizutani J, Tahara S (1999) Possible role of xanthobaccins produced by *Stenotrophomonas* sp. strain SB-K88 in suppression of sugar beet damping-off disease. Appl Environ Microbiol 65:4334–4339

Newton JA, Fray RG (2004) Integration of environmental and host-derived signals with quorum sensing during plant–microbe interactions. Cell Microbiol 6:213–224

Nielsen TH, Christophersen C, Anthoni U, Sørensen J (1999) Viscosinamide, a new cyclic depsipeptide with surfactant and antifungal properties produced by *Pseudomonas fluorescens*. J Appl Microbiol 86:80–90

Nielsen TH, Sørensen D, Tobiasen T, Andersen JB, Christophersen C, Givskov M, Sørensen J (2002) Antibiotic and biosurfactant properties of cyclic lipopeptides produced by fluorescent *Pseudomonas* spp. from the sugar beet rhizosphere. Appl Environ Microbiol 68:3416–3423

Nowak-Thompson B, Gould SJ, Kraus J, Loper JE (1994) Production of 2,4-diacetylphloroglucinol by the biocontrol agent *Pseudomonas fluorescens* Pf-5. Can J Microbiol 40:1064–1066

Omura S, Tanaka Y, Takahashi Y, Chia I, Inoue M, Iwai Y (1984) Irumamycin, an antifungal 20-membered macrolide produced by a *Streptomyces*. Taxonomy, fermentation and biological properties. J Antibiot (Tokyo) 37:1572–1578

Paulitz TC, Belanger RR (2001) Biological control in greenhouse systems. Annu Rev Phytopathol 39:103–133

Pfeferle W, Anke H, Bross M, Steffan B, Vianden R, Steglich W (1990) Asperfuran, a novel antifungal metabolite from *Aspergillus oryzae*. J Antibiot 43:648–654

Pierson LS III, Pierson EA (1996) Phenazine antibiotic production in *Pseudomonas aureofaciens*: role in rhizosphere ecology and pathogen suppression. FEMS Microbiol Lett 136:101–108

Prabavathy VR, Mathivanan N, Murugesan K (2006) Control of blast and sheath blight diseases of rice using antifungal metabolites produced by *Streptomyces* sp. PM5. Biol Control 39:313–319

Prabavathy VR, Vajayanandraj VR, Malarvizhi K, Mathivanan N, Mohan N, Murugesan K (2008) Role of actinomycetes and their metabolites in crop protection. In: Khachatourian GC, Arora DK, Rajendran TP, Srivastava AK (eds) Agriculturally important microorganisms. Academic World International, Bhopal, India, pp 243–255

Prashanth S (2007) Biological control of *Macrophomina* root rot and plant growth promotion in groundnut by *Bacillus licheniformis* MML2501, an azole compound producing rhizobacterium. Ph.D. thesis, University of Madras, Madras, India

Raaijmakers JM, Weller DM (1998) Natural plant protection by 2,4-diacetylphloroglucinol-producing *Pseudomonas* spp. in take-all decline soils. Mol Plant Microbe Interact 11:144–152

Ramamoorthy V, Viswanathan R, Raguchander T, Prakasam V, Samiyappan R (2001) Induction of systemic resistance by plant growth promoting rhizobacteria in crop plants against pests and diseases. Crop Prot 20:1–11

Ramesh S (2009) Marine actinomycetes diversity in Bay of Bengal, India: isolation and characterization of bioactive compounds from *Streptomyces fungicidicus* MML1614. Ph.D. thesis, University of Madras, Madras, India

Sandra AI, Wright CH, Zumoff LS, Steven VB (2001) *Pantoea agglomerans* strain EH318 produces two antibiotics that inhibit *Erwinia amylovora* in vitro. Appl Environ Microbiol 67:282–292

Shanmugaiah V (2007) Biocontrol potential of phenazine-1-carboxamide producing plant growth promoting rhizobacterium, *Pseudomonas aeruginosa* MML2212 against sheath blight disease of rice. Ph.D. thesis, University of Madras, Madras, India

Shanmugaiah V, Mathivanan N, Varghese B (2010) Purification, crystal structure and antimicrobial activity of phenazine-1-carboxamide produced by a growth-promoting biocontrol bacterium, *Pseudomonas aeruginosa* MML2212. J Appl Microbiol 108:703–711

Shen YC (1997) Recent progress on the research and development in agricultural antibiotic. Plant Prot Technol Ext 17:35–37

Shibata M, Iwasa T, Wakae O, Matsuura K, Yamamoto H, Asai M, Mizuno K (1970) DE Patent 1954110, Takeda Chemical Industries, Ltd., 1970 (prior 29 Oct 1968)

Siddiqui IA, Shaukat SS, Khan GH, Ali NA (2003) Suppression of *Meloidogyne javanica* by *Pseudomonas aeruginosa* IE-6S$^+$ in tomato: the influence of NaCl, oxygen and iron levels. Soil Biol Biochem 35:1625–1634

Simon A, Dunlop AW, Ghisalberti EL, Sivasithamparam K (1988) *Trichoderma koningii* produces a pyrone compound with antibiotic properties. Soil Biol Biochem 20:263–264

Sivasithamparam K, Ghisalberti EL (1998) Secondary metabolism in *Trichoderma* and *Gliocladium*. In: Kubicek CP, Harman GE (eds) *Trichoderma* and *Gliocladium*, vol 1. Taylor and Francis, London, pp 139–191

Smith KP, Havey MJ, Handelsman J (1993) Suppression of cottony leak of cucumber with *Bacillus cereus* strain UW85. Plant Dis 77:139–142

Stephen OD, Lydon J (1987) Herbicides from natural compounds. Weed Technol 1:122–128

Strange RN, Scott PR (2005) Plant disease: a threat to global food security. Annu Rev Phytopathol 43:83–116

Suzni T (1992) Biological control of soil borne diseases with antagonistic microbes. In: Kim SU (ed) New biopesticides: proceedings of the agricultural biotechnology symposium. The Research Center of New Bio-Materials in Agriculture, Suweon, Korea, pp 55–76

Suzuki S, Isono K, Nagatsu J, Mizutani T, Kawashima Y, Mizuno T (1965) A new antibiotic, polyoxin A. J Antibiot (Tokyo) 18:131

Szekeres A, Leitgeb B, Kredics L, Antal Z, Hatvani L, Manczinger L, Vagvolgyi C (2005) Peptaibols and related peptaibiotics of *Trichoderma*: a review. Acta Microbiol Immunol Hung 52:137–168

Taga ME, Bassler BL (2003) Chemical communication among bacteria. Proc Natl Acad Sci USA 100:14549–14554

Takeuchi S, Hirayama K, Ueda K, Sakai H, Yonehara H (1958) Blasticidin S, a new antibiotic. J Antibiot A 11:1–5

Tanaka Y, Omura S (1993) Agroactive compounds of microbial origin. Annu Rev Microbiol 47:57–87

Thomashow LS, Weller DM, Bonsall RF, Pierson LS III (1990) Production of the antibiotic phenazine-1-carboxylic acid by fluorescent *Pseudomonas* in the rhizosphere of wheat. Appl Environ Microbiol 56:908–912

Thrane C, Nielsen TH, Nielsen MN, Olsson S, Sørensen J (2000) Viscosinamide-producing *Pseudomonas fluorescens* DR54 exerts biocontrol effect on *Pythium ultimum* in sugar beet rhizosphere. FEMS Microbiol Ecol 33:139–146

Tsujibo H, Sato T, Inui M, Yamamoto H, Inamori Y (1988) Intracellular accumulation of phenazine antibiotics produced by an alkalophilic actinomycete. I. Taxonomy, isolation and identification of the phenazine antibiotics. Agric Biol Chem 52:301–306

Umezawa H, Okami Y, Hashimoto T, Sukara Y, Hamada M, Takeuchi M (1965) A new antibiotic, kasugamycin. J Antibiot (Tokyo) A 18:101–103

Urbasch J (1992) Hyperparasitic attack of *Trichothecium roseum* on conidia of *Pestalotia funerea*. J Phytopathol 136:231–237

van Loon LC, Bakker PAHM, Pietrse CMJ (1998) Systemic resistance induced by rhizosphere bacteria. Annu Rev Phytopathol 36:453–483

Vasudevan P, Kavitha S, Priyadarisini VB, Babujee L, Gnanamanickam SS (2002) Biological control of rice diseases. In: Gnanamanickam SS (ed) Biological control of crop diseases. Dekker, New York, pp 11–32

Voisard C, Keel C, Haas D, Defago G (1989) Cyanide production by *Pseudomonas fluorescens* helps suppress black root of tobacco under gnotobiotic conditions. EMBO J 8:351–358

Vyas RK, Mathur K (2002) *Trichoderma* spp. in cumin rhizosphere and their potential in suppression of wilt. Indian Phytopathol 55:455–457

Wan W, Li C, Zhang J, Jiang D, Huan HC (2008) Effect of volatile substances of *Streptomyces platensis* F-1 on control of plant fungal diseases. Biol Control 46:552–559

Weindling R, Emerson OH (1936) The isolation of a toxic substance from the culture of a *Trichoderma*. Phytopathology 26:1068–1070

Weller DM, Raaijmakers J, McSpadden Gardener B, Thomashow LM (2002) Microbial populations responsible for specific soil suppressiveness to plant pathogens. Annu Rev Phytopathol 40:309–348

Wiest A, Grzegorski D, Xu BW, Goulard C, Rebuffat S, Ebbole DJ, Bodo B, Kenerley C (2002) Identification of peptaibols from *Trichoderma virens* and cloning of a peptaibol synthetase. J Biol Chem 277:20862–20868

Woo JH, Kitamura E, Myouga H, Kamei Y (2002) An antifungal protein from the marine bacterium *Streptomyces* sp. strain AP77 is specific for *Pythium porphyrae*, a causative agent of red rot disease in *Porphyra* spp. Appl Environ Microbiol 68:2666–2675

Yamaguchi I (1996) Pesticides of microbial origin and applications of molecular biology. In: Copping LG (ed) Crop protection agents from nature: natural products and analogues. The Royal Society of Chemistry, Cambridge, pp 27–49

Zhang J, Banko G, Wolfe S, Demain AL (1987) Methionine induction of ACV synthetase in Cephalosporium acremonium. J Ind Microbiol 2:251–255

Zhang L, Yan K, Zhang Y, Huang R, Bian J, Zheng C, Sun H, Chen Z, Sun N, An R, Min F, Zhao W, Zhuo Y, You J, Song Y, Yu Z, Liu Z, Yang K, Gao H, Dai H, Zhang X, Wang J, Fu C, Pei G, Liu J, Zhang S, Goodfellow M, Jiang Y, Kuai J, Zhou G, Chen X (2007) High-throughput synergy screening identifies microbial metabolites as combination agents for the treatment of fungal infections. Proc Natl Acad Sci USA 104:4606–4611

Zhang J, Wang XJ, Yan YJ, Jiang L, Wang JD, Li BJ, Xiang WS (2010) Isolation and identification of 5-hydroxyl-5-methyl-2-hexenoic acid from *Actinoplanes* sp. HBDN08 with antifungal activity. Bioresour Technol 101:8383–8388

Zhao Z, Wang Q, Wang K, Brian K, Liu C, Gu Y (2010) Study of the antifungal activity of *Bacillus vallismortis* ZZ185 in vitro and identification of its antifungal components. Bioresour Technol 101:292–297

Wiest A, Grzegorski D, Xu BW, Goulard C, Rebuffat S, Ebbole DJ, Bodo B, Kenerley C (2002) Identification of peptaibols from *Trichoderma virens* and cloning of a peptaibol synthetase. J Biol Chem 277:20862–20868

# Chapter 11
# The Role of 2,4-Diacetylphloroglucinol- and Phenazine-1-Carboxylic Acid-Producing *Pseudomonas* spp. in Natural Protection of Wheat from Soilborne Pathogens

**Dmitri V. Mavrodi, Olga V. Mavrodi, James A. Parejko, David M. Weller, and Linda S. Thomashow**

## 11.1 Antibiotics and Antibiosis in Plant-Associated Bacteria

Considerable research on plant growth-promoting rhizobacteria (PGPR) over the past half century has focused on antibiosis, "the inhibition or destruction of one organism by a metabolic product of another." By the 1950s, workers already had recognized the inhibitory activity of antibiotics and shown that purified substances or extracts prepared from cultures of antibiotic producers could inhibit plant pathogens (Stallings 1954). Biological control studies in the 1970s and 1980s turned to fluorescent *Pseudomonas* spp. and seminal work with *Pseudomonas fluorescens* Pf-5 (Howell and Stipanovic 1979, 1980) sparked renewed interest in the role of antibiotic production in biocontrol. Strain Pf-5, isolated from the rhizosphere of cotton, produces 2,4-diacetylphloroglucinol (2,4-DAPG), pyrrolnitrin (Prn), and pyoluteorin (Plt) and suppresses damping-off of cotton caused by *Pythium ultimum* and *Rhizoctonia solani*. Prn and Plt purified from cultures of Pf-5 provided the same protection against *Rhizoctonia* and *Pythium* damping-off, respectively, as did the bacterium. More recently, biotechnological and molecular advances have bridged the gap from the laboratory to the field, and it has become routine practice not only to identify genes involved in the regulation and synthesis of antibiotics, but also to demonstrate antibiotic production in situ. Among the fluorescent *Pseudomonas* spp., these studies have focused mainly on Plt, Prn, 2,4-DAPG, and the phenazines, a large group of heterocyclic, nitrogen-containing

D.V. Mavrodi (✉) • O.V. Mavrodi
Department of Plant Pathology, Washington State University, Pullman, WA 99164-6430, USA
e-mail: mavrodi@mail.wsu.edu

J.A. Parejko
School of Molecular Biosciences, Washington State University, Pullman, WA 99164-4234, USA

D.M. Weller • L.S. Thomashow
USDA-ARS, Root Disease and Biological Control Research Unit, Washington State University, Pullman, WA 99164-6430, USA

D.K. Maheshwari (ed.), *Bacteria in Agrobiology: Plant Nutrient Management*,
DOI 10.1007/978-3-642-21061-7_11, 

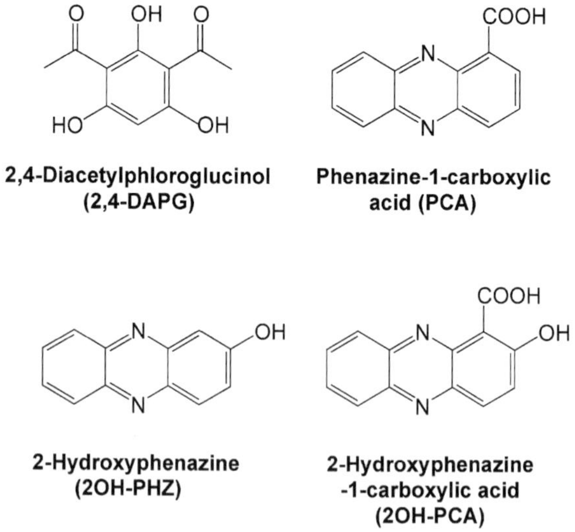

**Fig. 11.1** Chemical structures of 2,4-diacetylphloroglucinol (2,4-DAPG) and some common phenazines produced by fluorescent *Pseudomonas* spp. inhabiting rhizosphere of cereal crops

**2,4-Diacetylphloroglucinol (2,4-DAPG)**

**Phenazine-1-carboxylic acid (PCA)**

**2-Hydroxyphenazine (2OH-PHZ)**

**2-Hydroxyphenazine -1-carboxylic acid (2OH-PCA)**

redox-active compounds derived from phenazine-1-carboxylic acid (PCA) (Fig. 11.1). We focus here on the phenazines and 2,4-DAPG. We briefly review past research that established the importance of these antibiotics in the suppression of fungal wheat root pathogens, and show how the foundations laid by this work have opened the door to studies that address their broader roles in the physiology of the organisms that produce them as well as their interactions in the environment not only with soilborne pathogens, but also with the host plant and other rhizosphere inhabitants.

## 11.2 Biochemistry and Regulation of Synthesis of Phenazines and 2,4-DAPG

Although the phenazine compound pyocyanin was first described in 1859 because of its colorful pigmentation, it was not until over 100 years later that phenazine synthesis was shown to proceed via the shikimic acid pathway, with chorismic acid as the branch point intermediate and glutamine the source of the nitrogen in the tricyclic moiety (Fig. 11.1) (Mentel et al. 2009; Turner and Messenger 1986). Elucidation of subsequent steps in the pathway followed the cloning and characterization of a conserved core operon, *phzABCDEFG*, responsible for synthesis of PCA (Pierson et al. 1995; Mavrodi et al. 1998, 2001), and additional genes involved in the derivatization of PCA to phenazine-1-carboxamide, 2-hydroxyphenazine-1-carboxylic acid, and pyocyanin (Chin-A-Woeng et al. 2001; Delaney et al. 2001; Mavrodi et al. 2001; Parsons et al. 2007). Within the core operon, *phzC* encodes 3-deoxy-D-arabinoheptulosonate-7-phosphate synthase, the branch point enzyme of the shikimate pathway, which presumably directs intermediates from primary metabolism into phenazine synthesis. The remaining genes encode enzymes required for efficient phenazine synthesis in pseudomonads. Of these, PhzE and PhzD sequentially catalyze the synthesis of 2-amino-4-deoxychorismic acid and 2,3-dihydro-3-hydroxyanthranilic acid (DHHA), respectively (McDonald et al. 2001; Parsons et al. 2003). PhzF, a homodimer, then catalyzes the isomerization of DHHA to a ketone, two molecules of which react further, in a reaction facilitated by PhzA/B, to form a tricyclic phenazine precursor that apparently then undergoes a spontaneous oxidative decarboxylation reaction (Blankenfeldt et al. 2004; Ahuja et al. 2008; Mentel et al. 2009). Further oxidation of this reactive intermediate may be facilitated by PhzG, an FMN-dependent oxidase (Parsons et al. 2004; Mentel et al. 2009), with the resulting phenazine product then going on to function as a redox-active electron shuttle or a substrate for derivatization by phenazine-modifying enzymes. In the wheat rhizosphere-associated strain *P. aureofaciens* 30-84, derivatization is accomplished by the product of *phzO*, which is situated downstream of *phzG* and encodes an aromatic monooxygenase that hydroxylates PCA to form 2-hydroxyphenazine and 2-hydroxyphenazine-1-carboxylic acid, both of which are potent bactericidal and fungicidal antibiotics (Delaney et al. 2001).

The physical and chemical environment strongly influences phenazine synthesis in vitro (Slininger and Jackson 1992; Slininger and Shea-Wilbur 1995) and presumably also in the rhizosphere, where biocontrol activity by *P. fluorescens* 2-79 was correlated with a variety of soil characteristics (Ownley et al. 1992, 2003). Regulation at the environmental level presumably is integrated at the molecular level via global mechanisms, at least some of which are known to maintain homeostasis in *Pseudomonas* spp. These include the GacS/GacA and RpeA/RpeB two-component signal transduction systems (Chancey et al. 1999; Whistler and Pierson 2003; Pierson and Pierson 2010). Direct control of phenazine biosynthesis in *P. fluorescens* 2-79 and *P. aureofaciens* 30-84, which control take-all disease of wheat, is regulated by quorum sensing, an elegant mechanism by which bacteria modulate gene expression in response to their population density (Pierson et al. 1994, 1998). In both strains a pair of genes, *phzI* and *phzR*, is located adjacent to one another and convergently oriented, with *phzR* just upstream and in opposite orientation to the phenazine biosynthesis operon. The *phzI* gene encodes for the synthesis of low molecular weight *N*-acyl-homoserine lactone signal molecules that accumulate in the environment of the growing population until reaching a threshold concentration, whereupon they bind to the product of *phzR*, triggering a conformational change that enables activation not only of the promoter of the phenazine biosynthesis operon, but also of *phzI* and *phzR* themselves, making expression of the circuit autoinducible (Pierson et al. 1994; Wood and Pierson 1996; Chancey et al. 1999; Khan et al. 2007). Such regulation is ideally suited to environments like the rhizosphere, where nutrients are likely to be transiently available and populations need to be able to regulate the flow of nutrients into pathways not essential for growth and maintenance.

As in the case of the phenazines, elucidation of the mechanism of synthesis of 2,4-DAPG followed from the cloning and characterization of the biosynthesis operon *phlACBDE* in several biocontrol strains of *P. fluorescens* (Bangera and

Thomashow 1999; Schnider-Keel et al. 2000). Of particular interest is *phlD*, which encodes a novel type III polyketide synthase responsible for the synthesis of phloroglucinol from malonyl-CoA (Achkar et al. 2005; Zha et al. 2006). The *phlD* gene is conserved among all known 2,4-DAPG-producing fluorescent pseudomonads, but exhibits sequence polymorphism sufficient that is useful as a marker of strain diversity within populations of 2,4-DAPG producers (Mavrodi et al. 2007). Of the remaining genes in the operon, *phlACB* encode enzymes involved in the acetylation of phloroglucinol to monoacetylphloroglucinol and 2,4-DAPG (Achkar et al. 2005; Zha et al. 2006) and *phlE* encodes a membrane protein thought to function in transport or resistance (Bangera and Thomashow 1999; Abbas et al. 2004).

Like the synthesis of phenazines, that of 2,4-DAPG is tightly controlled and integrated with other cellular processes. Directly involved in modulating levels of 2,4-DAPG are three genes, *phlF*, *phlG*, and *phlH*, that lie upstream of the biosynthesis operon. Of these, *phlF* and *phlH* encode transcriptional regulators that repress and activate, respectively, expression of the biosynthesis genes and *phlG* encodes a hydrolase that degrades 2,4-DAPG (Bangera and Thomashow 1999; Schnider-Keel et al. 2000; Abbas et al. 2004; Bottiglieri and Keel 2006). Also as in the production of phenazines, synthesis of 2,4-DAPG is autoregulated, albeit not by acylhomoserine lactones but by 2,4-DAPG itself, which binds to and dissociates the PhlF repressor from the promoter of *phlA* (Schnider-Keel et al. 2000; Abbas et al. 2004). Further control is achieved at the global level by transcriptional regulators such as *mvaT* and *mvaV* (Baehler et al. 2006), the relative levels of RNA polymerase sigma factors (Sarniguet et al. 1995; Schnider et al. 1995; Pechy-Tarr et al. 2005), and by a GacS/GacA signal transduction cascade in which the response regulator GacA promotes the transcription of small regulatory RNAs that interfere with the negative effects of translational repressors to modulate expression of the *phl* operon post-transcriptionally (Dubuis and Haas 2007).

Expression of the *phl* genes and production of 2,4-DAPG are influenced by a variety of abiotic and biotic factors. Using liquid cultures, Duffy and Défago (1997) determined that $Zn^{2+}$, $NH_4Mo^{2+}$, and glucose stimulated and inorganic phosphate generally reduced accumulation, but the effect varied considerably from strain to strain. Numerous investigators have observed differences in *phl* gene expression or the accumulation of 2,4-DAPG in response to host plant species, cultivar, and age, with monocots or younger plants generally supporting greater synthesis than dicots or older plants (Notz et al. 2001; Bergsma-Vlami et al. 2005; de Werra et al. 2008; Okubara and Bonsall 2008; Jamali et al. 2009; Rochat et al. 2010). Cross-talk between producers of 2,4-DAPG can occur in vitro and the rhizosphere, such that 2,4-DAPG produced by one strain can trigger the autoregulatory circuit of another, stimulating synthesis (Maurhofer et al. 2004). Other metabolites synthesized by 2,4-DAPG producers themselves, or by heterologous rhizosphere inhabitants, also can influence *phl* gene expression. For example, in 2,4-DAPG-producing strains that produce the antibiotic pyoluteorin, synthesis of the two is balanced such that either antibiotic represses the synthesis of the other (Schnider-Keel et al. 2000; Baehler et al. 2005). Phytopathogenic fungi such as *Fusarium oxysporum* can produce metabolites that strongly repress *phl* gene expression both in vitro and on roots, whereas expression on roots was stimulated in the presence of pathogens such as *P. ultimum* or *R. solani*, perhaps due to nutrients released during fungal infection (Duffy and Défago 1997; Notz et al. 2002; Jamali et al. 2009). In contrast, foliar infection or physical stress lowered *phl* gene expression in the rhizosphere (de Werra et al. 2008). Finally, recent studies indicate that bacteriovorous rhizosphere inhabitants such as *Acanthamoeba* can also influence *phl* gene expression. Soluble compounds released by this predator led to induction of *phl* gene expression, whereas direct contact with the predator led to reduced expression of these and other genes involved in the bacterial defense response (Jousset et al. 2010).

## 11.3 Diversity of Phenazine- and 2,4-DAPG-Producing Bacteria in the Rhizosphere of Wheat

Current knowledge of wheat-associated phenazine-producing bacteria is largely devoted to the *Gammaproteobacteria* class of Gram-negative bacteria (Mavrodi et al. 2010). The most studied and most frequently isolated phenazine producers fall into the pyoverdine-producing subset of the genus *Pseudomonas* termed the fluorescent *Pseudomonas* spp. complex (Mavrodi et al. 2006). This complex comprises several closely related species, with *P. fluorescens* encompassing the best-characterized biocontrol agents. *P. fluorescens* 2-79 was originally isolated from the rhizosphere of dryland wheat and is a model PCA producer that has been studied in detail for its PCA-dependent biocontrol properties (Thomashow and Weller 1988; Thomashow et al. 1990). Until a recent study in which 15 new PCA producers were described, no strains closely related to *P. fluorescens* 2-79 had been isolated from wheat (Mavrodi et al. 2010). All 15 new strains fell into the *P. fluorescens* complex, with two separate groups based on 16S rDNA. One group, more closely related to *P. fluorescens* 2-79, included *P. synxantha* and the lesser known *P. gessardii* and *P. libanensis* (Saini et al. 2008; Mavrodi et al. 2010) while the other, larger group was closely related to *P. orientalis*, a little known species not previously reported to produce phenazines (Dabboussi et al. 1999). Phenazine-producing pseudomonads also are found within *P. chlororaphis*, which includes strains formerly classified as *P. aureofaciens*. *P. chlororaphis* 30-84, originally isolated from the roots of wheat in a take-all suppressive field, produces 2-OH-PHZ, 2-OH-PCA, and PCA.

Unlike the diversity among phenazine producers, that of 2,4-DAPG producers colonizing the wheat rhizosphere is restricted to a small subset of *P. fluorescens* strains, likely due to the ancestral nature of the *phl* gene cluster (Moynihan et al. 2009). However, the inter-strain diversity spans at least 22 BOX-PCR and *phlD* restriction fragment length polymorphism (RFLP) genotypes (genotypes A to T, PfY, and PfZ) that fall into three coarsely defined amplified ribosomal DNA restriction analysis (ARDRA) lineages (Weller et al. 2007). Currently, only ARDRA group 2 has been isolated from wheat and within that group, genotype D has been found as the dominant wheat colonizing genotype (Weller et al. 2007). The classical representative of the D genotype, *P. fluorescens* Q8r1-96, has certain colonization and competition properties (such as pyocin-like bacteriocin production) that make it a superior biocontrol strain to other closely related genotypes of 2,4-DAPG-producers (Raaijmakers and Weller 2001; Validov et al. 2005). Dominance by a single genotype, often genotype D, is common to suppressive soils upon which wheat has been grown in monoculture for multiple years (Gardener et al. 2000; Weller et al. 2002; Bergsma-Vlami et al. 2005). For example, in a single field cropped continuously to wheat for 115 years, 2,4-DAPG producers of the D-genotype comprised 77% of the isolates (Landa et al. 2006). Although the D-genotype has been found to be dominant on wheat, representatives of 17 of the 22 genotypes of 2,4-DAPG producers have been isolated from wheat grown in soils worldwide (De La Fuente et al. 2006).

Considering the relatively high intra- and inter-species diversity discovered thus far in 2,4-DAPG- and phenazine-producing species, it is highly likely that future studies using high throughput culture-independent techniques will uncover further taxonomic and allelic diversity among phenazine and 2,4-DAPG-producing populations of *Pseudomonas* spp. inhabiting the rhizosphere of cereal crops.

## 11.4 Role of 2,4-DAPG and Phenazines in Suppression of Soilborne Diseases of Wheat

The polyketide metabolite 2,4-DAPG is active in the control of numerous root and seedling diseases (Stutz et al. 1986; Vincent et al. 1991; Keel et al. 1992; Laville et al. 1992; Shanahan et al. 1992; Harrison et al. 1993; Cronin et al. 1997; Duffy and Défago 1997; de Souza et al. 2003a) and is a key component of soils that have become suppressive to take-all disease of wheat, a phenomenon known as take-all decline (TAD). Take-all, caused by *Gaeumannomyces graminis var. tritici*, is a serious root disease of wheat (Hornby 1998) and TAD is the spontaneous decrease in take-all incidence and severity induced by monoculture wheat or barley after a severe outbreak of the disease (Cook and Weller 1987). TAD occurs worldwide (Hornby 1998), typically after five or six consecutive wheat or barley crops, and involves microbiological changes in the soil or rhizosphere that suppress the pathogen (Weller et al. 2002). The specific suppression associated with TAD is transferable (1–10% TAD soil) to conducive soil and is eliminated by soil pasteurization (60°C, 30 min) or fumigation (Cook and Rovira 1976), and by rotation with non-cereal crops (Cook 1981).

Beginning in the 1970s, antagonistic *Pseudomonas* spp. were implicated in TAD by studies in our lab and elsewhere (Weller et al. 2002). We later hypothesized that fluorescent *Pseudomonas* spp. producing DAPG or PCA were important in the suppressiveness of some TAD soils (Weller et al. 2002). This idea was prompted by earlier findings that some of the most effective biocontrol strains from TAD soils produce either 2,4-DAPG (Vincent et al. 1991; Harrison et al. 1993) or PCA (Thomashow and Weller 1988), and the antibiotics are responsible for the biocontrol activity of the strains. *G. graminis var. tritici* is highly sensitive to 2,4-DAPG (the 90% effective dose is 3.1–11.1 μg $ml^{-1}$), and isolates insensitive to PCA or 2,4-DAPG were suppressed to a lesser extent by producer biocontrol strains than were sensitive isolates of the fungus (Mazzola et al. 1992). Our recent findings have revealed that under field conditions, even prolonged exposure of *G. graminis var. tritici* to 2,4-DAPG in TAD soils does not lead to the widespread selection of tolerant or resistant strains of the pathogen (Kwak et al. 2010). The development of primers and probes specific to PCA and 2,4-DAPG biosynthesis genes allowed quantification of indigenous antibiotic producers in the wheat rhizosphere (Raaijmakers et al. 1997). Several lines of evidence (Weller et al. 2002, 2007) demonstrated that 2,4-DAPG-producing strains of *P. fluorescens* play a key role in TAD in Washington State, USA: (1) 2,4-DAPG producers were present on roots from TAD soils at densities above the threshold level ($10^5$ CFU $g^{-1}$ root) (Raaijmakers and Weller 1998) required for take-all control, but were below the threshold or not detected on roots from conducive soils (Raaijmakers et al. 1997); (2) inverse relation exists between the densities of indigenous 2,4-DAPG producers and take-all severity (Raaijmakers and Weller 1998); (3) specific suppression in TAD soil was lost when 2,4-DAPG producers were eliminated by soil pasteurization (Raaijmakers and Weller 1998); (4) adding TAD soil to conducive soil transferred suppressiveness and established population densities of 2,4-DAPG producers above the threshold required for disease control (Raaijmakers and Weller 1998); (5) cultivation of oats, a non-host crop that eliminated suppressiveness, reduced the population density of 2,4-DAPG producers below the threshold level; (6) introduction of DAPG producers from TAD soils into conducive soils rendered the soils as suppressive as TAD soil (Raaijmakers and Weller 1998); (7) 2,4-DAPG was detected on roots of wheat grown in TAD soil, but not on roots from conducive soil (Raaijmakers et al. 1999); and (8) 2,4-DAPG producers were above the threshold level on wheat collected from TAD fields, but were not detectable on wheat from nearby conducive fields (Raaijmakers et al. 1999).

The role of 2,4-DAPG producers in TAD is not restricted to Washington. 2,4-DAPG producers were above the threshold density on wheat grown in monoculture soils from Fargo, ND (116 years), Hallock, MN (10 years) (Weller et al. 2007) and Woensdrecht, The Netherlands (14 and 27 years) (de Souza et al. 2003b) but were at or below the detection limit on wheat grown in soils from adjacent non-monoculture fields (de Souza et al. 2003b; Weller et al. 2007). Studies by the group of Raaijmakers showed a key role for 2,4-DAPG in the Dutch TAD soils (de Souza et al. 2003b).

Under non-irrigated dryland conditions in the Pacific Northwest, USA, take-all disease is less severe and root and crown rots caused by *F. culmorum* and *F. pseudograminearum*, and *Rhizoctonia* root rot caused by *R. solani* AG-8 and *R. oryzae*, become more important soilborne diseases of wheat (Cook and Veseth 1991). Recently, we reported large populations of indigenous phenazine-producing

*Pseudomonas* strains on cereals grown in non-irrigated dryland fields in the Columbia Plateau of central Washington State (Mavrodi et al. 2010). Many isolates of these phenazine-producing pseudomonads are strongly inhibitory to *R. solani* AG-8, and our preliminary results suggest that in dryland agroecosystems these indigenous phenazine-producing *Pseudomonas* spp. may contribute to the natural protection of wheat from Rhizoctonia root rot (Unpublished results).

## 11.5 Mechanism of Action of 2,4-DAPG and Phenazines and Physiological Role in Bacterial Producers

2,4-DAPG and phenazines possess broad-range antibiotic activity and, depending on the amount applied, will inhibit the growth of bacteria, oomycetes, fungi, nematodes, plants and animal cells (Haas and Défago 2005; Mavrodi et al. 2006). This broad spectrum activity suggests that 2,4-DAPG and phenazines interfere with some common and important cellular functions, and the identification of some of these functions has been carried out using a *Saccharomyces cerevisiae* model system. An extensive screening of a yeast deletion library (Ran et al. 2003) revealed the capacity of the phenazine antibiotic pyocyanin to disrupt multiple basic cellular pathways including the cell cycle, apoptosis, respiration, and the oxidative stress response. In addition, pyocyanin interfered with the assembly and functioning of vacuolar ATPases, a diverse group of ATP-driven proton pumps that acidify a wide array of intracellular compartments and function in processes such as endocytosis, protein sorting and vesicle transport, intracellular targeting of lysosomal enzymes, and the coupled transport of small molecules (Nishi and Forgac 2002). Much of the biological activity of phenazines is attributed directly to their redox properties (Hassan and Fridovich 1980; Hassett et al. 1992; Denning et al. 2003; Giddens and Bean 2007). Phenazines can undergo cellular redox cycling in the presence of oxygen and reducing agents including NADH and NADPH, causing the accumulation of toxic superoxide ($O_2^-$) and hydrogen peroxide ($H_2O_2$) (Hassan and Fridovich 1980). In contrast to phenazines, the exact mechanism of action of 2,4-DAPG remains to be elucidated. However, a recent study by Gleeson et al. (2010) demonstrated that 2,4-DAPG interferes with mitochondrial function in *S. cerevisiae*.

Though traditionally considered "secondary metabolites" of no direct benefit to the cells that produce them, phenazines synthesized by bacteria in their native habitats clearly contribute to competitiveness and long-term survival. Thus, phenazine-producing pseudomonads are more competitive and survive longer on the roots of wheat than do phenazine-nonproducing mutants (Mazzola et al. 1992). The contribution of phenazines to the survival of pseudomonads in the plant rhizosphere is consistent with the emerging link between phenazine production and biofilm formation. For example, cultures of *P. aureofaciens* did not establish biofilms in the absence of phenazines, and biofilm architecture and bacterial dispersal rates were dependent on the identity and ratios of the phenazines produced

(Maddula et al. 2006, 2008). Similarly, although phenazine-deficient mutants of *P. aeruginosa* formed biofilms, the identity and amounts of amended phenazines influenced biofilm architecture and cell swarming activity (Ramos et al. 2010). Price-Whelan et al. (2006) suggested that because the diffusion rate of oxygen through biofilms is thought to be slow, phenazines could help maintain the redox homeostasis of cells embedded in the film by acting as electron acceptors for the reoxidation of accumulating NADH. Indeed, in oxygen-limited stationary-phase cultures of *P. aeruginosa*, a decrease in intracellular $NADH/NAD^+$ was correlated with the presence of pyocyanin in the culture (Price-Whelan et al. 2007), and phenazine-facilitated electron transfer promoted anaerobic survival but not growth under conditions of oxidant limitation (Wang et al. 2010). Interestingly, other electron shuttles that were reduced but not made by *P. aeruginosa* did not facilitate survival, suggesting that sophisticated systems are needed to control the reactivity of these molecules within the cell and that mechanisms have evolved in pseudomonads to be specific for the phenazines they produce (Wang et al. 2010).

In contrast to phenazines, 2,4-DAPG does not seem to contribute to the ecological competence of producing strains, and 2,4-DAPG-deficient mutants colonize the plant rhizosphere on par with corresponding wild-type strains (Carroll et al. 1995; de Souza et al. 2003b). However, from the standpoint of evolutionary diversity, it has been speculated that 2,4-DAPG producers comprise a group of strains that evolve significantly more slowly than their closely related 2,4-DAPG-nonproducing counterparts (Moynihan et al. 2009). This suggests that 2,4-DAPG production is an important trait to the ecology of certain *P. fluorescens* strains and that retention of the *phl* gene cluster is of evolutionary advantage (Moynihan et al. 2009).

## 11.6 Plant Responses to Phenazines and 2,4-DAPG

Rhizosphere-dwelling *Pseudomonas* spp. may utilize phenazines and 2,4-DAPG to interfere with processes of nutrient exudation from roots of the host plant. Stable isotope probing experiments (Phillips et al. 2004) revealed that 2,4-DAPG blocked the uptake of exogenous Ala by axenic alfalfa roots and caused a concurrent increase in the net efflux of $^{15}N$-labeled compounds from root tissues. The exposure of alfalfa roots to 200 μM phenazine or 2,4-DAPG increased the efflux of amino acids by 200 and 1,600%, respectively, and similar effects were observed in other plant species, including wheat.

Due to their redox-active properties, phenazines may also play a significant role in the mobilization and plant uptake of Fe and Mn, particularly in dry calcareous soils. In arid regions, soils typically have high pH and pE levels because the soils are less weathered and oxygen, a strong electron acceptor, can diffuse freely into the soil. Under such conditions, iron and manganese are thermodynamically stable as Mn(III), Mn(IV), and Fe(III) insoluble oxy-hydroxides [(hydr)oxides] (Stumm and Morgan 1981). The solubilities of the Fe- and Mn-(hydr) oxides also decrease under higher pH, which can lead to Fe and Mn deficiency in crops and microorganisms (Marschner 1995). Phenazine-producing rhizosphere bacteria may participate in the reduction of Fe(III) and Mn (III, IV) to their soluble counterparts, probably by producing biofilms and limiting $O_2$ diffusion and thus creating an environment where Fe and Mn reduction can occur (Hernandez et al. 2004; Price-Whelan et al. 2006; Wang and Newman 2008). Wang and Newman (2008) have shown that the rate of Fe (III) reduction by phenazines depends on the reduction potential of the phenazine, the solubility of the Fe (hydr) oxide, and pH. They also found that phenazines show different rates of oxidation by $O_2$: pyocyanin > 1-hydroxyphenazine > PCA, the opposite order of their Fe reducing rates. These findings suggest that indigenous antibiotic-producing rhizosphere bacteria may contribute to plant nutrition by bringing recalcitrant metal ions into solution (Rengel et al. 1996; Marschner et al. 2003; Robin et al. 2008).

Finally, there is mounting evidence suggesting a role of antibiotics produced by PGPR in elicitation of the induced systemic resistance (ISR) response in plants. During ISR, microbe-associated microbial patterns (MAMPs) or secreted metabolites of rhizosphere bacteria are recognized by the plant, resulting in activation of defense responses (De Vleesschauwer and Hofte 2009). The onset of ISR does not involve direct activation of defense genes, but rather, results in priming of the plant innate immune system for a rapid response (Pieterse et al. 2009). The rhizobacteria-mediated ISR is often, but not always, regulated by the jasmonic acid (JA) and ethylene (ET) signaling pathways and provides protection against herbivorous insects and necrotrophic pathogens that are sensitive to defense responses controlled by JA and ET. However, recent studies have also revealed a great deal of variation and cross-communication between the hormone signaling pathways that accompany induction of ISR by different rhizobacteria (De Vleesschauwer and Hofte 2009).

Both 2,4-DAPG and phenazines are known to induce ISR in different plant species, although the molecular mechanism of this phenomenon is slightly better understood in the case of 2,4-DAPG. 2,4-DAPG was first implicated in the elicitation of ISR by *P. fluorescens* CHA0, where it induces resistance in *Arabidopsis* and tomato against the oomycete *Hyaloperonospora arabidopsis* (Iavicoli et al. 2003) and the root knot nematode *Meloidogyne javanica* (Siddiqui and Shaukat 2004), respectively. 2,4-DAPG was also identified as an ISR elicitor in *P. fluorescens* Q2-87 (Weller et al. 2004). In all of the aforementioned studies, the ability to induce ISR was abolished by mutations in the 2,4-DAPG biosynthesis pathway and was restored upon complementation of the mutant. Although the precise details of ISR induction by 2,4-DAPG are unknown, it has been suggested that this bacterial metabolite may interfere with auxin signaling in the host plant. For example in *Arabidopsis*, the elicitation of ISR was dependent on the presence of functional EIR1 protein, which is involved in ET signaling and has also been proposed to have a role in auxin transport in roots (Iavicoli et al. 2003). A recent study by Brazelton et al. (2008) also demonstrated that 2,4-DAPG alters the morphology of roots in tomato seedlings and that the effect was blocked in the auxin-resistant mutant *diageotropica*.

The capacity of phenazines to elicit ISR was demonstrated in a rhizosphere isolate of *P. aeruginosa* 7NSK2, which produces the redox-active phenazine antibiotic pyocyanin (De Vleesschauwer et al. 2006). In 7NSK2, pyocyanin acts as a prime elicitor of resistance in rice against the rice blast pathogen *Magnaporthe oryzae*. The protective effect of 7NSK2 treatment could be replicated by exposure of rice seedlings to purified pyocyanin, whereas colonization of plants with pyocyanin-nonproducing isogenic mutants did not reduce rice blast disease symptoms (De Vleesschauwer et al. 2006). Curiously, the exposure of rice to pyocyanin also rendered plants more susceptible to colonization by *R. solani*. The molecular mechanism of ISR induction in rice by pyocyanin appears to involve the transient generation of low levels of reactive oxygen species, and the application of free radical scavengers interfered with the ability of this phenazine compound to protect rice against *M. oryzae* (De Vleesschauwer et al. 2006).

## 11.7 Conclusion and Future Prospects

Fluorescent *Pseudomonas* species that synthesize phenazine antibiotics and 2,4-diacetylphloroglucinol inhabit the rhizosphere of wheat and have important roles in the suppression of take-all and a variety of other cereal diseases caused by soilborne fungal plant pathogens. These antibiotics exhibit broad-spectrum activity consistent with the idea that they interfere in target fungi with important cellular activities such as mitochondrial function, the cell cycle, apoptosis, respiration, and the oxidative stress response. Effects of the antibiotics on the host plant are still relatively unexplored, but in some systems they include the induction of systemic resistance and alterations in the secretion and uptake of metabolites by roots. Considering the relatively high intra- and inter-species diversity, respectively, discovered thus far among producers of 2,4-DAPG- and phenazines, it is highly likely that future studies will uncover further taxonomic and allelic diversity among populations of *Pseudomonas* spp. producing these antibiotics. Genomic and functional genetic analyses of well-characterized model strains also will reveal additional complexity in the repertoire and targets of the bioactive metabolites they produce. Genetic resources are readily available that can be used to identify phenazine and 2,4-DAPG biosynthesis operons in new isolates with biocontrol potential, and these resources will be useful in dissecting the basis of naturally suppressive soils, particularly in sites where wheat has been grown over long periods of time. Strains producing these and other natural antibiotics also are well-suited as introduced biological control agents, particularly as they are compatible with organic and sustainable agricultural production systems, but their use ultimately will depend on wider public acceptance of microbial biopesticides.

# References


Abbas A, McGuire JE, Crowley D, Baysse C, Dow M, O'Gara F (2004) The putative permease PhlE of *Pseudomonas fluorescens* F113 has a role in 2,4-diacetylphloroglucinol resistance and in general stress tolerance. Microbiology 150:2443–2450

Achkar J, Xian M, Zhao H, Frost JW (2005) Biosynthesis of phloroglucinol. J Am Chem Soc 127:5332–5333

Ahuja EG, Janning P, Mentel M, Graebsch A, Breinbauer R, Hiller W, Costisella B, Thomashow L, Mavrodi D, Blankenfeldt W (2008) PhzA/B catalyzes the formation of the tricycle in phenazine biosynthesis. J Am Chem Soc 130:17053–17061

Baehler E, Bottiglieri M, Pechy-Tarr M, Maurhofer M, Keel C (2005) Use of green fluorescent protein-based reporters to monitor balanced production of antifungal compounds in the biocontrol agent *Pseudomonas fluorescens* CHA0. J Appl Microbiol 99:24–38

Baehler E, de Werra P, Wick LY et al (2006) Two novel MvaT-like global regulators control exoproduct formation and biocontrol activity in root-associated *Pseudomonas fluorescens* CHA0. Mol Plant Microbe Interact 19:313–329

Bangera MG, Thomashow LS (1999) Identification and characterization of a gene cluster for synthesis of the polyketide antibiotic 2,4-diacetylphloroglucinol from *Pseudomonas fluorescens* Q2-87. J Bacteriol 181:3155–3163

Bergsma-Vlami M, Prins ME, Raaijmakers JM (2005) Influence of plant species on population dynamics, genotypic diversity and antibiotic production in the rhizosphere by indigenous *Pseudomonas* spp. FEMS Microbiol Ecol 52:59–69

Blankenfeldt W, Kuzin AP, Skarina T et al (2004) Structure and function of the phenazine biosynthetic protein PhzF from *Pseudomonas fluorescens*. Proc Natl Acad Sci USA 101: 16431–16436

Bottiglieri M, Keel C (2006) Characterization of PhlG, a hydrolase that specifically degrades the antifungal compound 2,4-diacetylphloroglucinol in the biocontrol agent *Pseudomonas fluorescens* CHA0. Appl Environ Microbiol 72:418–427

Brazelton JN, Pfeufer EE, Sweat TA, Gardener BB, Coenen C (2008) 2,4-diacetylphloroglucinol alters plant root development. Mol Plant Microbe Interact 21:1349–1358

Carroll H, Moenne-Loccoz Y, Dowling DN, O'Gara F (1995) Mutational disruption of the biosynthesis genes coding for the antifungal metabolite 2,4-diacetylphloroglucinol does not influence the ecological fitness of *Pseudomonas fluorescens* F113 in the rhizosphere of sugarbeets. Appl Environ Microbiol 61:3002–3007

Chancey ST, Wood DW, Pierson LS III (1999) Two-component transcriptional regulation of *N*-acyl-homoserine lactone production in *Pseudomonas aureofaciens*. Appl Environ Microbiol 65:2294–2299

Chin-A-Woeng TFC, Thomas-Oates JE, Lugtenberg BJJ, Bloemberg GV (2001) Introduction of the *phzH* gene of *Pseudomonas chlororaphis* PCL1391 extends the range of biocontrol ability of phenazine-1-carboxylic acid-producing *Pseudomonas* spp. strains. Mol Plant Microbe Interact 14:1006–1015

Cook RJ (1981) The influence of rotation crops on take-all decline phenomenon. Phytopathology 71:189–192

Cook RJ, Rovira AD (1976) The role of bacteria in the biological control of *Gaeumannomyces graminis* by suppressive soils. Soil Biol Biochem 8:269–273

Cook RJ, Veseth RJ (1991) Wheat health management. APS, St. Paul, MN

Cook RJ, Weller DM (1987) Management of take-all in consecutive crops of wheat or barley. In: Chet I (ed) Innovative approaches to plant disease control. Wiley, New York, NY, pp 41–76

Cronin D, Moenne-Loccoz Y, Fenton A, Dunne C, Dowling DN, O'Gara F (1997) Role of 2,4-diacetylphloroglucinol in the interactions of the biocontrol pseudomonad strain F113 with the potato cyst nematode *Globodera rostochiensis*. Appl Environ Microbiol 63:1357–1361

Dabboussi F, Hamze M, Elomari M, Verhille S, Baida N, Izard D, Leclerc H (1999) Taxonomic study of bacteria isolated from Lebanese spring waters: proposal for *Pseudomonas cedrella* sp. nov. and *P. orientalis* sp. nov. Res Microbiol 150(5):303–316
De La Fuente L, Mavrodi DV, Landa BB, Thomashow LS, Weller DM (2006) *phlD*-based genetic diversity and detection of genotypes of 2,4-diacetylpholorogucinol-producing *Pseudomonas fluorescens*. FEMS Microbiol Ecol 56:64–78
de Souza JT, Arnould C, Deulvot C, Lemanceau P, Gianinazzi-Pearson V, Raaijmakers JM (2003a) Effect of 2,4-diacetylphloroglucinol on *Pythium*: cellular responses and variation in sensitivity among propagules and species. Phytopathology 93:966–975
de Souza JT, Weller DM, Raaijmakers JM (2003b) Frequency, diversity, and activity of 2,4-diacetylphloroglucinol-producing fluorescent *Pseudomonas* spp. in Dutch take-all decline soils. Phytopathology 93:54–63
De Vleesschauwer D, Hofte M (2009) Rhizobacteria-induced systemic resistance. Adv Bot Res 51:223–281
De Vleesschauwer D, Cornelis P, Hofte M (2006) Redox-active pyocyanin secreted by *Pseudomonas aeruginosa* 7NSK2 triggers systemic resistance to *Magnaporthe grisea* but enhances *Rhizoctonia solani* susceptibility in rice. Mol Plant Microbe Interact 19:1406–1419
de Werra P, Baehler E, Huser A, Keel C, Maurhofer M (2008) Detection of plant-modulated alterations in antifungal gene expression in *Pseudomonas fluorescens* CHA0 on roots by flow cytometry. Appl Environ Microbiol 74:1339–1349
Delaney SM, Mavrodi DV, Bonsall RF, Thomashow LS (2001) *phzO*, a gene for biosynthesis of 2-hydroxylated phenazine compounds in *Pseudomonas aureofaciens* 30-84. J Bacteriol 183:318–327
Denning GM, Iyer SS, Reszka KJ, O'Malley Y, Rasmussen GT, Britigan BE (2003) Phenazine-1-carboxylic acid, a secondary metabolite of *Pseudomonas aeruginosa*, alters expression of immunomodulatory proteins by human airway epithelial cells. Am J Physiol 285:L584–L592
Dubuis C, Haas D (2007) Cross-species GacA-controlled induction of antibiosis in pseudomonads. Appl Environ Microbiol 73:650–654
Duffy BK, Défago G (1997) Zinc improves biocontrol of Fusarium crown and root rot of tomato by *Pseudomonas fluorescens* and represses the production of pathogen metabolites inhibitory to bacterial antibiotic biosynthesis. Phytopathology 87:1250–1257
Gardener BBM, Schroeder KL, Kalloger SE, Raaijmakers JM, Thomashow LS, Weller DM (2000) Genotypic and phenotypic diversity of *phlD*-containing *Pseudomonas* strains isolated from the rhizosphere of wheat. Appl Environ Microbiol 66:1939–1946
Giddens SR, Bean DC (2007) Investigations into the in vitro antimicrobial activity and mode of action of the phenazine antibiotic D-alanylgriseoluteic acid. Int J Antimicrob Agents 29:93–97
Gleeson O, O'Gara F, Morrissey JP (2010) The *Pseudomonas fluorescens* secondary metabolite 2,4 diacetylphloroglucinol impairs mitochondrial function in *Saccharomyces cerevisiae*. Antonie Van Leeuwenhoek 97:261–273
Haas D, Défago G (2005) Biological control of soil-borne pathogens by fluorescent pseudomonads. Nat Rev Microbiol 3:307–319
Harrison LA, Letendre L, Kovacevich P, Pierson E, Weller D (1993) Purification of an antibiotic effective against *Gaeumannomyces graminis var. tritici* produced by a biocontrol agent, *Pseudomonas aureofaciens*. Soil Biol Biochem 25:215–221
Hassan HM, Fridovich I (1980) Mechanism of the antibiotic action of pyocyanine. J Bacteriol 141:156–163
Hassett DJ, Charniga L, Bean K, Ohman DE, Cohen MS (1992) Response of *Pseudomonas aeruginosa* to pyocyanin: mechanisms of resistance, antioxidant defenses, and demonstration of a manganese-cofactored superoxide dismutase. Infect Immun 60:328–336
Hernandez ME, Kappler A, Newman DK (2004) Phenazines and other redox-active antibiotics promote microbial mineral reduction. Appl Environ Microbiol 70:921–928
Hornby D (1998) Take-all of cereals: a regional perspective. CAB International, Wallingford, UK

Howell CR, Stipanovic RD (1979) Control of *Rhizoctonia solani* in cotton seedlings with *Pseudomonas fluorescens* and with an antibiotic produced by the bacterium. Phytopathology 69:480–482

Howell CR, Stipanovic RD (1980) Suppression of *Pythium ultimum* induced damping-off of cotton seedlings by *Pseudomonas fluorescens* and its antibiotic pyoluteorin. Phytopathology 70:712–715

Iavicoli A, Boutet E, Buchala A, Metraux JP (2003) Induced systemic resistance in *Arabidopsis thaliana* in response to root inoculation with *Pseudomonas fluorescens* CHA0. Mol Plant Microbe Interact 16:851–858

Jamali F, Sharifi-Tehrani A, Lutz MP, Maurhofer M (2009) Influence of host plant genotype, presence of a pathogen, and coinoculation with *Pseudomonas fluorescens* strains on the rhizosphere expression of hydrogen cyanide- and 2,4-diacetylphloroglucinol biosynthetic genes in *P. fluorescens* biocontrol strain CHA0. Microb Ecol 57:267–275

Jousset A, Rochat L, Scheu S, Bonkowski M, Keel C (2010) Predator-prey chemical warfare determines the expression of biocontrol genes by rhizosphere-associated *Pseudomonas fluorescens*. Appl Environ Microbiol 76:5263–5268

Keel C, Schnider U, Maurhofer M et al (1992) Suppression of root diseases by *Pseudomonas fluorescens* CHA0 – importance of the bacterial secondary metabolite 2,4-diacetylphloroglucinol. Mol Plant Microbe Interact 5:4–13

Khan SR, Herman J, Krank J et al (2007) *N*-(3-hydroxyhexanoyl)-L-homoserine lactone is the biologically relevant quormone that regulates the *phz* operon of *Pseudomonas chlororaphis* strain 30-84. Appl Environ Microbiol 73:7443–7455

Kwak YS, Bakker PA, Glandorf DC, Rice JT, Paulitz TC, Weller DM (2010) Isolation, characterization, and sensitivity to 2,4-diacetylphloroglucinol of isolates of *Phialophora* spp. from Washington wheat fields. Phytopathology 100:404–414

Landa BB, Mavrodi OV, Schroeder KL, Allende-Molar R, Weller DM (2006) Enrichment and genotypic diversity of *phlD*-containing fluorescent *Pseudomonas* spp. in two soils after a century of wheat and flax monoculture. FEMS Microbiol Ecol 55:351–368

Laville J, Voisard C, Keel C, Maurhofer M, Défago G, Haas D (1992) Global control in *Pseudomonas fluorescens* mediating antibiotic synthesis and suppression of black root rot of tobacco. Proc Natl Acad Sci USA 89:1562–1566

Maddula VSRK, Zhang Z, Pierson EA, Pierson LS III (2006) Quorum sensing and phenazines are involved in biofilm formation by *Pseudomonas chlororaphis* (*aureofaciens*) strain 30-84. Microb Ecol 52:289–301

Maddula VSRK, Pierson EA, Pierson LS III (2008) Altering the ratio of phenazines in *Pseudomonas chlororaphis (aureofaciens)* strain 30-84: effects on biofilm formation and pathogen inhibition. J Bacteriol 190(8):2759–2766

Marschner H (1995) Mineral nutrition in higher plants. Academic Press, London, UK

Marschner P, Kandeler E, Marschner B (2003) Structure and function of the soil microbial community in a long-term fertilizer experiment. Soil Biol Biochem 35:453–461

Maurhofer M, Baehler E, Notz R, Martinez V, Keel C (2004) Cross talk between 2,4-diacetylphloroglucinol-producing biocontrol pseudomonads on wheat roots. Appl Environ Microbiol 70:1990–1998

Mavrodi DV, Ksenzenko VN, Bonsall RF, Cook RJ, Boronin AM, Thomashow LS (1998) A seven-gene locus for synthesis is of phenazine-1-carboxylic acid by *Pseudomonas fluorescens* 2-79. J Bacteriol 180:2541–2548

Mavrodi DV, Bonsall RF, Delaney SM, Soule MJ, Phillips G, Thomashow LS (2001) Functional analysis of genes for biosynthesis of pyocyanin and phenazine-1-carboxamide from *Pseudomonas aeruginosa* PAO1. J Bacteriol 183:6454–6465

Mavrodi DV, Blankenfeldt W, Thomashow LS (2006) Phenazine compounds in fluorescent *Pseudomonas* spp.: biosynthesis and regulation. Annu Rev Phytopathol 44:417–445

Mavrodi OV, Mavrodi DV, Thomashow LS, Weller DM (2007) Quantification of 2,4-diacetylphloroglucinol-producing *Pseudomonas fluorescens* strains in the plant rhizosphere by real-time PCR. Appl Environ Microbiol 73:5531–5538

Mavrodi DV, Peever TL, Mavrodi OV, Parejko JA, Raaijmakers JM, Lemanceau P, Mazurier S, Heide L, Blankenfeldt W, Weller DM, Thomashow LS (2010) Diversity and evolution of the phenazine biosynthesis pathway. Appl Environ Microbiol 76:866–879

Mazzola M, Cook RJ, Thomashow LS, Weller DM, Pierson LS (1992) Contribution of phenazine antibiotic biosynthesis to the ecological competence of fluorescent pseudomonads in soil habitats. Appl Environ Microbiol 58:2616–2624

McDonald M, Mavrodi DV, Thomashow LS, Floss HG (2001) Phenazine biosynthesis in *Pseudomonas fluorescens*: branchpoint from the primary shikimate biosynthetic pathway and role of phenazine-1,6-dicarboxylic acid. J Am Chem Soc 123:9459–9460

Mentel M, Ahuja EG, Mavrodi DV, Breinbauer R, Thomashow LS, Blankenfeldt W (2009) Of two make one: the biosynthesis of phenazines. Chembiochem 10:2295–2304

Moynihan JA, Morrissey JP, Coppoolse ER, Stiekema WJ, O'Gara F, Boyd EF (2009) Evolutionary history of the *phl* gene cluster in the plant-associated bacterium *Pseudomonas fluorescens*. Appl Environ Microbiol 75:2122–2131

Nishi T, Forgac M (2002) The vacuolar ($H^+$)-ATPases – nature's most versatile proton pumps. Nat Rev Mol Cell Biol 3:94–103

Notz R, Maurhofer M, Schnider-Keel U, Duffy B, Haas D, Défago G (2001) Biotic factors affecting expression of the 2,4-diacetylphloroglucinol biosynthesis gene *phlA* in *Pseudomonas fluorescens* biocontrol strain CHA0 in the rhizosphere. Phytopathology 91:873–881

Notz R, Maurhofer M, Dubach H, Haas D, Défago G (2002) Fusaric acid-producing strains of *Fusarium oxysporum* alter 2,4-diacetylphloroglucinol biosynthetic gene expression in *Pseudomonas fluorescens* CHA0 in vitro and in the rhizosphere of wheat. Appl Environ Microbiol 68:2229–2235

Okubara PA, Bonsall RF (2008) Accumulation of *Pseudomonas*-derived 2,4-diacetylphloroglucinol on wheat seedling roots is influenced by host cultivar. Biol Control 46:322–331

Ownley BH, Weller DM, Thomashow LS (1992) Influence of in situ and in vitro pH on suppression of *Gaeumannomyces graminis var. tritici* by *Pseudomonas fluorescens* 2-79. Phytopathology 82:178–184

Ownley BH, Duffy BK, Weller DM (2003) Identification and manipulation of soil properties to improve the biological control performance of phenazine-producing *Pseudomonas fluorescens*. Appl Environ Microbiol 69:3333–3343

Parsons JF, Calabrese K, Eisenstein E, Ladner JE (2003) Structure and mechanism of *Pseudomonas aeruginosa* PhzD, an isochorismatase from the phenazine biosynthetic pathway. Biochemistry 42:5684–5693

Parsons JF, Calabrese K, Eisenstein E, Ladner JE (2004) Structure of the phenazine biosynthesis enzyme PhzG. Acta Crystallogr D 60:2110–2113

Parsons JF, Greenhagen BT, Shi K, Calabrese K, Robinson H, Ladner JE (2007) Structural and functional analysis of the pyocyanin biosynthetic protein PhzM from *Pseudomonas aeruginosa*. Biochemistry 46:1821–1828

Pechy-Tarr M, Bottiglieri M, Mathys S et al (2005) RpoN ($\sigma^{54}$) controls production of antifungal compounds and biocontrol activity in *Pseudomonas fluorescens* CHA0. Mol Plant Microbe Interact 18:260–272

Phillips DA, Fox TC, King MD, Bhuvaneswari TV, Teuber LR (2004) Microbial products trigger amino acid exudation from plant roots. Plant Physiol 136:2887–2894

Pierson LS III, Pierson EA (2010) Metabolism and function of phenazines in bacteria: impacts on the behavior of bacteria in the environment and biotechnological processes. Appl Microbiol Biotechnol 86:1659–1670

Pierson LS, Keppenne VD, Wood DW (1994) Phenazine antibiotic biosynthesis in *Pseudomonas aureofaciens* 30-84 is regulated by PhzR in response to cell density. J Bacteriol 176:3966–3974

Pierson LS, Gaffney T, Lam S, Gong FC (1995) Molecular analysis of genes encoding phenazine biosynthesis in the biological control bacterium *Pseudomonas aureofaciens* 30-84. FEMS Microbiol Lett 134:299–307

Pierson LS, Wood DW, Pierson EA (1998) Homoserine lactone-mediated gene regulation in plant-associated bacteria. Annu Rev Phytopathol 36:207–225
Pieterse CM, Leon-Reyes A, Van der Ent S, Van Wees SC (2009) Networking by small-molecule hormones in plant immunity. Nat Chem Biol 5:308–316
Price-Whelan A, Dietrich LE, Newman DK (2006) Rethinking 'secondary' metabolism: physiological roles for phenazine antibiotics. Nat Chem Biol 2:71–78
Price-Whelan A, Dietrich LE, Newman DK (2007) Pyocyanin alters redox homeostasis and carbon flux through central metabolic pathways in *Pseudomonas aeruginosa* PA14. J Bacteriol 189:6372–6381
Raaijmakers JM, Weller DM (1998) Natural plant protection by 2,4-diacetylphloroglucinol-producing *Pseudomonas* spp. in take-all decline soils. Mol Plant Microbe Interact 11:144–152
Raaijmakers JM, Weller DM (2001) Exploiting genotypic diversity of 2,4-diacetylphloroglucinol-producing *Pseudomonas* spp.: characterization of superior root-colonizing *P. fluorescens* strain Q8r1-96. Appl Environ Microbiol 67:2545–2554
Raaijmakers JM, Weller DM, Thomashow LS (1997) Frequency of antibiotic-producing *Pseudomonas* spp. in natural environments. Appl Environ Microbiol 63:881–887
Raaijmakers JM, Bonsall RE, Weller DM (1999) Effect of population density of *Pseudomonas fluorescens* on production of 2,4-diacetylphloroglucinol in the rhizosphere of wheat. Phytopathology 89:470–475
Ramos I, Dietrich LE, Price-Whelan A, Newman DK (2010) Phenazines affect biofilm formation by *Pseudomonas aeruginosa* in similar ways at various scales. Res Microbiol 161:187–191
Ran HM, Hassett DJ, Lau GW (2003) Human targets of *Pseudomonas aeruginosa* pyocyanin. Proc Natl Acad Sci USA 100:14315–14320
Rengel Z, Gutteridge R, Hirsch P, Hornby D (1996) Plant genotype, micronutrient fertilization and take-all infection influence bacterial populations in the rhizosphere of wheat. Plant Soil 183:269–277
Robin A, Vansuyt G, Hinsinger P, Meyer JM, Briat JF, Lemanceau P (2008) Iron dynamics in the rhizosphere: consequences for plant health and nutrition. Adv Agron 99:183–225
Rochat L, Pechy-Tarr M, Baehler E, Maurhofer M, Keel C (2010) Combination of fluorescent reporters for simultaneous monitoring of root colonization and antifungal gene expression by a biocontrol pseudomonad on cereals with flow cytometry. Mol Plant Microbe Interact 23:949–961
Saini HS, Barragan-Huerta BE, Lebron-Paler A et al (2008) Efficient purification of the biosurfactant viscosin from *Pseudomonas libanensis* strain M9-3 and its physicochemical and biological properties. J Nat Prod 71:1011–1015
Sarniguet A, Kraus J, Henkels MD, Muehlchen AM, Loper JE (1995) The sigma factor $\sigma^s$ affects antibiotic production and biological control activity of *Pseudomonas fluorescens* Pf-5. Proc Natl Acad Sci USA 92:12255–12259
Schnider U, Keel C, Blumer C, Troxler J, Défago G, Haas D (1995) Amplification of the housekeeping sigma factor in *Pseudomonas fluorescens* CHA0 enhances antibiotic production and improves biocontrol abilities. J Bacteriol 177:5387–5392
Schnider-Keel U, Seematter A, Maurhofer M, Blumer C, Duffy B, Gigot-Bonnefoy C, Cornelia R, Notz R, Défago G, Haas D, Keel C (2000) Autoinduction of 2,4-diacetylphloroglucinol biosynthesis in the biocontrol agent *Pseudomonas fluorescens* CHA0 and repression by the bacterial metabolites salicylate and pyoluteorin. J Bacteriol 182:1215–1225
Shanahan P, O'Sullivan DJ, Simpson P, Glennon JD, O'Gara F (1992) Isolation of 2,4-diacetylphloroglucinol from a fluorescent pseudomonad and investigation of physiological parameters influencing its production. Appl Environ Microbiol 58:353–358
Siddiqui IA, Shaukat SS (2004) Systemic resistance in tomato induced by biocontrol bacteria against the root-knot nematode, *Meloidogyne javanica* is independent of salicylic acid production. J Phytopathol 152:48–54
Slininger PJ, Jackson MA (1992) Nutritional factors regulating growth and accumulation of phenazine-1-carboxylic acid by *Pseudomonas fluorescens* 2-79. Appl Microbiol Biotechnol 37:388–392

Slininger PJ, Shea-Wilbur MA (1995) Liquid culture pH, temperature, and carbon (not nitrogen) source regulate phenazine productivity of the take-all biocontrol agent *Pseudomonas fluorescens* 2-79. Appl Microbiol Biotechnol 43:794–800

Stallings JH (1954) Soil produced antibiotics-plant disease and insect control. Bacteriol Rev 18:131–146

Stumm W, Morgan JJ (1981) Aquatic chemistry. John Wiley and Sons, New York, NY

Stutz EW, Défago G, Kern H (1986) Naturally occurring fluorescent pseudomonads involved in suppression of black root rot of tobacco. Phytopathology 76:181–185

Thomashow LS, Weller DM (1988) Role of a phenazine antibiotic from *Pseudomonas fluorescens* in biological control of *Gaeumannomyces graminis var. tritici*. J Bacteriol 170:3499–3508

Thomashow LS, Weller DM, Bonsall RF, Pierson LS (1990) Production of the antibiotic phenazine-1-carboxylic acid by fluorescent *Pseudomonas* species in the rhizosphere of wheat. Appl Environ Microbiol 56:908–912

Turner JM, Messenger AJ (1986) Occurrence, biochemistry and physiology of phenazine pigment production. Adv Microb Physiol 27:211–275

Validov S, Mavrodi O, De La Fuente L, Boronin A, Weller D, Thomashow L, Mavrodi D (2005) Antagonistic activity among 2,4-diacetylphloroglucinol-producing fluorescent *Pseudomonas* spp. FEMS Microbiol Lett 242:249–256

Vincent MN, Harrison LA, Brackin JM, Kovacevich PA, Mukerji P, Weller DM, Pierson EA (1991) Genetic analysis of the antifungal activity of a soilborne *Pseudomonas aureofaciens* strain. Appl Environ Microbiol 57:2928–2934

Wang Y, Newman DK (2008) Redox reactions of phenazine antibiotics with ferric (hydr)oxides and molecular oxygen. Environ Sci Technol 42:2380–2386

Wang Y, Kern SE, Newman DK (2010) Endogenous phenazine antibiotics promote anaerobic survival of *Pseudomonas aeruginosa* via extracellular electron transfer. J Bacteriol 192: 365–369

Weller DM, Raaijmakers JM, Gardener BBM, Thomashow LS (2002) Microbial populations responsible for specific soil suppressiveness to plant pathogens. Annu Rev Phytopathol 40:309–348

Weller DM, van Pelt JA, Mavrodi DV, Pieterse CMJ, Bakker PAHM, van Loon LC (2004) Induced systemic resistance (ISR) in *Arabidopsis* against *Pseudomonas syringae* pv. tomato by 2,4-diacetylphloroglucinol (DAPG)-producing *Pseudomonas fluorescens*. Phytopathology 94:S108

Weller DM, Landa BB, Mavrodi OV et al (2007) Role of 2,4-diacetylphloroglucinol-producing fluorescent *Pseudomonas* spp. in plant defense. Plant Biol 9:4–20

Whistler CA, Pierson LS III (2003) Repression of phenazine antibiotic production in *Pseudomonas aureofaciens* strain 30-84 by RpeA. J Bacteriol 185:3718–3725

Wood DW, Pierson LS III (1996) The *phzI* gene of *Pseudomonas aureofaciens* 30-84 is responsible for the production of a diffusible signal required for phenazine antibiotic production. Gene 168:49–53

Zha W, Rubin-Pitel SB, Zhao H (2006) Characterization of the substrate specificity of PhlD, a type III polyketide synthase from *Pseudomonas fluorescens*. J Biol Chem 281:32036–32047

# Chapter 12
# Plant Root Associated Biofilms: Perspectives for Natural Product Mining

**Salme Timmusk and Eviatar Nevo**

## 12.1 Introduction

For many years microbes in nature have been viewed as simple life forms growing as individual cells. This has enabled the characterization of the microorganisms. Most of our understanding of microbiology originates from experiments in liquid culture-free living bacteria. However, planktonic growth is not the natural situation for microorganisms and care needs to be taken then to interpret these results in their natural state. During the last decades an intensive research has been conducted in the area of biofilms: medical-industrial and plant associated biofilms. Usually biofilms are defined as complex microbial communities attached to the surface or interface enclosed in an extracellular matrix of microbial and host origin to produce a spatially organized three-dimensional structure (Costerton et al. 1995). It should also be noted that phenotypic variation in the biofilm forming bacteria is included (Branda et al. 2005; Lemon et al. 2008; Lopez et al. 2009, 2010). Genotypically identical biofilm bacteria are inherently different from the planktonic bacteria. Individual cells within a population control their gene expression to ensure that regulation of cell differentiation will occur (Lopez et al. 2009; O'Toole et al. 2000). There are complete reviews in the literature covering biofilm biology and genetics (Branda et al. 2005; Furukawa et al. 2006; Hogan and Kolter 2002; Langer et al. 2006; Lloyd et al. 2007; Lopez et al. 2010; Moons et al. 2009; Nunez et al. 2005; Watnick and Kolter 2000; Zegans et al. 2002). Biofilm is a normal common existence in bacterial ecosystems. Within the biofilms bacteria have cooperative behavior and they may be susceptible to harsh environmental conditions. It is the

S. Timmusk (✉)
Department of Forest Mycology and Pathology, Uppsala BioCenter, SLU, Uppsala, Box 7026, SE-75007, Sweden
e-mail: salme.timmusk@mykopat.slu.se

E. Nevo
Institute of Evolution, University of Haifa, Mt. Carmel, Haifa, Israel

D.K. Maheshwari (ed.), *Bacteria in Agrobiology: Plant Nutrient Management*,
DOI 10.1007/978-3-642-21061-7_12, 

preferred state of existence, because bacterial community adds defenses and multiple mechanism of bacterial survival, and enhances its fitness. Microorganisms also gain access to resources and niches that require critical mass and cannot effectively be utilized by isolated cells. Acquisition of new genetic traits, nutrient availability and metabolic cooperation have also been suggested as means for optimization of population survival in biofilms (Anderson and O'Toole 2008; Lemon et al. 2008; Lopez et al. 2009, 2010; Monds and O'Toole 2009).

In several areas of medical and industrial biofilms, the microorganisms have relatively little to do with the surface quality. In the area of plant associated microorganisms, it is generally accepted that plant roots live in firm teamwork with the surrounding microorganisms forming a unique self-regulating complex system (Deutschbauer et al. 2006; Stahl and Wagner 2006). Microorganisms are not only the most abundant organisms in natural systems, but are also key players in ecological processes. Among other plant-associated bacteria, the aerobic endospore-forming bacteria, mainly those belonging to *Bacillus* and related genera, are ubiquitous in agricultural systems due to their multilayer cell wall structure, ability to form stress resistant endospores, and to produce a wide variety of antibiotic substances. Exploiting these abilities, the bacteria can inhabit diverse niches in agro-ecosystems and out complete other microorganisms on the plant root. Therefore, the colonization niches for the bacteria are more reproducibly stable, and these bacteria are likely to be used in precision management of agroecosystems. For example, it was shown that an endospore forming species *Paenibacillus polymyxa* colonizes as biofilms the regions around root tips (Timmusk et al. 2005) (Fig. 12.1). The bacterial biofilms can protect plants against pathogens, as well as against abiotic stress conditions (Haggag and Timmusk 2007; Timmusk et al. 2009; Timmusk and Wagner 1999).

In this review we highlight themes regarding the nature and diversity of the bacterial biofilms and elucidate their potential as a rich source of novel biologically active compounds. The underground resources of plant rhizosphere could provide insights associated with global climate change. So far these resources have been neglected to large extent, but hopefully with the help of new technologies we will be able to understand and employ the natural potential of biofilms for our agroecosystems.

## 12.2 Structure

Biofilms formation is a dynamic sequence of events that has been carefully studied in *Vibrio cholerae* in Kolter's laboratory (Watnick and Kolter 1999, 2000). Four general biofilm formation stages have been described. The first stage is initiated as an attachment stage. Here bacteria grow as planktonic cells and approach the surface so closely that motility is slowed as a result. The bacterium may form then a transient association with the surface and with other microbes that previously attached to the surface. The transient association refers to the search for a place to

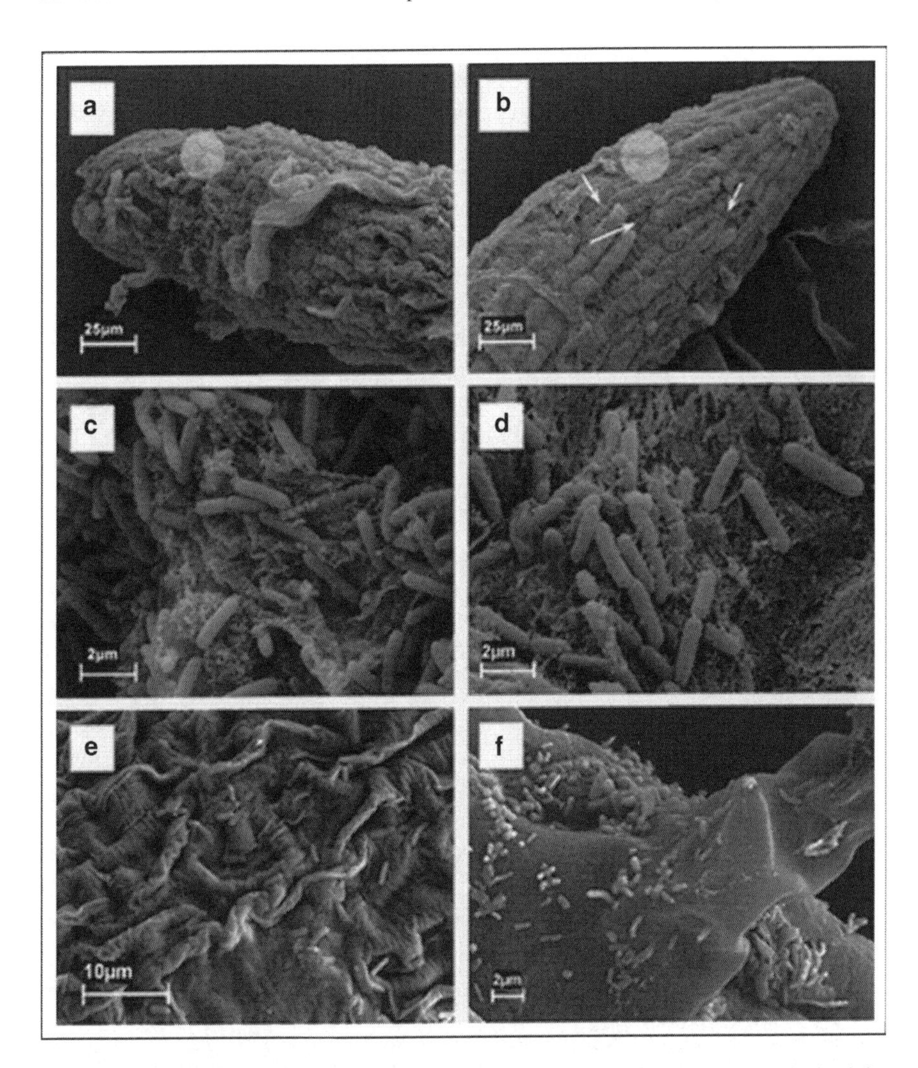


**Fig. 12.1** Scanning electron microscopy (SEM) micrographs of plant roots colonized by *Paenibacillus polymyxa*. *P. polymyxa* B1 colonization and biofilm formation on *plant* roots in the gnotobiotic system (**a**, **c**, and **e**), and in soil assays after 1 week of colonization (**b**, **d**, **f**). Roots were prepared and analyzed as described in Timmusk et al. (2005). Images were taken from the root tips (**a**–**d**) and from tip-distal regions (**e** and **f**). Note the biofilm formation on root tips (**a**–**d**). Much fewer bacteria colonize the regions behind root tip (**e** and **f**). In the non-sterile system only *P. polymyxa* was present at the biofilm-covered regions (**d**), whereas *P. polymyxa* cells mixed with indigenous bacteria were found on the distant regions of the plant root (**f**)

settle and is followed by a stable association. Stage two includes binding to the surface resulting in monolayer formation. After adhering to the surface the bacteria begin to multiply while emitting chemical signals that inter-communicate between bacterial cells and root. Once the signal intensity exceeds a certain level the genetic mechanisms underlying extracellular matrix production are activated. During this stage the cell motility is decreased and microcolonies are formed (O'Toole et al. 2000; O'Toole and Kolter 1998; Pratt and Kolter 1999). The cell layers are progressively added by extracellular matrix production (Branda et al. 2005, 2006; Nadell et al. 2008), and the biofilm three-dimensional structure is formed. Finally, the bacteria return to the planktonic stage (Watnick and Kolter 2000). Recently, a number of studies described the vast diversity in biofilm structure (Kolter and Greenberg 2006). Are there any principles of general nature? One feature that seems to apply to biofilms is that they all seem to create matrix. What is inside a matrix? An extracellular matrix can provide an almost infinite range of macromolecules. It was suggested that in the model bacterium *Bacillus subtilis* polysaccharides and a protein Tas A are the major components of its biofilm. Mutations that eliminate Tas A and extracellular polysaccharides (EPS) production have a severe effect on biofilm production (Branda et al. 2006; Kolter and Greenberg 2006). The sugars in biofilms can be divided into simple sugars (monosaccharides, oligosaccharides, polysaccharides), and complex sugars: all of which can play various roles in host microbe interactions (Lloyd et al. 2007; Vu et al. 2009). Water retention varies with the type of polysaccharides, but EPS water retention capacity may exceed 70 g water per g polysaccharide (Chenu 1993; Sutherland 2001; Vu et al. 2009; Zhang et al. 1998). Our experiments show that bacteria can engineer their own microenvironment in a form of porous EPS mixed soil particles. The environment immediately interacts with plant root providing buffered and predictable hydration, and transport properties (Fig. 12.2, Timmusk manuscript in preparation). The EPS producing *Paenibacillus* sp. strains significantly increased soil aggregation in comparison to the null mutants of the strains (Timmusk manuscript in preparation). The EPS may also contribute to mechanical stability of the biofilm, and interact with other macromolecules and low molecular mass solutes, providing a multitude of microenvironments within the biofilm (Vu et al. 2009). Currently many of these effects can only be speculated. Due to

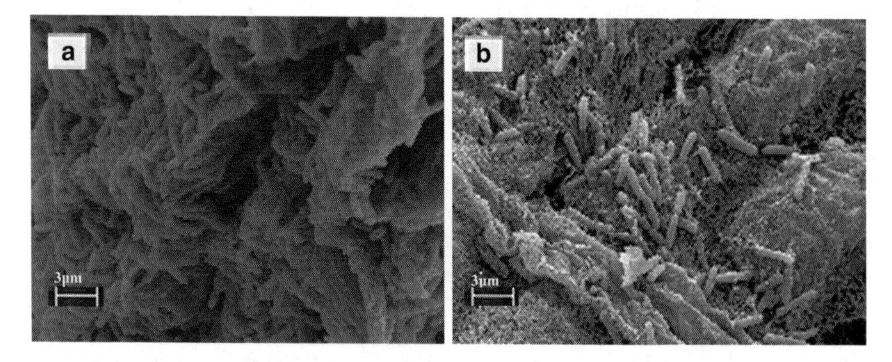


**Fig. 12.2** SEM micrographs of wild barley *Hordeum spontaneum* roots colonized by biofilm forming bacteria. Typical pattern of bacterial biofilm formation on wild barley root tips at AS (**a**) and ES (**b**). Wild barley plants were sampled, prepared and analyzed as described in Timmusk et al. (2009). Note that wild barley root tips at AS (**a**) are well colonized with mainly rod-shaped biofilm forming bacilli. Significantly less biofilm is formed on ES wild barley root tips (**b**)

their abundance in nature it is tempting to suggest polysaccharides as thc vehicle for biofilm manipulation.The diverse structural variations of EPS produced by bacteria of different taxonomic lineages makes the task hardly realistic.

## 12.3 Signaling

Quorum sensing (QS) is a well-known relatively conserved general communication mechanism. Since the initial discovery of Davies et al. (1998) the QS involvement in biofilm formation has been shown in variety of species. The cell to cell communication in this process is based on utilization signal molecules – the messengers that transform information across the space. QS is regulation of gene expression in response to cell population density. Gram positive and gram negative bacteria use QS to regulate diverse physiological activities. It has been shown that such activity occurs both inside and between the species. In general gram negative bacteria use homoserine lactones, and gram positive bacteria use small peptides. QS nature and potential applications are reviewed (Choudhary and Schmidt-Dannert 2010; Decho et al. 2010; Dickschat 2010; Thoendel and Horswill 2010). Kevin Foster and colleagues (Nadell et al. 2008) recently published a study examining the evolution of QS within biofilms. They illustrated how in the process of gaining fitness some bacterial species activate EPS production, whereas other species repress EPS synthesis upon QS activation.

There is growing evidence that in addition to the well documented quorum sensing systems other molecules act as signal molecules (Shank and Kolter 2009). Initially it was shown by the Davies group that the subinhibitory concentration of various antibiotics may function as signals. Surprisingly, these small molecules have the activity to modulate global gene transcription. There are bacteria in plant rhizospheres that produce the antibiotics in concentrations that are capable of killing other microbial cells. However, most attempts to detect the high antibiotic concentrations produced under natural conditions have limited success. Hence, besides being weapons fighting against competitors they are also considered signaling molecules that regulate the homeostasis of microbial communities. Strangely enough, it was shown that some antibiotics at low concentrations may even be beneficial to the bacteria in natural environments (Davies 2009; Fajardo and Martinez 2008; Goh et al. 2002; Linares et al. 2006; Mesak et al. 2008; Mlot 2009; Skindersoe et al. 2008; Yim et al. 2006, 2007). If the antibiotics are handled as signaling compounds, it gives also a totally new view to antibiotic resistance in the natural systems. In this case antibiotic resistance may serve as protection against new signals in environment in order to maintain the biofilm community (Davies 2009; Yim et al. 2006, 2007). Beside antibiotics several other secondary metabolites (SM) are known to be involved in microbial signaling Shank and Kolter (2009).

The environmental signals, such as nutrient sources, local PH, temperature, and oxygen surface properties evoke changes in biofilms in order to be able to gain optimal nutrition and colonize the environment efficiently (Davey and O'Toole

2000; Moons et al. 2009). As mentioned above, biofilm formation has four steps surface attachment, micro colony formation, maturation, and architecture formation. The initial steps, attachment and microcolony formation are regulated by the signals that differ from bacteria to bacteria, and reflect the natural habitat. The steps that follow are relatively more conserved and mainly reveal the physiology of cells inside the biofilm (Stanley and Lazazzera 2004). It was shown in Kolter's laboratory that bacteria initiate biofilm formation through different pathways depending on environmental conditions (O'Toole et al. 2000). Hence, the bacterial strain can achieve biofilm phenotype under different conditions through different mechanisms (Pratt and Kolter 1999). Studies on wild barley *Hordeum spontaneum* biofilms show that different types of biofilms are formed on the root tips from the "Evolution Canyons," "African," and "European" slopes (Fig. 12.2) (detailed below) (Timmusk et al. 2011). Since bacteria cannot escape stressful environmental conditions, their sensitive mechanisms must be evolved to allow the rapid perception of stress and homeostasis maintenance. This adds more dimensions to the complexity of biofilms and draws our attention to the necessity to study biofilms under contrasting environmental conditions, e.g., stress and non-stress environments.

## 12.4 "Evolution Canyon"

Insights into microbial biofilms biological and evolutionary significance necessitates the study of coevolution with the host plant, ideally under contrasting environmental stresses. The "Evolution Canyon" (EC) model (Fig. 12.3) is a natural laboratory focusing on the study of the evolution of biodiversity and adaptation at a microsite. The project is navigated by the Institute of Evolution at the Haifa University in Israel. The model present sharp interslope ecological contrasts caused by interslope microclimate divergence (Pavlicek et al. 2003). Both the geology and macro-climate are similar for both slopes. Since the canyon runs east–west, the canyon slopes display opposite orientations. The south-facing "African" slope, AS or SFS, receives 200–800% more solar radiation than the north-facing "European" slope, ES or NFS. Consequently, the savannoid AS is warmer and drier, and more drought-stressed than the cooler and more humid, ES. The opposite slopes are separated at bottom by 100 m and at top by 400 m, averaging 200 m (Nevo 2006).

**Fig. 12.3** Cross section of the "Evolution Canyon" indicating the collection sites on "African Slope" (AS) 1 and 2 and "European Slope" (ES) 5 and 7

The EC model reveals evolution in action across life at a microscale involving biodiversity divergence, adaptation and incipient sympatric ecological speciation (Nevo 1997, 2006; Nevo et al. 2002; Nunez et al. 2005). The model highlights diverse taxa species richness, genomics, proteomics, and phenomics phenomena by exploring genetic polymorphisms at protein and DNA levels. Four EC's are currently being investigated in Israel in the Carmel, Galilee, Negev, and Golan Mountains (EC I-IV), respectively. We identified 2,500 species in EC I (Carmel) from bacteria to mammals in an area of 7,000 $m^2$. Local biodiversity patterns parallel global patterns (Nevo 2006). Higher terrestrial species richness was found on the AS. Aquatic species richness prevails on the ES. In 9 out of 14 (64%) model organisms across life, we identified a significantly higher genetic polymorphism on the more drought-stressful AS (Nevo 1997). Likewise, in some model taxa, we found largely higher levels of mutation rates, gene conversion, recombination, DNA repair, genome size, small sequence repeats (SSRs), single nucleotide polymorphism (SNPs), retrotransposons, transposons, and candidate gene diversity on the more stressful AS. Remarkably, interslope incipient sympatric ecological speciation was found across life from bacteria to mammals. The EC model could potentially highlight many mysteries of evolutionary biology, including the genetic basis of adaptation and speciation, especially now with the rapid high-throughput techniques of whole genome analysis (Joly et al. 2004; Nevo 1995, 1997, 2006, 2009).

Among other model organisms, wild progenitors of cereals, emmer wheat (*Triticum dicoccoides*), and wild barley (*Hordeum spontaneum*) have been studied at the "EC" for more than 30 years. The work has produced more than 200 publications (see the full list at http://enevo.haifa.ac.il and at http://evolution.haifa.ac.il) and the book, "Evolution of Wild Emmer and Wheat Improvement" (Nevo et al. 2002). This book contains interdisciplinary studies on the ecological, genetic, genomic, agronomic, and evolutionary aspects of wild emmer, conducted at the Institute of Evolution from 1980 to 2002. Wild emmer and wild barley are the progenitors of most cultivated wheat and barley, and thus are important sources of wheat and barley improvement. It is known that plants have coevolved together with biofilm-forming rhizobacteria over millennia. It is not clear, however, whether the modern cropping systems have retained all the beneficial components that are present in the naturally coevolved systems. *Paenibacillus polymyxa* as a representative of the wild progenitors rhizobacteria has been thoroughly studied. This bacterium is capable of imparting resistance to pathogens and improve drought tolerance (Timmusk and Wagner 1999). A model system to study and compare the bacterial biofilm formation in soil was developed (Timmusk et al. 2005). To investigate bacterial interactions in natural systems, real-time PCR for the biofilm forming bacterial rapid detection was also developed (Timmusk et al. 2007). *P. polymyxa* antagonism studies in interaction with agricultural plants against different pathogens, e.g., *Aspergillus niger*, *Pythium* and *Phytophthora* spp. highlighted the importance of biofilms in biocontrol initiation (Haggag and Timmusk 2007) (Fig. 12.4).

Biofilm formation is a complex phenomenon and is affected by physicochemical environment. For example, nutrient resources, attachment efficiency, cyclic stage

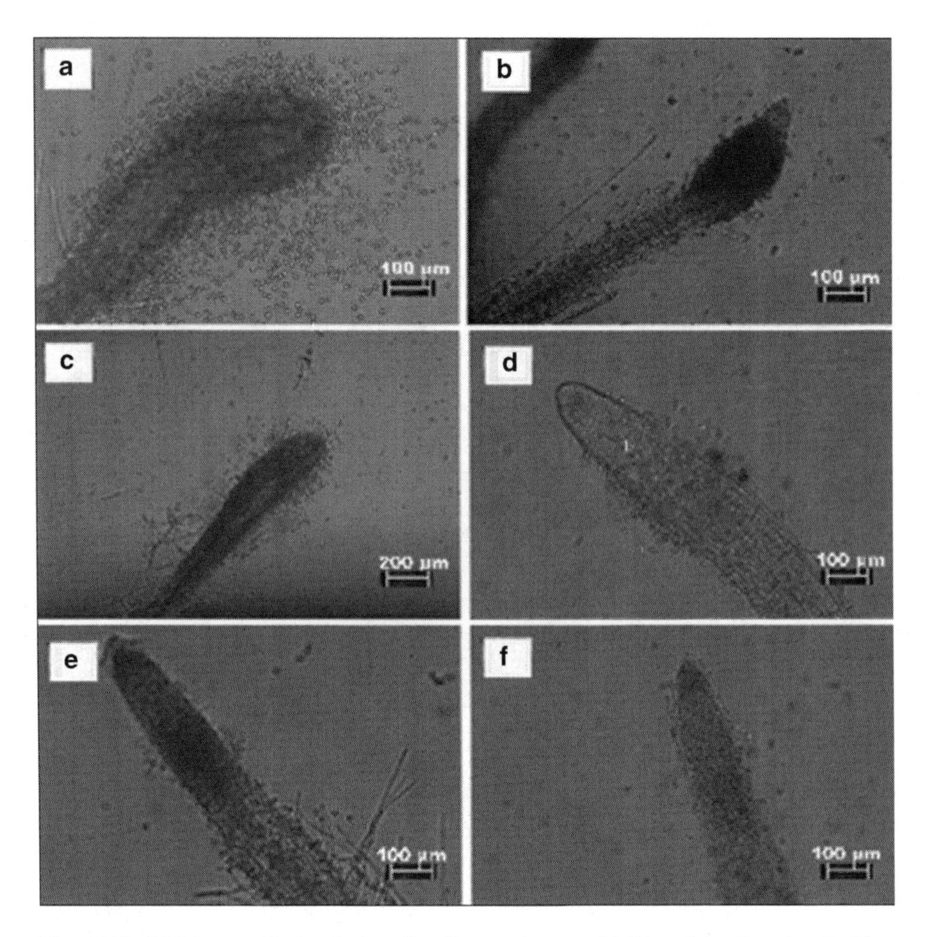


**Fig. 12.4** Inhibitory effect of *Paenibacillus polymyxa* biofilm formation to *Pythium aphanidermatum* and *Phytophthora palmivora* root colonization. *Arabidopsis thaliana* seedlings were grown and inoculated with the *P. polymyxa* and pathogens as described in Timmusk et al. (2009). The pattern of *P. aphanidermatum* (**a**) and *P. palmivora* (**b**) zoospore colonization on plant root is affected by *P. polymyxa* pre-inoculation (**c–f**). *P. polymyxa* relatively poor biofilm forming strain caused somewhat reduced *P. aphanidermatum* (**c**) and *P. palmivora* (**d**) zoospore colonization. Efficient biofilm forming *P. polymyxa* strains pretreated sample showed significantly less *P. aphanidermatum* [typical example on (**e**)] and *P. palmivora* (**f**) zoospore colonization

of the bacteria are factors that affect crosstalk between bacteria and plant roots (Aparna et al. 2008). Using SEM it was shown that wild barley seedlings from AS and ES have different types of biofilms formed around their root tips (Fig. 12.2). Both AS and ES biofilms are formed mainly by rod-shaped bacilli. Significantly more EPS containing biofilm is formed on the stressful AS (Fig. 12.2, Timmusk manuscript in preparation). The EPS role in protection against desiccation was shown by Tamaru et al. (2005). Their results confirm that EPS directly contributes to desiccation resistance enhancement. Bacteria from the biofilm forming regions of both slopes were isolated and screened for their metabolic properties (Timmusk et al. 2011). The drought-stressful AS slope contains significantly higher population of 1-aminocyclopropane-1-carboxylate deaminase (ACCd) producing, phosphorus solubilizing, osmotic stress tolerant bacteria (Timmusk et al. 2011). The features are likely to have provided a selective advantage for the plant-bacterial biofilm complex survival, and the bacteria may have helped the plant to tolerate various stresses using one or more of those mechanisms. These results suggest that bacterial biofilms on the plant root behave much like a multicellular organism. They excrete the "matrix" to provide a buffer against the environment and hold themselves in place. Whatever is produced inside the biofilm has a suitable environment and higher probability to get through to the target. This indicates that the rhizosphere bacteria, together with the plant roots at the AS wild barley rhizosphere, might function as communities with elevated complexity and plasticity, which in aggregate, have afforded the plant the adaptability to the harsh conditions encountered. The bacteria that coevolved with their hosts, over millennia, are likely to control, to a large extent, plant adaptation to the environment and have a huge potential for application in our agricultural systems enhancing plant stress tolerance.

## 12.5 New Perspectives

Biofilm research is currently one of the most topical research issues of molecular microbial ecology. First, it is expected that an improved understanding of the bacterial behavior will lead to develop agents that control the biology of biofilms. Secondly, biofilms are a rich source for novel natural products. Natural products are chemical compounds that usually exhibit biological activity and are presumed to have an ecological function. The compounds underwent an evolutionary process during which they were optimized for specific purposes. One of the most promising resources for new drugs, signaling compounds and plant growth promoting substances are biofilm SM (Wang and Tan 2009). There are millions of these compounds produced in the microbial world and several of them successfully applied. The biosynthetic pathways of SM are rather complex (Singh and Pelaez 2008).

The two most common classes of SMs are the nonribosomal peptides (NRP) and the polyketides (PK) (Koglin and Walsh 2009; McIntosh et al. 2009; Yagasaki and Hashimoto 2008; Zhang et al. 2008). PK synthetases (PKS) and NRP synthetases (NRPS) are both multienzyme multimodular biocatalysts containing numerous enzymatic domains organized into functional units (Powell et al. 2007a, b; Yagasaki and Hashimoto 2008; Wilkinson and Micklefield 2007). The vast structural diversity is due to a wide range of available substrates compared to 20 amino acids (AA) available for ribosomal synthesis. There are over 300 different amino, hydroxy or carboxy acid substrates that have been identified in NRP compounds (Kleinkauf and von Dohren 1990). Additionally NRP compounds also include fatty acid chains, macrocyclic, and heterocyclic rings. NRP usually contain between 2 and 20 AA. However, exceptionally the longest NRP known so far contains 48 AA (Hamada et al. 2010). The evolution of nonribosomal expression systems has allowed evolving the peptide based compounds with relatively low ATP cost. It is suggested to be sixfold lower in cost than the consumption for ribosomal synthesis, where ATP is required for aminoacyl-tRNA synthesis proofreading, elongation, and translation (Kallow et al. 1998, 2002). Both PKS and NRPS contain conserved domains. These domains are used in the overall assembly process. Three types of domains adenylation (A), thiolation (T), and condensation (C) domains are essential for the compound synthesis. A domain activates the corresponding AA and aminoacy-adenylates are subsequently transferred to 4-phospho-pantheinyl cofactors attached to downstream T-domains. During the stepwise elongation formation of the peptide bond between two adjacent aminoacyl intermediates bound to T domain is carried out by the intervening C domain. In some cases there is an additional epimerization (E) domain, which catalyzes the racemization of L amino acid to D amino acid.

How does one identify the compounds and correspondence in complex mixtures of microbes? The conserved domains have been valuable in predicting the metabolites in the structurally difficult to characterize PKS and NRPS groups. Usually the cosmid libraries from the microbial isolates are constructed, the libraries are screened with radioactive, degenerate DNA probes, or PCR primers, which target conserved regions of PKS or NRPS gene clusters. Then chromosome walking is used from identified genes to retrieve the sequence of the entire gene. Gene knockouts coupled with comparative metabolic profiling of wild type and mutant strains are then used as tool to identify the actual products (Yin et al. 2007). Yet, it is also known that there is an heterologous expression of the single biosynthetic genes. This can be found out by Northern blotting, DNA microarray analysis, or RT-PCR. The pleiotropic SM regulator manipulation at the cellular level is a good strategy to find and activate the silent cryptic pathways.

Taking into account that 99% of the microorganisms from most environments on earth cannot be grown under laboratory conditions, DNA based technologies should also be applied in the process of compound isolation and identification. Microbe and community genome sequences have revealed many genes and gene clusters encoding compounds similar to the ones known to be involved in the biosynthesis of biologically active compounds (Corre and Challis 2009) (Fig. 12.5). Often the gene clusters represent biosynthesis of novel natural products. Significant advances have been made in the past 20 years through the application of metagenomics, also referred as environmental and community genomics. Metagenomics is the genomic analysis of microorganisms by direct extraction and cloning of DNA from an assemblage of organisms (Handelsman 2004). Comprehensive reviews have been written on the area (Ferrer et al. 2005, 2007; Gabor et al. 2007; Li and Qin 2005; Rajendhran and Gunasekaran 2008; Singh et al. 2009; Singh and Pelaez 2008; Taupp et al. 2009; Valenzuela et al. 2006; Ward and Fraser 2005). It became apparent that metagenomic approach could allow the isolation of genes encoding novel compounds from any environment (Daniel 2005; Langer et al. 2006; Lorenz and Eck 2005). It was proposed that if the gene clusters could be expressed in heterologous hosts, it would provide a direct route to the production of bioactive compounds. Hence, it was hoped that characterization of the communication

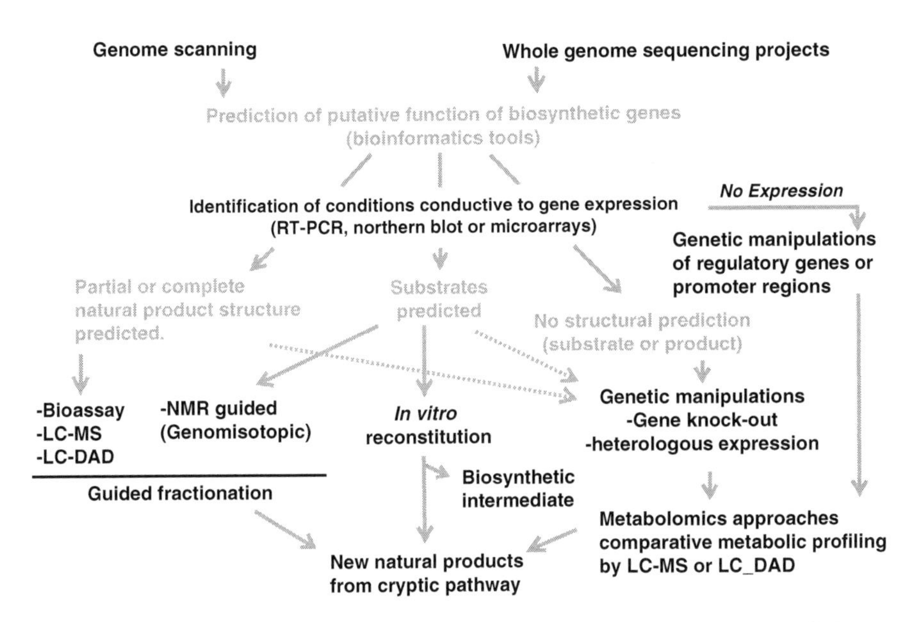


**Fig. 12.5** Strategies for discovery of novel natural products by genome mining (modified from Corre and Challis 2009)

networks and the natural roles of SMs was an available task. Even though several of the initial efforts encountered shortage of suitable techniques and tools for the natural product discovery, it was a necessary platform to reach the current stage. Nowadays, protocols have been developed to capture unexplored microbial diversity to overcome the existing barriers in estimation of diversity. New screening methods have been designed to select specific functional genes within metagenomic libraries to detect novel biocatalysts, as well as other bioactive molecules (Singh and Pelaez 2008). To study the complete gene or operon clusters, various vectors including cosmid, fosmid, or bacterial artificial chromosomes are being developed (Taupp et al. 2009). Bioinformatics tools and databases have added enormously to the study of microbial diversity (Singh et al. 2009).

If the compound is identified and isolated, then atomic force microscopy (AFM) can be used as a tool to study its production and performance under complex microbial associations. The earlier works mainly focused on gaining morphological and topographic information of the biofilm surface (Steele et al. 1998). The components of biofilm forming bacterial metabolism can be visualized in real time assays. One way to do it is immobilization of molecules at AFM probes. The AFM cantilever tips can then measure breakaway forces between biomolecules. With the specific antibodies on the cantilevers researchers have measured antibody–antigen interactions, and at the same time imagined their target antigens (Hinterdorfer and Dufrene 2006). The molecular recognition force (Hinterdorfer and Dufrene 2006) is applicable to study the biomolecule localization and function on the surface of biofilms. Single molecule studies have elucidated the important parameters of microbial protein folding and rupture. For example, the AFM imaging and force measurements studies have been performed on surface polysaccharides of *Lactobacillus* sp. Lecithin modified tips were used to study individual polysaccharide molecules on the surface of biofilms (Francius et al. 2008). In order to understand their function in biofilms, polysaccharides were characterized with single molecule force spectroscopy (Sletmoen et al. 2003). Glucans were characterized on the *Streptococcus mutans* biofilms and their possible role in biofilms was studied (Cross et al. 2007). The study was conducted with various mutants which ability to synthesize glucans was affected. The technique also provides the possibility for microbial surface molecular recognition using specific binding, such as antibody–antigen interaction. Employing AFM it is possible to study properties of attachment to the surfaces under natural conditions. The studies of pathogens were performed and structural details of Hif-typ pili at the early stage of biofilm were described (Ahimou et al. 2007). Force measurements of chemically fixed planktonic cells and native biofilm cells showed major difference in physical properties, such as elasticity and adhesion (Volle et al. 2008a, b). It has been also shown that biofilm formation is strongly dependent on the characteristics of substrate material (Oh et al. 2009). AFM was used to image *Bdellovibrio bacteriovorus* attack on *Escherichia coli* biofilms. The morphological changes in nanoscale of *E. coli* cells were monitored while attacked by the predator (Nunez et al. 2005). AFM studies are even more efficient when combined with other methods. As such AFM cannot produce information about the chemical composition of the biofilm under the surface. Hence, it can be used in combination of florescent and confocal microscopy (Lulevich et al. 2006; Martin-Cereceda et al. 2010). Raman spectroscopy would also facilitate to identify the materials. It uses a nondestructive laser to identify the components peaks of the Raman spectra (McEwen et al. 2010).

In sum, we are just beginning to understand the complexity and potential of biofilms. Yet, it is already clear that much is to be gained from studying this area. Intelligent biofilm engineering will be crucial in meeting the needs of handling the biofilms in agro-ecological systems. The contrasting environmental study locations where plants have coevolved with microbial representatives under stress over long period of time, such as the contrasting opposite slopes of "Evolution Canyon" (AS and ES), are especially good source for microbial representatives in order to study the biofilm structure, properties as well as production and composition of biologically active compounds.

## References

Ahimou F, Semmens MJ, Novak PJ, Haugstad G (2007) Biofilm cohesiveness measurement using a novel atomic force microscopy methodology. Appl Environ Microbiol 73:2897–2904

Anderson GG, O'Toole GA (2008) Innate and induced resistance mechanisms of bacterial biofilms. Curr Top Microbiol Immunol 322:85–105

Aparna M, Sharma P, Yadav S (2008) Biofilms: microbes and disease. Braz J Infect Dis 2:526–530

Branda SS, Vik S, Friedman L, Kolter R (2005) Biofilms: the matrix revisited. Trends Microbiol 13:20–26
Branda SS, Chu F, Kearns DB, Losick R, Kolter R (2006) A major protein component of the *Bacillus subtilis* biofilm matrix. Mol Microbiol 59:1229–1238
Chenu C (1993) Clay polysaccharide or sand polysaccharide associations as models for the interface between microorganisms and soil-water related properties and microstructure. Geoderma 56:143–156
Choudhary S, Schmidt-Dannert C (2010) Applications of quorum sensing in biotechnology. Appl Microbiol Biotechnol 86:1267–1279
Corre C, Challis GL (2009) New natural product biosynthetic chemistry discovered by genome mining. Nat Prod Rep 26:977–986
Costerton JW, Lewandowski Z, Caldwell DE, Korber DR, Lappin-Scott HM (1995) Microbial biofilms. Annu Rev Microbiol 49:711–745
Cross SE, Kreth J, Zhu L, Sullivan R, Shi W, Qi F, Gimzewski JK (2007) Nanomechanical properties of glucans and associated cell-surface adhesion of *Streptococcus mutans* probed by atomic force microscopy under in situ conditions. Microbiology 153:3124–3132
Daniel R (2005) The metagenomics of soil. Nat Rev Microbiol 3:470–478
Davies DG, Parsek MR, Pearson JP, Iglewski BH, Costerton JW & Greenberg EP (1998) The involvement of cell-to-cell signals in the development of a bacterial biofilm. Science 280:295–298
Davey ME, O'Toole GA (2000) Microbial biofilms: from ecology to molecular genetics. Microbiol Mol Biol Rev 64:847–867
Davies J (2009) Everything depends on everything else. Clin Microbiol Infect 15(Suppl 1):1–4
Decho AW, Norman RS, Visscher PT (2010) Quorum sensing in natural environments: emerging views from microbial mats. Trends Microbiol 18:73–80
Deutschbauer AM, Chivian D, Arkin AP (2006) Genomics for environmental microbiology. Curr Opin Biotechnol 17:229–235
Dickschat JS (2010) Quorum sensing and bacterial biofilms. Nat Prod Rep 27:343–369
Fajardo A, Martinez JL (2008) Antibiotics as signals that trigger specific bacterial responses. Curr Opin Microbiol 11:161–167
Ferrer M, Martinez-Abarca F, Golyshin PN (2005) Mining genomes and 'metagenomes' for novel catalysts. Curr Opin Biotechnol 16:588–593
Ferrer M, Golyshina O, Beloqui A, Golyshin PN (2007) Mining enzymes from extreme environments. Curr Opin Microbiol 10:207–214
Francius G, Lebeer S, Alsteens D, Wildling L, Gruber HJ, Hols P, De Keersmaecker S, Vanderleyden J, Dufrene YF (2008) Detection, localization, and conformational analysis of single polysaccharide molecules on live bacteria. ACS Nano 2:1921–1929
Furukawa S, Kuchma SL, O'Toole GA (2006) Keeping their options open: acute versus persistent infections. J Bacteriol 188:1211–1217
Gabor E, Liebeton K, Niehaus F, Eck J, Lorenz P (2007) Updating the metagenomics toolbox. Biotechnol J 2:201–206
Goh EB, Yim G, Tsui W, McClure J, Surette MG, Davies J (2002) Transcriptional modulation of bacterial gene expression by subinhibitory concentrations of antibiotics. Proc Natl Acad Sci USA 99:17025–17030
Haggag W, Timmusk S (2007) Colonization of peanut roots by biofilm forming *Paenibacillus polymyxa* initiates biocontrol against crown rot disease. J Appl Microbiol 104:961–969
Hamada T, Matsunaga S, Fujiwara M, Fujita K, Hirota H, Schmucki R, Guntert P, Fusetani N (2010) Solution structure of polytheonamide B, a highly cytotoxic nonribosomal polypeptide from marine sponge. J Am Chem Soc 132:12941–12945
Handelsman J (2004) Metagenomics: application of genomics to uncultured microorganisms. Microbiol Mol Biol Rev 68:669–685
Hinterdorfer P, Dufrene YF (2006) Detection and localization of single molecular recognition events using atomic force microscopy. Nat Methods 3:347–355

Hogan D, Kolter R (2002) Why are bacteria refractory to antimicrobials? Curr Opin Microbiol 5:472–477

Joly D, Korol A, Nevo E (2004) Sperm size evolution in *Drosophila*: inter- and intraspecific analysis. Genetica 120:233–244

Kallow W, von Dohren H, Kleinkauf H (1998) Penicillin biosynthesis: energy requirement for tripeptide precursor formation by delta-(L-alpha-aminoadipyl)-L-cysteinyl-D-valine synthetase from *Acremonium chrysogenum*. Biochemistry 37:5947–5952

Kallow W, Pavela-Vrancic M, Dieckmann R, von Dohren H (2002) Nonribosomal peptide synthetases-evidence for a second ATP-binding site. Biochim Biophys Acta 1601:93–99

Kleinkauf H, von Dohren H (1990) Nonribosomal biosynthesis of peptide antibiotics. Eur J Biochem 192:1–15

Koglin A, Walsh CT (2009) Structural insights into nonribosomal peptide enzymatic assembly lines. Nat Prod Rep 26:987–1000

Kolter R, Greenberg EP (2006) Microbial sciences: the superficial life of microbes. Nature 441:300–302

Langer M, Gabor EM, Liebeton K, Meurer G, Niehaus F, Schulze R, Eck J, Lorenz P (2006) Metagenomics: an inexhaustible access to nature's diversity. Biotechnol J 1:815–821

Lemon KP, Earl AM, Vlamakis HC, Aguilar C, Kolter R (2008) Biofilm development with an emphasis on *Bacillus subtilis*. Curr Top Microbiol Immunol 322:1–16

Li X, Qin L (2005) Metagenomics-based drug discovery and marine microbial diversity. Trends Biotechnol 23:539–543

Linares JF, Gustafsson I, Baquero F, Martinez JL (2006) Antibiotics as intermicrobial signaling agents instead of weapons. Proc Natl Acad Sci USA 103:19484–19489

Lloyd DH, Viac J, Werling D, Reme CA, Gatto H (2007) Role of sugars in surface microbe-host interactions and immune reaction modulation. Vet Dermatol 18:197–204

Lopez D, Vlamakis H, Kolter R (2009) Generation of multiple cell types in *Bacillus subtilis*. FEMS Microbiol Rev 33:152–163

Lopez D, Vlamakis H, Kolter R (2010) Biofilms. Cold Spring Harb Perspect Biol 2(7):a000398

Lorenz P, Eck J (2005) Metagenomics and industrial applications. Nat Rev Microbiol 3:510–516

Lulevich V, Zink T, Chen HY, Liu FT, Liu GY (2006) Cell mechanics using atomic force microscopy-based single-cell compression. Langmuir 22:8151–8155

Martin-Cereceda M, Roberts EC, Wootton EC, Bonaccorso E, Dyal P, Guinea A, Rogers D, Wright CJ, Novarino G (2010) Morphology, ultrastructure, and small subunit rDNA phylogeny of the marine heterotrophic flagellate *Goniomonas aff. amphinema*. J Eukaryot Microbiol 57:159–170

McEwen GD, Wu Y, Zhou A (2010) Probing nanostructures of bacterial extracellular polymeric substances versus culture time by Raman microspectroscopy and atomic force microscopy. Biopolymers 93(2):171–177

McIntosh JA, Donia MS, Schmidt EW (2009) Ribosomal peptide natural products: bridging the ribosomal and nonribosomal worlds. Nat Prod Rep 26:537–559

Mesak LR, Miao V, Davies J (2008) Effects of subinhibitory concentrations of antibiotics on SOS and DNA repair gene expression in *Staphylococcus aureus*. Antimicrob Agents Chemother 52:3394–3397

Mlot C (2009) Microbiology. Antibiotics in nature: beyond biological warfare. Science 324:1637–1639

Monds RD, O'Toole GA (2009) The developmental model of microbial biofilms: ten years of a paradigm up for review. Trends Microbiol 17:73–87

Moons P, Michiels CW, Aertsen A (2009) Bacterial interactions in biofilms. Crit Rev Microbiol 35:157–168

Nadell CD, Xavier JB, Levin SA, Foster KR (2008) The evolution of quorum sensing in bacterial biofilms. PLoS Biol 6:e14

Nevo E (1995) Asian, African and European biota meet at "Evolution Canyon" Israel: local tests of global biodiversity and genetic diversity patterns. Proc R Soc Lond (Biol) 262:149–155

Nevo E (1997) Evolution in action across phylogeny caused by microclimatic stresses at "Evolution Canyon". Theor Popul Biol 52:231–243
Nevo E (2006) 'Evolution Canyon': a microcosm of life's evolution focusing on adaptation and speciation. Israel J Ecol Evolut 52:485–506
Nevo E (2009) Ecological genomics of natural plant populations: the Israeli perspective. Methods Mol Biol 513:321–344
Nevo E, Korol AB, Beiles A, Fahima T (2002) Evolution of wild emmer and wheat improvement. Springer, Berlin
Nunez ME, Martin MO, Chan PH, Spain EM (2005) Predation, death, and survival in a biofilm: *Bdellovibrio* investigated by atomic force microscopy. Colloids Surf B Biointerfaces 42:263–271
O'Toole GA, Kolter R (1998) Flagellar and twitching motility are necessary for *Pseudomonas aeruginosa* biofilm development. Mol Microbiol 30:295–304
O'Toole G, Kaplan HB, Kolter R (2000) Biofilm formation as microbial development. Annu Rev Microbiol 54:49–79
Oh YJ, Lee NR, Jo W, Jung WK, Lim JS (2009) Effects of substrates on biofilm formation observed by atomic force microscopy. Ultramicroscopy 109:874–880
Pavlicek T, Sharon D, Kravchenko LV, Saaroni H, Nevo E (2003) Microclimatic interslope differences underlying biodiversity contrasts in "Evolution Canyon", Mt. Carmel, Israel. Isr J Earth Sci 52:1–9
Powell A, Nakeeb MAl, Wilkinson B, Micklefield J (2007a) Precursor-directed biosynthesis of nonribosomal lipopeptides with modified glutamate residues. Chem Commun (Camb) 2683–2685
Powell A, Borg M, Amir-Heidari B, Neary JM, Thirlway J, Wilkinson B, Smith CP, Micklefield J (2007b) Engineered biosynthesis of nonribosomal lipopeptides with modified fatty acid side chains. J Am Chem Soc 129:15182–15191
Pratt LA, Kolter R (1999) Genetic analyses of bacterial biofilm formation. Curr Opin Microbiol 2:598–603
Rajendhran J, Gunasekaran P (2008) Strategies for accessing soil metagenome for desired applications. Biotechnol Adv 26:576–590
Shank EA, Kolter R (2009) New developments in microbial interspecies signaling. Curr Opin Microbiol 12:205–214
Singh SB, Pelaez F (2008) Biodiversity, chemical diversity and drug discovery. Prog Drug Res 65:143–174
Singh J, Behal A, Singla N, Joshi A, Birbian N, Singh S, Bali V, Batra N (2009) Metagenomics: concept, methodology, ecological inference and recent advances. Biotechnol J 4:480–494
Skindersoe ME, Ettinger-Epstein P, Rasmussen TB, Bjarnsholt T, de Nys R, Givskov M (2008) Quorum sensing antagonism from marine organisms. Mar Biotechnol (NY) 10:56–63
Sletmoen M, Maurstad G, Sikorski P, Paulsen BS, Stokke BT (2003) Characterisation of bacterial polysaccharides: steps towards single-molecular studies. Carbohydr Res 338:2459–2475
Stahl DA, Wagner M (2006) The knowledge explosion in environmental microbiology offers new opportunities in biotechnology. Curr Opin Biotechnol 17:227–228
Stanley NR, Lazazzera BA (2004) Environmental signals and regulatory pathways that influence biofilm formation. Mol Microbiol 52:917–924
Steele A, Goddard D, Beech IB, Tapper RC, Stapleton D, Smith JR (1998) Atomic force microscopy imaging of fragments from the Martian meteorite ALH84001. J Microsc 189:2–7
Sutherland I (2001) Biofilm exopolysaccharides: a strong and sticky framework. Microbiology 147:3–9
Tamaru Y, Takani Y, Yoshida T, Sakamoto T (2005) Crucial role of extracellular polysaccharides in desiccation and freezing tolerance in the terrestrial cyanobacterium *Nostoc commune*. Appl Environ Microbiol 71:7327–7333
Taupp M, Lee S, Hawley A, Yang J, Hallam SJ (2009) Large insert environmental genomic library production. J Vis Exp 23(31):1387

Thoendel M, Horswill AR (2010) Biosynthesis of peptide signals in gram-positive bacteria. Adv Appl Microbiol 71:91–112

Timmusk S, Wagner EG (1999) The plant-growth-promoting rhizobacterium *Paenibacillus polymyxa* induces changes in *Arabidopsis thaliana* gene expression: a possible connection between biotic and abiotic stress responses. Mol Plant Microbe Interact 12:951–959

Timmusk S, Grantcharova N, Wagner EGH (2005) *Paenibacillus polymyxa* invades plant roots and forms biofilms. Appl Environ Microbiol 11:7292–7300

Timmusk S, Paalme V, Lagercratz U, Nevo E (2007) Detection and quantification of plant drought tolerance enhancing bacterium *Paenibacillus polymyxa* in the rhizosphere of wild barley (*Hordeum spontaneum*) with real-time PCR. J Appl Microbiol 107:736–745

Timmusk S, van West P, Gow CN, Huffstutler RP (2009) Biofilm forming *Paenibacillus polymyxa* antagonizes oomycete plant pathogens *Phytophthora palmivora* and *Pythium aphanidermatum*. J Appl Microbiol 106:1473–1481

Timmusk S, Paalme V, Pavlicek T, Bergquist J, Vangala A, Danilas T, Nevo E (2011) Bacterial distribution in the rhizosphere of wild barley under contrasting microclimates. PloS One 6(3): e17968

Valenzuela L, Chi A, Beard S, Orell A, Guiliani N, Shabanowitz J, Hunt DF, Jerez CA (2006) Genomics, metagenomics and proteomics in biomining microorganisms. Biotechnol Adv 24:197–211

Volle CB, Ferguson MA, Aidala KE, Spain EM, Nunez ME (2008a) Quantitative changes in the elasticity and adhesive properties of *Escherichia coli* ZK1056 prey cells during predation by *Bdellovibrio bacteriovorus* 109J. Langmuir 24:8102–8110

Volle CB, Ferguson MA, Aidala KE, Spain EM, Nunez ME (2008b) Spring constants and adhesive properties of native bacterial biofilm cells measured by atomic force microscopy. Colloids Surf B Biointerfaces 67:32–40

Vu B, Chen M, Crawford RJ, Ivanova EP (2009) Bacterial extracellular polysaccharides involved in biofilm formation. Molecules 14:2535–2554

Wang L, Tan H (2009) Molecular regulation of microbial secondary metabolites – a review. Wei Sheng Wu Xue Bao 49:411–416

Ward N, Fraser CM (2005) How genomics has affected the concept of microbiology. Curr Opin Microbiol 8:564–571

Watnick PI, Kolter R (1999) Steps in the development of a *Vibrio cholerae* El Tor biofilm. Mol Microbiol 34:586–595

Watnick P, Kolter R (2000) Biofilm, city of microbes. J Bacteriol 182:2675–2679

Wilkinson B, Micklefield J (2007) Mining and engineering natural-product biosynthetic pathways. Nat Chem Biol 3:379–386

Yagasaki M, Hashimoto S (2008) Synthesis and application of dipeptides; current status and perspectives. Appl Microbiol Biotechnol 81:13–22

Yim G, Wang HH, Davies J (2006) The truth about antibiotics. Int J Med Microbiol 296:163–170

Yim G, Wang HH, Davies J (2007) Antibiotics as signalling molecules. Philos Trans R Soc Lond B Biol Sci 362:1195–1200

Yin J, Straight PD, Hrvatin S, Dorrestein PC, Bumpus SB, Jao C, Kelleher NL, Kolter R, Walsh CT (2007) Genome-wide high-throughput mining of natural-product biosynthetic gene clusters by phage display. Chem Biol 14:303–312

Zegans ME, Becker HI, Budzik J, O'Toole G (2002) The role of bacterial biofilms in ocular infections. DNA Cell Biol 21:415–420

Zhang XQ, Bishop PL, Kupferle MJ (1998) Measurement of polysaccharides and proteins in biofilm extracellular polymers. Water Sci Technol 37:345–348

Zhang H, Wang Y, Pfeifer BA (2008) Bacterial hosts for natural product production. Mol Pharm 5:212–225

# Chapter 13
# Bacterial, Fungal, and Plant Volatile Compounds in Reducing Plant Pathogen Inoculums

**W.G. Dilantha Fernando and Vidarshani Nawalage**

## 13.1 Introduction

Biocontrol broadly refers the use of one living organism to curtail the growth and proliferation of another undesirable one. Biocontrol can be defined as "any condition under which a practice where by survival or activity of a pathogen is reduced through the activity of another living organisms except by man himself with the result there is a reduction in incidence of disease caused by pathogens" (Garrette 1965). Currently, biological control is becoming an increasingly effective alternative control measure to replace postharvest fungicide applications or to be implemented as part of integrated control programs (Conway et al. 1999). Biocontrol applications for postharvest disease control is now directed more towards the use of natural volatile compounds (VCs) produced by microorganisms that are biodegradable; that do not leave toxic residues on the fruit surface and displays as effective disease control as conventional fungicides (Mercier and Smilanick 2005). In biocontrol various mechanisms operate such as antibiosis, pre-emptive colonization of target site, antifungal volatiles, and induced systemic resistance (Fernando et al. 2007). Among these mechanisms, even though the antifungal volatile-mediated biocontrol plays an important role, this area was not broadly studied.

The rhizosphere of plants is the habitat of a community comprised of many different organisms. Soil bacteria often possess traits that enable them to act as antagonists by suppressing soil-borne plant diseases. For example, certain bacteria excrete antifungal metabolites that directly or indirectly support plant growth (Bilgrami et al. 1991; Gupta et al. 2000; Haas and Defago 2005; Handelsman and Stabb 1996). Many of these specialized compounds, such as antibiotics, are either liquid or solid at room temperature. Little is known about volatiles (with molecular masses <300 Da, low polarity, and a high vapor pressure) that can act as

W.G.D. Fernando (✉) • V. Nawalage
Department of Plant Science, University of Manitoba, Winnipeg, MB, Canada R3T 2N2
e-mail: D_Fernando@umanitoba.ca

D.K. Maheshwari (ed.), *Bacteria in Agrobiology: Plant Nutrient Management*,
DOI 10.1007/978-3-642-21061-7_13, 

antimicrobials and cause growth inhibition or have more deleterious effects on organisms. The microbial world synthesizes and emits many volatiles.

Pathogenic microorganisms affecting plant health are a major and chronic threat to food production and ecosystem stability worldwide. As agricultural production intensified over the past few decades, producers became more and more dependent on agrochemicals as a relatively reliable method of crop protection helping with economic stability of their operations. However, increasing use of chemical inputs causes several negative effects, i.e., development of pathogen resistance to the applied agents and their nontarget environmental impacts (McKee and Robinson 1988; Gerhardson 2002; Whipps 2001). Furthermore, the growing cost of pesticides, particularly in less-affluent regions of the world, and consumer demand for pesticide-free food has led to a search for substitutes for these products. There are also a number of fastidious diseases for which chemical solutions are few, ineffective, or nonexistent (Gerhardson 2002). Biological control is thus being considered as an alternative or a supplemental way of reducing the use of chemicals in agriculture.

In a previous publication Fiddaman and Rossall (1994) reported that another potential mode of action may lie with the production of antifungal volatile metabolites. The production of antifungal volatile metabolites has been shown to occur in several bacterial species (Hora and Baker 1972; Moore-Landecker and Stotzky 1972; Herrington et al. 1987), yeasts (Glen and Hutchinson 1969; Saksena and Tripathi 1987), fungi (Robinson et al. 1968; Dennis and Webster 1971; Upadhyay 1980), and from decomposing plant tissue (Lewis 1976). The production of such volatiles in the soil is likely to have a profound effect on the microbial ecology of that environment (Fiddaman and Rossall 1994).

*Bacillus subtilis* produces very potent AFVs and it is possible that these may contribute to this organism's potential as an effective biological control agent. For example, *B. subtilis* 12376 produces a novel high molecular weight non-volatile antibiotic (Bennett 1988), and the role of the antifungal volatiles may well be supplementary in nature. Localized concentration of these compounds as a result of solubilization and trapping within the soil biosphere may lead to greater in situ antifungal activity than would be predicted from in vitro assays (Fiddaman and Rossall 1994). This chapter broadly discusses the work of researchers on plant, fungal, and bacterial volatile metabolites which were successful in controlling the pathogen inoculums.

## 13.2 Plant Volatiles

It is interesting to know that plants have the potentiality of producing a vast number of VCs having a different spectrum of function. So far with the spotlight of recently advanced analytical techniques 1,700 volatile compounds were recognized and even extracted from more than 90 plant families (Dudareva et al. 2004). These volatiles are essentially plant secondary metabolites and Dudareva et al. have stated that they constitute about 1% of plant secondary metabolites released from leaves, flowers, and fruits into the atmosphere and from roots into the soil (Dudareva et al. 2004). When considering their functional roles in plants, these compounds assist plants in their communication and interaction as a language between the surrounding environment and plants. Also the fragrance of flowers and fruits is due to VCs consisting of small organic molecules with high vapor pressures which provide a reproductive advantage by attracting pollinators and seed dispersers. Also some of the plant volatile plant compounds are evolved to repel herbivores such as insects, which detect volatiles through the antennae on their heads or using certain mouthparts called axillary palps.

### *13.2.1 Plant Volatiles in Reducing the Pathogen Inoculums*

Although there are a number of naturally occurring VCs available in various plant tissues such as aroma component of fruits and vegetables and essential oil in spices and herbs, very little is known about their antimicrobial effects against harmful plant pathogens (Tripathi and Dubey 2004). This section focuses on a few studies carried out on some plant volatiles which were proven to be given highly promising effects against some of bacterial and fungal pathogen inoculums.

#### 13.2.1.1 Control of *Penicillium expansum* by Plant Volatile Compounds (Neri et al. 2006)

This work was focused in testing the efficacy of some plant VCs against pear fruit pathogens. Here the aldehydes (hexanal, *trans*-2-hexenal, citral, *trans*-cinnamaldehyde, *p*-anisaldehyde), the phenols (carvacrol and eugenol), and the ketones 2-nonanone (−)-carvon were screened for their ability to control *P. expansum* conidia germination and mycelial growth. The most active compounds in vitro against *P. expansum* were found to be *trans*-2-hexenal, carvacrol, *trans*cinnamaldehyde, and citral, measured from effects on conidial germination or mycelial growth.

#### 13.2.1.2 The Effect of Selected Cotton-Leaf Volatiles on Growth, Development, and Aflatoxin Production of *Aspergillus parasiticus* (Greene-McDowelle et al. 1999)

In this study Wright and co-workers have screened the ability of cotton leaf volatiles against pathogenic *Aspergilllus flavus* and *Aspergilllus parasiticus*. These two strains are the two prominent producers of aflatoxins. Aflatoxins are naturally occurring mycotoxins which are toxic and among the most carcinogenic substances known. In this particular study, these two aflatoxigenic strains of *A. flavus* and *A. parasiticus* were grown in the presence of specific cotton-leaf volatiles using the two alcohols [3-methyl-1-butanol (3-MB) and nonanol] and two terpenes (camphene and limonene) which are the representatives of the cotton leaf volatiles. In the presence of the above volatiles it was noted that a variation in the aflatoxin production was indicated by the altered growth patterns of the fungi when compared to the normal growth. Among the VCs examined, 3-MB has given the most promising effect by decreasing the radial growth of the fungi. Interestingly, the radial growth was directly proportional to the volatile dose used.

#### 13.2.1.3 Phytotxic and Fungitoxic Activities of the Essential Oil of Kenaf (*Hibiscus cannabinus* L.) Leaves (Kobaisy et al. 2001)

Another fascinating example of the plant volatiles is reported from the essential oil of Kenaf (*Hibiscus cannabinus* L.) leaves. Kenaf belongs to the family Malvaceae in which cotton (*Gossypium hirsutum* L.) and okra [*Abelmoschus esculentus* (L.) Moench] are the well-known members. Among the several uses of Kenaf, it has been evaluated for its allelopathic properties. For example, Kenaf mulch in vegetable production reduced weed populations. Apart from this study there are records of leaf volatiles of *H. cannabinus* biotype collected in Egypt where they have recorded the presence of ten components: ethyl alcohol, isobutyl alcohol, limonene, phellandrene, R-terpenyl acetate, citral, and four unidentified components. In this particular study it was noticed that the essential oil has phytotoxic, algicidal, and antifungal activities which could be useful for new, eco-friendly natural products in agriculture. They extracted a light yellow essential oil of *H. cannabinus* and successfully identified 57 components responsible for more than 79% of the total oil composition. Among the identified volatiles there are four monoterpene hydrocarbons (2.84%) and five sesquiterpene hydrocarbons (2.84%). Among the oxygenated compounds, alcohols (39.34%) were most predominant in the oil, followed by carbonyl compounds (30.79%). Diterpenes (36.18%) constituted most of the alcohol compounds; the main components were (*E*)-phytol (28.16%) and (*Z*)-phytol (8.02%). Aldehydes (20.32%) constituted most of the carbonyl compounds; the main components were *n*-nonanal (5.70%), benzene acetaldehyde (4.39%), (*E*)-2-hexenal (3.10%), and 5-methylfurfural (3.00%). Potentiality of the oil as a cyanobactericide was recommended after testing it against one cyanobacterium and one green alga.

However the most remarkable property of the oil of *H. cannabinus* is that of its inhibitory activity to the growth of certain species of the plant pathogenic fungi *Colletotrichum*. Genus *Colletotrichum* and its teleomorph *Glomerella* cause typical symptoms of anthracnose, in which watery, sunken patches surrounded by a red margin is visible in flowers, leaves, petioles, stolons, and crowns in a vast number of crops. From this particular study it was proven that the essential oil in Kenaf has the capacity to inhibit the growth of *C. gloeosporioides*, *C. fragariae*, and *C. accuhtatum*.

#### 13.2.1.4 Effect of Volatile Metabolites of Dill, Radish, and Garlic on Growth of Bacteria (Tirranen et al. 2001)

This particular study was targeted to screen the potentiality of dill, radish, and garlic in the evolving of antimicrobial volatile substances against certain bacteria. Here the plants were grown hydroponically and the antimicrobial activity of volatile metabolites was studied on seven bacterial strains from *Escherichia* (*E. coli* – two strains), *Staphylococcus* (*S. aureus*), *Haphnia* (*H. alvei*), *Bacillus* (*B. cereus*, *B. brevis*), and *Nocardia* (*Nocardia species*) genera.

The growth responses of test-microbes in the presence of volatile metabolites of dill, radish, and garlic were investigated by measuring the difference in the size of bacterial colonies in experimental and in control dishes. Control dishes of test-cultures were kept without exposure to the effect of volatile, biologically active substances from the investigated plants, i.e., grown in the atmosphere of clean air. To study the effectiveness of volatile metabolites produced by the above plants, experimental data was compared with background values which were found in each experiment separately.

It has been noted that the antimicrobial activity of volatile metabolites from radish was the poorest among the three plants as compared to all test-bacteria under study. Here the effect of volatile substances emitted by 38-day-old radish plants was considered zero, as the test-bacteria in experiments with radish volatiles did not differ much in size from the control.

However when compared to radish, antimicrobial effect of 30-day-old dill plants was prominent against some of the test bacteria. Volatile substances emitted by dill have strongly inhibited growth of two strains of bacteria of the coliform group and spore bacteria (*B. cereus*) and have stimulated growth of other Gram-positive bacteria (*B. brevis* and *Nocardia* species); but they were unable to control the growth of staphylococci and other Gram-negative bacteria.

Among the three plants tested, garlic demonstrated the most promising effect in controlling the test bacteria. From earlier times it has been known that the garlic herb is an effective herbal remedy to treat viral, bacterial, fungal, and other parasitic infections in the human body. Interestingly, a volatile substance released by crushed raw garlic called allicin, is known to be much more potent as an antibiotic than the common antibiotics such as penicillin and tetracycline issued in most standard medical treatments.

Volatile substances emitted from 33- to 50-day-old garlic plants successfully inhibited the test-bacteria with varying capacities. However it was observed that volatile emissions of younger garlic plants were able to stimulate the growth of *S. aureus*, *B. brevis*, *H. alvei* and slightly slow down the growth of *E. coli* (two strains) and *B. cereus*. The study concluded that the volatile substances emitted by dill and garlic were found to have a different spectrum of action on the growth of bacteria. The sensitivity spectrum of bacteria to the effect of volatile metabolites of dill and garlic differed with bacterial species.

#### 13.2.1.5 Antimicrobial Activities of Leaf Essential Oil from *Agastache rugosa* (Kim 2008)

*Agastache rugosa* Kuntz is an aromatic plant species with a perennial growth habit and is a member of the Lamiaceae family, native to Korea, China, and Japan. In this particular study, chemical composition of leaves from *A. rugosa* was analyzed by gas chromatography–mass spectrometry (GC–MS) in an effort to identify volatile constituents. Phytotoxic and antimicrobial activities of this essential oil were studied with 31 components, primarily methylchavicol (80.24%), DL-limonene (3.50%), linalool (4.23%), 5-methyl-2-(1-methylethylidene)-cyclohexanone (3.84%), and β-caryophyllene (2.39%).

#### 13.2.1.6 Antifungal Activity of Strawberry Fruit Volatile Compounds Against *Colletotrichum acutatum* (Moreno et al. 2007)

This interesting study concentrated on eight VCs which are characteristic of the scent of strawberry and originated by the oxidative degradation of linoleic and linolenic acids by a lipoxygenase (LOX) pathway. The focus was to test whether they have any potential in destroying or inhibiting the growth of the *Colletotrichum acutatum*, one of the causal agents of strawberry anthracnose. A range of volatile organic compounds was examined, including aldehydes, alcohols, and esters and these were tested on mycelial growth and conidia development. Among them (*E*)-Hex-2-enal was found to be the best inhibitor of mycelial growth [MID (minimum inhibitory doses) = 33.65 μL $L^{-1}$] and of spore germination (MID = 6.76 μL $L^{-1}$), while hexyl acetate was the least effective of all VCs tested (MID = 6441.89 μL $L^{-1}$ for mycelial growth and MID = 1351.35 μL $L^{-1}$ for spore germination). It was obvious that these compounds were capable in preventing the appearance of the anthracnose symptoms when brought in touch with the strawberry fruits inoculated with a spore suspension of *C. acutatum*. The promising effect of the antifungal ability 9891 of the (*E*)-hex-2-enal on conidial cells of *C. acutatum* was further proven by observations under the transmission electron microscopy. The distorted structure of the cell wall and plasma membrane was clearly visualized due to disorganization and lysis of organelles.

## 13.3 Volatile Fungal Metabolites in Reducing the Pathogen Inoculums

It is apparent that antifungal volatiles are widespread throughout the microbial community, although some volatiles seem to be more active than others. Identification of the volatile metabolite(s) responsible for reduction of pathogen inoculums and must await for further investigations.

### *13.3.1 Biological Activities of Volatile Fungal Metabolites (Hutchinson 1973)*

Outside the cell, the distinction between volatility and non-volatility is only qualitative, in the magnitude of the vapor pressures in particular conditions; within the cell solutions the distinction serves no purpose. In practice it has become a stimulating focus for thoughts and that this has led to useful new work and knowledge and the following six attributes seem to affect this closely.

(a) The capability of approaching an organism through the gaseous phase is an important aspect for fungi as most of the fungi grow part or their entire reproductive structure in moist air.
(b) Many compounds having an affinity in dissolving in lipids (lipophilic) and almost insoluble in water are directly produced upon the relief from the donor source and collect faster in the plasma membrane of an acceptor cell if it transfers via the gas phase than the usual liquid phase.
(c) When these volatile metabolites move into the gas phase they will be exposed to gaseous and physical inactivating or stimulating factors. Those diffusing in complex liquid solutions are likely to be exposed to more concentrated chemical inactivating or stimulating factors, and their movements will be limited by discontinuities in water films.
(d) As these volatile metabolites are active gases they are simple molecules with low molecular weights. Therefore their identification is much easier than their measurements and control and encourage exploring their chemical nature.
(e) The very low concentrations in which some of the identified ones are active suggests a comparison with antibiotics, growth factors, and vitamins, all areas of knowledge in which a loosely defined concept has promoted inquiry and discovery.
(f) Among all the other aspects, the most important point is that some of the volatile nature of these fungal metabolites leads to impermanence in particular situations, and therefore there is a risk of not giving the proper observation.

### *13.3.2 New Biocontrol Method for Parsley Powdery Mildew by the Antifungal Volatiles-Producing Fungus Kyu-W63 (Koitabashi 2005)*

Research on biocontrol of powdery mildew has mainly focused on studies of its hyperparasites (Abo-Foul et al. 1996; Bèlanger and Labbè 2002; Urquhart et al. 1994), but little is known about the usefulness of filamentous fungi that produce antifungal volatiles. The objective of this research was to develop a new biocontrol technique that would reduce labor compared with spraying an aqueous chemical to control the disease. In previous reports (Koitabashi et al. 2002, 2004), a filamentous fungus, Kyu-W63 was isolated which inhibited the development of wheat powdery mildew. Kyu-W63 treatment also suppressed parsley powdery mildew under greenhouse conditions similar to growers' conditions in Kyushu, in the southern part of Japan.

The inhibitory effect of Kyu-W63 was ascribed to its antifungal volatiles, which had a wide fungistatic spectrum and suppressed the growth of other pathogenic fungi in vitro (Koitabashi et al. 2002). The two types of antifungal volatiles produced by Kyu-W63 were determined to be 5-pentyl-2-furaldehyde and 5-(4-pentenyl)-2-furaldehyde (Koitabashi et al. 2004). Although known as an antinematode substance, 5-(4-pentenyl)-2-furaldehyde was first confirmed to have antifungal activity in this study.

### *13.3.3 Influence of Volatile Metabolites from* Geotrichum candidum *on Other Fungi (Tariq and Campbell 1991)*

There are a number of reports describing the fungi which produce volatile metabolites capable of suppressing or stimulating both on spore germination and mycelial extension, in the producer organism as well as in other species. But the previous studies of the effects of volatile metabolites produced by *Geotrichum candidum* have been limited to the role of these compounds in the self-regulation of growth and differentiation in this fungus. The results of this study prove that one or more volatile metabolites produced by self-inhibitory arthrospore suspensions of *G. candidum* in sterile distilled water are capable of inhibiting or stimulating germination and hyphal extension in some pathogenic fungi.

The phenomenon of self-inhibition of germination in arthrospores of *G. candidum* link was first reported by Park and Robinson (1970). They also observed that the addition of nutrients inhibited the self-inhibition phenomenon, proving that there is a close relationship between spore concentration and the level of nutrients required to overcome this auto-inhibition. This work was further extended by McKee and Robinson who reported that colonies and dense arthrospore suspensions of *G. candidum* produce one or more volatile metabolites which inhibit arthrospore germination and reduce the rate of mycelial extension in the producer organism (McKee and Robinson 1988). The study focused on whether volatile metabolites from self-inhibitory arthrospore suspensions of *G. candidum* influence germination and hyphal extension in other fungi. For this purpose they cultured the fungi, *Aspergillus jlavus*, *Aspergillus fumigatus* Fresen. *Botrytis allii* Munn, *Fusarium oxysporum* f.sp. *lycopersici*, *Penicillium italicum*, and *Microsporum canis* Bodin (NCPF 742). They noticed that volatile metabolites produced by self-inhibitory arthrospore suspensions of *G. candidum* inhibited germination in conidia of *A. flavus*, *B. allii*, *P. italicum*, and *M. canis* in contrast, germination of the conidia of *A. fumigatus* were encouraged in the presence of the arthrospore suspension when compared to the control dishes. The germination of the microconidia of *F. oxysporum* f.sp. *lycopersici* showed a very slight inhibition when they were exposed to the volatile metabolites. They also studied the influence of volatile metabolites from *G. candidum* on hyphal extension. When the standard assay system was used, volatile metabolites from self-inhibitory arthrospore suspensions of *G. candidum* significantly reduced the rate of hyphal extension in *A. flavus* and *B. allii*, but did not influence hyphal extension in *F. oxysporum* f.sp. *lycopersici*, *M. canis* or *P. italicum.* Also when examining the relationship between the volatile metabolites and the rate of hyphal extension in the tested species, it was obvious that the rate of hyphal extension in test discs exposed to the volatile metabolites was significantly reduced in all these five test species. By contrast, the rate of hyphal extension in *A. fumigatus* increased in the presence of volatile metabolites in both the standard and sensitized assay systems.

The type and level of response is dependent upon the species tested and the nutrient status of the test discs. McKee and Robinson (1988) reported that a reduction in mycelial extension in *G. candidum*, induced by self-produced volatile metabolites, was apparent only when the nutrient status of the test disc was reduced and the extension was already declining due to nutrient limitation. The results suggested that responses to these volatiles by some other species are similarly nutrient-dependent. In the standard assay, significant differences in hyphal extension between control and treatment discs for *A. flavus* and *A. fumigatus* were not apparent during the initial 8-h period of observations but were observed 24 h later. Many previous reports of the effects of volatile fungal metabolites on growth and development have implicated the primary metabolites carbon dioxide, ethanol, and acetaldehyde as the compounds responsible for the inhibitory and stimulatory effects observed (Hutchinson 1973). Analysis of air samples from Petri dishes containing self-inhibitory arthrospore suspension of *G. candidum* in malt broth and SOW have revealed the presence of oxygen, carbon dioxide, ethanol, acetaldehyde, and trimethylamine (McKee 1988; McKee and Robinson 1988; Robinson et al. 1989). However, McKee (1988) reported that self-inhibitory arthrospore suspensions of *G. candidum* in SOW, trimethylamine was the only volatile metabolite present in concentrations significantly higher than levels detected in control dishes. Robinson et al. (1989) observed that samples of pure trimethylamine inhibited arthrospore germination and hyphal extension in *G. candidum* and that the level of inhibition increased when nutrients in the assay discs were limiting (sensitized assay). In contrast, other reports suggested that growth in some fungi is stimulated in the presence of trimethylamine, possibly due to utilization of this compound as a source of nitrogen (Fujii et al. 1978; Yamada et al. 1976).

### 13.3.4 Effect of Volatile Phylloplane Fungal Metabolites on the Growth of a Foliar Pathogen (Upadhyay 2005)

When paying attention to the plant surfaces having the direct contact with air (aerial plant surfaces), volatile metabolites of fungi play an important role in the development of an oncoming pathogen (spore). Based on this principle, this study investigated the effect of volatile metabolites of various micro fungi inhabiting leaves on the growth of *Pestalotiopsis funerea* Desm, which is the causal agent of the leaf spot disease of *Eucalyptus globulus* Labill. They isolated the leaf-inhabiting micro-fungi by using different cultural techniques and then maintained them on yeast extract agar. The different fungal species were then tested on yeast extract agar in Petri dishes. Each plate was inoculated centrally with an agar disc of 6 mm diameter cut along the margin of a freshly growing culture of antagonistic fungus inhabiting the leaf. After incubation the lid of each dish was replaced by a bottom containing the agar medium and inoculated centrally with a 6-mm agar disc of *P. funerea*. The dishes were affixed together with an adhesive tape.

In their results, they observed that on control plates, incubated for 96 h, the average colony diameter was 47 $\pm$ 5 mm. The 5 mm or 10% difference in inhibition or stimulation of growth could be due to the variability of the growth of the pathogen. Any variation beyond 10% was due to the volatile substances emitted from the corresponding fungus. As they observed, the maximum inhibition which was more than 20% was recorded for *Trichodern viride* and *Trichothecium roseum* followed by *Aspergillus chevalieri*, *Curvularia lunata*, and *Penicillium chrysogenum*. Pathogenic fungi *A. flavus*, *Aspergillus terreus*, *Cladosporium cladosporiodes*, *Penicillium oxalicum*, and *Penicillium purpurogenum* were found to be least effective or not effective at all. In a previous study, Hutchinson and Cowan reported that inhibition in growth and sporulation of a culture of *Pestalotiopsis rhododendri* was due to the volatile secondary metabolites such as HCN, ethylene, ammonia, and acetaldehyde. However in this study, the growth of the fungal pathogen was encouraged due to the volatile production of *Cladosporium herbarum*, *Nigrospora spaerica*, and *Papu spora* sp., due to low concentration of $CO_2$, which is known to be a fungal growth stimulant. In addition, microscopic investigations show that volatile metabolites from *Trichoderma viride*, *T. roseum*, and *C. lunata* cause deformity, vacuolation of the hyphal cytoplasm and stunted mycelial growth. These hyphae were more highly branched than the normal ones. These effects were not permanent as they grew normally when placed on fresh nutrient agar. In addition, these observations also suggest that the volatiles are effective on the youngest portion, i.e., the tip of the fungal hypha which ceased to grow and started branching just below the tip.

## 13.4 Volatile Bacterial Metabolites in Reducing the Pathogen Inoculums

A large number of bacterial and fungal genera are known for producing chemical metabolites both diffusible and volatile in nature that have role to play in combating various phytopathogens.

### 13.4.1 *Diffusible and Volatile Compounds Produced by an Antagonistic* **Bacillus subtilis** *Strain Cause Structural Deformations in Pathogenic Fungi In Vitro (Chaurasia et al. 2005)*

This particular study was targeted at the soil bacteria inhabiting the immediate surroundings of the tea plant roots of tea gardens found in the Himalayan region in India. They encountered certain strains of *Bacillus* which were isolated and tested against a range of pathogenic fungi such as *Alternaria alternata*, *Cladosporium oxysporum*, *F. oxysporum*, *Paecilomyces lilacinus*, *Paecilomyces variotii*, and *Pythium afertile*. In this study authors have specially focused at the potential of the bacterium *B. subtilis* in producing diffusible and VCs which could damage the usual structure of the mycelia of the pathogenic fungi collected from the temperate and alpine forests of Uttaranchal Himalaya. The test fungi are known to cause diseases (*F. oxysporum*: *Fusarium* wilt and rots, *P. afertile*: damping off of seedlings and *Pythium* blight, *A. alternata*: leaf spot and leaf blight, *C. oxysporum*: fruit and crop rots, *P. lilacinus* and *P. variotii*: various human diseases) (Bilgrami et al. 1991; Dhindsa et al. 1995; Fletcher et al. 1998). In order to measure the antagonistic ability of the VCs produced by the *B. subtilis*, they measured the radial growth of the test fungi grown on a carrot potato agar plate in the presence of a sterilized, 5 mm diameter paper disc dipped in yeast extract broth containing the bacterial culture ($10^8$ cfu $mL^{-1}$). After 24 h it was observed that due to the VCs, radial growth of the fungi was reduced by values of $15.90 \pm 0.14\%$ (*C. oxysporum*) to $34.40 \pm 0.04\%$ (*F. oxysporum*). In addition to the growth suppression, the structural abnormality of the fungal mycelium of the all tested fungi induced by the VCs emitted by the bacterium was also noted. The entire disappearance of the transverse as well as longitudinal septae of *A. alternata* was observed, as well the conidia became thick walled and spherical or irregular in shape.

In another study, Fiddaman and Rossall (1994) investigated the effect of the substrate on the production of antifungal volatiles from *B. subtilis*. They focused on a strain of *B. subtilis* having a remarkable ability of producing a potent antifungal volatile and proceeded testing it with a range of growth media. Through in vitro assays it was confirmed that the antifungal volatile activity is more profound on nutrient agar enriched with D-glucose, along with some carbohydrates and peptones whereas the activity is reduced in the presence of L-glucose in the medium. Interestingly they also noticed significantly less antifungal volatile activity of this particular bacterium when inoculated on canola roots.

### 13.4.2 *Identification and Use of Potential Bacterial Organic Antifungal Volatiles in Biocontrol (Fernando et al. 2005)*

This appears to be the first work reported on the identification and the use of the bacterial antifungal organic volatiles in bio control of plant pathogens. The study focused on bacteria isolated from canola and soybean plants and was initiated to observe the role of bacterial antifungal organic volatiles in the bio control of plant disease. Prior to this study, the potential of *Pseudomonas* spp. producing organic volatiles against *Pytophthora vignae* in cowpea was already screened (Fernando and Linderman 1997).

*Sclerotinia sclerotiorum* has a broad spectrum of hosts that it can infect and cause disease. This pathogen causes disease conditions in over 400 crop species. It is responsible for diseases in two major economically important crop species; sunflower and canola, causing the stem rot of canola and head, stem and basal rot of sunflower. It is apparent that the most effective way of controlling *S. sclerotiorum* is the destruction of the overwintering sclerotia and in this case bacterial antifungal volatiles have the capacity to diffuse through the soil and kill the overwintering sclerotia. Therefore the antifungal volatile-producing bacteria isolated from canola and soybean plants were targeted to detect their specific mode of inhibition affecting mycelium, sclerotia or ascospores. Volatiles were isolated and identified by gas chromatography and mass spectroscopy. Individual organic VCs were screened for inhibition on mycelial growth and sclerotia. The study tested 197 bacterial isolates of which ten were identified as *Pseudomonas*. Their volatile production ability was tested using the divided plate method in a three compartment Petri plate where the first compartment contained Nutrient Agar (NA) or Tryptic Soy Agar (TSA) and in which the bacterial strain was streaked. The second compartment which was left empty except in the control which was filled with activated charcoal for the adsorption of any volatile produced by the bacteria and to investigate the effect on mycelial growth. The growth of the fungus was measured every 24 h as compared to the control over a period of 7 days. In addition to the divided plate method, volatile production was also tested using the sealed plate method. Bacteria were streaked on NA or TSA in the bottom dish of a Petri plate and the mycelial disc of the pathogen was placed in the center of the bottom dish of a second Petri plate containing PDA. The dish containing the mycelia plug was inverted over the bacterial plate and the dishes were sealed with parafilm. The plates were incubated at room temperature and the growth of the fungus was measured every 24 h, as compared to the control, over a period of 7 days. The control plates used only the mycelial discs. The sealed plate method was also used to check for the inhibition of surface-sterilized sclerotial germination. After 7 days, all mycelial plugs and sclerotia were removed from the plates and tested for viability in a fresh Petri plate containing PDA.

The volatile production in soil was screened by examining the mycelial and sclerotial inhibition of *S. sclerotiorum* by testing the ability of the volatiles to penetrate through soil. The ability of volatiles to inhibit ascospores was also tested. In 197 isolates screened for the production of volatile antifungal compounds in divided plates, 14 isolates consistently produced antifungal volatiles, and inhibited mycelial growth and sclerotial germination of *S. sclerotiorum*. Among the tested isolates, five *Pseudomonas chlororaphis* (Biotype D), three *P. fluorescens* (Biotype G), one *P. corrugata*, and one *P. aurantiaca* were identified using the Biolog© identification procedure.

The effect of the antifungal volatiles was clearly observed in the NA and TSA divided plates and in the control plate in which the third compartment was filled with activated charcoal. It was found that none of the bacteria had any inhibitory effect, allowing complete growth of the mycelial plug and germination of the sclerotia (Fig. 13.1).

Among the isolates, isolate DF35, *P. fluorescens* (Biotype-G), was able to inhibit sclerotial germination by 100%, but the mycelial growth only by 50%. Other bacteria screened caused 100% inhibition of both the growth of the mycelia and germination of the sclerotia. In the NA sealed plate assays, all the bacteria had higher inhibition of sclerotial germination than mycelial germination, except isolate DF209, *P. chlororaphis* (Biotype-D), which had high inhibition of both. The isolates DF200, *P. aurantiaca* and DF209 *P. chlororaphis* (Biotype-D) both had high inhibition of ascospore germination at 88 and 90%, respectively. To identify the volatiles emitted by the bacteria screened in this study, 14 were analyzed using GC–MS. This analysis yielded 23 organic compounds which included a range of aldehydes, ketones, alcohols, aliphatic alkanes, and organic acids. Of the 14 bacteria tested, 12 isolates found to have medium-to-high levels of antifungal volatile production were isolated from the soil and soil-associated plant parts. If volatile antibiotics are to be effective in controlling the pathogen, the most effective place would be in the soil where these volatiles would come in direct contact with

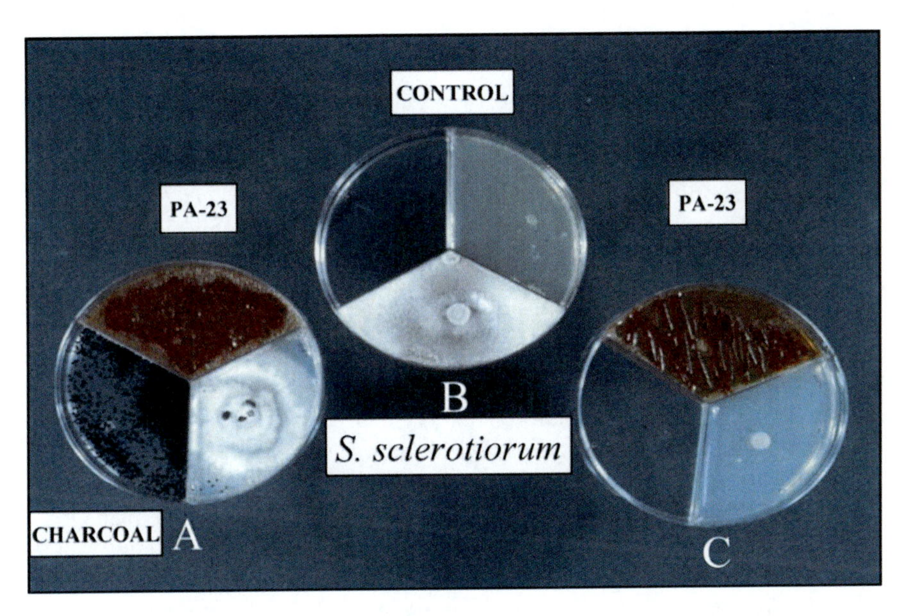


**Fig. 13.1** Bacterial antifungal-volatile activity in divided plates: Mycelial plug growth was completely inhibited in the presence of the bacteria streaked in a different compartment (**c**), as compared to the control (**b**), which had no bacteria. Mycelial growth was unaffected by the presence of volatile producing bacteria, when the third compartment of the plates was amended with activated charcoal (**a**). Charcoal adsorbs volatiles as soon as they are produced, and hence no inhibitory effect was observed (Fernando et al. 2005)

the sclerotinia propagules. Nitrogen sources also seemed to enhance the volatile production, where TSA, richer in carbon and nitrogen sources than NA, seemed to induce more volatile production and antifungal activity. This proved again that volatile production in soil is high in organic niches. Studies in greenhouse and under field conditions continue to look at the efficiency of bacteria to produce these antifungal volatiles in natural soil conditions and to reduce sclerotial viability of *S. sclerotiorum*; their continued existence and population dynamics, and their role in induction of the lipoxygenase pathway in canola and production of antifungal plant volatiles in biocontrol of *S. sclerotiorum*.

### *13.4.3 Cyanide Production by* **Pseudomonas fluorescens** *Helps Suppress Black Root Rot of Tobacco Under Gnotobiotic Conditions (Voisard et al. 1989)*

It has long been known that, fluorescent *Pseudomonas* which aggressively colonize root surfaces, are capable of fighting against certain plant diseases caused by the pathogens present in soil (Schroth and Hancock 1982; Davison 1988; Weller 1988; Defago and Haas 1990). The common feature shared by these plant-beneficial Pseudomonads is that they are antagonists of phytopathogenic fungi and bacteria (De Weger et al. 1995). To study this antagonistic effect, *P. fluorescens* strain CHAO was first isolated from a suppressive soil near Payerne in Switzerland. It was known that this strain suppresses black root rot of tobacco, a disease caused by the fungus *Thielaviopsis basicola*. The inhibition of this disease mostly occurs in iron-rich natural and artificial soils containing vermiculite clay (Stutz et al. 1986; Keel et al. 1989). This strain is known to inhibit the disease by producing pyoverdine, antibiotics, and volatile hydrogen cyanide (HCN) (Ahl et al. 1986). To assess the importance of these compounds, an aseptic environment was developed which consisted of artificial soil, a tobacco plant and known amounts of *T. basicola* and *P. fluorescens* cells (Keel et al. 1989). The antagonistic ability of the both wild type and mutant strains of *P. fluorescens* could be measured quantitatively and reproducibly using this particular system. This study tested the ability of volatile hydrogen cyanide in inhibiting the pathogenic *T. basicola* endo-conidia. Nearly 800 plant species of the plant kingdom evolve cyanide when they are wounded or attacked by fungi and hence successfully avoid the cyanide-sensitive pathogens. Certain successful fungal pathogens of cyanogenic plants however, are tolerant to cyanide as they can detoxify it by converting it to formamide (VanEtten and Kistler 1984). *T. basicola* was observed to be susceptible for the cyanide concentrations produced by *P. fluorescens* CHAO growing on nutrient agar or on pieces of cotton roots (Ahl et al. 1986), but it was not determined whether *P. fluorescens* produce cyanide in soil. From experiments conducted in the gnotobiotic system, it was proven that the inhibitory action of *P. fluorescens* is most probably due to the production of cyanide in situ. In 1977 it was found that *P. fluorescens* CHAO produced cyanide in the chemically defined medium of Castric. Both Glycine, which is the precursor of cyanide in *Pseudomonas aeruginosa* (Castric 1977), and $Fe^{3+}$ ions stimulated cyanide production in strain CHAO. It was also discovered that the cyanide generation (cyanogenesis) occurs at the end of exponential growth and at the beginning of stationary phase in batch cultures of *P. fluorescens*.

### *13.4.4 Production of Ammonia by* Enterobacter cloacae *and Its Possible Roles in the Biological Control of* Pythium *Pre-emergence Damping-off (Howell et al. 1988)*

*Enterobacter cloacae* is a well-known nitrogen-fixing bacterium which has shown excellent results towards the disease of pre-emergence damping off in pea, beet, cotton, and cucumber seedlings susceptible to *Pythium* species. A study of the mechanisms involved in the biological control of *E. cloaceae* of seedling disease induced by *Pythium* showed that there was a tight relationship between attachment of the bacterium to the hyphae of the fungus and inhibition of fungal growth and the control of the particular disease. In previous studies it was noted that some of the sugars available in the environment have suppressed the attachment of the bacterium to the fungal hyphae and thereby achieving the inhibition of the growth of the fungus and promoting the disease control. However, in earlier reports on *E. cloaceae*, the presence of the detectable antibiotic productivity or hyper-parasitic activity of the bacterium and the exact mechanisms of biological control was not properly reported. Hence the purpose of this particular study was to isolate and identify the inhibitory factors observed in the dual cultures and to determine its possible role in biological control.

The results suggested that the mechanisms involved in the biological control of *Pythium* pre-emergence damping off by *E. cloacae* was more complicated than earlier thought as other studies have not reported the presence of any antibiotic excluded by *E. cloacae*. From these results it was obvious *E. cloacae* produced a strong antifungal compound inhibitory to many fungi and that this compound was in fact, volatile ammonia. There are reports that ammonia is toxic to fungi in low concentrations which occur at fungistatic levels.

Ammonia released from various nitrogenous compounds or in the form of anhydrous ammonia has been used a fungicide to reduce disease severity. Nelson et al. also reported that "*E. cloacae* is effective in disease control only on seeds that do not exude large amounts of sugar in to the spermosphere during germination." It was also found that these sugars interfered with bacterial attachment to fungi as well as ammonia production by *E. cloacae*. It was thought that under conditions of low concentration of readily metabolized sugars in the spermosphere, *E. cloacae* deaminates amino acids generated by the germinating seeds to obtain a carbon source and thereby ammonia is produced as an antifungal by-product of this process. An important discovery showed that although *E. cloacae* had poor inhibition (50% in their study) it was shown to effectively control *Fusarium* wilt of cucumber. Although not clear, it is probably due to the form of wilt pathogen in soil. Germination of chlamydospores and conidia is often suppressed at a much lower inhibitor concentration than it is required to inhibit the further growth of the mycelium, which is growing rapidly. However according to the study, *E. cloacae* is not an effective bio control agent of seedling diseases caused by *Rhizoctonia solani*. This may be due to *R. solani* being less sensitive to ammonia than *P. ultimum* and infection by the pathogen usually occurs later in the development of the seedling stage when the bacterial population is diminished below a critical threshold level.

"As ammonia is highly toxic to *P. ultimum* at low concentrations, *E. cloacae* actively produces ammonia under conditions that might easily occur in spermosphere, and that the pattern of sugar inhibition of ammonia production is consistent with that observed for suppression of biocontrol activity by *E. cloacae*, it appears that ammonia production may well be the part of the mechanism by which *E. cloacae* controls *Pythium* preemergence damping off." In order to confirm this hypothesis studies are under way which will test the removal and restoration of the capacity to produce ammonia in mutants of *E. cloacae*.

## 13.5 Conclusions

In nature there are arrays of plant, fungal, and bacterial volatiles that inhibit other microbes in its niche either as a mechanism of competition, and/or antagonism. Most suppressive soils go unnoticed as the plants are healthy, or perform to the satisfaction of the grower. It is important to identify these soils that harbor beneficial microbes or plant metabolites to manipulate the systems in order to improve the plant health. More robust methods such as 454-pyro-sequencing and other molecular techniques available will assist in finding these organisms and their metabolites rapidly. As the world population is growing, and arable land is becoming increasingly limited, and the potential hazards of extensive use of chemicals are becoming obvious, the need for such environmentally friendly disease management strategies is prudent.

## References

Abo-Foul S, Raskin VI, Sztejnberg A, Marder JB (1996) Disruption of chlorophyll organization and function in powdery mildew-diseased cucumber leaves and its control by the hyperparasite *Ampelomyces quisqualis*. Phytopathology 86:195–199

Ahl P, Voisard C, Defago G (1986) Iron bound-siderophores, cyanic acid, and antibiotics involved in suppression of *Thielaviopsis basicola* by a *Pseudomonas fluorescens* strain. J Phytopathol 116:121–134

Bèlanger RR, Labbè C (2002) Control of powdery mildews without chemicals: prophylactic and biological alternatives for horticultural crops. In: Bèlanger RR, Bushnell WR, Dik AJ, Carver TLW (eds) The powdery mildews. APS, St. Paul, MN, pp 256–267

Bennett JM (1988) Biological control of chocolate spot on *Vicia faba* by *Bacillus subtilis*. Ph.D. thesis, University of Nottingham

Bilgrami BS, Jammaluddine S, Rizwi MA (1991) Fungi of India, list and references. Todays and Tomorrow's Printer and Publishers, New Delhi, p 798

Castric PA (1977) Glycine production by *Pseudomonas aeruginosa*. Hydrogen cyanide biosynthesis. J Bacteriol 130:826–831

Chaurasia B, Pandey A, Palni LM, Trivedi P, Kumar B, Colvin N (2005) Diffusible and volatile compounds produced by an antagonistic Bacillus subtilis strain cause structural deformations in pathogenic fungi in vitro, Microbiol Res.160(1):75–81

Conway WS, Janisieswicz WJ, Klein JD, Sams CE (1999) Strategy for combining heat treatment, calcium infiltration and biological control to reduce postharvest decay of Gala apple. Hortscience 34:700–704

Davison J (1988) Plant beneficial bacteria (review). Biotechnology 6:282–286

De Weger LA, van der Bij AJ, Dekkers LC, Simons M, Wijffelman CA, Lugtenberg BJJ (1995) Colonization of the rhizosphere of crop plants by plant-beneficial pseudomonads. FEMS Microbiol Ecol 17:221–228

Defago G, Haas D (1990) Pseudomonads as antagonists of soilborne plant pathogens: modes of action and genetic analysis. In: Bollag JM, Stotzky G (eds) Soil biochemistry, vol 6. Dekker, New York, pp 249–291

Dennis C, Webster J (1971) Antagonistic properties of species-groups of *Trichoderma*. 11. Production of volatile antibiotics. Trans Br Mycol Soc 57:41–48

Dhindsa MK, Naidu J, Singh SM, Jain SK (1995) Chronic supprative otitis media caused by *Paecilomyces varioti*. J Med Vet Mycol 33:59–61

Dudareva NE, Gershenzon J (2004) Biochemistry of plant volatiles. Plant Physiol 135:1893–1902

Dudareva N*, Pichersky E, Gershenzon J, (2004) Biochemistry of Plant Volatiles Plant Physiol. Vol. 135:1898–1902

Fernando WGD, Ramarathnam R, Krishnamoorthy AS, Savchuk SC (2005) Identification and use of potential bacterial organic antifungal volatiles in biocontrol. Soil Biol Biochem 37: 955–964

Fernando WGD, Linderman RG (1997) Effects of the mycorrhizal (Glomus intraradices) colonization on the development of root and stem rot (Phytophthora vignae) of cowpea. J. Nat. Sci. Council of Sri Lanka 25:39–47

Fernando WGD, Ramarathnam R, Kievit T (2007) Bacterial weapons of fungal destruction: phyllosphere targeted biological control of Sclerotinia stem rot and Blackleg diseases in Canola (*Brassica napus* L.). In: Esser K, Kubicek CP, Druzhinina IS (eds) The mycota, vol IV, A comprehensive treatise on fungi as experimental systems for basic and applied research. Springer, Berlin, pp 189–200

Fiddaman PJ, Rossall S (1994) Effect of substrate on the production of antifungal volatiles from *Bacillus subtilis*. J Appl Bacteriol 76:395–405

Fletcher CL, Hay RJ, Midgley G, Moore M (1998) Onychomycosis caused by infection with *Paecilomyces lilacinus*. Br J Dermatol 139:1133–1135

Francisco T Arroyo, Moreno J, Daza P, Boianova L, Romero F (2007) Antifungal Activity of Strawberry Fruit Volatile Compounds against Colletotrichum acutatum. J Agric Food Chem 55(14):5701–5707

Fujii T, Ishida Y, Kadota H (1978) Consumption of trimethylamine by molds in salted fish during storage at low temperature. Bull Jpn Soc Sci Fish 44:39–43

Garrette SD (1965) Towards the biological control of soil borne plant pathogens. In: Baker KF, Synder WC (eds) Ecology of soil borne plant pathogens. University of California Press, Los Angeles, CA, pp 4–17, p 571

Gerhardson B (2002) Biological substitutes for pesticides. Trends Biotechnol 20:338–343

Glen AT, Hutchinson SA (1969) Some biological effects of volatile metabolites from cultures of *Saccharomyces cerevisiae*. J Gen Microbiol 55:19–27

Greene-McDowelle DM, Ingber B, Wright MS, Zeringue HJ Jr, Bhatnagar D, Cleveland TE (1999) The effects of selected cotton-leaf volatiles on growth, development and aflatoxin production of Aspergillus parasiticus, Toxicon. 1999 Jun;37(6):883–93

Gupta AM, Gopal KVB, Tilak R (2000) Mechanism of plant growth promotion by rhizobacteria. Indian J Exp Biol 38:856–862

Haas D, Defago G (2005) Biological control of soil-borne pathogens by fluorescent pseudomonads. Nat Rev Microbiol 3:307–319

Handelsman J, Stabb EV (1996) Biocontrol of soil borne plant pathogens. Plant Cell 8:1855–1869

Herrington PR, Craig JT, Sheridan JE (1987) Methyl vinyl ketone: a volatile fungistatic inhibitor from *Streptomyces griseoruber*. Soil Biol Biochem 19:509–512

Hora TS, Baker R (1972) Soil fungistasis: microflora producing a volatile inhibitor. Trans Br Mycol Soc 59:491–500

Howell CR, Beier RC, Stipanovic RD (1988) Production of ammonia by *Enterobacter cloacae* and its possible role in the biological control of *Pythium* preemergence damping-off by the bacterium. Phytopathology 78:1075–1078

Hutchinson SA (1973) Biological activities of volatile fungal metabolites. Annu Rev Phytopathol 11:223–246

Keel C, Voisard C, Berling CH, Kahr G, Defago G (1989) Iron sufficiency, a prerequisite for the suppression of tobacco black root rot by *Pseudomonas fluorescens* strain CHAO under gnotobiotic conditions. Phytopathology 79:584–589

Kim J (2008) Phytotoxic and antimicrobial activities and chemical analysis of leaf essential oil from *Agastache rugosa*. J Plant Biol 51:276–283

Kobaisy M, Tellez MR, Webber CL, Dayan FE, Schrader KK, Wedge DE (2001) Phytotoxic and fungitoxic activities of the essential oil of Kenaf (*Hibiscus cannabinus* L.) leaves and its composition. J Agric Food Chem 49:3768–3771

Koitabashi M (2005) New biocontrol method for parsley powdery mildew by the antifungal volatiles-producing fungus Kyu-W63. J Gen Plant Pathol 71:280–284

Koitabashi M, Iwano M, Tsushima S (2002) Aromatic substances inhibiting wheat powdery mildew produced by a fungus detected with a new screening method for phylloplane fungi. J Gen Plant Pathol 68:183–188

Koitabashi M, Kajitani Y, Hirashima K (2004) Antifungal substances produced by fungal strain Kyu-W63 from wheat leaf and its taxonomic position. J Gen Plant Pathol 70:124–130

Lewis JA (1976) Production of volatiles from decomposing plant tissues and effect of these volatiles on *Rhizoctonia solani* in culture. Can J Microbiol 22:1300–1306

McKee ND (1988) Volatile inhibitors of growth and development produced by fungi. Ph.D. thesis, The Queen's University of Belfast

McKee ND, Robinson PM (1988) Production of volatile inhibitors of germination and hyphal extension by *Geotrichum candidum*. Trans Br Mycol Soc 91:157–160

Mercier J, Smilanick JL (2005) Control of green mold and sour rot of stored lemon by biofumigation with *Muscodor albus*. Biol Control 32:401–407

Moore-Landecker E, Stotzky G (1972) Inhibition of fungal growth and sporulation by volatile metabolites from bacteria. Can J Microbiol 18:957–962

Neri F, Mari M, Brigati S (2006) Control of *Penicillium expansum* by plant volatile compounds. Plant Pathol 55:100–105

Park D, Robinson PM (1970) Germination studies with *Geotrichum candidum*. Trans Br Mycol Soc 54:83–92

Robinson PM, Park D, Garrett MK (1968) Sporostatic products of fungi. Trans Br Mycol Soc 51:113–124

Robinson PM, McKee ND, Thompson LAA, Harper DB, Hamilton JTG (1989) Autoinhibition of germination and growth in *Geotrichum candidum*. Mycol Res 93:214–222

Saksena N, Tripathi HHS (1987) Effect of organic volatiles from *Saccharomyces* on the spore germination of fungi. Acta Microbiol Hung 34:255–257

Schroth MN, Hancock JG (1982) Disease-suppressive soil and root-colonizing bacteria. Science 216:1376–1381

Stutz EW, Defago G, Kern H (1986) Naturally occurring fluorescent pseudomonads involved in suppression of black root rot of tobacco. Phytopathology 76:181–185

Tariq VN, Campbell VM (1991) Influence of volatile metabolites from *Geotrichum candidum* on other fungi. Mycol Res 95:891–893

Tirranen LS, Borodina EV, Ushakova SA, Rygalov VY, Gitelson JI (2001) Effect of volatile metabolites of dill, radish and garlic on growth of bacteria. Acta Astronaut 49:105–108

Tripathi P, Dubey NK (2004) Exploitation of natural products as an alternative strategy to control postharvest fungal rotting of fruit and vegetables. Postharvest Biol Technol 32:235–245

Upadhyay RK (1980) Effect of volatile phylloplane fungal metabolites on the growth of a foliar pathogen. Experientia 37:707–708

Upadhyay RK (2005) Effect of volatile phylloplane fungal metabolites on the growth of a foliar pathogen. Cell Mol Life Sci 37(7):707–708

Urquhart EJ, Menzies JG, Punja ZK (1994) Growth and biological control activity of *Tilletiopsis* species against powdery mildew (*Sphaerotheca fuliginea*) on greenhouse cucumber. Phytopathology 84:341–351

VanEtten HD, Kistler HC (1984) Molecular and Genetic Perspectives. In: Kosuge T, Nester EW (eds) Plant–microbe interactions. Macmillan, NewYork, pp 42–68

Voisard C, Keel C, Haas D, Defago G (1989) Cyanide production by *Pseudomonas fluorescens* helps suppress black root rot of tobacco under gnotobiotic conditions. EMBO J 8:351–358

Weller DM (1988) Biological control of soil-borne plant pathogens in the rhizosphere with bacteria. Annu Rev Phytopathol 26:379–407

Whipps JM (2001) Microbial interaction and biocontrol in the rhizosphere. J Exp Bot 52:487–511

Yamada H, Kishimoto N, Kumagai H (1976) Metabolism of *N* substituted amines by yeasts. J Ferment Technol 54:726–737

# Chapter 14
# Denitrification Activity in Soils for Sustainable Agriculture

**Leticia A. Fernández, Eulogio J. Bedmar, Marcelo A. Sagardoy, María J. Delgado, and Marisa A. Gómez**

## 14.1 Introduction

Denitrification is a microbial process in the nitrogen cycle in which oxidized nitrogen compounds are used as alternative electron acceptors for energy production. It consists of four reaction steps in which nitrate is reduced to dinitrogen gas by metalloenzymes. These enzymes are induced sequentially under anaerobic conditions (Philippot 2002). The ability to denitrify is observed in a range of genera among both bacteria and archaea, and the widespread distribution is likely due to a common ancestor that existed before the prokaryotes split into two domains. Denitrifying bacteria occur in practically every sort of environmental niche: they represent 10–15% of the bacterial population in soil, water, and sediment (Casella and Payne 1996).

From an environmental point of view, denitrification has both positive and negative effects. One of the positive effect is that it decreases the leaching of nitrate to ground and surface waters, thus diminishing consequences derived from abuse of nitrogen fertilizers during agriculture practices. Aquifer contamination by nitrates and nitrites is of great impact in public health as it origins may lead to cancer in digestive apparatus tract and metahemoglobinemia. Other positive effect is that it is the main biological process responsible for the return of fixed nitrogen to the atmosphere. In contrast, the negative effects are that denitrification contributes to the greenhouse effects, the destruction of the ozone layer, and is a loss of nitrogen otherwise available for the growth of plants (Munch and Velthof 2007).

L.A. Fernández (✉) • M.A. Sagardoy • M.A. Gómez
Departamento de Agronomía, Universidad Nacional del Sur, San Andrés s/n (8000), Bahía Blanca, Provincia de Buenos Aires, República Argentina
e-mail: fernandezletic@yahoo.com.ar

E.J. Bedmar • M.J. Delgado
Departamento de Microbiología del Suelo y Sistemas Simbióticos, Estación Experimental del Zaidin, CSIC, Apartado Postal 419, 18080 Granada, España

D.K. Maheshwari (ed.), *Bacteria in Agrobiology: Plant Nutrient Management*,
DOI 10.1007/978-3-642-21061-7_14, 

The subject of this review is to focus on denitrification activity in soils. Investigation into sustainable agriculture is trying to improve productivity while taking care of the environment. Thus, the relevance of denitrification process must be considered as an important part of the nitrogen cycle in soil agroecosystems. On the other hand, the genus *Bradyrhizobium* has been always studied for its role in biological nitrogen fixation. However, several approaches have considered the relevance of this genus in denitrification, because it is, along with *Azorhizobium caulinodans*, the only rhizobia that have been shown to grow under denitrifying conditions. Consequently, we will intent to explain the reasons for considering this group of bacteria in relation with denitrification process. This review will also include the latest advances in molecular biology that have been made with regard to denitrification.

## 14.2 Soil Nitrogen Transformations

Nitrogen was discovered by the Scottish chemist and physician Daniel Rutherford in 1772 by removing oxygen and carbon dioxide from air. At the same time, the French chemist, Antoine Laurent Lavoisier, isolated what he called *azote*, meaning without life because it did not support life or combustion. Nitrogen is the fourth most common element in many biomolecules, some of them are essential for life, such as proteins and nucleic acids (Philippot and Germon 2005).

Nitrogen can occur in numerous oxidation states and has stable valences ranging from −3, as in ammonia ($NH_4^+$) to +5, as in nitrate ($NO_3^-$) (Kuenen and Roberson 1987; Poth and Focth 1985). Interconversions of these nitrogen species constitute the global biogeochemical nitrogen cycle, which is sustained by biological processes with bacteria playing a predominant role.

The atmosphere is the largest reservoir of nitrogen. This nitrogen enters the cycle through the action of several unique types of microorganisms that convert nitrogen gas to inorganic forms available to plants. This is called nitrogen fixation. The product of this process is generally used directly by free nitrogen-fixing bacteria or exported to the plants by symbiotic nitrogen-fixing bacteria.

Free-living diazotrophs fix nitrogen for their own benefit and may do so under aerobic, anaerobic, or microaerobic conditions, and they may be chemotrophs or phototrophs (Newton 2007). Symbiotic diazotrophs almost always live and fix nitrogen under microaerobic conditions inside a specialized structure on plant roots called nodules and on the stems of some aquatic legumes. The genera *Allorhizobium*, *Azorhizobium*, *Bradyrhizobium*, *Mesorhizobium*, *Rhizobium,* and *Sinorhizobium*, collectively referred to as rhizobia, are members, among others, of the bacterial order Rhizobiales of the α-proteobacteria (Delgado et al. 2007). Rhizobia are bacteria with the unique ability to establish a $N_2$-fixing symbiosis. We are interested in rhizobia, more precisely, in the genera *Bradyrhizobium* sp., because this bacterial species both fix nitrogen and are able to denitrify.

After the death of an organism (plant, animal, fungi, bacteria, etc.) organic nitrogen is degraded to ammonia by microorganisms. This decomposition of dead organic matter by saprophytic microorganisms is termed ammonification or mineralization. Once ammonia is produced, it can be fixed by clay or by soil organic matter, volatilized as ammonia, assimilated by plants and microorganisms, and finally it can be converted to nitrate by highly specialized bacteria during a two-step process called nitrification. The conversion by nitrification of the relatively immobile nitrogen form ammonia to the highly mobile form nitrate provides opportunities for nitrogen losses from soil. The nitrate formed during this process can be assimilated by plant roots and bacteria or used as terminal electron acceptor by microorganisms when the oxygen is limited. Therefore, nitrification can be considered to be the central for the flow, transfer, or loss of nitrogen in soil. The reduction of nitrate into gaseous nitrogen in a microaerobic/anaerobic environment is performed by a four-step reaction respiratory process called denitrification. Alternatively, nitrate produced by nitrification can be also reduced into ammonium by other either respiratory or dissimilatory process (Martínez Toledo 1992).

### 14.2.1 Symbiotic Nitrogen Fixation

The importance of symbiotic nitrogen fixation for sustainable agricultural systems cannot be underestimated, and great potential exists for increasing the usefulness of leguminous crops through breeding and palatability of legumes and through use of legumes as "green" fertilizers. Furthermore, the association of free-living diazotrophic bacteria with plants could also possibly contribute to the N economy of a crop. This would be especially important in organic agriculture and for small-scale farmers who cannot afford the extensive application of fertilizers (Miller and Cramer 2004).

Among rhizobia, *Bradyrhizobium japonicum* is one of the most agriculturally relevant species because it has the ability to form root nodules on soybean (*Glycine max*). This leguminous plant is of extraordinary importance because of its nutritional properties and is a significant crop component in many countries as the production area has steadily increased over the past 20 years. Moreover, *B. japonicum* is the rhizobial strain where the denitrification process has been more extensively studied. However, this bacterium can fix atmospheric nitrogen in soybean plants but also can carry out denitrification with production of nitrous oxide (Ciampitti et al. 2008).

Chemically, biological nitrogen fixation as symbiotic is essentially the conversion of dinitrogen to ammonia catalyzed by the enzyme nitrogenase. The reaction can be represented as follows:

| | | | | |
|---|---|---|---|---|
| 16Mg ATP | | | $\longrightarrow$ | 16Mg ADP + 16Pi |
| $N_2^+$ | 8H+ + | 8e− | $\longrightarrow$ | $2NH_3 + H_2$ |

Symbiotic nitrogen fixation by grain legumes and nitrous oxide emissions, the consequence of denitrification, are the two major processes of nitrogen transformation in agroecosystems. However, the relationship between these two processes is not very well understood (Zhong et al. 2009).

## 14.3 Denitrification

Denitrification is a part of the nitrogen cycle and transforms nitrate ($NO_3^-$) into $N_2$ gas. It is a reductive process and is a form of respiration which occurs in four stages: ($NO_3^-$) to nitrite ($NO_2^-$), ($NO_2^-$) to nitric oxide (NO), NO to nitrous oxide ($N_2O$), and $N_2O$ to $N_2$. All steps within this metabolic pathway are catalyzed by complex multisite metalloenzymes. In general, the proteins required for denitrification are only produced under anoxic conditions, and if anaerobically grown cells are exposed to oxygen, the activities of the proteins are inhibited. Thus, for denitrifying organisms, respiration of oxygen usually occurs in preference to the use of N-oxides or oxyanions (van Spanning et al. 2007).

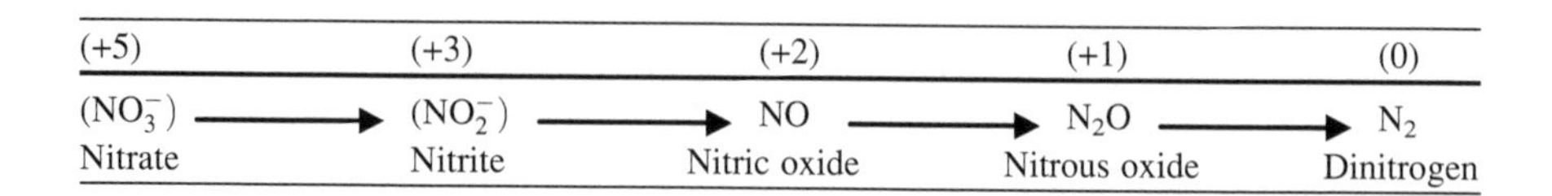


Although denitrification is initiated by respiratory (dissimilatory) nitrate reduction, this reaction is not unique to denitrification since it also occurs in ammonification and nitrification (see Sect. 14.3.1). Therefore, it is considered that the defining reaction in the process is the one-electron reduction of nitrite to the first gaseous intermediate nitric oxide. Thus, the production of nitric oxide as an obligate intermediate in respiratory denitrification is the condition at least for most denitrifiers. Numerous evidences confirm this hypothesis: isolation and characterization of nitric oxide reductases, studies with mutants preparations, measurements of different levels of nitric oxide under various conditions and trapping studies have shown that this N-oxide possesses the properties of a kinetically component intermediate (Ye et al. 1994). On the other hand, the two criteria that must be met to claim an organism as a respiratory denitrifier are: dinitrogen or nitrous oxide are produced from nitrate or nitrite in a higher rate than that of other mechanisms and that such reductions are coupled to a growth yield increase (Mahne and Tiedje 1995).

### *14.3.1 Nitrate Reducing Processes*

Historically, denitrification was not associated with any particularly mechanism. However, there are several types of processes that reduce nitrate. While respiratory denitrification is often of the greatest quantitative significance to nitrogen budgets, awareness of the other processes is necessary (Tiedje 1988, 1994). Therefore, the following paragraphs consider them.

Mulder et al. (1995) demonstrated that nitrification and denitrification were not necessarily separate phenomena. They showed an anaerobic ammonium oxidation in which nitrate was serving as the electron acceptor. The reaction taking place was:

$$5NH_4^+ + 3NO_3^- \rightarrow 4N_2 + 9H_2O + 2H^+$$

Further work showed that the oxidation of ammonium was actually coupled to the reduction of nitrite rather than to nitrate (Trimmer et al. 2003):

$$NO_2^- + NH_4^+ \rightarrow N_2 + 2H_2O$$

The phenomenon was studied in a denitrifying reactor and the responsible organism was identified as a new lithotrophic planctomycete. Also, it was shown that the reaction was energy yielding (Strous et al. 1999).

### *14.3.2 Diversity of Denitrifying Microbiota*

Denitrification is called completed, when denitrifier organism has the complete set of denitrification genes (*nar*/*nap*, *nir*, *nor*, *nos*) which conferred the capacity to reduce nitrate up to molecular nitrogen. In this case, a large part of the nitrous oxide emitted from soils comes from the actions of denitrifying bacteria which in the absence of oxygen, use nitrate as a terminal electron acceptor reducing it sequentially to nitrite, nitric oxide, nitrous oxide, and dinitrogen gas. However, not all nitrate-respiring bacteria possess the full complement of reductase enzymes required to produce dinitrogen gas: some accumulate nitrite and others nitrous oxide as the end product. This process is called incomplete denitrification (Zumft 1997; Cheneby et al. 2000).

Bacteria capable of denitrification are frequently isolated from sediment, soil, and aquatic environments and denitrifying ability is present in many phylogenetically diverse groups of bacteria (Cheneby et al. 2000). The most common denitrifiers in nature are species of *Pseudomonas* followed by the closely related *Alcaligenes*. Denitrification is present in strains contained in ten different prokaryotic families: Rhodospirillaceae, Cytophagaceae, Spirileaceae, Pseudomonaceae, Rhizobiaceae, Halobacteriaceae, Neisseriaceae, Nitrobacteraceae, and Bacillaceae (Tiedje 1988). Among the biogeochemical cycles on earth, there are no inorganic biotransformations that are carried out by a wider distribution and biodiversity than in the case of denitrification.

Nowadays, it is well established that denitrification is not exclusively a bacterial process. There are some works in filamentous fungi like *Fusarium*, in yeasts and actinomycetes. All of these novel organisms produce $N_2O$ as the major denitrification product (Shoun et al. 1992; Kumon et al. 2002; Laughlin and Stevens 2002). For several fungi, the system is localized at the respiring mitochondria, where the cytochrome P450 (P450nor) is located, which acts as nitric oxide reductase (Nor) (Kumon et al. 2002). However, as Cheneby et al. (2000) pointed out in their work: as fungi biomass is comparatively inferior to the bacteria biomass, denitrification process in this eukaryotic group has less importance from an ecological point of view.

### *14.3.3 Biochemistry and Molecular Biology*

The advantages of genome sequencing projects provide the opportunity to accelerate the knowledge on the genetic, evolution, and diversity of denitrification genes. This information could allow the identification of new potential denitrifying organisms which cannot be found by traditional studies. Nowadays, the information obtained by molecular approaches could not be depreciated and it must be included in the investigation programs.

Jones et al. (2008) combined the use of phylogenetic network analyses, statistical comparison of functional gene tree topologies, and examination of genome features to better understand the evolution of denitrification pathway. They observed that although horizontal gene transfer cannot be ruled out as a factor in the evolution of denitrification genes, the analysis suggests other phenomenon such as duplication or divergence that may have influenced the evolution.

Moreover, Demaneche et al. (2009) characterized clusters of denitrification genes from soil DNA extracts using a metagenomic approach which combines molecular screening and pyrosequencing. Interestingly, they identified nine clusters and presented the physical maps which show the *nosZ* clusters, the *nirS* clusters and genes from a family involved in the expression control of the denitrification process, the *nirK* clusters and other genes not directly involved in the process.

Finally, Philippot (2002) suggested that the exploration and study of molecular biology of denitrification genes is required for a more accurate understanding and modeling of the nitrogen fluxes.

The following paragraphs describe the important knowledge which has been accumulated for all genes coding for enzymes of denitrification (Fig. 14.1).

#### 14.3.3.1 Nitrate Reductases

Nitrate reduction in bacteria has three functions: the utilization of nitrate as a nitrogen source for growth (assimilation), the generation of metabolic energy by using nitrate as a terminal electron acceptor (respiration), and the dissipation of excess reducing power for redox balancing (dissimilation). In these three processes are involved, respectively, three different types of nitrate reductase: assimilatory nitrate reductase (Nas), and two dissimilatory nitrate reductase, which differ in their location: a membrane-bound (Nar) and a periplasmic-bound (Nap) (Moreno et al. 1999).

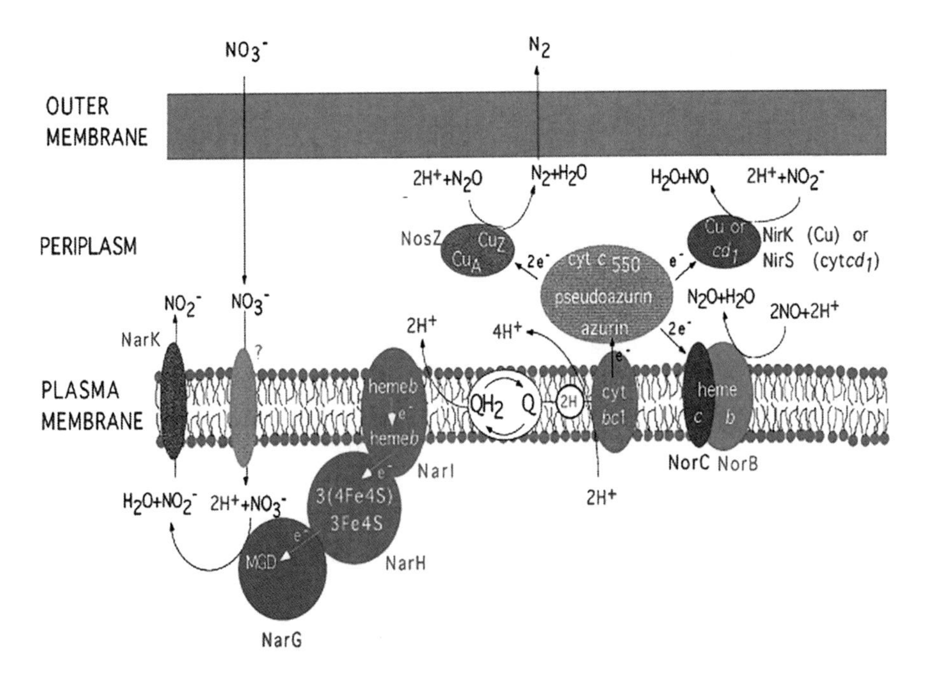


**Fig. 14.1** Denitrification pathways in bacteria

Nar enzymes are composed of three subunits: a catalytic $\alpha$ subunit (NarG) of 112–140 kDa, a soluble $\beta$ subunit (NarH) of 52–64 kDa, and both of them are anchored to subunit $\gamma$, NarI. Nap are 3-subunit enzymes composed of a catalytic subunit (90 kDa, NapA), a cytochrome c with two heme groups (15 kDa, NapB), and a membrane protein with 4 *c*-types heme (25 kDa, NapC) (Moreno et al. 1999).

#### 14.3.3.2 Nitrite Reductases

Two distinct types of Nir are found in denitrifying bacteria: one contains a *c*-type and a $d_1$-type heme as the redox active center, and the other contains Cu as the redox active transition metal. Both of them are periplasmic proteins which accept electrons from a cytochrome *c* and catalyzed the reduction of nitrite to nitric oxide. The *nirS* and *nirK* gene encode the $cd_1$ and copper-containing nitrite reductase, respectively, and with the two enzymes existing with differing levels of species diversity between them (Moreno et al. 1999).

Bremer et al. (2007) studied the influence of eight nonleguminous grassland plant species on the composition of soil denitrifier communities by T-RFLP analysis of the *nirK* gene. The results showed that the plants affected the *nirK*-type denitrifier community composition directly, for example, by root exudates. Moreover, the molecular technique revealed that environmental condition such as the sampling time had additional significance effect.

#### 14.3.3.3 Nitric Oxide Reductases

Three nitric oxide reductases (Nors) in bacteria: cNor, qNor, and qCuANor. cNor had been the better studies in Gram-negative bacteria like *Pseudomonas stutzeri*, *P. denitrificans,* and *Paracoccus halodenitrificans* (Bedmar et al. 2005).

#### 14.3.3.4 Nitrous Oxide Reductases

The last step of denitrification pathway is reduction of nitrous oxide to nitrogen which is catalyzed by nitrous oxide enzyme (Nos). The enzyme of periplasmic localization has been biochemically characterized in *P. stutzeri*, *P. denitrificans,* and *Paracoccus pantotrophus*. It is a homodimeric enzyme formed by two subunits of 65 kDa which contains copper in its active site (Bedmar et al. 2005).

## 14.4 Denitrification by *Bradyrhizobium* sp.

Bradyrhizobial and rhizobial denitrification has been known and widely discussed from a long time ago (O'Hara and Daniel 1985; van Berkum and Keyser 1985; Smith and Smith 1986; Breitenbeck and Bremner 1989). Some of these authors considered that denitrifying ability was a common trait within some rhizobia and bradyrhizobia. Nowadays, it has been well established that *B. japonicum* and *A. caulinodans* are the only rhizobia which are true denitrifiers. It has been shown that they reduces nitrate simultaneously to ammonia and nitrogen when cultured microanaerobically with nitrate as the terminal electron acceptor and the sole source of nitrogen (Delgado and Bedmar 2006).

Few workers have studied the occurrence and the denitrification activity of bradyrhizobia in natural populations. Asakawa (1993) described the ability of 103 *B. japonicum* isolates from Japanese soils to denitrify. Only one strain, *B. japonicum* S107 produced 30 nmol of $N_2O$ permicrogram of cell. The other strains were classified in two groups: 58 produced about 1.33 nmol of $N_2O$ permicrogram of cell and 44 produced 0.005 nmol of $N_2O$. On the other hand, Sameshima-Saito et al. (2004) evaluated the denitrification ability of 65 *Bradyrhizobium* isolates from other Japanese soils with a new 15N-labeled $N_2$ detection methodology. These authors divided the isolates in three categories: (1) 28 were full denitrifiers (up to $N_2$); (2) 18 belonged to the group of truncated denitrifiers (up to $N_2O$); and (3) 19 were nondenitrifiers.

### *14.4.1 Molecular Aspects*

At the molecular level, denitrification in *B. japonicum* has been extensively studied during the last decade. The following paragraphs comprehend a brief description of the relevant knowledge.

Denitrification in *B. japonicum* depends on the *napEDABC*, *nirK*, *norCBQD*, and *nosRZDFYLX* gene clusters encoding nitrate-, nitrite-, nitric oxide-, and nitrous oxide reductase, respectively (Fig. 14.2), which are dispersed over the chromosome (Kaneko et al. 2002). *B. japonicum* has the periplasmic nitrate reductase Nap as the only enzyme responsible for nitrate respiration under anaerobic growth conditions. Although it has been considered that membrane-bound nitrate reductase (Nar) catalyzes the first step of anaerobic denitrification (see Sect. 14.3.2), and that the Nap system was more important for aerobic denitrification or in redox balancing, the *B. japonicum* Nap enzyme can support anaerobic growth by reducing nitrate to nitrite, allowing the bacterium to grow in denitrification conditions (Bedmar et al. 2005).

Five genes were identified because they presented homology with *nap* genes from other microorganisms: *napE*, *napD*, *napA*, *napB*, and *napC* which codified proteins of 6.6, 11.8, 94.5, 16.9, and 23.9 kDa, respectively. The *nirK* gene encodes the copper nitrite reductase, a protein of 34.4 kDa, whose deduced primary sequence has greater than 68% identity with translated sequences of *nirK* genes from other denitrifiers. *NorC* and *norB* encode the cytochrome *c*-containing subunit II and the cytochrome *b*-containing subunit I, respectively. *NorQ* encodes a protein with an ATP/GTP-binding motif, and the predicted *norD* gene product is of unknown function. Mutational analysis indicated that the two structural *norC* and *norB* genes are required for microaerobic growth under nitrate-respiring conditions. The genes *nosR*, *nosZ*, *nosD*, *nosF*, *nosY*, *nosL*, and *nosX* codified proteins of 85, 72, 49, 33, 28, 19, and 38.5 kDa, respectively. It is worth noting that *B. japonicum* strains carrying either a *nosZ* or a *nosR* mutation grow well when cultured microaerobically with nitrate as the final electron acceptor (Bedmar et al. 2005).

#### 14.4.1.1 Regulation of Denitrification Genes

Like in many other denitrifiers, gene expressions in *B. japonicum* occur under oxygen limited conditions in the presence of nitrate. Regulatory studies using transcriptional *lacZ* fusions to the *napEDABC*, *nirK*, *norCBQD*, and *nosRZDFYLX* promoter regions indicated that microaerobic induction of denitrification genes is dependent on the *fixLJ* and *fixK2* genes whose products form the *FixLJ*/*FixK2* regulatory cascade. Another transcriptional regulator, the *NnrRprotein*, is responsible for N-oxide regulation of the *B. japonicum nirK* and *norCBQD* genes. Thus, the FixLJ-FixK2-NnrR cascade integrates both oxygen limitation and the presence of an N-oxide that are critical for maximal induction of the *B. japonicum* denitrification genes (Bedmar et al. 2005; Delgado et al. 2007).

More recently, it has been confirmed that maximal expression of denitrification genes requires the NifA protein. Moreover, the work of Bueno et al. (2010) re-affirmed previous results which indicated that *FixK2* is absolutely required for *nirK*, *napEDABC*, and *norCBQD* expression in addition to the classical nitrogen fixation genes.

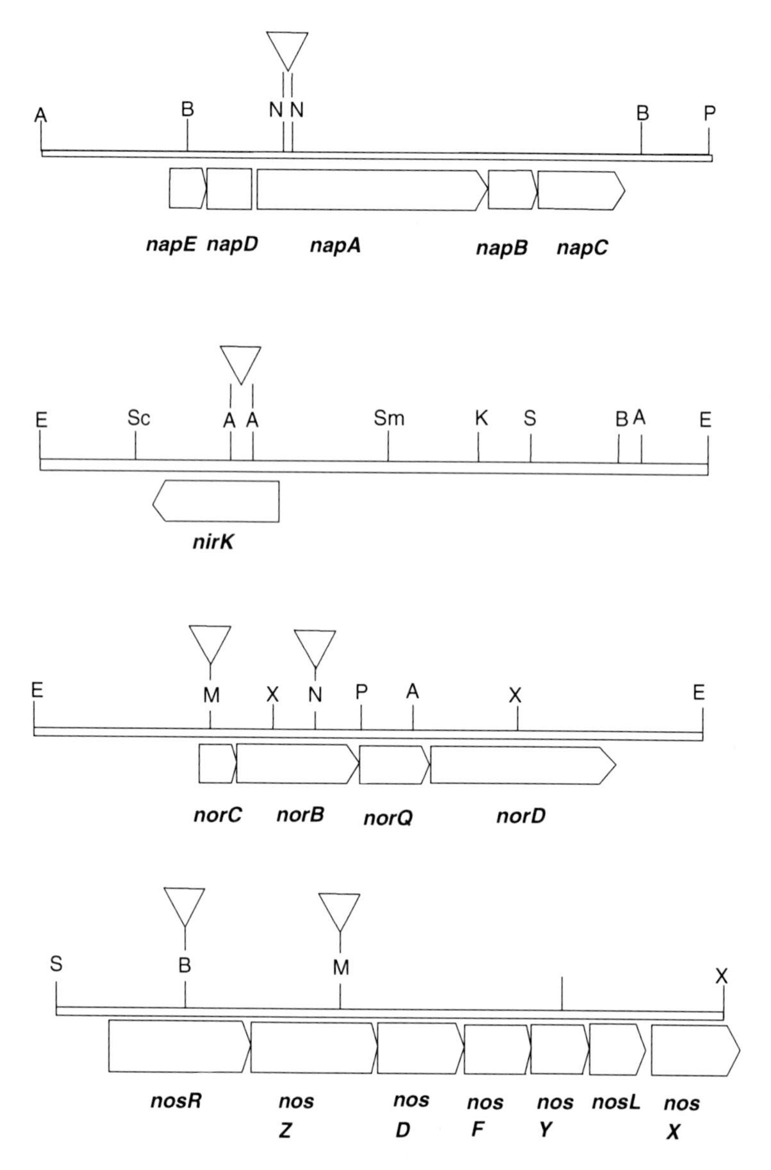


**Fig. 14.2** Organization of the *B. japonicum napEDABC*, *nirK*, *norCBQD,* and *nosRZDEFYLX* genes

## *14.4.2 PGPR Activity*

The beneficial effects of the symbiotic association between *Bradyrhizobium* and legumes are well known. However, many studies show that rhizobia can form associations with other economically important grain crops, such as maize, rice, and wheat and with vegetable crops like lettuce and radishes. These new associations can be beneficial for nonleguminous plants but it can also have a deleterious effect. Therefore, it is very important when crop rotation or intercrops systems are used to select strains of rhizobia that will have PGPR effects on both the plant involved (Antoun and Prévost 2000). Antoun et al. (1998) studied in a greenhouse trial the effect of inoculation of radishes with 266 collection strains of rhizobia and bradyrhizobia. Some of the strains examined were found to have deleterious effect while others were neutral or displayed PGPR activity on radishes.

## 14.5 Denitrification in Agricultural Soils

Agriculture has an important potential role in mitigating greenhouse gas (GHG) emissions, and it is known that in addition to carbon dioxide, agricultural soils are typically a major source to the atmosphere of the GHG, nitrous oxide (Hernandez-Ramirez et al. 2009).

Carbon dioxide production results from organic matter mineralization while nitrous oxide emissions are due to nitrification and denitrification processes. Atmospheric concentrations of both gases have increased during the last 250 years: from 278 to 365 for carbon dioxide production and from 0.270 to 0.314 $\mu L\ L^{-1}$ for nitrous oxide (Intergovernmental Panel on Climate Change 2001). As a matter of fact, soil emissions of nitrous oxide are quantitatively small (9.5 × 1012 g $N_2O$-N $year^{-1}$) when compared with $CO_2$ release (Schlesinger 1997; Yang et al. 2009). However, nitrous oxide in the lower atmosphere acts as a GHG that is more than 100 times more powerful in the warming potential than carbon dioxide. Moreover, this gas is chemically stable in the troposphere and reaches the stratosphere where it forms nitric oxide radicals in photochemical reactions which are involved in the destruction of ozone layer (Xing et al. 2009). In addition, arable soils are responsible for almost 60% of the global anthropogenic emissions of nitrous oxide (Smith et al. 2007). Therefore, it can be considered that the worst negative effect of agricultural denitrification is nitrous oxide emissions from soils.

### *14.5.1 Nitrous Oxide Emissions*

In soils, the rate of nitrous oxide production can be affected by different soil factors such as temperature, $N - -NO_3^-$, pH, water content, and organic C availability. In cultivated soils, the major nitrogen fluxes include inputs from fertilizer, fixation, atmospheric deposition, and manure; outputs including crop harvest, nitrate leaching, and denitrification, and the internal transformations of nitrogen including mineralization, nitrification, immobilization, and crop residue decomposition (Mahmood et al. 1998; Ciarlo et al. 2007; Ciampitti et al. 2008).

Because of the high nitrogen inputs, agricultural soils are considered as critical locations for denitrification. Therefore, the factors which control denitrification also affect nitrous oxide emissions. However, there is not always a positive linear relationship between denitrification and the gas emissions, because the ratio between the end products of denitrification, for example nitrous oxide and nitrogen, is also affected by different controlling factors such as temperature, nitrate, organic carbon compounds, and pH (Philippot and Germon 2005).

Although it is known that fluxes of nitrogen and nitrous oxide are difficult to measure under agricultural field conditions (David et al. 2009), there are numerous investigations which consider this subject. Yang et al. (2009) studied paddy soils in Taiwan during 2000–2006 and demonstrated that nitrous oxide emission coefficient at different locations was between 0.010 and 0.174 mg $m^{-2}$ $h^{-1}$ and that $N_2O$ emission increased with green manure amendment. The manure stimulated nitrous oxide emission rate due to the increase of soil organic matter and nitrogen content. Moreover, nitrous oxide derived from N fertilizer in paddy field varied between 0.05 and 0.28% in the central and southern Taiwan. The authors observed that nitrous oxide emission increased with the increasing of nitrogen fertilizer application and decreased when one slow-release N fertilizer was applied to the soils. They concluded that appropriate fertilization could reduce the nitrous oxide emission from the soils. In Madagascar, Baudoin et al. (2009) studied the differences between two soil management strategies [direct seeding with mulched crop (DMC) residues vs. tillage without incorporation of crop residues] along with a fertilization gradient (no fertilizer, organic fertilizer, organic plus mineral fertilizers). Denitrification activity and total C and N content in the soil were significantly increased by DMC. Denitrification enzyme activity was more closely correlated with C content than with N content in the soil. Principal component analysis confirmed that soil management had the strongest impact on the soil denitrifier community and total C and N content and further indicated that changes in microbial and chemical soil parameters induced by the use of fertilizer were favored in DMC plots. Carvalho et al. (2009) studied soil C sequestration as well as the GHG fluxes ($N_2O$ and $CH_4$) during the process of conversion of Cerrado into agricultural land in the southwestern Amazon region (Brazil), comparing no tillage (NT) and conventional tillage (CT) systems. Data showed that the $N_2O$ emissions in the wet season were different from those measured in the dry season. The $N_2O$ emissions were higher in the areas under NT and in the wet season. In regions with marked dry and wet seasons, the higher soil water content has a positive effect on $N_2O$ emissions to the atmosphere (Davidson et al. 1993). According to Groffman (1985), NT, as compared to CT, increases soil C and N stocks, improving soil aggregate stability, leading to a rise in $N_2O$ emissions from denitrification. The better soil aggregation, higher soil microporosity, and water content favor the formation of anoxic microsites.

### 14.5.2 Nitric Oxide Emissions

It was mentioned previously that denitrification by rhizobacteria diminishes the amount of nitrate available for plant nutrition. However, it may have positive effects on root development by means of nitric oxide production (NO), which is a key signal molecule that controls root growth and nodulation, stimulates seed germination, and is involved in plant defense responses against pathogens. Furthermore, NO can interact with other plant hormone signaling networks including that for indoleacetic acid (IAA). Bacterial denitrification with the production of NO by *Azospirillium brasilense* has been demonstrated on wheat roots (Richardson et al. 2009). Therefore, denitrification in agricultural soils has another benefit: promotion of root growth is linked to the ability of PGPR to produce phytohormones like NO.

### 14.5.3 Symbiotic Fixation and Denitrification

The relationship between soil nitrous oxide emissions and BNF by grain legumes is not very well understood and the contribution of actively $N_2$-fixing plants to $N_2O$ emissions has rarely been reported. However, its comprehension will allow us to understand both nitrogen fluxes in soil and emission of nitrous oxide from legumes.

The use of grain legumes in rotation with cereal and oilseed crops is a well-established practice to increase soil fertility and crop yields. Since indigenous rhizobia are incapable of, or are ineffective in, supporting an adequate level of $N_2$ fixation if grain legumes have not been grown previously, it is recommended that grain legumes be inoculated with rhizobia, particularly when introducing these grain legumes for the first time into the crop rotation (Zhong et al. 2009).

The consequence of denitrification by free-living rhizobia could result in a significant loss of soil nitrate (Delgado et al. 2007) because there are vast areas all over the world which are cultivated with legumes. However, as it was mentioned previously, *B. japonicum* and *A. caulinodans* are the only rhizobia which are true denitrifiers. Fernández et al. (2008) characterized the denitrification activity of 250 strains, all of them representatives of native *Bradyrhizobium* sp., isolated from argentine soils cultivated with soybean. The results showed that 73 were scored as probably denitrifiers by a preliminary screening method. Only 41 were considered denitrifiers because they produced gas bubbles in Durham tubes, cultures reached an absorbance of more than 0.1, and $NO_3$ and $NO_2$ were not present. Ten of these isolates were selected to confirm denitrification based on colony diameter and utilization of carbohydrates. According to $N_2O$ production with $NO_3$ and cell protein concentration (Table 14.1), the isolates could be differentiated into three categories of denitrifiers: group I included the collection strains USDA 110 and MSDJ G 49 which produced 7.06 and 7.30 nmol of $N_2O$ per microgram of protein, respectively; group II incorporated four isolates soils which produced between 2 and 5 nmol of

**Table 14.1** Denitrifying activity of indigenous *Bradyrhizobium* sp. isolates from argentine soybean cultivated soils

| Strains | Group[a] | Denitrifying activity[b] | Denitrification genes[c] |
|---|---|---|---|
| MSDG G49 | | 7.30* ± 0.22d** | |
| USDA 110 | I (1)[D] | 7.06 ± 0.15d | *napA*, *nirK*, *norC*, *nosZ* |
| Per3 64 | | 3.26 ± 0.19b | *napA*, *nirK*, *norC*, *nosZ* |
| Per1 12 | | 4.20 ± 0.25c | *napA*, *nirK*, *norC* |
| Per1 31 | | 3.36 ± 0.18bc | *nirK*, *norC* |
| Per3 34 | II (1) | 2.4 ± 0.09b | *norC* |
| Per1 64 | | 0.60 ± 0.03a | – |
| Per1 1 | | 0.82 ± 0.03a | – |
| Per3 45 | | 0.49 ± 0.05a | *napA*, *norC* |
| Per3 61 | III (1) | 1.03 ± 0.08a | *napA*, *nirK*, *norC*, *nosZ* |
| Man1 18 | | 0 | |
| Man1 34 | (2) | 0 | – |

*Data are average values of three replicates ± SE

**Means with different letters in the same column differ significantly at $P \leq 0.05$ according to Fisher LSD

[a]Groups of isolates according to $N_2O$ production and cell protein concentration with $NO_3^-$

[b]Denitrifying activity in nmol $N_2O$ $\mu g^{-1}$ of cellular protein

[c]Screening of presence/absence of denitrification genes

$N_2O$ per microgram of protein; whereas group III included four *Bradyrhizobium* sp. which yielded $\leq 1$ nmol of $N_2O$ per microgram of protein. Clearly, this classification indicates that there is a different denitrification activity in indigenous *Bradyrhizobium* sp. from soils cultivated with soybean. On the other hand, Ciampitti et al (2008) studied the effects of the inoculation of soybean with *B. japonicum* on $N_2O$ evolution during all phenological stages and during the stubbles decomposition period in the presence of nitrogen fertilizers. They found that nitrous oxide emissions increased during the soybean growing season, with the highest accumulation from grain filling until commercial maturity. Nitrogen fertilization affected $N_2O$ losses especially with inoculated soybean plants. Significant correlation was observed between $N_2O$ emissions and soil nitrate contents with inoculated plants, suggesting that the main controlling variable of $N_2O$ emissions was nitrate contents until harvest.

Several authors have worked with other species which also have capacity to denitrify. In Canada, Zhong et al. (2009) quantified $N_2O$ emissions associated with $N_2$ fixation by grain legumes under controlled conditions and the denitrifying capability of two *Rhizobium leguminosarum* biovar viciae strains. Results indicated that (1) neither *R. leguminosarum* strain, 99A1 or RGP2 was capable of denitrification in pure culture, nor in symbiosis with lentil and pea in sterile Leonard jars, suggesting that introducing these *Rhizobium* into soils through rhizobial inoculation onto lentil and pea will not increase denitrification or $N_2O$ emissions, and (2) soil-emitted $N_2O$ from well-nodulated lentil and pea crops grown under controlled conditions was not significantly different than that from the check treatments. They conclude that $N_2O$ emissions may not be directly related to biological nitrogen fixation by grain legumes under conditions comparable to those provided in their experiments.

#### 14.5.3.1 Denitrification in Soybean Nodules

When a rhizobial species is used to inoculate the corresponding legume host, bacteroids within the nodules are also able to express the denitrification pathway. The process of denitrification in nodules is an energy-producing mechanism as well as a nitrite and nitric oxide detoxifying mechanism. Consequently, this capacity is of great interest in symbiotic association as it leads bacteroids to survive in the nodules.

Bacteroids isolated from nodules of $N_2$-dependent plants that were incubated with nitrate may produce large amounts of nitrite, nitrous oxide, and nitrogen, depending on whether the species used for inoculation contains a complete or incomplete set of denitrification genes. Emission of $NO_x$ from legume nodules also contributes to the release of GHG into the atmosphere. Soil nitrate has restricted access to the bacteroid-infected zone within the nodule; consequently, despite the presence of an active denitrification system, the process within the nodules could be limited by substrate availability (Delgado et al. 2007).

## 14.6 Concluding Remarks

The relevance of denitrification process as an important part of the nitrogen cycle in soil agroecosystems is absolutely without discussion. The positive effects are far superior as compared to what it could be considered as negative effects. Sustainable agriculture needs that every microbiological process, such as in this case denitrification, must be preserved.

On other hand, *Bradyrhizobium* naturalized population in the soybean cultivated soils was quantitatively important with respect to other groups of microorganisms, as these bacteria may have a relevant influence on the relative denitrification potential of soils. Further investigation is needed to determine whether the denitrifying activity of these bacteria significantly influences the extent or the products of denitrification.

## References

Antoun H, Prévost D (2000) PGPR activity of *Rhizobium* with nonleguminous plants. In: Proceedings of the 5th International PGPR workshop, Villa Carlos Paz, Córdoba

Antoun H, Beauchamp CJ, Goussard N, Chabot R, Lalande R (1998) Potential of *Rhizobium* and *Bradyrhizobium* species as plant growth promoting rhizobacteria on non-legumes: effect on radishes (*Raphanus sativus* L.). Plant Soil 204:57–67

Asakawa S (1993) Denitrifying ability of indigenous strains of *Bradyrhizobium japonicum* isolated from fields under paddy-upland rotation. Biol Fert Soils 15:196–200

Baudoin E, Philippot L, Cheneby D, Chapuis-Lardy F, Fromin N, Bru D, Rabary B, Brauman A (2009) Direct seeding mulch-based cropping increases both the activity and the abundance of denitrifier communities in a tropical soil. Soil Biol Biochem 41:1703–1709

Bedmar E, Robles EF, Delgado MJ (2005) The complete denitrification pathway of the symbiotic, nitrogen-fixing bacterium *Bradyrhizobium japonicum*. 10th nitrogen cycle meeting 2004. Biochem Soc Trans 33(Part 1):141–145

Bedmar EJ, Delgado MJ (2006). Metabolismo anaerobio del nitrato en bacterias simbióticas: respiración y desnitrificación. In: Bedmar EJ, González J, Lluch C, Rodelas B (eds) Fijación de Nitrógeno: Fundamentos y Aplicaciones. Editorial SEFIN, Granada, Spain. ISBN: 84-61-1198-5. pp. 92–101

Breitenbeck GA, Bremner JM (1989) Ability of free-living cells of *Bradyrhizobium japonicum* to denitrify in soils. Biol Fert Soils 7:219–224

Bremer C, Braker G, Matthies D, Reuter A, Engels C, Conrad R (2007) Impact of plant functional group, plant species, and sampling time on the composition of *nirK*-type denitrifier communities in soil. Appl Environ Microbiol 73:6876–6884

Bueno E, Mesa S, Sanchez C, Bedmar E, Delgado MJ (2010) *NifA* is required for maximal expression of denitrification genes in *Bradyrhizobium japonicum*. Environ Microbiol 12:393–400

Carvalho JLN, Cerri CEP, Feigl BJ, Pccolo MC, Godinho VP, Cerri CC (2009) Carbon sequestration in agricultural soils in the Cerrado region of the Brazilian Amazon. Soil Till Res 103: 342–349

Casella S, Payne WJ (1996) Potential of denitrifiers for soil environment protection. FEMS Microbiol Lett 140:1–8

Cheneby D, Philippot L, Hartmann A, Hénault C, Germon J-C (2000) 16S DNA analysis for characterization of denitrifying bacteria isolated from three agriculture soils. FEMS Microbiol Ecol 34:121–128

Ciampitti IA, Ciarlo EA, Conti ME (2008) Nitrous oxide emissions from soil during soybean (*Glycine max* L. Merrill) crop phenological stages and stubbles decomposition period. Biol Fert Soils 44:581–588

Ciarlo E, Conti M, Bartoloni N, Rubio G (2007) The effect of moisture on nitrous oxide emissions from soil and the $N_2O/(N_2O+N_2)$ ratio under laboratory conditions. Biol Fert Soils 43:675–681

David MB, Del Grosso SJ, Hu X, Marshall EP, McIsaac GF, Parton WJ, Tonitto C, Youssef MA (2009) Modeling denitrification in a tile-drained, corn and soybean agroecosystem of Illinois, USA. Biogeochemistry 93:7–30

Davidson EA, Matson PA, Vitousek PM, Riley R, Dunkin K, Garcia-Mendez G, Maass JM (1993) Process regulating soil emissions of NO and $N_2O$ in a seasonally dry tropical forest. Ecology 74:130–139

Delgado MJ, Casella S, Bedmar EJ (2007) Denitrification in rhizobia-legume symbiosis. In: Bothe H, Ferguson SJ, Newton WE (eds) Biology of the nitrogen cycle. Elsevier, Amsterdam, pp 84–92

Demaneche S, Philippot L, David MM, Navarro E, Vogel TM, Simonet P (2009) Characterization of denitrification gene clusters of soil bacteria via a metagenomic approach. Appl Environ Microbiol 75:534–537

Fernández LA, Perotti EB, Sagardoy MA, Gómez MA (2008) Denitrification activity of *Bradyrhizobium* sp. isolated from argentine soybean cultivated soils. World J Microbiol Biotechnol 24:2577–2585

Groffman PM (1985) Nitrification and denitrification in conventional and notillage soils. Soil Sci Soc Am J 49:329–334

Hernandez-Ramirez G, Brouder SM, Smith DR, Van Scoyoc GE (2009) Greenhouse gas fluxes in an Eastern corn belt soil: weather, nitrogen source, and rotation. J Environ Qual 38:841–854

Intergovernmental Panel on Climate Change (2001) The third assessment report. "Climate Change 2001". Cambridge University Press, Cambridge, UK

Jones CM, Stres B, Rosenquist M, Hallin S (2008) Phylogenetic analysis of nitrite, nitric oxide, and nitrous oxide respiratory enzymes reveal a complex evolutionary history for denitrification. Mol Biol Evol 25:1955–1966

Kaneko T, Nakamura Y, Sato S, Minamisawa K, Uchiumi T, Sasamoto S, Watanabe A, Idesawa K, Iriguchi M, Kawashima K, Kohara M, Matsumoto M, Shimpo S, Tsuruoka H, Wada T, Yamada M, Tabata S (2002) Complete genomic sequence of nitrogen-fixing symbiotic bacterium *Bradyrhizobium japonicum* USDA110. DNA Res 9:189–197

Kuenen JG, Roberson LA (1987) In: Cole JA, Ferguson S (eds) The nitrogen and sulfur cycles. Cambridge University Press, UK, pp 162–218

Kumon Y, Sasaki Y, Kato I, Takaya N, Shoun H, Beppu T (2002) Codenitrification and denitrification are dual metabolic pathways through which dinitrogen evolves from nitrate in *Streptomyces antibioticus*. J Bacteriol 184:2963–2968

Laughlin RJ, Stevens RJ (2002) Evidence for fungal dominance of denitrification and codenitrification in a grassland soil. Soil Sci Soc Am J 66:1540–1548

Mahmood T, Malik KA, Shamsi SRA, Sajjad MI (1998) Denitrification and total N losses from an irrigated sandy-clay loam under maize-wheat cropping system. Plant Soil 199:239–250

Mahne I, Tiedje JM (1995) Criteria and methodology for identifying respiratory denitrifiers. Appl Environ Microbiol 61:1110–1115

Martínez Toledo MV (1992) Biología del Nitrógeno. In: López JG, Lluch C (eds) Interacción Planta-Microorganismo: Biología del Nitrógeno. Editorial Rueda, Alcorcón, Madrid, Spain. ISBN: 84-7207-065-4

Miller AJ, Cramer MD (2004) Root nitrogen acquisition and assimilation. Plant Soil 274:1–36

Moreno VC, Cabello P, Martínez-Luque M, Blasco R, Castillo F (1999) Prokaryotic nitrate reduction: molecular properties and functional distinction among bacterial nitrate reductases. J Bacteriol 181:6573–6584

Mulder A, van de Graaf AA, Robertson LA, Kuenen JG (1995) Anaerobic ammonium oxidation discovered in a denitrifying fluidized bed reactor. FEMS Microbiol Ecol 16:177–183

Munch JCH, Velthof GL (2007) Denitrification and agriculture. In: Bothe H, Ferguson SJ, Newton WE (eds) Biology of the nitrogen cycle. Elsevier, Amsterdam, pp 331–341

Newton WE (2007) Physiology, biochemistry, and molecular biology of nitrogen fixation. In: Bothe H, Ferguson SJ, Newton WE (eds) Biology of the nitrogen cycle. Elsevier, Amsterdam, pp 109–127

O'Hara GW, Daniel RM (1985) Rhizobial denitrifiation: a review. Soil Biol Biochem 17:1–9

Philippot L (2002) Denitrifying genes in bacterial and archaeal genomes. Biochim Biophys Acta 1577:355–376

Philippot L, Germon JC (2005) Contribution of bacteria to initial input and cycling of nitrogen in soils. In: Varma A, Buscot F (eds) Microorganisms in soils: roles in genesis and functions. Springer, Berlin, pp 159–176

Poth M, Focth DD (1985) $^{15}N$ kinetic analysis of $N_2O$ production by *Nitrosomonas europea*: an examination of nitrifier denitrification. Appl Environ Microbiol 49:1134–1141

Richardson AE, Barea JM, McNeill AM, Prigent-Combaret C (2009) Acquisition of phosphorus and nitrogen in the rhizosphere and plant growth promotion by microorganisms. Plant Soil 321:305–339

Sameshima-Saito R, Chiba K, Minamisawa K (2004) New method of denitrification analysis of *Bradyrhizobium* field isolates by gas chromatographic determination of $^{15}N$-labeled $N_2$. Appl Environ Microbiol 70:2886–2891

Schlesinger WH (1997) Biogeochemistry: an analysis of global change, 2nd edn. Academic, San Diego, CA

Shoun H, Kima D-H, Uchiyamab H, Sugiyamac J (1992) Denitrification by fungi. FEMS Microbiol Lett 94:277–281

Smith GB, Smith MS (1986) Symbiotic and free-living denitrification by *Brady japonicum*. Soil Sci Soc Am J 50:349–354

Smith P, Martino D, Cai Z, Gwary D, Janzen H, Kumar P, McCarl B, Ogle S, Oma Rice C, Scholes B, Sirotenko O (2007) Agriculture. In: Metz B, Davidson OR, Bosch PR, Dave R, Meyer LA (eds) Climate Change 2007: Mitigation. Contribution of Working Group III to the Fourth

Assessment Report of the Intergovernmental Panel on Climate Change. Cambridge University Press, Cambridge, United Kingdom and New York, NY, USA. pp. 499–532
Strous M, Fuerst JA, Kramer EHM, Logemann S, Muyzer G, Vam de Pas-Schoonen KT, Webb R, Kuenen JG, Jetten MSM (1999) Missing lithotrophic identified as a new planctomycete. Nature 400:446–449
Tiedje JM (1988) Dissimilatory nitrate-reducing bacteria. In: Zehnder AJB (ed) Biology of anaerobic microorganisms. Wiley, New York, pp 179–243
Tiedje JM (1994) Denitrifiers. In: Klute A (ed) Methods of soil analysis, part 2: microbiological and biochemical properties, 2nd edn. SSSA, Madison, pp 245–265
Trimmer M, Nicholls JC, Deflandre B (2003) Anaerobic ammonium oxidation measured in sediments along the Thames estuary, United Kingdom. Appl Environ Microbiol 69:6447–6454
van Berkum P, Keyser H (1985) Anaerobic growth and denitrification among different serogroups of soybean rhizobia. Appl Environ Microbiol 49:772–777
van Spanning RJM, Richardson DJ, Ferguson SJ (2007) Introduction to the biochemistry and molecular biology of denitrification. In: Bothe H, Ferguson SJ, Newton WE (eds) Biology of the nitrogen cycle. Elsevier, Amsterdam, pp 4–20
Xing G, Zhao X, Xiong Z, Yan X, Xu H, Xie Y, Shi S (2009) Nitrous oxide emission from paddy fields in China. Acta Ecol Sin 29:45–50
Yang S-S, Lai C, Chang H, Chang E, Wei C (2009) Estimation of methane and nitrous oxide emissions from paddy fields in Taiwan. Renew Energ 34:1916–1922
Ye RW, Averill BA, Tiedje JM (1994) Denitrification: production and consumption of nitric oxide. Appl Environ Microbiol 60:1053–1058
Zhong Z, Lemke RL, Nelson LM (2009) Nitrous oxide emissions associated with nitrogen fixation by grain legumes. Soil Biol Biochem 41:2283–2291
Zumft WG (1997) Cell biology and molecular basis of denitrification. Microbiol Mol Biol Rev 4:533–616

# Erratum to: Role of PGPR in Integrated Nutrient Management of Oil Seed Crops

**Sandeep Kumar, R.C. Dubey and D.K. Maheshwari**

Erratum to:
Chapter 1 in:
D.K. Maheshwari (ed.), *Bacteria in Agrobiology: Plant Nutrient Management*,
DOI 10.1007/978-3-642-21061-7_1

---

*In this chapter, the name of one author,* ***Sandeep Kumar****, has erroneously been deleted during the proofs corrections. The corrected version is:*

## Chapter 1
## Role of PGPR in Integrated Nutrient Management of Oil Seed Crops

**Sandeep Kumar, R.C. Dubey and D.K. Maheshwari**

Sandeep Kumar • R.C. Dubey • D.K. Maheshwari (✉)
Department of Botany and Microbiology, Faculty of Life Sciences, Gurukul Kangri University, Haridwar 249404, Uttarakhand, India
e-mail: maheshwaridk@gmail.com

The online version of the original chapter can be found under
DOI 10.1007/978-3-642-21061-7_1

**Sandeep Kumar** • R.C. Dubey • D.K. Maheshwari (✉)
Department of Botany and Microbiology, Faculty of Life Sciences, Gurukul Kangri University, Haridwar 249404, Uttarakhand, India
e-mail: maheshwaridk@gmail.com

D.K. Maheshwari (ed.), *Bacteria in Agrobiology: Plant Nutrient Management*,
DOI 10.1007/978-3-642-21061-7_15, 

# Index


**A**
ABA. *See* Abscisic acid
ABC transporters FecCDE, 128
*Abisida cylindrospora*, 57–59
Abscisic acid (ABA)
biosynthesis, 165–166
structure, 165
ACC. *See* 1-Aminocyclopropane–1-carboxylic acid
*Acidiphillum acidophilum*, 89, 91, 92
*Acidithiobacillus*, 90–93, 95–96, 98
Acinetobactin, 114–115
Actinomycetes, 239, 246, 253–254
*N*-acyl-homoserine lactone (AHL), 209–228, 256–257
Adenosine–5'-phosphosulfate (APS), 85
Aerobactin, 120
Affinities, 122–123
Aflatoxins, 303
*Agrobacterium*, 71, 72
Agrochemicals, 3, 238, 239
AHL. *See N*-acyl-homoserine lactone
AIPs. *See* Autoinducing polypeptides
Albumin, 112
Alcaligin, 126
Alfalfa, 87
Allicin, 305
1-Aminocyclopropane–1-carboxylic acid (ACC)
deaminase, 18, 20, 26, 27, 32, 34, 183–202
gene, 190, 193, 195–199, 201–202
protein, 188, 192, 193
Ammonia, 322, 323, 328
*Amycolatopsis orientalis*, 53
Anaerobic conditions, 84
Anguibactin, 114–115
Antagonism, 240
Anthracnose, 304
Antibiosis, 239, 240, 243, 267–268
Antibiotics, 209–210, 214, 216–222, 224, 225, 227, 228, 238, 240–241, 243, 245, 246, 251, 253–255, 258, 267–270, 273, 274, 276, 277, 301
peoluteorin (Plt), 221
phenazines (PHZ), 214, 217–219, 224
pyrrolnitrin (PRN), 216, 217, 219–221
Antifungal volatiles, 301
Antimicrobial peptides, 241
*Aphanothece*, 90
APS. *See* Adenosine–5'-phosphosulfate
*Araucaria* sp., 57
Arthrospore suspension, 308
*Aspergillus awamori*, 96, 98
*Aspergillus niger*, 48, 58–59
*Aspergillus terreus*, 98
ATP-binding cassette (ABC) transporter, 125
Autoinducers, 256–257
Autoinducing polypeptides (AIPs), 256
Auxins
IAA, 186
*Azospirillum*, 70–72
*Azotobacter*, 67, 70
*Azotobacter chroococcum*, 58

**B**
*Bacillus* sp., 66, 70–72
*B. edaphicus*, 58
*B. megaterium*, 58, 66
*B. subtilis*, 52, 54
Bacterial metabolites, 310–316
BCAs. *See* Biocontrol agents
*Beauveria caledonica*, 59, 60
Bicarbonate–extractable-P, 95


D.K. Maheshwari (ed.), *Bacteria in Agrobiology: Plant Nutrient Management*,
DOI 10.1007/978-3-642-21061-7, 

Biocontrol, 214, 216–227, 301
antibiotics, 33, 34
ISR, 33
lytic enzymes, 18
root colonization, 33
Biocontrol agents (BCAs), 238–242, 244, 245, 254–257
Biodegradability, 243, 258
Biofertilizers, 1–5, 7–10
Biofilm, 217, 220, 222–224, 227, 228
Bioinoculants, 8
Biological control, 238–240, 252, 257, 258
Biologically activated sulfur, 96, 98
Biopesticides, 2
*Bosea thiooxidans*, 92
*Bradyrhizobium* sp., 50, 322, 328, 330, 333–335
*japonicum*, 323
*Brassica*, 5–9
*Brassica oleracea*, 86
*Burkholderia cepaciae*, 58
*Burkholderia kururiensis* subsp. *thiooxydans*, 91

C
Calcareous soil, 88
Calcium hydroxide, 88
*Candidda utilils*, 50
Canola, 96, 101, 102
Carbon source, 67
Catecholate, 113, 114
*Catenococcus thiocyclus*, 89
C-bonded S, 83, 84
Cellular communication, 256
Cellular respiration, 249
Chemiosmotic, 48
Chemolithotrophic, 92–93
Chemolithotrophs, 89, 90, 92–93
Chitinases, 217, 220, 225
*Chlorella vulgaris*, 50
*Chlorobium* spp., 90
*Chloroflexus aurantiacus*, 90
Chlorosis, 50
Chlorosulfolipids, 85
*Chromatium* spp., 90
Citrate, 115
*Citreicella thiooxidans*, 89
CK. *See* Cytokinins
Clay minerals, 82, 84
Climate change, 237
Co-inoculation, 57, 97, 98
Competition, 238–241, 244
Complex S-forms, 83–84
Conjugation, 257
Coprogen, 120
Cotton-leaf volatiles, 303–304
Crop
productivity, 237–239
protection, 237–238, 243–245, 255, 256
*Cupriavidus eutrophs* CH34, 50
*Cupriavidus metallidurans*, 53, 54
Cyanide, 314
Cyanobactericide, 304
Cyanogenic plants, 314
Cysteine, 81, 84–86
Cytokinins (CK)
biosynthesis, 30, 142
effects, 30
kinetin, 161–162, 170
physiology, 161, 163, 173
structure, 29
zeatin, 142, 161, 163, 170

**D**
Denitrification, 321–335
bacteria, 321, 325–327, 333, 335
enzymes, 326, 328, 329, 332
genes, 325–327, 329–330, 334, 335
Desferal, 114
Desferrioxamine, 114
Desferrioxamine B, 121
D-genotype, 272
2,4-Diacetylphloroglucinol (DAPG), 267–277
Dihydroxybenzoic, 120
Divided plate method, 312
*Dyella ginsengisoli*, 93, 95, 96, 98, 101

**E**
Ectomycorrhizae, 59
*Ectothiorhodospira* spp., 90
Electron transport chain, 248
Elicitors, 256
Enter–Doudoroff pathway, 91
*Enterobacter*, 66, 70, 72, 74
Enterobactin, 121, 127, 132–133
Entrobactin, 120
Ericoid mycorrhizae, 59
*Escherichia coli*, 52, 54
Ethylene
ACC deaminase, 142, 165
biosynthesis, 26, 141–142, 164, 165, 169, 171

effects, 27, 28, 30–32
structure, 31–32
Evolution Canyon (EC) model
"African" slope (AS), 290
biofilm formation, 291–292
cross section, 290
drought-stressful AS slope, 293
EC I (Carmel), 291
EPS role, 292
inhibitory effect, 292
*Paenibacillus polymyxa*, 291

**F**
*fec*A, 119–120, 128
*feg*A, 126
*feg*AB, 124, 127
*feg*B, 124
Fenton type chemistry, 128
*fep*A (ferric-enterobactin receptor), 120
*fep*C, 119–120
Ferrichrome, 113, 120, 121, 124, 126, 127
Ferrioxamine, 113, 114
Ferritins, 111
*fhu*A, 119–120, 124–126
*fhu*CDB, 124–125
Fluorescent pseudomonads (FPs), 241, 252
Food security, 237
Formulations, 242, 257
FPs. *See* Fluorescent pseudomonads
*fpv*A, 120, 127
Fungal metabolites, 306–310
Fungal Zn solubilization, 58–60
Fungistatic spectrum, 308
Fur (ferric uptake regulator), 128

**G**
*Gaeumannomyces graminis* var. *tritici*, 272
*Gamma proteobacteria*, 89, 92
GAs. *See* Gibberellins
Geomicrobiologically, 102
Gibberellins (GAs), 18, 20, 26, 27, 29–31
biosynthesis, 154–159
conjugates, 153–154, 156, 159
dwarf, 155–157
gibberellic acid, 142, 171, 173
retardants, 157–159
structure, 153–155
*Gluconacetobacter diazotrophicus*, 58
Glucose dehydrogenase, 58
Glucose metabolic pathway, 91
Glycine betaine, 86
Greenhouse effect, 321
Groundnut (*Arachis hypogaea* L.), 2, 4, 6

**H**
*Haemophilus ducreyi*, 54
*Haemophilus influenzae*, 54
*Halothiobacillus* sp., 95, 96, 101
HCN. *See* Hydrogen cyanide
Heme, 111
Hemopexin, 111
Hemophores, 112
Hemoproteins, 112
Hepatoglobin–Hemoglobin, 111
Herbicides, 253
Heterologous siderophores, 123–126
Heterotrophs, 56, 89–91
High throughput screening, 257
*hu*A (ferrichrome receptor), 120
Hydrogen cyanide (HCN), 241, 248, 252, 254–255
Hydroxamate, 113, 126
Hydroxamate siderophores, 132
Hydroxamic acid, 113
Hydroxyapatite, 66–68
Hyperparasites, 307
Hyphal extension, 309

**I**
IAA. *See* Indole–3-acetic acid
Indole–3-acetic acid (IAA), 225
biosynthesis, 27, 28
degradation, 29
effects, 27–29
Induced systemic resistance (ISR), 217, 222–223, 225, 240, 276, 277
INM. *See* Integrated nutrient management
Inoculants, 167, 170–173
Inorganic fertilization, 5, 7–10
Inorganic S, 84
Integrated nutrient management (INM), 1–10
Integrated plant nutrition system (IPNS), 1
*In-vitro* test, 74
*In-vivo* test, 75
IPNS. *See* Integrated plant nutrition system
Iron
acquisition, 117–128
fluorescent pseudomonads, 22
siderophores, 18, 22–23
Iron-siderophore affinity, 121
Iron–sulfur enzymes, 85
ISR. *See* Induced systemic resistance

**J**
Jasmonic acid (JA)
  biosynthesis, 142, 168
  jasmonates, 167–168
  OPDA, 168

**K**
Kaolinite, 82
Korean soils, 93

**L**
Lactoferrin, 120–121
Lead molecules, 256, 258
Leguminous plants, 323
*Leptosphaeria maculans*, 101
Lipid-S, 83–84
Lipid synthesis, 85
*Lyngbya* spp., 90

**M**
Macronutrients, 57–58, 87
Membrane
  de-stability, 247
  disruption, 247
  permeability, 251
  transport, 248–249
*Mesorhizobium thiogangeticum*, 92
Metabolic engineering, 257–258
Metabolic system biology, 257
Metabolites
  agroactive, 244, 255, 257
  antibiotic, 238, 241, 245, 254, 255, 258
  microbial, 237–258
  volatile, 246, 252
Metal inorganic transport, 48
Metalloenzyme, 48
Methionine, 81, 84–86
*Methylobacterium fujisawaense*, 91
*Methylobacterium goesingense*, 91
*Methylobacterium oryzae*, 91
*Methylobacterium thiocyanatum*, 92
*Microbacterium phyllosphaerae*, 93, 95, 96, 98, 101
*Microbacterium saperdae*, 56
Microbial communication, 257
Microbial metabolism, 257
*Micrococcus* spp., 31
Mitochondrial function, 247, 249
Mitochondrial respiration, 249, 251
Mixotrophs, 89–91, 102
Montmorillonite, 82
Mugineic acid, 121, 129
Mycelial growth, 246, 252, 254, 255
Mycoherbicidal activity, 255
Mycorrhizal fungus, 70, 72, 74, 75
*Myrothecium* sp., 98

**N**
$NH_4$ assimilation, 66
Nitrate, 321–326, 328, 329, 331–335
  reductases, 87, 326–327, 329
Nitric oxide, 324–329, 331, 333, 335
  biosynthesis, 167
  reductases, 324, 326, 328
Nitrite, 321, 324, 325, 327, 329, 335
  reductases, 327, 329
Nitrogen, 323, 326, 328, 329, 331, 332, 334
  cycle, 321, 322, 324, 335
  fixation, 18, 21, 24–26, 85
  *nif* genes, 24, 25
  *Rhizobia*, 24
Nitrogenase complex, 85
Nitrogen-fixing bacterium, 315
*Nitrosomonas*, 66
Nitrous oxide, 323–325, 328, 329, 331–335
  reductases, 328, 329
Non C-bonded S, 83, 84
Nonribosomal peptides (NRP), 293
Non-volatile antibiotics, 243, 255
NRP synthetases (NRPS), 293

**O**
Organic acid
  citric acid (citrate), 66, 67
  gluconic acid, 67
  oxalic acid (oxalate), 66, 67
Organic fertilization, 5, 8, 9
Organic volatiles, 311
*Oscillatoria* sp., 90
Oxidative damage, 248

**P**
*Pandoraea* sp., 95, 96, 101
PAPS. *See* 3'-Phosphoadenosine–5'-phosphosulfate
*Paracoccus bengalensis*, 92
Paracoccus sulfur oxidation (PSO) pathway, 102
*Paracoccus thiocyanatus*, 92
Parasitism, 237–240, 242, 247

PCA. *See* Phenazine–1-carboxylic acid
*Penicillium aurantiogriseum*, 57
*Penicillium simplicissimum*, 58–59
Pentanoic acid, 86
Pentose phosphate pathway, 91
*pfe*A, 127
PGPB. *See* Plant growth-promoting bacteria
PGPR. *See* Plant growth promoting rhizobacteria
Phenazine–1-carboxylic acid (PCA), 267–277
*Phormidium* spp., 90
Phosphate fixation, 65, 66, 69, 70, 72, 73
  precipitation, 65
  $P_{0.2}$-value, 65
Phosphate fixation-sorption, 65–66
Phosphate solubilization
  mixed inocula, 21
  phosphate-solubilizing bacteria (PSB), 20, 21
Phosphatidylsulfocholine, 85
3'-Phosphoadenosine–5'-phosphosulfate (PAPS), 85
Phosphobacterin, 69
Photolithotrophs, 89, 90
Phytohormones
  auxins
    biosynthesis, 147–149
    biosynthetic pathways, 147
    catabolism, 150
    conjugation, 173
    IAA, 143, 145
    IPDC, 145, 146, 148
    physiology, 150–152
    structures, 150, 151
Phytopathogens, 245, 252, 254, 310
Phytosiderophores, 115
*Pinus radiate*, 57
PK synthetases (PKS), 293
Plant diseases, 237–258
Plant growth-promoting bacteria (PGPB), 17–35
Plant growth promoting rhizobacteria (PGPR), 130
  biofertilizers, 184, 199
  crop yield, 199–200
  nutrient uptake, 186, 200
  plant fitness, 200–202
Plant growth-promoting rhizobacteria (PGPR), 1–10, 209–228, 267, 276
Plant growth promotion, 131–132
Plant root associated biofilms
  AFM cantilever tips, 295
  ATP cost, 293–294
  *Bacillus*, 286
  defined, 285
  EC model, 290–293
  *Lactobacillus* sp., 296
  metagenomics, 294
  multienzyme multimodular biocatalysts, 293
  natural products discovery, 295
  *Paenibacillus polymyxa*, 286
  PKS and NRPS, 294
  screening methods, 295
  signaling
    bacterial strain, 290
    environmental signals, 289–290
    quorum sensing (QS) activation, 289
  *Streptococcus mutans*, 296
  structure
    extracellular polysaccharides (EPS) production, 288
    formation stages, 286–288
    *Hordeum spontaneum* roots, 288
    *P. polymyxa* B1 colonization, 287
    *Vibrio cholerae*, 286
Plant sulfolipid, 85
Plant volatiles, 302–305
Polyamines
  cadaverine, 144, 166–167
Polyketides (PK), 293
  synthase, 270
Polythionate, 91, 102
Post harvest pathogens, 246
Protein synthesis, 247, 250–252
Provide$^{TM}$, 73
*Pseudaminobacter salicylatoxidans*, 92
Pseudobactin, 121
*Pseudomonas* sp., 66–68, 70–72
  *P. aeruginosa*, 71, 275, 277
  *P. chlororaphis*, 271
  *P. fluorescens*, 56, 57, 71, 267, 269, 271–273, 275, 276
  *P. monteilli*, 56
  *P. putida*, 57, 71
  *P. simplicissium*, 57
  *P. sputorum*, 95, 96, 101
  *P. striata*, 71
PSO pathway. *See* Paracoccus sulfur oxidation pathway
Pyochelin, 114–115
Pyocyanin, 268, 274–277
Pyoverdines, 115, 127
Pyrite, 82, 84
*Pythium* blight, 311

**Q**

Quorum quenching (QQ), 221, 226–228, 256–257
Quorum sensing (QS), 209–228, 256–257, 269

**R**

Radish volatiles, 305
Reduction-oxidation, 82, 88
Rhizobacteria, 240, 245, 256
  *Achromobacter*, 156
  *Azospirillum* sp.
    *A. brasilense*, 148, 149, 151–152, 157, 165–167, 169–171
    *A. brasilense* AZ39, 170–173
    *A. lipoferum*, 155–157, 169, 170
  *Bacillus*, 146, 147, 150, 156, 157, 165, 166
  biocontrol-PGPB, 141
  *Bradyrhizobium*, 145, 146, 150, 162–163
  *Bradyrhizobium japonicum*
    *B. japonicum* E109, 171
  PGPR, 141–142, 144–168
  *Pseudomonas*, 145–147, 149
  PSHR, 141, 142, 171
  *Rhizobium*, 145–147, 150, 151, 162, 164, 168
Rhizobactin, 125
Rhizobia, 322, 323, 325, 328, 330, 331, 333, 334
Rhizobios
  *Azorhizobium*, 71
  *Bradyrhizobium japonicum*, 71
  *Mesorhizobium mediterraneum*, 71
  *Rhizobium*, 70, 71
*Rhizobium leguminosarum*, 50
*Rhizoctonia solani*, 101, 267, 271, 273, 274, 277
Rhizoremediation, 56, 60
Rhizosphere, 141–144, 150, 156, 159–163, 267–277
  competency, 257
*Rhodopseudomonas* spp., 90
Rock phosphate (RP), 67, 68, 70, 72–76, 93–96, 102
Root
  elongation, 184, 186, 187, 191, 199, 201
  ethylene, 184, 190–193
  exudates, 47, 184, 190
  rhizosphere, 184
RP. *See* Rock phosphate

**S**

Salicylate-mediated systemic acquired resistance (SAR), 222
Salmochelin, 120
SAR. *See* Salicylate-mediated systemic acquired resistance
*Scolecobasidium constrictum*, 98
S-containing compounds, 83
S-dimethyl sulfonium, 86
Sealed plate method, 312
Secondary metabolites, 238–256, 258
Self-inhibition, 308
Semi arid, 83
*Serratia marcescens*, 67
Sesame, 2–5
Sesame oil, 4
Siderophore affinities, 120–121, 133
Siderophores, 132, 240, 241, 245
Sinigrin, 85–86
S4 intermediate (S4I) pathway, 102
Soil minerals, 65, 66, 68
Soil order, 65, 69
Soluble S-forms, 84
*Staphylococcus aureus*, 53, 54
*Starkeya novella*, 89–91
*Stenotrophomonas maltophila*, 54
Sterol sulfur ester, 85
Strawberry fruit volatile compounds, 306
*Streptococcus pneumonia*, 52, 54
Stress
  abiotic, 183, 184, 189, 200, 202
  biotic, 183, 184, 189, 200, 202
Sulfur oxidizing bacteria, 81–102
Sunflower (*Helianthus annus* L.), 2, 4, 7, 9
Symbiotic nitrogen fixation, 323
Synthetic biology, 257

**T**

Take-all, 269, 271–273, 277
Take-all decline (TAD), 272, 273
*Tetrathiobacter intermedius*, 90, 91
*Tetrathiobacter kashmirensis*, 92
*Tetrathiobacter thermosulfatus*, 92
Tetrathionate, 84, 102
Thermodynamics, 112, 195
*Thermotaga maritima*, 54
*Thermothrix thiopara*, 90
*Thilaspi* sp.
  *T. arvensi*, 56
  *T. caerulscens*, 56–57
*Thioalkalimicrobium*, 92
Thioalkalivibrio, 92

*Thiobacillus* sp., 66
    *T. acidophilus*, 91
    *T. denitrificans*, 90
    *T. ferroxidans*, 56, 90
    *T. intermedius*, 90–91
    *T. thermosulfatus*, 92
    *T. thioparus*, 90, 91
*Thioclava pacifica*, 90
Thiosulfate-oxidising organisms, 91, 92
*Thiovirga sulfuroxydans*, 90–91
*Thlaspi caeulescens*, 60
Transferrin, 120–121
Transferrin and lactoferrin, 111
Transporter proteins, 50, 53
*Trichoderma harzianum*, 57
Trithionate, 84, 102

**V**
Vibriobactin, 114–115
Volatile organic compounds (VOCs), 245, 246, 249, 252
Volatile phylloplane fungal metabolites, 309–310
Volatiles, 216, 217, 220, 222, 224–225, 228
    antibiotics, 243, 313
    dose, 304
    metabolites, 305

**W**
Weedicides, 252, 253

**X**
*Xanthobacter tagetidis*, 92

**Y**
Yersiniabactin, 114–115

**Z**
Zinc
    binding proteins, 48, 50
    carbonate, 57
    compounds, 54, 55
    contaminated areas, 50
    deficiency, 47, 49–51, 57
    dust powder, 55
    efflux systems, 49–54
    homeostasis, 51, 53
    hyper accumulator, 56
    levels, 47, 49, 52, 53, 57
    limiting conditions, 53
    mines, 49–50
    oxide, 55, 57, 59
    pollution, 49
    requirements, 51
    resistance, 50–51
    solubilization, 47–60
    sulfate, 55
    sulfide, 49, 55, 56
    toxicity, 47–50
    uptake, 47, 50–54, 57
Zincophilic plants, 60
Zincophore, 53, 60